Mohammed Hichem Mortad

Counterexamples in Operator Theory

 Birkhäuser

Mohammed Hichem Mortad
Department of Mathematics
University of Oran 1
Oran, Algeria

ISBN 978-3-030-97816-7   ISBN 978-3-030-97814-3   (eBook)
https://doi.org/10.1007/978-3-030-97814-3

Mathematics Subject Classification: 47-01, 47A, 47B

This book is published under the imprint Birkhäuser, www.birkhauser-science.com by the registered company Springer Nature Switzerland AG
The registered company address is: Gewerbestrasse 11, 6330 Cham, Switzerland

Preface

Counterexamples undeniably play an outstanding role in mathematics. As is known, counterexamples are patently examples but with a different purpose. The commonly agreed usage of counterexamples is linked with demonstrating the boundaries of possible assertions. In some sense, we may say that there are two types of counterexamples (though some might not totally agree with me). The easiest ones perhaps are those which show that some particular assumptions in certain results may not just be dropped. The hardest counterexamples are usually the ones connected to long-standing open problems or conjectures. In general, the longer a problem resists solutions, the harder it is to solve it (in particular, to find a counterexample in the case there is any). Indeed, there are many cases where the counterexample has to be constructed in several steps with sometimes highly non-trivial constructions. As a result, many counterexamples have been around as initially discovered for quite a while, and without any alterations or simplifications whatsoever. That is, we usually speak of "the counterexample" rather than "a counterexample". Note in passing that some of these examples are just unreadable. Thankfully, there are always curious and brilliant mathematicians who constantly discover "nicer examples" which replace the "ugly ones". Nonetheless, as time goes on, as well as new tools and results become available to us, we realize that, in many cases, the difficulty was not in the counterexample per se but rather in the fact that we are not under the same pressure anymore, i.e. we are not now wondering whether we should look for proof or a counterexample?

That being said, be that as it may, we should always be crediting the effort made by the pioneer of a given counterexample. By way of illustration, we will always remember P. Enflo [104] for pointing out the main idea of the counterexample to the invariant subspace problem on Banach spaces, then effectively constructing it (with the understandable variant by B. Beauzamy [15]), even though we are quite happy with the fairly simple counterexample by G. Sirotkin in [342] (which was a variant of C. J. Read's [301] different construction of such a counterexample). This applies to other famous counterexamples, e.g. by E. Nelson and then K. Schmüdgen on (strong) commutativity of unbounded self-adjoint operators. This also applies to

M. Naimark then to P. R. Chernoff on the triviality of the domain of the square of closed symmetric operators, then the one in [86], etc...

The simplification of counterexamples is also of some pedagogical interest. To elaborate more, if, for instance, we find a yet simpler counterexample to the invariant subspace problem, then it could be included in our lectures (after all, who is disposed to teach Enflo's 100-page counterexample in their lecture?).

In this modest work, I have collected counterexamples (and examples) about many topics on linear bounded and unbounded operators on *Hilbert spaces*. Obviously, it is impossible to cover all topics given that operator theory has become such a vast area of mathematics. In addition, the fact that the title is somewhat attractive gives the mathematical community the entire right to expect to see much in it, and I hope that they would not be too disappointed to see some missing notions or counterexamples. Notwithstanding, I hope that readers will not find that I have dealt ad nauseam with some other notions.

Throughout this book, readers will observe that some examples are pretty easy to find, others are somewhat harder, and some are very much involved. Put differently, questions are extremely varied in their degree of difficulty. For this reason, I have marked the hard problems with "⊛" while the hardest problems are indicated by "⊛⊛". Apart from the very hard questions, it has not been an easy task to decide which questions should be indicated by "⊛". Indeed, several problems marked with "⊛" amply have their place in the class of "⊛⊛" questions. But, because they are based on other counterexamples already available in the book or because similar cases were treated in this volume, they have been classified as "⊛". Notice that in the case of unbounded operators, even the unmarked questions might be viewed as a little hard by neophytes.

Many counterexamples in this book are made by me. I have also cited most of the relevant references (and any omission is purely made by inadvertence). One of my main aims has been to include as simple examples as possible following the French saying "Pourquoi faire compliqué quand on peut faire simple?", translated into English as "Why complicate things when they can be simple?". It is worth alluding that when trying to discover new counterexamples, I have been well assisted with Albert Einstein's quote: "Failure is success in progress".

To the best of my knowledge, there are not any books wholly devoted to counterexamples in operator theory. I could mention, e.g. [123, 148], and [206] which certainly contain many counterexamples, but they are not exclusively dedicated to this matter. Additionally, they do not cover unbounded operators. Still additionally, many interesting counterexamples have appeared in the literature (as well as improvements of already existing ones) since those last three references were published, and so they very much deserve to be added. In the end, I wish that the present book will be as successful as *Counterexamples in Analysis* [128] and *Counterexamples in Topology* [347].

The book is divided into two main parts: "Bounded Linear Operators" and "Unbounded Linear Operators". The two parts are similarly structured. Most chapters have a "Basics" section in order that the book be readily usable. Notice that these parts are rudimentary and cannot replace any detailed course on the subject.

For bounded operators, some well-established references are, e.g., [66, 136, 148, 314, 341]. For unbounded operators, I recommend the textbooks [326] and [375], and also [28, 66, 190, 302, 303, 314] and [341] (not to forget the Dunford-Schwartz trilogy [95, 96] and [97]).

Each counterexample is associated with a titled subsection to allow readers to find what they are looking for just by consulting the table of contents. When necessary, the questions are preceded by comments and results (and also definitions in some cases) to emphasize the connection to the counterexample. I may say that I have presented some parts of the book à la Halmos even though we all agree that Halmos' mathematical and expository skills are just inimitable.

Among the diverse potential audience for whom this book is designed (starting from senior undergraduate students in mathematics and other fields), the main audience of this book remains graduate students. It should also be useful to researchers and lecturers. Despite the fact that the main needed notions and results have been recalled, to use this book, readers must have knowledge of advanced linear algebra, operator theory, functional analysis, ℓ^p and L^p-spaces (and their famous dense subspaces), and some measure theory. As for the Fourier transform, distributions, and basic Sobolev spaces, two small chapters are devoted to them at the end of the book.

With all the pressure authors are usually under when writing a book, I may say that I have overall greatly enjoyed writing this one, and I hope readers will equally enjoy reading it.

Finally, I am indebted to my PhD supervisor (2000–2003) Professor Alexander M. Davie who has made some interesting and useful comments and suggestions. *Sandy, thank you for everything!*

Thanks are also due to Professors Robert B. Israel, Konrad Schmüdgen, and Jan Stochel. Their appreciated help is indicated in the text.

I must likewise thank my fellow student Dr. S. Dehimi with whom I discussed many problems in the last 2 years. I also thank my students for their support (in particular, Ms. Imene Lehbab), who have never let me give up this project.

Lastly, even though I believe that I have done my best, as a human work, the book is definitely not perfect. So, I will be happy to hear from readers about any other counterexamples, possible errors, typos, and suggestions, which can be sent to me via e-mail: mhmortad@gmail.com.

Oran, Algeria   Mohammed Hichem Mortad
October 2021

Contents

Part I   Bounded Linear Operators

1   Some Basic Properties 3
   1.1   Basics 3
   1.2   Questions 7
         1.2.1   Does the "Banachness" of $B(X, Y)$ Yield That
               of Y? 7
         1.2.2   An Operator $A \neq 0$ with $A^2 = 0$ and So
               $\|A^2\| \neq \|A\|^2$ 7
         1.2.3 $A, B \in B(H)$ with $ABAB = 0$ but $BABA \neq 0$ 8
         1.2.4   An Operator Commuting with Both $A + B$ and
               AB, But It Does Not Commute with Any of A
               and B 9
         1.2.5   The Non-transitivity of the Relation
               of Commutativity 9
         1.2.6   Two Operators A, B with $\|AB - BA\| = 2\|A\|\|B\|$ ... 9
         1.2.7   Two Nilpotent Operators Such That Their Sum
               and Their Product Are Not Nilpotent 10
         1.2.8   Two Non-nilpotent Operators Such That Their
               Sum and Their Product Are Nilpotent 10
         1.2.9   An Invertible Operator A with $\|A^{-1}\| \neq 1/\|A\|$ 10
         1.2.10   An $A \in B(H)$ Such That $I - A$ Is Invertible
               and Yet $\|A\| \geq 1$ 11
         1.2.11   Two Non-invertible $A, B \in B(H)$ Such That
               AB Is Invertible 12
         1.2.12   Two A, B Such That $A + B = AB$ but $AB \neq BA$ 12
         1.2.13   Left (Resp. Right) Invertible Operators with
               Many Left (Resp. Right) Inverses 12
         1.2.14   An Injective Operator That Is Not Left Invertible 13

	1.2.15	An $A \neq 0$ Such That $\langle Ax, x \rangle = 0$ for All $x \in H$	14
	1.2.16	The Open Mapping Theorem Fails to Hold True for Bilinear Mappings	14
2	**Basic Classes of Bounded Linear Operators**		**21**
	2.1	Basics	21
	2.2	Questions	22
		2.2.1 A Non-unitary Isometry....................................	22
		2.2.2 A Nonnormal A Such That $\ker A = \ker A^*$	23
		2.2.3 Do Normal Operators A and B Satisfy $\|ABx\| = \|BAx\|$ for All x?	23
		2.2.4 Do Normal Operators A and B Satisfy $\|ABx\| = \|AB^*x\|$ for All x?	23
		2.2.5 Two Operators B and V Such That $\|BV\| \neq \|B\|$ Where V Is an Isometry	23
		2.2.6 An Invertible Normal Operator That Is not Unitary	24
		2.2.7 Two Self-Adjoint Operators Whose Product Is Not Even Normal...	24
		2.2.8 Two Normal Operators A, B Such That AB Is Normal, but $AB \neq BA$	24
		2.2.9 Two Normal Operators Whose Sum Is Not Normal	25
		2.2.10 Two Unitary U, V for Which $U + V$ Is Not Unitary....	25
		2.2.11 Two Anti-commuting Normal Operators Whose Sum Is Not Normal	25
		2.2.12 Two Unitary Operators A and B Such That AB, BA, and $A + B$ Are All Normal yet $AB \neq BA$ ...	26
		2.2.13 A Non-self-adjoint A Such That A^2 Is Self-Adjoint	26
		2.2.14 Three Self-Adjoint Operators A, B, and C Such That ABC Is Self-Adjoint, Yet No Two of A, B, and C Need to Commute.......................	26
		2.2.15 An Orthogonal Projection P and a Normal A Such That PAP Is Not Normal	26
		2.2.16 A Partial Isometry That Is Not an Isometry..............	26
		2.2.17 A Non-partial Isometry V Such That V^2 Is a Partial Isometry...	27
		2.2.18 A Partial Isometry V Such That V^2 Is a Partial Isometry, but Neither V^3 Nor V^4 Is One................	27
		2.2.19 No Condition of $U = U^*$, $U^2 = I$ and $U^*U = I$ Needs to Imply Any of the Other Two	27
		2.2.20 A B Such That $BB^* + B^*B = I$ and $B^2 = B^{*2} = 0$	27
		2.2.21 A Nonnormal Solution of $(A^*A)^2 = A^{*2}A^2$	28
		2.2.22 An $A \in B(H)$ Such That $A^n = I$, While $A^{n-1} \neq I, n \geq 2$	28

	2.2.23	A Unitary A Such That $A^n \neq I$ for All $n \in \mathbb{N}$, $n \geq 2$	28
	2.2.24	A Normal Non-self-adjoint Operator $A \in B(H)$ Such That $A^* A = A^n$	28
	2.2.25	A Nonnormal A Satisfying $A^{*p} A^q = A^n$	29

3 Operator Topologies **39**
3.1 Questions 39
	3.1.1	Strong Convergence Does Not Imply Convergence in Norm, and Weak Convergence Does Not Entail Strong Convergence	39
	3.1.2	$s - \lim_{n \to \infty} A_n = A \not\Rightarrow s - \lim_{n \to \infty} A_n^* = A^*$	40
	3.1.3	$(A, B) \mapsto AB$ Is Not Weakly Continuous	40
	3.1.4	The Uniform Limit of a Sequence of Invertible Operators	41
	3.1.5	A Sequence of Self-adjoint Operators Such That None of Its Terms Commutes with the (Uniform) Limit of the Sequence	41
	3.1.6	Strong (or Weak) Limit of Sequences of Unitary or Normal Operators	41

4 Positive Operators **47**
4.1 Basics 47
4.2 Questions 48
	4.2.1	Two Positive Operators A, B Such That $AB = 0$	48
	4.2.2	Two A, B Such That $A \not\leq 0$, $A \not\geq 0$, $B \not\leq 0$, $B \not\geq 0$, yet $AB \geq 0$	48
	4.2.3	$KAK^* \not\leq A$ Where $A \geq 0$ and K Is a Contraction	48
	4.2.4	$KAK^* \leq A \not\Rightarrow AK = KA$ Where $A \geq 0$ and K Is an Isometry	49
	4.2.5	$KAK^* \leq A \not\Rightarrow AK^* = KA$ Where $A \geq 0$ and K Is Unitary	50
	4.2.6	The Operator Norm Is Not Strictly Increasing	50
	4.2.7	$A \geq B \geq 0 \not\Rightarrow A^2 \not\geq B^2$	50
	4.2.8	$A, B \geq 0 \not\Rightarrow AB + BA \not\geq 0$	51
	4.2.9	Two Non-self-adjoint A and B Such That $A^n + B^n \geq 0$ for All n	51
	4.2.10	Two Positive A, B (with $A \neq 0$ and $B \neq 0$) and Such That $AB \geq 0$ but $A^2 + B^2$ Is Not Invertible	51
	4.2.11	Two A, B Satisfying $\|AB - BA\| = 1/2\|A\|\|B\|$	52
	4.2.12	Two A, B Satisfying $\|AB - BA\| = \|A\|\|B\|$	52
	4.2.13	On Normal Solutions of the Equations $AA^* = qA^*A$, $q \in \mathbb{R}$	53

5   Matrices of Bounded Operators 59
   5.1   Basics 59
   5.2   Questions 62
      5.2.1   A Non-invertible Matrix Whose Formal
           Determinant Is Invertible 62
      5.2.2   An Invertible Matrix Whose Formal
           Determinant Is Not Invertible........................... 63
      5.2.3   Invertible Triangular Matrix vs. Left and Right
           Invertibility of Its Diagonal Elements 63
      5.2.4   Non-invertible Triangular Matrix vs. Left and
           Right Invertibility of Its Diagonal Elements 64
      5.2.5   An Invertible Matrix yet None of Its Entries Is
           Invertible... 64
      5.2.6   A Normal Matrix yet None of Its Entries Is Normal 64
      5.2.7   A Unitary Matrix yet None of Its Entries Is Unitary 65
      5.2.8   Two Non-comparable Self-Adjoint Matrices
           yet the Corresponding Entries Are Comparable 65
      5.2.9   An Isometry S Such That S^2 Is Unitarily
           Equivalent to $S \oplus S$ 65
      5.2.10   An Infinite Direct Sum of Invertible Operators
           Need Not Be Invertible.................................. 66
      5.2.11   The Similarity of $A \oplus B$ to $C \oplus D$ Does Not
           Entail the Similarity of A to C or That of B to D 66
      5.2.12   A Matrix of Operators T on H^2 Such That
           $T^3 = 0$ But $T^2 \neq 0$ 66
      5.2.13   Block Circulant Matrices Are Not Necessarily
           Circulant 67

6   (Square) Roots of Bounded Operators 75
   6.1   Basics 75
   6.2   Questions 76
      6.2.1   A Self-Adjoint Operator with an Infinitude of
           Self-Adjoint Square Roots 76
      6.2.2   An Operator Without Any Square Root 77
      6.2.3   A Nilpotent Operator with Infinitely Many
           Square Roots 77
      6.2.4   An Operator Having a Cube Root but Without
           Any Square Root 77
      6.2.5   An Operator Having a Square Root but Without
           Any Cube Root 78
      6.2.6   A Non-invertible Operator with Infinitely
           Many Square Roots 78
      6.2.7   An Operator A Without Any Square Root, but
           $A + \alpha I$ Always Has One ($\alpha \in \mathbb{C}^*$)...................... 78
      6.2.8 $A^2 \geq 0 \nRightarrow A \geq 0$ Even When A Is Normal 79

6.2.9 $A^3 \geq 0 \nRightarrow A \geq 0$ Even When A Is Normal   79
6.2.10   An Operator Having Only Two Square Roots   80
6.2.11   Can an Operator Have Only One Square Root?   80
6.2.12   Can an Operator Have Only Two Cube Roots?   80
6.2.13   A Rootless Operator.......................................   80
6.2.14   On Some Result By B. Yood on Rootless Matrices   80
6.2.15   A Non-nilpotent Rootless Matrix........................   81
6.2.16   Two (Self-Adjoint) Square Roots of a
         Self-Adjoint Operator Need Not Commute   81
6.2.17   A $B \in B(H)$ Commuting with A Need Not
         Commute with an Arbitrary Root of A..................   81
6.2.18   A Self-Adjoint Operator Without Any Positive
         Square Root...   82
6.2.19   Three Positive Operators $A, B, C \in B(H)$
         Such That $A \geq B \geq 0$ and C Is Invertible Yet
         $(CA^2C)^{\frac{1}{2}} \nsucceq (CB^2C)^{\frac{1}{2}}$ 82
6.2.20   Three Positive Operators $A, B, C \in B(H)$
         Such That $A \leq C$ and $B \leq C$ Yet $(A^2 + B^2)^{\frac{1}{2}}$
         $\nleq \sqrt{2}C$...   82
6.2.21   On Some Result by F. Kittaneh on Normal
         Square Roots   83
6.2.22   On the Normality of Roots of Normal
         Operators Having Co-prime Powers   83
6.2.23   An Isometry Without Square or Cube Roots............   83
6.2.24   Two Operators A and B Without Square Roots,
         Yet $A \oplus B$ Has a Square Root   84

7   **Absolute Value, Polar Decomposition**   95
7.1   Basics   95
7.2   Questions   96
7.2.1   An A Such That $|\operatorname{Re} A| \nleq |A|$ and $|\operatorname{Im} A| \nleq |A|$ 96
7.2.2   A Weakly Normal T Such That T^2 Is Not Normal   97
7.2.3   Two Self-Adjoints A, B Such That $|A + B|$
         $\nleq |A| + |B|$...   97
7.2.4   Two Self-Adjoint Operators A, B That Do Not
         Satisfy $|A||B| + |B||A| \geq AB + BA$....................   98
7.2.5   Two Self-Adjoint Operators A and B Such
         That $\||A| - |B|\| \nleq \|A - B\|$ 98
7.2.6   Two Non-commuting Operators A and B That
         Are Not Normal and Yet $|A + B| = |A| + |B|$ 99
7.2.7   Two Positive Operators A and B with
         $|A - B| \nleq A + B$ 99
7.2.8   Two Self-adjoint Operators A and B Such That
         $I + |AB - I| \nleq (I + |A - I|)(I + |B - I|)$ 99

7.2.9   Two Self-Adjoints $A, B \in B(H)$ Such That
          $|AB| \neq |A||B|$ 100
7.2.10   Two Operators A and B Such That $AB = BA$,
          However, $|A||B| \neq |B||A|$ 100
7.2.11   A Pair of Operators A and B Such That
          $A|B| = |B|A$ and $B|A| = |A|B$, But
          $AB \neq BA$ and $AB^* \neq B^*A$ 100
7.2.12   An Operator A Such That $A|A| \neq |A|A$ 101
7.2.13   An A Such That $|A||A^*| = |A^*||A|$
          But $AA^* \neq A^*A$ 101
7.2.14   An Operator A Such That $|A^2| \neq |A|^2$ 101
7.2.15   A Non-surjective A Such That $|A|$ Is Surjective 102
7.2.16   Two Self-Adjoint Operators A, B with $B \geq 0$
          Such That $-B \leq A \leq B$ but $|A| \nleq B$ 102
7.2.17   The Failure of the Inequality $|\langle Ax, x \rangle| \leq \langle |A|x, x \rangle$ 103
7.2.18   On the Generalized Cauchy–Schwarz Inequality 103
7.2.19   On the Failure of Some Variants of the
          Generalized Cauchy–Schwarz Inequality 104
7.2.20   A Sequence of Self-Adjoint Operators (A_n)
          Such That $|||A_n| - |A||| \to 0$ But $\|A_n - A\| \nrightarrow 0$ 104
7.2.21   The Non-weakly Continuity of $A \mapsto |A|$ 105
7.2.22   A Sequence of Operators (A_n) Converging
          Strongly to A, but $(|A_n|)$ Does Not Converge
          Strongly to $|A|$ 105
7.2.23   An Invertible $A = U|A|$ with $U|A| \neq |A|U$,
          $UA \neq AU$, and $A|A| \neq |A|A$ 106
7.2.24   Left or Right Invertible Operators Do Not
          Enjoy a ("Unitary") Polar Decomposition 106
7.2.25   A Normal Operator Whose Polar
          Decomposition Is Not Unique 106
7.2.26   On a Result of the Uniqueness of the Polar
          Decomposition By Ichinose–Iwashita................... 106
7.2.27   An Operator A Expressed as $A = V|A|$ with
          $A^3 = 0$ but $V^3 \neq 0$ 107
7.2.28   An Invertible Operator A Expressed as
          $A = U|A|$ with $A^3 = I$ but $U^3 \neq I$ 107

8   Spectrum 121
    8.1   Basics 121
    8.2   Questions 123
          8.2.1   An $A \in B(H)$ Such That $\sigma(A) = \varnothing$ 123
          8.2.2   An A Such That $\sigma_p(A^*) \neq \{\bar{\lambda} : \lambda \in \sigma_p(A)\}$ 123
          8.2.3   An $A \in B(H)$ with $p[\sigma(A)] \neq \sigma[p(A)]$
                    Where p Is a Polynomial 124

8.2.4   A Non-self-adjoint Operator Having a Real
          Spectrum.. 124
8.2.5   A Non-positive Operator with a Positive Spectrum 125
8.2.6   A Non-orthogonal Projection A with $\sigma(A) = \{0, 1\}$ 125
8.2.7   A Self-Adjoint Operator Without Any Eigenvalue 125
8.2.8   A Self-Adjoint A Such That
          $\sigma(A) \cap \sigma(-A) = \varnothing$, Yet A Is Neither
          Positive Nor Negative 126
8.2.9   A Self-Adjoint A Such That
          $\sigma(A) \cap \sigma(-A) = \{0\}$, Yet A Is
          Neither Positive Nor Negative 126
8.2.10   A Self-Adjoint Matrix A Such That
          $\sigma(A) \cap \sigma(-A) = \{0\}$ and tr $A = 0$, Yet A Is
          Neither Positive Nor Negative 127
8.2.11   A Nilpotent $T = A + iB$ Such That
          $\sigma(A) \cap \sigma(-A) = \{0\}$ 127
8.2.12   A Nilpotent $T = A + iB$ Such That $\sigma(A) \cap$
          $\sigma(-A) = \varnothing$ 127
8.2.13   Two Operators A, B Such That $\sigma(AB) \neq \sigma(BA)$ 127
8.2.14   Two Self-Adjoints A, B with $\sigma(A) + \sigma(B) \not\supset$
          $\sigma(A + B)$ 128
8.2.15   Two Commuting A, B with $\sigma(A) + \sigma(B) \not\subset$
          $\sigma(A + B)$ 129
8.2.16   Two Self-Adjoints A and B with $\sigma(A)\sigma(B) \not\supset$
          $\sigma(AB)$ 129
8.2.17   Two Commuting A, B Such That $\sigma(A)\sigma(B) \not\subset$
          $\sigma(AB)$ 129
8.2.18   Two Commuting Self-Adjoint Operators A
          and B Such That $\sigma_p(AB) \not\subset \sigma_p(A)\sigma_p(B)$ and
          $\sigma_p(A + B) \not\subset \sigma_p(A) + \sigma_p(B)$ 129
8.2.19   A Unitary Operator A Such That
          $|\sigma(A)|^2 \neq \sigma(A^*)\sigma(A)$ and $\sigma(A^*)\sigma(A) \not\subset \mathbb{R}$............ 130
8.2.20   Two Operators A and B Such That
          $\sigma(A) \cap \sigma(-B) = \varnothing$, but $A + B$ Is Not Invertible 130
8.2.21   "Bigger" Operators Defined on Larger Hilbert
          Spaces Do Not Necessarily Have "Bigger" Spectra 131
8.2.22   An Operator A Such That $A \neq 0$ and $\sigma(A) = \{0\}$ 131
8.2.23   A Self-Adjoint Operator A Such That
          $\sigma_p(A) = \{0, 1\}$, $\sigma_c(A) = (0, 1)$, and $\sigma(A) = [0, 1]$..... 132
8.2.24   An Operator A with $\sigma(A) = \{1\}$, But A Is
          Neither Self-Adjoint Nor Unitary 132
8.2.25   Two Self-Adjoint Operators A, B Such That
          $\sigma(AB)$ Is Purely Imaginary............................. 132
8.2.26 $A \leq B$ Does Not Yield $\sigma(A) \subset \sigma(B)$ 132
8.2.27   An Operator A with $\overline{\sigma_p(A)} \subsetneq \sigma_a(A) \subsetneq \sigma(A)$ 133

8.2.28   A $T \in B(H)$ Such That $|\lambda| \in \sigma(|T|)$, but $\lambda \notin \sigma(T)$ 133
8.2.29   A $T \in B(H)$ Such That $\lambda \in \sigma(T)$, yet $|\lambda| \notin \sigma(|T|)$ 133
8.2.30   A Sequence of Normal Operators A_m
         Converging Strongly to A, but
         $\sigma(A) \neq \lim_{m \to \infty} \sigma(A_m)$ 134
8.2.31 $T = \bigoplus_{n \in \mathbb{N}} T_n$, but $\cup_{n \in \mathbb{N}} \sigma(T_n) \neq \sigma(T)$ 134
8.2.32 $A = \bigoplus_{n \in \mathbb{N}} A_n$ Where All (A_n) Are Nilpotent,
         but A Is Not Nilpotent 135

9   **Spectral Radius, Numerical Range** 145
    9.1   Basics 145
    9.2   Questions 146
          9.2.1   An $A \in B(H)$ Such That $r(A) < \|A\|$ 146
          9.2.2   A Quasinilpotent Operator Which Is Not Nilpotent 146
          9.2.3   A Non-normaloid Operator T Such That
                   $\mathrm{Re}\, T \geq 0$ 147
          9.2.4   An Operator A with $A \neq I$, $\sigma(A) = \{1\}$
                   and $\|A\| = 1$ 147
          9.2.5   No Inequality Between $r(AB)$ and $r(A)r(B)$
                   Need to Hold 148
          9.2.6   The Discontinuity of $A \mapsto r(A)$ with Respect
                   to the S.O.T. 148
          9.2.7   Two Commuting Normaloid Operators with a
                   Non-normaloid Product 149
          9.2.8   Two Commuting Self-Adjoint Operators A and
                   B Such That $W(A + B) \not\supset W(A) + W(B)$ 149
          9.2.9   A Bounded Operator Whose Numerical Range
                   Is Not a Closed Set in $\mathbb{C}$ 149
          9.2.10   Neither $\sigma(A) \subset W(A)$ Nor $W(A) \subset \sigma(A)$
                   Need to Hold, Even for Self-Adjoint Operators 150
          9.2.11   Two Self-Adjoint $A, B \in B(H)$ with Equal
                   Spectra but Different Numerical Ranges 150
          9.2.12   On the Failure of $\sigma(AB) \subset \overline{W(A)} \cdot \overline{W(B)}$ 150
          9.2.13   On the Failure of $\sigma(A^{-1}B) \subset \overline{W(B)}/\overline{W(A)}$
                   for An Invertible A 151
          9.2.14   A Nonnormal Operator A with $\overline{W(A)} = \mathrm{conv}(\sigma(A))$ ... 151

10 **Compact Operators** 159
    10.1 Basics 159
    10.2 Questions 160
          10.2.1   A Non-compact A Such That A^2 Is Compact 160
          10.2.2   A Non-compact Operator A Such That
                   $\|Ae_n\| \to 0$ as $n \to \infty$ Where (e_n) Is An
                   Orthonormal Basis 160
          10.2.3   Does the Shift Operator Commute with a
                   Nonzero Compact Operator? 161

10.2.4   A Compact Operator Which Is Not Hilbert-Schmidt....   161
10.2.5   A Compact Operator Without Eigenvalues   162
10.2.6   A Non-positive Integral Operator Having a
         Positive Kernel   162
10.2.7   On λ-Commutativity Related to Compact and
         Finite Rank Operators....................................   163
10.2.8   The Infinite Direct Sum of Compact Operators
         Need Not Be Compact   164
10.2.9   A Compact Quasinilpotent Operator A with
         $0 \notin W(A)$..   164

11   Functional Calculi   169
11.1   Basics   169
11.2   Questions   171
         11.2.1   A Bounded Borel Measurable Function f
                  and a Self-Adjoint $A \in B(H)$ Such That
                  $\sigma(f(A)) \neq f(\sigma(A))$....................................   171
         11.2.2   A Normal Operator A Having Only Real
                  Eigenvalues But A Is Not Self-Adjoint   171
         11.2.3   A Non-self-adjoint Operator Whose Square
                  Root Is Its Adjoint..   172
         11.2.4 $A, B, C \in B(H)$, A Being Self-Adjoint,
                  $AB = CA$ but $f(A)B \neq Cf(A)$ for Some
                  Continuous Function f   172
         11.2.5   A Positive A Such That $BA = AB^*$ and
                  $B\sqrt{A} \neq \sqrt{A}B^*$ 172
         11.2.6   An $A \in B(H)$ Such That $\|A\| \neq \sup_{\|x\|\leq 1} |\langle Ax, x \rangle|$....   173
         11.2.7   A Normal $A \in B(H)$ (Where H Is Over the
                  Field $\mathbb{R}$) Such That $\|A\| \neq \sup_{\|x\|\leq 1} |\langle Ax, x \rangle|$ 173
         11.2.8   An A Such That $p(A) \neq A^*$ for Any
                  Polynomial p   173

12   Fuglede-Putnam Theorems and Intertwining Relations   179
12.1   Basics   179
12.2   Questions   180
         12.2.1   Does $TA = AT$ Give $TA^* = A^*T$ When
                  $\ker A = \ker A^*$?   180
         12.2.2   If $T, N, M \in B(H)$ Are Such That
                  N and M Are Normal, Do We Have
                  $\|TN - MT\| = \|TN^* - M^*T\|$? How About
                  $|TN - MT| = |TN^* - M^*T|$?........................   180
         12.2.3   A Self-Adjoint A and a Normal B Such That
                  $AB = BA$ but $B^*A \neq AB$ 181
         12.2.4   A Unitary $U \in B(H)$ and a Self-Adjoint
                  $A \in B(H)$ Such That $AU^* = UA$ and $AU \neq UA$......   181
         12.2.5   Two Nonnormal Double Commuting Matrices   181

    12.2.6 $AN = MB \nRightarrow AN^* = M^*B$ Even When All
              of M, N, A and B Are Unitary 182

    12.2.7 $AN = MB \nRightarrow BN^* = M^*A$ Where M Is an
              Isometry or N Is a Co-isometry 182

    12.2.8   Positive Invertible Operators A, B, N, M Such
              That $AN = MB$ and $AB = BA$ Yet $BN^* \neq M^*A$ 182

    12.2.9   On Some Generalization of the
              Fuglede-Putnam Theorem Involving
              Contractions 183

    12.2.10 A, B, $C \in B(H)$ with A Being Self-Adjoint
              Such That $AB = \lambda CA$ and λ Is Arbitrary 183

    12.2.11 Are There Two Normal Operators A and B
              Such That $AB = 2BA$? 184

    12.2.12 A Normal A and a Self-Adjoint B Such That
              AB Is Normal but BA Is Not Normal 184

    12.2.13 About Some Embry's Theorem 185

13 Operator Exponentials 191
  13.1 Basics 191
  13.2 Questions 191
    13.2.1   An Invertible Matrix Which is Not the
              Exponential of Any Other Matrix 191

    13.2.2   A Compact $A \in B(H)$ for Which e^A is Not
              Compact 192

    13.2.3 $e^A = e^B \nRightarrow A = B$ 192

    13.2.4   A Non-Self-Adjoint $A \in B(H)$ Such That e^{iA}
              Is Unitary 192

    13.2.5   A Normal A Such That $e^A = e^B$ But $AB \neq BA$ 192

    13.2.6   Two Self-Adjoint A, B Such That $A \geq B$
              and $e^A \ngeq e^B$ 193

    13.2.7   A Self-Adjoint $A \in B(H)$ Such That $\|e^A\| \neq e^{\|A\|}$ 193

    13.2.8   A Nonnormal Operator A Such That e^A Is Normal 194

    13.2.9 A, $B \in B(H)$, $AB \neq BA$ and $e^{A+B} \neq e^A e^B$
              $\neq e^B e^A$ 194

    13.2.10 A, $B \in B(H)$, $AB \neq BA$ and $e^{A+B} = e^A e^B$
              $= e^B e^A$ 194

    13.2.11 A, $B \in B(H)$, $AB \neq BA$ and e^{A+B}
              $\neq e^A e^B = e^B e^A$ 194

    13.2.12 A, $B \in B(H)$, $AB \neq BA$ and $e^{A+B} = e^A e^B$
              $\neq e^B e^A$ 195

    13.2.13 Real Matrices A and B Such That $AB \neq BA$
              and $e^{A+B} = e^A e^B \neq e^B e^A$ 195

    13.2.14 An Operator A with $e^A e^{A^*} = e^{A^*} e^A \neq e^{A+A^*}$ 195

    13.2.15 An Operator T Such That $|e^T| \neq e^{\operatorname{Re} T}$
              and $|e^T| \nleq e^{|T|}$ 195

14   Nonnormal Operators.. 207
  14.1   Basics 207
  14.2   Questions 210
      14.2.1   A Hyponormal Operator Which Is Not Normal 210
      14.2.2   An Invertible Hyponormal Operator Which Is
             Not Normal 210
      14.2.3   A Hyponormal Operator with a
             Non-Hyponormal Square 210
      14.2.4   A Hyponormal Operator $A \in B(H)$ Such That
             $A + \lambda A^*$ is Not Hyponormal for Some $\lambda \in \mathbb{C}$.......... 210
      14.2.5   On the Failure of Some Property on the
             Spectrum for Hyponormal Operators 211
      14.2.6   On the Failure of Some Friedland's Conjecture 211
      14.2.7   On the Failure of the ("Unitary") Polar
             Decomposition for Quasinormal Operators 211
      14.2.8   The Inclusions Among the Classes of
             Quasinormals, Subnormals, Hyponormals,
             Paranormals, and Normaloids are Proper 212
      14.2.9   A Subnormal (Or Quasinormal) Operator
             Whose Adjoint Is Not Subnormal 212
      14.2.10 A Paranormal Operator Whose Adjoint Is Not
             Paranormal.. 212
      14.2.11 A Semi-Hyponormal Operator Which Is Not
             Hyponormal 213
      14.2.12 Embry's Criterion for Subnormality:
             $A = A^* A^2$ Does Not Even Yield the
             Hyponormality of $A \in B(H)$ 213
      14.2.13 An Invertible Hyponormal Operator Which Is
             Not Subnormal 214
      14.2.14 Two Commuting Hyponormal Operators A and
             B Such That $A + B$ Is Not Hyponormal................ 214
      14.2.15 Two Quasinormal Operators $A, B \in B(H)$
             Such That $AB = BA = 0$ Yet $A + B$ Is Not
             Even Hyponormal 215
      14.2.16 Two Commuting Subnormal Operators S and
             T Such That Neither ST nor $S + T$ Is Subnormal 215
      14.2.17 The Failure of the Fuglede Theorem for
             Quasinormals et al.. 215
      14.2.18 A Unitary A and a Quasinormal B with
             $TB = AT$ but $TB^* \neq A^*T$ 216
      14.2.19 Two Double Commuting Nonnormal
             Hyponormal Operators................................... 216
      14.2.20 Two Hyponormal (Or Quasinormal) Operators
             A and B Such That $AB^* = B^*A$ but $AB \neq BA$ 216

14.2.21   Two Commuting (Resp. Double Commuting)
          Paranormal Operators Whose Tensor (Resp.
          Usual) Product Fails to Remain Paranormal   217
14.2.22   A Subnormal Operator Such That $|A^2| \neq |A|^2$ 218
14.2.23   A Subnormal A Such That $A|A| \neq |A|A$ 218
14.2.24   A Non-Quasinormal T Such That $T^{*2}T^2 = (T^*T)^2$ ...   218
14.2.25   A Non-Subnormal Operator Whose Powers
          Are All Hyponormal   219
14.2.26   A Hyponormal Operator $T \in B(H)$ Such That
          All T^n, $n \geq 2$ Are Subnormal but T Is Not............   220
14.2.27   Two Quasinormal Operators A and B Such
          That $AB = BA$ and $|AB| \neq |A||B|$.....................   220
14.2.28   Two Quasinormal Operators A and B Such
          That $AB = BA$ Yet $A|B| \neq |B|A$ 221
14.2.29   Two Quasinormal Operators A and B Such
          That $AB^* = B^*A$ and $B|A| = |A|B$ Yet
          $A|B| \neq |B|A$ 221
14.2.30   The Failure of Some Kaplansky's Theorem for
          Hyponormal Operators..................................   221
14.2.31   The Inequality $|\langle Ax, x \rangle| \leq \langle |A|x, x \rangle$ for All x
          Does Not Yield Hyponormality   222
14.2.32   The Failure of Reid's Inequality for
          Hyponormal Operators..................................   222
14.2.33   The Weak Limit of Sequences of Hyponormal
          Operators   222
14.2.34   Strong (and Weak) Limit of Sequences of
          Quasinormal Operators   223
14.2.35   A Quasinormal Operator $A \in B(H)$ Such That
          e^A Is Not Quasinormal   223
14.2.36   An Invertible Subnormal Operator A Without
          Any Bounded Square Root   224
14.2.37   An Invertible Operator Which Is Not the
          Exponential of Any Operator   224
14.2.38   A Positive Answer to the Curto-Lee-Yoon
          Conjecture About Subnormal and Quasinormal
          Operators, and a Related Question   224
14.2.39   A Subnormal Operator T Such That $p(T)$
          Is Quasinormal for Some Non-Constant
          Polynomial p yet T Is Not Quasinormal................   226
14.2.40   A Binormal Operator Which Is Not Normal   226
14.2.41   An Invertible Binormal Operator T Such That
          T^2 Is Not Binormal   226
14.2.42   Two Double Commuting Binormal Operators
          Whose Sum Is Not Binormal   227

14.2.43 A Non-Binormal Operator T Such That
$T^n = 0$ for Some Integer $n \geq 3$ 227
14.2.44 A Subnormal Operator Which Is Not Binormal 227
14.2.45 A Hyponormal Binormal Operator Whose
Square Is Not Binormal 227
14.2.46 Binormal Operators and the Polar Decomposition 228
14.2.47 Does the θ-Class Contain Subnormals?
Hyponormals? 228
14.2.48 An Operator in the θ-Class Which Is Not Even
Paranormal 229
14.2.49 Is the θ-Class Closed Under Addition? 229
14.2.50 Is the θ-Class Closed Under the Usual Product
of Operators? 229
14.2.51 Posinormal Operators 229
14.2.52 (α, β)-Normal Operators 230

15 Similarity and Unitary Equivalence 263
15.1 Questions 263
15.1.1 Two Similar Operators Which Are Not
Unitarily Equivalent 263
15.1.2 Two Operators Having Equal Spectra but They
Are Not Metrically Equivalent 264
15.1.3 Two Metrically Equivalent Operators yet They
Have Unequal Spectra 264
15.1.4 A Non-Self-Adjoint Normal Operator Which Is
Unitarily Equivalent to Its Adjoint 264
15.1.5 Two Commuting Self-Adjoint Invertible
Operators Having the Same Spectra yet They
Are Not Unitarily Equivalent 264
15.1.6 A Matrix Which Is Not Unitarily Equivalent to
Its Transpose 265
15.1.7 On Some Similarity Result by J. P. Williams 265
15.1.8 Two Self-Adjoint Operators A and B Such
That AB Is Not Similar to BA 265
15.1.9 Two Self-Adjoint Matrices A and B Such That
AB Is Not Unitarily Equivalent to BA 266
15.1.10 Two Normal Matrices A and B Such That AB
Is Not Similar to BA 266
15.1.11 Two Self-Adjoint A, B with
$\dim \ker AB = \dim \ker BA$ yet AB Is
Not Similar to BA 267
15.1.12 Two Similar Congruent Operators A and B
Which Are Not Unitarily Equivalent 267
15.1.13 Two Quasi-Similar Operators A, B with
$\sigma(A) \neq \sigma(B)$ 267

          15.1.14 Two Quasi-Similar Operators A and B Such
                   That One Is Compact and the Other Is Not 268
          15.1.15 Two Similar Subnormal Operators Which Are
                   Not Unitarily Equivalent 268
          15.1.16 Two Quasi-Similar Hyponormal Operators
                   Which Are Not Similar 269
          15.1.17 A Quasinilpotent Operator Not Similar to Its
                   Multiple 269

16 **The Sylvester Equation** 281
    16.1   Basics 281
    16.2   Questions 281
          16.2.1   The Condition $\sigma(A) \cap \sigma(B) = \varnothing$ Is Not
                   Necessary for the Existence of a Solution to
                   Sylvester's Equation 281
          16.2.2   An Equation $AX - XB = C$ Without a
                   Solution X for Some $C \in B(H)$ Where
                   $\sigma(A) \cap \sigma(B) \neq \varnothing$ 282
          16.2.3   Unitary Equivalence and Sylvester's Equation 282
          16.2.4   Schweinsberg's Theorem and the Sylvester
                   Equation for Quasinormal Operators.................... 282

17 **More Questions and Some Open Problems** 285
    17.1   More Questions... 285
    17.2   Some Open Problems 288

Part II   Unbounded Linear Operators

18 **Basic Notions** 293
    18.1   Basics 293
    18.2   Questions 298
          18.2.1   An Unbounded Linear Functional That Is
                   Everywhere Defined 298
          18.2.2   An Unbounded Linear Operator That Is
                   Everywhere Defined from H into H 299
          18.2.3   A Non-densely Defined $T \neq 0$ with
                   $\langle Tx, x \rangle = 0$ for All $x \in D(T)$............................ 299
          18.2.4   A Densely Defined Linear Operator A
                   Satisfying $D(A) = D(A^2) = \cdots = D(A^n) \neq H$ 300
          18.2.5   A Densely Defined Operator A Such That
                   $D(A^2) = \{0\}$... 300
          18.2.6   A Bounded B (Not Everywhere Defined) and a
                   Densely Defined A Such That $D(BA) \neq D(A)$ 300
          18.2.7   A Densely Defined T with ran $T \cap D(T) = \{0\}$
                   yet $D(T^2) \neq \{0\}$ 301

18.2.8   Two Densely Defined Operators A and B Such
          That $D(A + B) = D(A) \cap D(B) = \{0\}$ 301
18.2.9   The Same Symbol T with Two Nontrivial
          Domains D and D' Such That $D \cap D' = \{0\}$ 302
18.2.10   The Sum of a Bounded Operator and an
          Unbounded Operator Can Be Bounded 302
18.2.11   Two Densely Defined T and S Such That
          $T - S \subset 0$ But $T \not\subset S$ 302
18.2.12   Three Densely Defined Operators A, B, and C
          Satisfying $A(B + C) \not\subset AB + AC$ 302

19 Closedness 307
19.1   Basics 307
19.2   Questions 316
        19.2.1   A Closed Operator Having Any Other Operator
                  as an Extension 316
        19.2.2   A Bounded Operator That Is Not Closed, and
                  an Unbounded Operator That Is Closed 317
        19.2.3   A Left Invertible Operator That Is Not Closed 317
        19.2.4   A Right Invertible Operator That Is Not Closed 317
        19.2.5   A Closable A That Is Injective but $\overline{A}$ Is Not Injective ... 317
        19.2.6   A Closed Densely Defined Unbounded
                  Operator A and a $V \in B(H)$ Such That AV Is
                  Not Densely Defined 318
        19.2.7   Two Closed Operators A and B Such That BA
                  Is Not Closed 318
        19.2.8   A Compact $B \in B(H)$ and a Closed Operator
                  A Such That BA Is Not Closed......................... 318
        19.2.9   The Closedness of AB with $B \in B(H)$ Need
                  Not Yield the Closedness of A 319
        19.2.10   The Closedness of BA with $B \in B(H)$ Need
                  Not Yield the Closability of A 319
        19.2.11   A Left Invertible Operator A and a Closed
                  Operator B Such That AB Is Unclosed 319
        19.2.12   A Right Invertible Operator A and a Closed
                  Operator B Such That AB Is Unclosed 319
        19.2.13   Two Closable A and B Such That $\overline{AB} \neq \overline{A}\,\overline{B}$ 320
        19.2.14   Three Densely Defined Closed Operators A, B,
                  and C Satisfying $\overline{ABC} \neq A\overline{BC}$ 320
        19.2.15   A Densely Defined Unbounded Closed
                  (Nilpotent) Operator A Such That A^2 Is
                  Bounded and Unclosed 320
        19.2.16   Another Densely Defined Unbounded Closed
                  Operator T Such That T^2 Is Bounded and Unclosed.... 321

19.2.17 A Densely Defined Unbounded and Closed
Operator A Such That
$D(A) = D(A^2) = \cdots = D(A^n) \neq H$ 321

19.2.18 A Non-closable Unbounded Nilpotent Operator......... 321

19.2.19 Unbounded Closed (Resp., Unclosable)
Idempotent Operators 321

19.2.20 An Unclosed (but Closable) A Such That A^2
Is Closed 322

19.2.21 An Unclosed and Closable T Such That T^2 Is
Closed (Such That $D(T^2) \neq \{0\}$) 322

19.2.22 A Densely Defined Unclosed Closable A Such
That A^2 Is Closed and Densely Defined 322

19.2.23 An Unbounded Densely Defined Closed
Operator A Such That $A^2 \neq 0$ and $A^3 = 0$ 323

19.2.24 A Non-closable Operator A Such That $A^2 = 0$
Everywhere on H 323

19.2.25 An Unclosable Operator T Such That $T^2 = I$
Everywhere on H 324

19.2.26 An Everywhere Defined Unbounded Operator
That Is Neither Injective Nor Surjective 324

19.2.27 An Everywhere Defined Unbounded Operator
That Is Injective but Non-surjective 324

19.2.28 An Everywhere Defined Unbounded Operator
That Is Surjective But Non-injective 325

19.2.29 A Non-closable Unbounded Operator T Such
That $T^n = I$ Everywhere on H and $T^{n-1} \neq I$ 325

19.2.30 An Unclosable Unbounded Operator T Such
That $T^n = 0$ Everywhere on H While $T^{n-1} \neq 0$ 325

19.2.31 Three Densely Defined Closed Operators A, B,
and C Satisfying $A(B + C) \not\subset AB + AC$ 325

19.2.32 A Closed A and a Bounded (Not Everywhere
Defined) B Such That $A + B$ Is Unclosed 326

19.2.33 Two Closed Operators with an Unclosed Sum........... 326

19.2.34 Closable Operators A and B with $\overline{A + B} \neq \overline{A} + \overline{B}$ 326

19.2.35 Three Densely Defined Closed Operators A, B,
and C Satisfying: $A + \overline{B + C} \neq \overline{A + B} + C$ 326

19.2.36 Two Non-closable Operators Whose Sum Is
Bounded and Self-Adjoint 327

19.2.37 Two Closed Operators Whose Sum Is Not Even
Closable... 327

19.2.38 Two Densely Defined Closed Operators S and
T Such That $D(S) \cap D(T) = \{0\}$ 327

19.2.39 An Unbounded Nilpotent N Such That $I + N$
Is Not Boundedly Invertible 327

20 Adjoints, Symmetric Operators 345
   20.1 Basics 345
   20.2 Questions 356
       20.2.1 Is There Some Densely Defined Non-closable
             Operator A Such That $A^*A = I$? 356
       20.2.2 A Densely Defined T Such That $H \neq D(T)$
             $+ \operatorname{ran}(T^*)$ 357
       20.2.3 A Densely Defined Symmetric Operator A
             Such That $\ker(A) \neq \ker(A^*)$ 357
       20.2.4 The Failure of Some Maximality Relations.............. 357
       20.2.5 A Non-closable Symmetric Operator 358
       20.2.6 A Densely Defined Operator T Such That
             $D(T^*) = \{0\}$ 358
       20.2.7 A Densely Defined Operator T on a Hilbert
             Space Such That $T^* = 0$ on $D(T^*) \neq \{0\}$.............. 358
       20.2.8 An Everywhere Defined Linear Operator T
             Such That $D(T^*) = \{0\}$ 358
       20.2.9 A Densely Defined T with
             $D(TT^*) = D(T^*T) = D(T)$ 359
       20.2.10 Two Densely Defined Operators A and B Such
             That $(A + B)^* \neq A^* + B^*$ 359
       20.2.11 Two Densely Defined A and B (One of Them
             Is Bounded but Non-everywhere Defined) with
             $(A + B)^* \neq A^* + B^*$ 359
       20.2.12 Two Densely Defined Operators A and B Such
             That $(BA)^* \neq A^*B^*$ 360
       20.2.13 A Densely Defined Left Invertible Operator T
             and a Densely Defined Operator S Such That
             $(ST)^* \neq T^*S^*$ 360
       20.2.14 A Densely Defined Operator A with $(A^2)^* \neq (A^*)^2$ 361
       20.2.15 A Densely Defined Operator T with $(T^2)^* = (T^*)^2$ 361
       20.2.16 Two Bounded (Non-everywhere Defined) or
             Unbounded Operators A and B Such That
             $A \subset B$ and $A^* = B^*$ 361
       20.2.17 A Non-closed Densely Defined Operator T
             Such That $\operatorname{ran} T \subset D(T)$ and $\operatorname{ran} T^* \subset D(T^*)$ 361
       20.2.18 A Closed Densely Defined Operator T Such
             That $\operatorname{ran} T \subset D(T)$ and $\operatorname{ran} T^* \subset D(T^*)$ 362
       20.2.19 A Densely Defined Unbounded Closed
             Operator A Such That $\operatorname{ran} A \subset D(A)$ But
             $\operatorname{ran} A^* \not\subset D(A^*)$ 362
       20.2.20 A Densely Defined Closed Operator A with
             $A + A^*$ Being Densely Defined But Unclosed 362
       20.2.21 A Closed Operator A Satisfying $AA^* = A^*A + I$ 362

20.2.22   A Closed T with $D(T) = D(T^*)$ but
          $D(TT^*) \neq D(T^*T)$ 363
20.2.23   An Unbounded Closed Symmetric and Positive
          Operator A Such That $D(A^2) = \{0\}$ 363

21   Self-Adjointness... 375
21.1   Basics 375
21.2   Questions 382
       21.2.1   Self-Adjoint Operators A with $D(A) \not\subset D(A^2)$ 382
       21.2.2   Symmetric (Closed) Operators Are Not
                 Maximally Self-Adjoint.................................. 382
       21.2.3 $A^*A \subset AA^* \not\Rightarrow A^*A = AA^*$ 382
       21.2.4   Two Self-Adjoint Operators (T, D) and
                 (T, D') Such That T Is Not Self-Adjoint on $D \cap D'$.... 383
       21.2.5   A Dense Domain $D(A) \subset L^2(\mathbb{R})$ for
                 $Af(x) = xf(x)$ on which A Is Closed and
                 Symmetric but Non-self-adjoint 383
       21.2.6   A Dense Domain $D(B) \subset L^2(\mathbb{R})$ for
                 $Bf(x) = if'(x)$ on which B Is Only Closed
                 and Symmetric 383
       21.2.7   A Closed Densely Defined Non-self-Adjoint A
                 Such That A^2 Is Self-Adjoint........................... 384
       21.2.8   A Bounded Densely Defined Essentially
                 Self-Adjoint Operator A Such That $D(A^2) = \{0\}$....... 384
       21.2.9   An Unbounded Densely Defined Essentially
                 Self-Adjoint Operator A Such That $D(A^2) = \{0\}$....... 384
       21.2.10   A Densely Defined Closed Symmetric Operator
                 Which Has Many Self-Adjoint Extensions 384
       21.2.11   A Closed and Symmetric Operator Without
                 Self-Adjoint Extensions................................. 385
       21.2.12   An Unbounded Operator A with Domain $D(A)$
                 Such That $\langle Ax, x \rangle \geq 0$ for All $x \in D(A)$, yet
                 A Is Not Self-Adjoint 385
       21.2.13   A Densely Defined Operator A Such That
                 Neither A^*A nor AA^* Is Self-Adjoint 385
       21.2.14   A Non-closed A Such That A^*A Is Self-Adjoint 385
       21.2.15   Two Closed Operators A and B Such That
                 $A + B$ Is Unclosed and Unbounded..................... 386
       21.2.16   Two Unbounded Self-Adjoint Operators A and
                 B Such That $D(A) \cap D(B) = \{0\}$ 386
       21.2.17   Two Unbounded Self-Adjoint, Positive, and
                 Boundedly Invertible Operators A and B with
                 $D(A) \cap D(B) = \{0\}$ 386

21.2.18   An Unbounded Self-Adjoint Operator
          A Defined on $D(A) \subset H$ Such That
          $D(A) \cap D(U^*AU) \neq \{0\}$ for any unitary
          $U \in B(H)$ 387
21.2.19 $D(A) \cap D(B) = \{0\}$ Is Equivalent to
          $D(S) \cap D(T) = \{0\}$ where A and B are
          Positive Boundedly Invertible Self-Adjoint
          Operators, and S and T Are Closed   387
21.2.20   Three Unbounded Self-Adjoint Operators
          R, S, and T with $D(R) \cap D(S) \neq \{0\}$,
          $D(R) \cap D(T) \neq \{0\}$, and $D(S) \cap D(T) \neq \{0\}$,
          yet $D(R) \cap D(S) \cap D(T) = \{0\}$ 388
21.2.21   Self-Adjoint Positive Operators C and B Such
          That $D(C) \cap D(B) = \{0\}$, yet $D(C^\alpha) \cap D(B^\alpha)$
          Is Dense for All $\alpha \in (0, 1)$..............................   388
21.2.22   Invertible Unbounded Self-Adjoint
          Operators A and B Such That
          $D(A) \cap D(B) = D(A^{-1}) \cap D(B^{-1}) = \{0\}$ 388
21.2.23   Positive Operators T and S with Dense
          Ranges Such That $\operatorname{ran}(T^{\frac{1}{2}}) = \operatorname{ran}(S^{\frac{1}{2}})$ and
          $\operatorname{ran}(T) \cap \operatorname{ran}(S) = \{0\}$ 389
21.2.24   An Unbounded Self-Adjoint Positive
          Boundedly Invertible A and an Everywhere
          Defined Bounded Self-Adjoint B Such That
          $D(AB) = \{0\}$ and $D(BA) \neq \{0\}$.........................   389
21.2.25   An Unbounded Self-Adjoint Positive C and a
          Positive $S \in B(H)$ Such That $D(CS) = \{0\}$,
          $D(CS^\alpha) \neq \{0\}$, and $D(C^\alpha S^\alpha) \neq \{0\}$ for Each
          $0 < \alpha < 1$ 389
21.2.26   An Unbounded Self-Adjoint Positive Operator
          A on $L^2(\mathbb{R})$ Such That $D(A) \nsubseteq D(\mathcal{F}^*A^\alpha\mathcal{F})$
          for Any $0 < \alpha < 1$ Where $\mathcal{F}$ Is the
          $L^2(\mathbb{R})$-Fourier Transform   390
21.2.27   Another Densely Defined Linear Operator T
          Such That $D(T^*) = \{0\}$ 390
21.2.28   A T with $D(T^2) = D(T^*) = D(TT^*) =$
          $D(T^*T) = \{0\}$ 390
21.2.29   An Unclosed Operator T Such That TT^* and
          T^*T Are Closed, but Neither TT^* Nor T^*T Is
          Self-Adjoint   390
21.2.30   Yet Another Densely Defined Operator T Such
          That $D(T^*)$ Is Not Dense.................................   391
21.2.31   On the Operator Equation $A^*A = A^2$ 391

21.2.32 A Closed Operator A with $A^*A \subset A^2$ (or
         $A^2 \subset A^*A$), yet A Is Not Self-Adjoint................... 391
21.2.33 A Rank One Self-Adjoint Operator B and
         an Unbounded, Self-Adjoint, Positive, and
         Invertible A Such That BA Is Not Even Closable....... 391
21.2.34 Two Unbounded Self-Adjoint Operators A and
         B Such That $A + B$ Is Not Self-Adjoint................. 392
21.2.35 On the Failure of the Essential Self-Adjointness
         of $A + B$ for Some Closed and Symmetric A and B.... 392
21.2.36 A Densely Defined Closed A Such That
         $A + A^*$ Is Closed, Densely Defined, Symmetric
         but Non-self-adjoint....................................... 393
21.2.37 A Densely Defined and Closed but
         Non-symmetric Operator A Such That $A + A^*$
         Is Self-Adjoint... 393
21.2.38 Values of λ for which $A + \lambda|A|$ Is (Not)
         Self-Adjoint or (Not) Closed, Where A Is
         Closed and Symmetric 393
21.2.39 An Unbounded Self-Adjoint Operator A Such
         That $|A| \pm A$ Are Not Even Closed 394
21.2.40 Two Unbounded Self-Adjoint and Positive
         Operators A and B Such That $AB + BA$ Is Not
         Self-Adjoint 394
21.2.41 A Closed and Densely Defined Operator T
         Such That $D(T + T^*) = D(TT^*) \cap D(T^*T) = \{0\}$ ... 394
21.2.42 Closed S and T with $D(ST) = D(TS) =$
         $D(S + T) = \{0\}$ 395
21.2.43 Two Densely Defined Closed Operators A
         and B Such That $\overline{D(A^*) \cap D(B)} = H$ and
         $D(A) \cap D(B^*) = \{0\}$................................... 395
21.2.44 Two Unbounded Self-Adjoint Positive
         Invertible Operators A and B Such That
         $D(A^{-1}B) = D(BA^{-1}) = \{0\}$........................... 395
21.2.45 A Densely Defined Unbounded Closed
         Operator B Such That B^2 and $|B|B$ Are
         Bounded, Whereas $B|B|$ Is Unbounded and Closed 396
21.2.46 A Closed Operator T with $D(T^2) = D(T^{*2}) = \{0\}$ 396
21.2.47 A Densely Defined Closed T with $D(T^2) \neq \{0\}$
         and $D(T^{*2}) \neq \{0\}$ but $D(T^3) = D(T^{*3}) = \{0\}$......... 396
21.2.48 A Densely Defined Closed T Such That
         $\mathcal{D}(T^3) \neq \{0\}$ and $\mathcal{D}(T^{*3}) \neq \{0\}$ yet
         $\mathcal{D}(T^4) = \mathcal{D}(T^{*4}) = \{0\}$ 397

21.2.49 A Densely Defined Closed T Such That
$\mathcal{D}(T^5) \neq \{0\}$ and $\mathcal{D}(T^{*5}) \neq \{0\}$ While
$\mathcal{D}(T^6) = \mathcal{D}(T^{*6}) = \{0\}$ 397

21.2.50 For Each $n \in \mathbb{N}$, There Is a Closed Operator T
with $\mathcal{D}(T^{2^n-1}) \neq \{0\}$ and $\mathcal{D}(T^{*2^n-1}) \neq \{0\}$
but $\mathcal{D}\left(T^{2^n}\right) = \mathcal{D}\left(T^{*2^n}\right) = \{0\}$ 397

21.2.51 Two Unbounded Self-Adjoint Operators A and
B with $D(A) = D(B)$ While $D(A^2) \neq D(B^2)$ 398

21.2.52 Self-Adjoint Operators A and B with
$D(A) = D(B)$, but Neither $A^2 - B^2$ nor
$AB + BA$ Is Even Densely Defined 398

21.2.53 On the Impossibility of the Self-Adjointness
of AB and BA Simultaneously When B Is
Closable (Unclosed) and A Is Self-Adjoint 399

21.2.54 The Non-self-adjointness of PAP Where P Is
an Orthogonal Projection and A Is Self-Adjoint 400

21.2.55 The Unclosedness of PAP Where P Is an
Orthogonal Projection and A Is Self-Adjoint 400

21.2.56 Two Self-Adjoint Operators A and B
Such That A Is Positive, $B \in B(H)$, and
$D\left(A^{1/2}BA^{1/2}\right) = \{0\}$ 401

22 **(Arbitrary) Square Roots** 435
22.1 Basics 435
22.2 Questions 435
22.2.1 Compact Operators Having Unbounded Square
Roots 435
22.2.2 A Bounded Invertible Operator Without Any
Closed Square Root 436
22.2.3 A Non-closable Operator Without Any
Closable Square Root 436
22.2.4 An Operator S with a Square Root T but T^* Is
Not a Square Root of S^* 436
22.2.5 An Operator S with a Square Root T but $\overline{T}$ Is
Not a Square Root of $\overline{S}$ 436
22.2.6 A Densely Defined Closed Operator T Such
That T^2 Is Densely Defined and Non-closable 437
22.2.7 Square Roots of a Self-Adjoint Operator Need
Not Have Equal Domains 437

23 **Normality** 441
23.1 Basics 441
23.2 Questions 442
23.2.1 A Normal T^* Such That T Is Not Normal 442
23.2.2 A T Such That $TT^* = T^*T$ yet T Is Not Normal 442

23.2.3   An Unbounded Densely Defined Closed
         Nonnormal Operator T Such That
         $D(T) = D(T^*)$ and $D(TT^*) = D(T^*T)$ 443
23.2.4   A Nonnormal Densely Defined Closed S that
         Satisfies $T \subset S^*S$ and $T \subset SS^*$ 443
23.2.5   Two Unbounded Self-Adjoint Operators A and
         B, B Is Positive, Such That $\overline{AB}$ Is Normal
         Without Being Self-Adjoint 444
23.2.6   Two Unbounded Self-Adjoint Operators A and
         B Such That A Is Positive, AB is Normal But
         Non-self-adjoint 445

24   **Absolute Value. Polar Decomposition** 451
     24.1   Basics 451
     24.2   Questions 451
            24.2.1   Closed Densely Defined Operators S and T
                     Such That $S \subset T$ but $|S| \not\subset |T|$ 451
            24.2.2   A Closed Operator T and an Unclosed
                     Operator A Such That $|T^*| = |T| = |A|$ 452
            24.2.3   A Non-closed A with $D(A) \neq D(|A|)$ 452
            24.2.4   A Non-closed Densely Defined Operator A
                     Without Any Polar Decomposition 452
            24.2.5   A Non-closed Densely Defined Operator A and
                     a Unitary $U \in B(H)$ Such That $A \subset U|A|$ 452
            24.2.6   A Non-closed Densely Defined Operator A and
                     a Unitary $U \in B(H)$ Such That $A \subset U|A|$ and
                     $UA \subset |A|$ 453
            24.2.7   A Closed and Symmetric A Such That $|A^n| \neq |A|^n$ 453
            24.2.8   A Self-Adjoint $B \in B(H)$ and a Self-Adjoint
                     Positive A Such That BA Is Closed Without
                     Being Self-Adjoint 453

25   **Unbounded Nonnormal Operators** 457
     25.1   Questions 457
            25.1.1   An $S \in B(H)$ Such That $0 \notin \overline{W(S)}$, and an
                     Unbounded Closed Hyponormal T Such That
                     $ST \subset T^*S$ but $T \neq T^*$ 457
            25.1.2   A Closed A with $A^*AA \subset AA^*A$ Does Not
                     Yield the Quasinormality of A 458
            25.1.3   An Unbounded Paranormal Operator T Such
                     That $D(T^*) = \{0\}$ 458
            25.1.4   A Closable Paranormal Operator Whose
                     Closure Is Not Paranormal 459
            25.1.5   A Densely Defined Closed Operator T Such
                     That Both T and T^* Are Paranormal and
                     $\ker T = \ker T^* = \{0\}$ but T Is Not Normal 459

	25.1.6	Q-Normal Operators	460
	25.1.7	On the Operator Equations $A^*A = A^n$, $n \geq 3$	460
26	**Commutativity**		465
	26.1	Basics	465
	26.2	Questions	466
	26.2.1	$B\overline{A} \subset \overline{AB} \not\Rightarrow BA \subset AB$, $B \in B(H)$	466
	26.2.2	A Self-Adjoint $B \in B(H)$ and an Unclosed A with $BA \subset AB$ but $f(B)A \not\subset Af(B)$ for Some Continuous Function f	466
	26.2.3	A Self-Adjoint $S \in B(H)$ and a Non-closable T with $ST \subset TS$ but $f(S)T \not\subset Tf(S)$ for Some Continuous Function f	467
	26.2.4	An Unbounded Self-Adjoint Operator Commuting with the L^2-Fourier Transform	467
	26.2.5	A Closed Symmetric A and a Unitary U Such That $A \subsetneq U^*AU$	467
	26.2.6	A Self-Adjoint A and B Where B Is a Bounded Multiplication Operator and A Is a Differential Operator which Commute Strongly	468
	26.2.7	Is the Product of Two "Commuting" Unbounded Self-Adjoint (Respectively, Normal) Operators Always Self-Adjoint (Respectively, Normal)?	468
	26.2.8	Two Unbounded Self-Adjoint Operators A and B which Commute Pointwise on Some Common Core but A and B Do Not Commute Strongly: Nelson-Like Counterexample	469
	26.2.9	A Densely Defined Closed T Such That $T^2 = T^{*2}$ yet T^2 Is Not Self-Adjoint	470
	26.2.10	Is There a Normal T Such That $(T^2)^* \neq T^{*2}$? What About When T^2 Is Normal?	470
	26.2.11	$BA \subset T \not\Rightarrow BA = T$ Even if $B \in B(H)$, A and T Are All Self-Adjoint	471
	26.2.12	$T \subset AB \not\Rightarrow T = AB$ Where T, A, and B Are Normal	471
	26.2.13	Two Unbounded Self-Adjoint Operators A and B which Commute Strongly Where B Is a Multiplication Operator and A Is a Differential Operator	472
	26.2.14	On the Failure of the Ôta–Schmüdgen Criterion of Strong Commutativity for Normal Operators	473
	26.2.15	Anti-Commutativity and Exponentials	473
	26.2.16	A Densely Defined Unbounded Operator Without a Cartesian Decomposition	474

          26.2.17   Are There Unbounded Self-Adjoint Operators
                    A and B with $A + iB \subset 0$?   475
          26.2.18   Two Unbounded Self-Adjoint Operators A and
                    B Such That $\overline{A + iB} \neq \overline{A} + i\overline{B}$ 475
          26.2.19   A Closed Operator T Such That
                    $D(T) = D(T^*)$ but $(T + T^*)/2$ and
                    $(T - T^*)/2i$ Are Not Essentially Self-Adjoint   475
          26.2.20   An Unbounded Normal T Such That Both
                    $(T + T^*)/2$ and $(T - T^*)/2i$ Are Unclosed   476
          26.2.21   Closed Nilpotent Operators Having a Cartesian
                    Decomposition Are Always Everywhere
                    Defined and Bounded   476
          26.2.22   An Unbounded "Nilpotent" Closed Operator
                    Having Positive Real and Imaginary Parts   477
          26.2.23   Two Commuting Unbounded Normal
                    Operators Whose Sum Fails to Remain Normal   477
          26.2.24   A Formally Normal Operator Without Any
                    Normal Extension   477

27   **The Fuglede–Putnam Theorems and Intertwining Relations**   489
     27.1   Basics   489
     27.2   Questions   490
          27.2.1   A Boundedly Invertible Positive Self-Adjoint
                    Unbounded Operator A and an Unbounded
                    Normal Operator N Such That $AN^* = NA$ but
                    $AN \not\subset N^*A$ 490
          27.2.2   A Closed T and a Normal M Such That
                    $TM \subset MT$ but $TM^* \not\subset M^*T$ and $M^*T \not\subset TM^*$ 491
          27.2.3   A Normal $B \in B(H)$ and a Densely Defined
                    Closed A Such That $BA \subset AB$ Yet $B^*A \not\subset AB^*$ 491
          27.2.4   A Self-Adjoint T and a Unitary B with
                    $BT \subset TB^*$ but $B^*T \not\subset TB$ 491
          27.2.5   A Closed Operator Which Does Not Commute
                    with Any (Nontrivial) Everywhere Defined
                    Bounded Operator   492
          27.2.6   A Self-Adjoint and a Closed Operators Which
                    Are Not Intertwined by Any Bounded Operator
                    Apart from the Zero Operator   492
          27.2.7   A Self-Adjoint Operator A and a Closed
                    Symmetric Restriction of A Not Intertwined by
                    Any Bounded Operator Except the Zero Operator   493
          27.2.8   Two Densely Defined Closed Operators A and
                    B Not Intertwined by Any Densely Defined
                    Closed (Nonzero) Operator   493

	27.2.9	On the Failure of a Generalization to Unbounded Operators of Some Similarity Result by M.R. Embry	493
	27.2.10	Are There Two Normal Operators A and $B \in B(H)$ Such That $BA \subset 2AB$ with AB Is Normal?	494
	27.2.11	On a Result About Commutativity by C. R. Putnam	494

28 Commutators ... 503
28.1	Basics	503
28.2	Questions	503
	28.2.1	Two Operators A, B, One of Them Is Unbounded, with $BA - AB \subset I$	503
	28.2.2	On Some Theorem of F. E. Browder About Commutators of Unbounded Operators	504
	28.2.3	Are There Two Closable Unbounded Operators A and B Such That $AB - BA$ is Everywhere Defined?	505
	28.2.4	Two Self-Adjoint Operators A and B Such That $AB + BA$ Is Bounded, While $AB - BA$ Is Unbounded	505
	28.2.5	Two Self-Adjoint Positive Operators (One of Them Is Unbounded) A and B Such That $AB - BA$ Is Unbounded	506
	28.2.6	On the Positivity of Some Commutator	506
	28.2.7	Two Unbounded and Self-Adjoint Operators A and B Such That $D(A) = D(B)$ and $A^2 - B^2$ Is Bounded but $AB - BA$ Is Unbounded	506
	28.2.8	Two Densely Defined Unbounded and Closed Operators C and B Such That $CB - BC$ Is Bounded and Unclosed, While $|C|B - B|C|$ Is Unbounded and Closed	506
	28.2.9	Two Unbounded Injective Self-Adjoint Operators A and B Such That $AB - BA$ Is Bounded, While $|A|B - B|A|$ Is Unbounded	507
	28.2.10	Two Unbounded (Injective) Self-Adjoint Operators A and B Such That $AB - BA$ Is Bounded and Commutes with B (and A) Yet $|A|B - B|A|$ Is Unbounded	507
	28.2.11	Two Unbounded Self-Adjoint, Positive, and Boundedly Invertible Operators A and B Such That $AB - BA$ Is Bounded While $\sqrt{A}\sqrt{B} - \sqrt{B}\sqrt{A}$ Is Unbounded	508
	28.2.12	A Densely Defined Closed Operator T Such That $|T^*||T| - |T||T^*|$ Is Unbounded	508

28.2.13 A Densely Defined and Closed Operator T
         Such That $TT^* - T^*T$ Is Unbounded and
         Self-Adjoint but $|T||T^*| - |T^*||T|$ Is Bounded
         and Unclosed 509
28.2.14 A Densely Defined and Closed Operator T
         Such That $TT^* - T^*T$ Is Bounded Whereas
         $|T||T^*| - |T^*||T|$ Is Unbounded........................ 509
28.2.15 Self-Adjoint Operators A and B Such That
         $\frac{1}{2}|\langle \overline{[A,B]}f, f \rangle| \not\leq \|Af\|\|Bf\|$ 509
28.2.16 Unbounded Skew-Adjoint Operators Cannot
         Be Universally Commutable............................ 509

29 **Spectrum** 519
    29.1 Basics 519
    29.2 Questions 523
          29.2.1   A Densely Defined Closed Operator with an
                   Empty Spectrum... 523
          29.2.2   A Densely Defined Operator A with $\sigma(A) = \mathbb{C}$ 523
          29.2.3   A Densely Defined Closed Operator A
                   with $\sigma(A) = \mathbb{C}$ 523
          29.2.4   Two Unbounded Strongly Commuting
                   Self-Adjoint Operators A and B Such That
                   $\sigma(A + B) \not\subset \sigma(A) + \sigma(B)$ 523
          29.2.5   Two Densely Closed Unbounded Operators A
                   and B Such That $\sigma(A) + \sigma(B)$ Is Not Closed.......... 524
          29.2.6   Two Commuting Self-Adjoint Unbounded
                   Operators A and B Such That $\sigma(BA) \not\subset \sigma(B)\sigma(A)$ 524
          29.2.7   A Densely Defined Operator A with $\sigma(A^2)$
                   $\neq [\sigma(A)]^2$ 524
          29.2.8   A Normal Operator T with
                   $p[\sigma(T)] \neq \sigma[p(T, T^*)]$ for Some
                   Polynomial p 525
          29.2.9   Two Closed (Unbounded) Operators A and B
                   Such That $\sigma(AB) - \{0\} \neq \sigma(BA) - \{0\}$ 525
          29.2.10 A (Non-closed) Densely Defined Operator A
                   Such That $\sigma(AA^*) - \{0\} \neq \sigma(A^*A) - \{0\}$............. 525
          29.2.11 Two Self-Adjoint Operators A and B with
                   $B \in B(H)$ Such That B Commutes with
                   $AB - BA$ yet $\sigma(AB - BA) \neq \{0\}$ 526
          29.2.12 An Unbounded, Closed, and Nilpotent
                   Operator N with $\sigma(N) \neq \{0\}$............................ 526
          29.2.13 An $A \in B(H)$ and an Unbounded Closed
                   Nilpotent Operator N with $AN = NA$ but
                   $\sigma(A + N) \neq \sigma(A)$ 527

29.2.14   A Closed, Symmetric, and Positive Operator T
with $\sigma(T) = \mathbb{C}$ 528

29.2.15   The (Only) Four Possible Cases for the
Spectrum of Closed Symmetric Operators 528

29.2.16   Unbounded Densely Defined Closable
Operators A and B Such That $A \subset B^*$,
BA Is Essentially Self-Adjoint, and
$\sigma(\overline{BA}) = \sigma(\overline{A}\,\overline{B}) \neq \sigma(\overline{AB})$ 528

30   Matrices of Unbounded Operators 539

31.1   Questions 539

30.1.1   Equal Matrices and Pairwise Different Entries 539

30.1.2   A Bounded Matrix with All Unbounded Entries 540

30.1.3   The Failure of the Product Formula for Some
Matrices of Operators 540

30.1.4   A Closed Matrix Yet All Entries Are Unclosed 541

30.1.5   A Matrix Whose Closure Is Not Equal to the
Matrix of Closures 541

30.1.6   A Non-closed (Essentially Self-Adjoint)
Matrix Whose Entries Are All Self-Adjoint 541

30.1.7   A Non-closable Matrix Yet All Entries Are Closed 542

30.1.8   A Densely Defined Matrix Whose Formal
Adjoint Is Not Densely Defined 542

30.1.9   A Closed Matrix Whose Adjoint Does Not
Admit a Matrix Representation 542

30.1.10 A Matrix Whose Adjoint Differs from Its
Formal Adjoint 542

30.1.11 A Non-boundedly Invertible Matrix Yet All
Its Entries Pairwise Commute and Its Formal
Determinant Is Boundedly Invertible..................... 543

30.1.12 A Self-Adjoint Matrix Yet None of Its Entries
Is Even Closed... 543

30.1.13 A Matrix of Operators A of Size $n \times n$,
Where All of Its Entries Are Unclosable and
Everywhere Defined, Each A^p, $1 \leq p \leq n-1$,
Does Not Contain Any Zero Entry but $A^n = 0$ 543

31   Relative Boundedness 553

31.1   Questions 553

31.1.1   A Function $\varphi \in L^2(\mathbb{R}^2)$ Such That
$\frac{\partial^2 \varphi}{\partial x \partial y} \in L^2(\mathbb{R}^2)$ but $\varphi \notin L^\infty(\mathbb{R}^2)$ 553

31.1.2   A $u \in L^2(\mathbb{R}^{n+1})$ with
$(-i\partial/\partial t - \triangle_x)u \in L^2(\mathbb{R}^{n+1})$ yet
$u \notin L_t^q(L_x^r)$ for Given Values of q and r 555

31.1.3   A Function $u \in L^2(\mathbb{R}^2)$ with
$(\partial/\partial t + \partial^3/\partial x^3)u \in L^2(\mathbb{R}^2)$ yet
$u \notin L^p(\mathbb{R}^2)$ for Any $p > 8$ 557

31.1.4   A Negative V in L^2_{loc} Such That $-\Delta + V$ Is
Not Essentially Self-Adjoint on C_0^∞ 557

31.1.5   A Positive V in L^2_{loc} Such That $\frac{\partial^2}{\partial t^2} - \frac{\partial^2}{\partial x^2} + V$
Is Not Essentially Self-Adjoint on C_0^∞ 558

31.1.6   If B Is A-Bounded, Is B^2 A^2-Bounded? Is B^*
A^*-Bounded? 558

31.1.7   The Kato–Rellich Theorem for Three Operators? 558

31.1.8   The Kato–Rellich Theorem for Normal Operators? 559

32   **More Questions and Some Open Problems II** 567

32.1   More Questions 567

32.2   Some Open Problems 568

Appendix A: A Quick Review of the Fourier Transform 571

Appendix B: A Word on Distributions and Sobolev Spaces 575

Bibliography 579

Index 595

Part I
Bounded Linear Operators

Chapter 1
Some Basic Properties

1.1 Basics

Throughout this chapter, H and K denote two Hilbert spaces over $\mathbb{C}$ unless otherwise stated.

Definition 1.1.1 Let $A : H \to K$ be a linear operator, i.e. $A \in L(H, K)$. We say that A is bounded if

$$\exists M \geq 0, \forall x \in H : \ \|Ax\|_K \leq M\|x\|_H.$$

The set of all bounded linear operators from H to K is denoted by $B(H, K)$. In case $H = K$, then we write $B(H)$ instead of $B(H, H)$.

If A is not bounded, then we say that A is unbounded.

Definition 1.1.2 Let $A : H \to K$ be a linear mapping.

(1) The kernel of A (or the null-space of A) is the subspace

$$\ker A = \{x \in H : Ax = 0_K\}.$$

Its dimension is called the nullity of A, that is, the number $n(A) = \dim(\ker A)$.
(2) The range of A (or image of A) is the subspace

$$\operatorname{ran} A = A(H) = \{Ax : x \in H\}.$$

Its dimension is called the rank of A, that is, the number $\dim(\operatorname{ran} A)$.
(3) The graph of A, denoted by $G(A)$, is defined by

$$G(A) = \{(x, Ax) : x \in H\}.$$

© The Author(s), under exclusive license to Springer Nature Switzerland AG 2022
M. H. Mortad, *Counterexamples in Operator Theory*,
https://doi.org/10.1007/978-3-030-97814-3_1

Remark Let A be a linear operator (not necessarily bounded) defined in H. The restriction of A to a linear subspace $D \subset H$ is denoted by A_D or $A|_D$.

Definition 1.1.3 A linear operator $A : H \to \mathbb{F}$ is called a linear functional (where $\mathbb{F}$ denotes either $\mathbb{R}$ or $\mathbb{C}$).

Definition 1.1.4 Let $A \in B(H, K)$. Set

$$\|A\| = \inf\{\alpha \geq 0 : \|Ax\|_K \leq \alpha \|x\|_H, \ x \in H\}.$$

Then $\|A\|$ is called the norm of A.

The following result gives several choices of *equal* norms to use on $B(H, K)$ (with $H \neq \{0\}$).

$$\|A\| = \sup_{x \neq 0} \frac{\|Ax\|}{\|x\|} = \sup_{\|x\| \leq 1} \|Ax\| = \sup_{\|x\| < 1} \|Ax\| = \sup_{\|x\| = 1} \|Ax\|.$$

Examples 1.1.1

(1) Let H be a Hilbert space. The identity operator $I : H \to H$, defined by $Ix = x$ for all $x \in H$, plainly satisfies $\|I\| = 1$.
(2) Let φ be a continuous function on $[a, b]$ taking its values in $\mathbb{C}$. For all $f \in L^2([a, b])$, set

$$(Af)(x) = \varphi(x) f(x).$$

Then A is called a multiplication operator. Also

$$\|A\| = \|\varphi\|_\infty = \sup_{x \in [a,b]} |\varphi(x)| < \infty.$$

(3) Let (α_n) be a bounded complex sequence, and let $A : \ell^2 \to \ell^2$ be defined by

$$Ax = A(x_n) = (\alpha_n x_n).$$

Then $\|A\| = \|\alpha\|_\infty$ (the operator A too is called a multiplication operator).
(4) We define on $L^2[0, 1]$ the operator V (called the Volterra operator)

$$(Vf)(x) = \int_0^x f(t)dt.$$

Then (cf. Exercise 9.3.21 in [256])

$$\|V\| = \frac{2}{\pi}.$$

We make use of the following convention:

Definition 1.1.5 Let $A, B \in B(H)$. The composition $A \circ B$ is called the product of A with B and may be denoted just by AB.

Proposition 1.1.1 *Let $A, B \in B(H)$. Then $AB \in B(H)$ and*

$$\|AB\| \leq \|A\|\|B\|.$$

Next we define commutativity.

Definition 1.1.6 Let $A, B \in B(H)$. We say that A commutes with B (or that A and B commute) if

$$AB = BA, \text{ i.e. } ABx = BAx$$

for all $x \in H$.

The next result allows to introduce the notion of the adjoint of an operator.

Theorem 1.1.2 *Let H and K be two Hilbert spaces with inner products $\langle \cdot, \cdot \rangle_H$ and $\langle \cdot, \cdot \rangle_K$, respectively. Let $A \in B(H, K)$. Then there exists a unique operator in $B(K, H)$, denoted by A^*, such that*

$$\langle Ax, y \rangle_K = \langle x, A^*y \rangle_H$$

for all $x \in H$ and all $y \in K$.

Definition 1.1.7 Let $A \in B(H, K)$. Then the (unique) operator $A^* \in B(K, H)$, introduced in the previous theorem, is called the adjoint of A.

The next operator is so important that we just cannot imagine how operator theory would have been without it! Indeed, it throws light on many questions as it is a valuable source of counterexamples.

Example 1.1.2 Let $S \in B(\ell^2)$ be defined by

$$S(x_1, x_2, \cdots) = (0, x_1, x_2, \cdots).$$

Then S is called the (unilateral) shift operator. It is sometimes more useful to define an operator using its effect on the standard (or other) orthonormal basis. For instance, the shift acts on the standard basis $\{e_n : n \in \mathbb{N}\}$ as

$$Se_n = e_{n+1}, \ n \in \mathbb{N}.$$

We may also represent it as

$$S = \begin{pmatrix} 0 & 0 & 0 & 0 & \cdots \\ 1 & 0 & 0 & 0 & \cdots \\ 0 & 1 & 0 & 0 & \cdots \\ 0 & 0 & 1 & 0 & \cdots \\ \vdots & 0 & 0 & 1 & \ddots \\ \vdots & \vdots & \vdots & \ddots & \ddots \end{pmatrix}.$$

The adjoint of the shift operator is given by

$$S^*(x_1, x_2, x_3, \cdots) = (x_2, x_3, \cdots).$$

Its action on the Hilbert basis $\{e_n : n \in \mathbb{N}\}$ is then:

$$S^* e_1 = 0_{\ell^2}, \ S^* e_n = e_{n-1}, \ n \geq 2.$$

Remark If S is the shift operator on $\ell^2(\mathbb{N})$, then

$$S^* S = I \text{ and } SS^* \neq I.$$

Now, we introduce other "shift" operators:

Definition 1.1.8 On $\ell^2(\mathbb{Z})$, define an operator R by

$$R(x_n)_{n \in \mathbb{Z}} = R(\cdots, x_{-1}, \mathbf{x_0}, x_1, \cdots) = (\cdots, x_{-2}, \mathbf{x_{-1}}, x_0, \cdots),$$

where the "bold" indicates the zeros positions of the sequence. The operator R is called the bilateral (forward or right) shift.

Proposition 1.1.3 *Let R be the bilateral shift on $\ell^2(\mathbb{Z})$. Then $\|R\| = 1$, and its adjoint is given by*

$$R^*(\cdots, x_{-1}, \mathbf{x_0}, x_1, \cdots) = (\cdots, x_{-1}, x_0, \mathbf{x_1}, x_2, \cdots).$$

Definition 1.1.9 The adjoint of R, i.e. R^*, is called the bilateral backward (or left) shift.

Definition 1.1.10 A (unilateral) weighted shift is the operator A defined from ℓ^2 into ℓ^2 by

$$A(x_1, x_2, \cdots) = (0, \alpha_1 x_1, \alpha_2 x_2, \cdots)$$

where (α_n) is a bounded sequence (called a weight). In other words, $A = SM$, where S is the standard shift on ℓ^2 and M is the multiplication operator on ℓ^2 by (α_n).

In the end, we give basic properties of the adjoint operation:

Proposition 1.1.4 *Let A, $B \in B(H)$, and let $\alpha, \beta \in \mathbb{C}$. Then:*

(1) $(\alpha A + \beta B)^* = \overline{\alpha} A^* + \overline{\beta} B^*$.
(2) $(AB)^* = B^* A^*$.
(3) $(A^*)^* = A$.
(4) $\|A^*\| = \|A\|$.
(5) $\|A\|^2 = \|A^* A\|$.

Corollary 1.1.5 *Let $A \in B(H)$. Then*

$$\|A^* A\| = \|A A^*\| = \|A\|^2.$$

In particular, $A^* A = 0$ iff $A A^* = 0$ iff $A = 0$.

1.2 Questions

1.2.1 Does the "Banachness" of $B(X, Y)$ Yield That of Y?

Let X and Y be two normed vector spaces. A well known result states that if Y is a Banach space, then so is $B(X, Y)$ (see e.g. Exercise 2.3.5 in [256] for a detailed proof).

Question 1.2.1 Let X and Y be two normed vector spaces. If $B(X, Y)$ is a Banach space, is Y necessarily a Banach space?

1.2.2 An Operator $A \neq 0$ with $A^2 = 0$ and So $\|A^2\| \neq \|A\|^2$

Let $A \in B(H)$, and let $n \in \mathbb{N}$. We define the iterated operator, denoted by A^n, by composing A with itself n times. In other words, A^n is defined as follows:

$$A^0 = I, \ A^{n+1} = A^n A$$

where I is the identity operator on H.

Obviously, if $A \in B(H)$, then $A^n \in B(H)$ for any $n \in \mathbb{N}$. Moreover,

$$\|A^n\| \leq \|A\|^n$$

for all n.

More generally, if $p(x) = a_0 + a_1 x + \cdots + a_n x^n$ is a polynomial, then we define $p(A)$ symbolically as

$$p(A) = a_0 I + a_1 A + \cdots + a_n A^n$$

where I is the identity operator on H. Clearly, $p(A) \in B(H)$.

Question 1.2.2 Find an $A \in B(H)$ such that $A^2 = 0$ but $A \neq 0$, whereby

$$\|A^2\| \neq \|A\|^2.$$

1.2.3 $A, B \in B(H)$ with $ABAB = 0$ but $BABA \neq 0$

Can we find $A, B \in B(H)$ such that

$$AB = 0 \text{ but } BA \neq 0?$$

Obviously, to get a counterexample, one has to avoid taking e.g. $A = B^*$. As a counterexample, just take

$$A = \begin{pmatrix} 0 & 1 \\ 0 & 0 \end{pmatrix} \text{ and } B = \begin{pmatrix} 1 & 0 \\ 0 & 0 \end{pmatrix}.$$

Then

$$AB = \begin{pmatrix} 0 & 0 \\ 0 & 0 \end{pmatrix} \neq \begin{pmatrix} 0 & 1 \\ 0 & 0 \end{pmatrix} = BA.$$

Question 1.2.3 Find $A, B \in B(H)$ such that

$$ABAB = 0 \text{ but } BABA \neq 0.$$

Hint: According to [149], you need to look for a counterexample when $\dim H \geq 3$.

1.2.4 An Operator Commuting with Both $A + B$ and AB, But It Does Not Commute with Any of A and B

Question 1.2.4 (Cf. [102]) Let $A, B, T \in B(H)$. Find T that commutes with both $A + B$ and AB, but it commutes with none of A and B.

1.2.5 The Non-transitivity of the Relation of Commutativity

Define a binary relation $\mathcal{R}$ on $B(H)$ by

$$A\mathcal{R}B \iff AB = BA.$$

Clearly, $\mathcal{R}$ is reflexive and symmetric. Is it transitive?

Question 1.2.5 Find $A, B, C \in B(H)$ such that A commutes with B and B commutes with C, yet A does not commute with C.

1.2.6 Two Operators A, B with $\|AB - BA\| = 2\|A\|\|B\|$

We have the following notion related to commutativity: Let $A, B \in B(H)$. The commutator of A, B is defined by

$$[A, B] = AB - BA.$$

In particular, $AB = BA$ iff $[A, B] = 0$.

If A, B are in $B(H)$, then we easily obtain

$$\|[A, B]\| \leq 2\|A\|\|B\|.$$

Question 1.2.6 (Cf. Questions 4.2.11 and 4.2.12) Give an example of (non-trivial) A and B for which the equality

$$\|[A, B]\| = 2\|A\|\|B\|$$

holds.

1.2.7   Two Nilpotent Operators Such That Their Sum and Their Product Are Not Nilpotent

Let $A \in B(H)$. We say that A is nilpotent if $A^n = 0$ for some $n \in \mathbb{N}$. The smallest $k \in \mathbb{N}$ satisfying $A^k = 0$ is called the index of nilpotence of A.

While the operator $A = 0$ on H is not excluded from the definition of nilpotence, we will be more interested in finding nonzero nilpotent operators when it comes to counterexamples.

Question 1.2.7 Find two nilpotent operators $A, B \in B(H)$ such that $A + B$ is not nilpotent. The same question for the product AB.

1.2.8   Two Non-nilpotent Operators Such That Their Sum and Their Product Are Nilpotent

Question 1.2.8 Find two non-nilpotent operators $A, B \in B(H)$ such that $A + B$ is nilpotent. The same question for the product AB.

1.2.9   An Invertible Operator A with $\|A^{-1}\| \neq 1/\|A\|$

Let $A \in B(H, K)$. Say that A is invertible if there is a $B \in B(K, H)$ such that

$$AB = I_K \text{ and } BA = I_H,$$

where I_H (resp. I_K) designates the identity operator on H (resp. on K). The operator B is called the inverse of A, and we then write $B := A^{-1}$.

Example 1.2.1 Let φ be a continuous function on say $[a, b]$. Let A be the multiplication operator by φ, i.e.

$$(Af)(x) = \varphi(x)f(x)$$

where $f \in L^2([a, b])$. Then A is invertible if and only if $\varphi(x) \neq 0$ for all x.

Example 1.2.2 Let S be the shift operator on $\ell^2(\mathbb{N})$. Then S is not invertible, and neither is S^*.

Next, we recall some basic properties about invertibility.

Proposition 1.2.1 *Let $A, B \in B(H)$ be invertible. Then:*

(1) A^{-1} *is invertible and* $(A^{-1})^{-1} = A$.
(2) AB *is invertible and*

$$(AB)^{-1} = B^{-1}A^{-1}.$$

(3)

$$AB = BA \iff A^{-1}B^{-1} = B^{-1}A^{-1}.$$

(4) *The adjoint A^* is invertible and*

$$(A^*)^{-1} = (A^{-1})^*.$$

Now, let $A \in B(H)$ be invertible. Then

$$1 = \|I\| = \|AA^{-1}\| \le \|A\|\|A^{-1}\|.$$

So, there is a priori no reason why $\|A^{-1}\| = 1/\|A\|$ should hold true.

Question 1.2.9 Find an invertible operator $A \in B(H)$ such that

$$\|A^{-1}\| \ne \frac{1}{\|A\|}.$$

What if A satisfies $A^2 = I$? Are there cases when $\|A^{-1}\| = 1/\|A\|$ holds anyway?

1.2.10 An $A \in B(H)$ Such That $I - A$ Is Invertible and Yet $\|A\| \ge 1$

Recall a well known result: *If $A \in B(H)$ is a pure contraction, i.e. $\|A\| < 1$, then $I - A$ is invertible and*

$$(I - A)^{-1} = \sum_{n=0}^{\infty} A^n \text{ (the Neumann series).}$$

Question 1.2.10 Find an $A \in B(H)$ such that $I - A$ is invertible yet $\|A\| \ge 1$.

1.2.11   Two Non-invertible $A, B \in B(H)$ Such That AB Is Invertible

Question 1.2.11 Find non-invertible $A, B \in B(H)$ such that AB is invertible.

Hint: You need to avoid cases like: dim $H < \infty$, $AB = BA$, ker $A = $ ker A^*, ker $B = $ ker B^* (see e.g. [84] and [256]).

1.2.12   Two A, B Such That $A + B = AB$ but $AB \neq BA$

Let H be a finite-dimensional Hilbert space, and let I be the identity operator on H. Let $A, B \in B(H)$ be such that $A + B = AB$. Then $AB = BA$. Indeed, we have

$$(A - I)(B - I) = \underbrace{AB - A - B}_{=0} + I = I.$$

Since dim $H < \infty$, the previous actually means that $A - I$ and $B - I$ are both invertible and that also $(B - I)(A - I) = I$. Hence,

$$(A - I)(B - I) = (B - I)(A - I),$$

i.e.

$$AB = BA,$$

as wished.

Question 1.2.12 Find $A, B \in B(H)$ such that $A + B = AB$ but $AB \neq BA$.

1.2.13   Left (Resp. Right) Invertible Operators with Many Left (Resp. Right) Inverses

Recall the following definition: Let $A \in B(H)$. We say that A is left invertible if $BA = I$ for some $B \in B(H)$. Call A right invertible if $AC = I$ for some $C \in B(H)$.

The importance of the previous definition lies in the fact that if $A \in B(H)$ possesses both a left inverse B and a right inverse C, then $B = C$ and so A is invertible. The proof is pretty simple and holds even in the context of groups.

A natural question pops up. Are left inverses unique? That is, if A possesses two left inverses B and C, is it necessary that $B = C$? In general, this is not always the case as we will shortly see.

There are cases when this is true. For instance, when dim $H < \infty$, it is known that $A \in B(H)$ is invertible if and only if it is left (resp. right) invertible. As a matter of fact, we have an interesting characterization: *A real or complex vector space X is finite-dimensional if and only if right invertibility for linear operators on X implies invertibility.* This appeared in [220].

Also, if A is *onto* and having two left inverses say B and C, then $B = C$. Indeed, $BA = CA = I$ implies that $(B - C)A = 0$ and so $B = C$. Similarly, if A is *one-to-one* and has two right inverses B and C, then $B = C$.

Notice in the end that an $A \in B(H)$ is invertible if and only if A possesses a *unique* left (resp. right) inverse. See e.g. Exercise 14 on Page 293 in [10].

Question 1.2.13 Give an operator A on some infinite-dimensional space having an infinitude of left (resp. right) inverses, yet A is not invertible, i.e. it is not right (resp. left) invertible.

1.2.14   An Injective Operator That Is Not Left Invertible

Is left invertibility always equivalent to injectivity? The answer is yes over finite-dimensional spaces. This is not always the case, however, when we leave the setting of finite-dimensional spaces. An example will be given below. Obviously, a left invertible operator is always one-to-one. In fact, an $A \in B(H)$ is left invertible if and only if A is injective with a *closed* range. See e.g. 10.29 in [10]. It is worth noting in the end that an $A \in B(H)$ is right invertible if and only if A is surjective.

Question 1.2.14 Give an example of an injective operator that is not left invertible.

1.2.15   An $A \neq 0$ Such That $\langle Ax, x \rangle = 0$ for All $x \in H$

Recall the following:

Theorem 1.2.2 *Let $A \in B(H)$ where H is a complex Hilbert space. Then the following statements are equivalent:*

(1) $A = 0$.
(2) $\langle Ax, x \rangle = 0$ *for all $x \in H$.*
(3) $\langle Ax, y \rangle = 0$ *for all $x, y \in H$.*

Question 1.2.15 (Cf. Question 18.2.3) Find a nonzero $A \in B(H)$ such that

$$\langle Ax, x \rangle = 0 \text{ for all } x \in H.$$

Hint: By the result above, you should avoid a $\mathbb{C}$-Hilbert space (and on an $\mathbb{R}$-Hilbert space, you must also avoid an A such that $A = A^*$!).

1.2.16   The Open Mapping Theorem Fails to Hold True for Bilinear Mappings

It was asked by W. Rudin in [312] (Page 67) whether a continuous bilinear map from the product of two Banach spaces onto a Banach space must be open at the origin (see e.g. [256] for their definition). The first counterexample was found by P. J. Cohen in [65], and it was a fairly nontrivial construction on an infinite-dimensional space. Much simpler was the counterexample found by Ch. Horowitz [163]. Indeed, what is even more striking about this example is that spaces are finite-dimensional, and usually interesting questions about Banach spaces are rather trivial in finite dimensions.

Question 1.2.16 (⊛) Find a bilinear continuous surjective mapping $T : \mathbb{C}^3 \times \mathbb{C}^3 \to \mathbb{C}^4$ that is not open at the origin.

Answers

Answer 1.2.1 The answer is negative in general. For example, consider the trivial case $X = \{0\}$. Then $B(X, Y) = B(\{0\}, Y)$ reduces to the vector space constituted of the zero operator only, and so it is trivially a Banach space. At the same time, Y can be any normed vector space (in particular, a non-complete one).

If, however, $X \neq \{0\}$, then the result is true, i.e. $B(X, Y)$ being a Banach space does imply that Y is a Banach space. A proof based on the Hahn–Banach theorem may be consulted in Exercise 2.3.43 in [256].

Answer 1.2.2 On $\mathbb{C}^2$, define

$$A = \begin{pmatrix} 0 & 0 \\ 1 & 0 \end{pmatrix}$$

and so $A \neq \begin{pmatrix} 0 & 0 \\ 0 & 0 \end{pmatrix}$. However,

$$A^2 = \begin{pmatrix} 0 & 0 \\ 0 & 0 \end{pmatrix}.$$

Observe in the end that

$$\|A^2\| = 0 \neq 1 = \|A\|^2,$$

as wished.

Answer 1.2.3 ([149]) Consider

$$A = \begin{pmatrix} 1 & 0 & 0 \\ 0 & 1 & 0 \\ 0 & 0 & 0 \end{pmatrix} \text{ and } B = \begin{pmatrix} 0 & 0 & 0 \\ 1 & 0 & 0 \\ 0 & 1 & 0 \end{pmatrix}.$$

Hence,

$$AB = \begin{pmatrix} 0 & 0 & 0 \\ 1 & 0 & 0 \\ 0 & 0 & 0 \end{pmatrix} \text{ and so } (AB)^2 = \begin{pmatrix} 0 & 0 & 0 \\ 0 & 0 & 0 \\ 0 & 0 & 0 \end{pmatrix}.$$

On the other hand, $BA = B$ and so

$$(BA)^2 = B^2 = \begin{pmatrix} 0 & 0 & 0 \\ 0 & 0 & 0 \\ 1 & 0 & 0 \end{pmatrix} \neq \begin{pmatrix} 0 & 0 & 0 \\ 0 & 0 & 0 \\ 0 & 0 & 0 \end{pmatrix}.$$

Answer 1.2.4 Let

$$A = \begin{pmatrix} 0 & 0 \\ 0 & 1 \end{pmatrix} \text{ and } B = \begin{pmatrix} 1 & 0 \\ 0 & 0 \end{pmatrix}.$$

Then $A + B = I$ and $AB = 0$. Hence, *any* T commutes with both $A + B$ and AB. So, we only have to choose a T that commutes with none of A and B. For example,

$$T = \begin{pmatrix} 0 & 2 \\ 0 & 0 \end{pmatrix}$$

does the job.

Answer 1.2.5 Let A and C be any two *non-commuting* operators (e.g. $A = \begin{pmatrix} 0 & 1 \\ 1 & 0 \end{pmatrix}$ and $C = \begin{pmatrix} 2 & 0 \\ 0 & 1 \end{pmatrix}$ both defined on say $\mathbb{C}^2$). By setting $B = 0$, we see that A commutes with B and that B commutes with C, but A does not commute with C.

Since B above is not invertible, one could think that perhaps the relation is transitive on the set of invertible operators. This is still not sufficient as may be seen by letting $B = I$ and by taking A and C as above (which are two invertible operators).

Answer 1.2.6 Consider the two-dimensional examples

$$A = \begin{pmatrix} 0 & 1 \\ 0 & 0 \end{pmatrix} \text{ and } B = \begin{pmatrix} 1 & 0 \\ 0 & -1 \end{pmatrix}.$$

Then

$$\|AB - BA\| = 2, \ \|A\| = \|B\| = 1,$$

i.e.

$$\|AB - BA\| = 2\|A\|\|B\|,$$

as required.

Answer 1.2.7 Let

$$A = \begin{pmatrix} 0 & 1 \\ 0 & 0 \end{pmatrix} \text{ and } B = \begin{pmatrix} 0 & 0 \\ 1 & 0 \end{pmatrix}.$$

Then, both A and B are nilpotent, while $A + B$ is not nilpotent as

$$(A + B)^2 = \begin{pmatrix} 1 & 0 \\ 0 & 1 \end{pmatrix} \neq \begin{pmatrix} 0 & 0 \\ 0 & 0 \end{pmatrix}.$$

The same pair works for the product. Indeed,

$$AB = \begin{pmatrix} 0 & 1 \\ 0 & 0 \end{pmatrix} \begin{pmatrix} 0 & 0 \\ 1 & 0 \end{pmatrix} = \begin{pmatrix} 1 & 0 \\ 0 & 0 \end{pmatrix}$$

is obviously not nilpotent.

Answer 1.2.8 Let

$$A = \begin{pmatrix} 0 & 1 \\ -1 & 0 \end{pmatrix} \text{ and } B = \begin{pmatrix} 1 & 0 \\ 0 & -1 \end{pmatrix}.$$

Then neither A nor B is nilpotent. However,

$$A + B = \begin{pmatrix} 1 & 1 \\ -1 & -1 \end{pmatrix}$$

and so

$$(A + B)^2 = \begin{pmatrix} 0 & 0 \\ 0 & 0 \end{pmatrix},$$

i.e. $A + B$ is nilpotent.

As for the product, let

$$C = \begin{pmatrix} 0 & 1 \\ 1 & 0 \end{pmatrix} \text{ and } D = \begin{pmatrix} 1 & 0 \\ 0 & 0 \end{pmatrix}.$$

Then neither C nor D is nilpotent, but

$$CD = \begin{pmatrix} 0 & 0 \\ 1 & 0 \end{pmatrix}$$

is nilpotent.

Answer 1.2.9 Let

$$A = \begin{pmatrix} 1 & 0 \\ 0 & 2 \end{pmatrix}.$$

Then $\|A\| = 2$ and $\|A^{-1}\| = 1$. Therefore,

$$\|A^{-1}\| \neq \frac{1}{\|A\|}.$$

Let us give a counterexample in the case $A^2 = I$. Consider

$$A = \begin{pmatrix} 0 & 2 \\ 1/2 & 0 \end{pmatrix}.$$

Plainly, $A^2 = I$ whereby $A^{-1} = A$. Since $\|A\| = 2$, it results that $\|A^{-1}\| \neq 1/\|A\|$.

As for the last case, there are in effect operators A satisfying the property $\|A^{-1}\| = 1/\|A\|$. For instance, $A = \alpha I$ where $\alpha \in \mathbb{C}^*$ does the job, and so does $\begin{pmatrix} 1 & 0 \\ 0 & -1 \end{pmatrix}$.

Answer 1.2.10 Let

$$Af(x) = (e^x + 1)f(x) \text{ be defined for } f \in L^2(0, 1).$$

Then it is easy to see that $I - A$ is invertible with inverse given by

$$(I - A)^{-1} f(x) = -e^{-x} f(x)$$

whereas

$$\|A\| = \max_{0 \le x \le 1} |e^x + 1| = e + 1 \ge 1.$$

Another simpler example reads: Let $A = \alpha I$ for some constant $\alpha > 1$. Then $I - \alpha I = (1 - \alpha)I$ is invertible yet

$$\|A\| = \|\alpha I\| = \alpha > 1.$$

Answer 1.2.11 Simply, let $S \in B(\ell^2)$ be the shift operator. Then neither S nor S^* is invertible, yet $S^*S = I$ is invertible.

Answer 1.2.12 The counterexample is based on the shift operator S defined on ℓ^2. Recall that $S^*S = I$ and $SS^* \neq I$. Let

$$A = I - S^* \text{ and } B = I - S.$$

Then

$$A + B = I - S^* + I - S = 2I - (S^* + S)$$

and

$$AB = (I - S^*)(I - S) = I - S^* - S + S^*S = 2I - (S^* + S),$$

that is,

$$AB = A + B.$$

If BA were equal to AB, then this would mean that

$$SS^* = I$$

and this is the sought contradiction. Therefore, $AB \neq BA$.

Answer 1.2.13 Let H be an infinite-dimensional space, and let $A \in B(H)$ be a left invertible operator, i.e. $BA = I$ for some $B \in B(H)$. From say Page 63 in [134], we know how to produce infinitely many other left inverses. Such a family is given by $B + \lambda(I - AB)$ where $\lambda \in \mathbb{C}$. Indeed, for any λ in $\mathbb{C}$, we have

$$[B + \lambda(I - AB)]A = BA + \lambda A - \lambda ABA = I + \lambda A - \lambda A = I.$$

In a similar way, if A has a right inverse C, that is $AC = I$, then all the operators $C + \lambda(I - CA)$ $(\lambda \in \mathbb{C})$ are right inverse to A.

As an illustration, let $S \in B(\ell^2)$ be the usual shift operator. Then $S^*S = I$ making S left invertible. Then each of $S^* + \lambda(I - SS^*)$, $\lambda \in \mathbb{C}$, is again a left inverse to S. Also, each of $S + \lambda(I - SS^*)$, $\lambda \in \mathbb{C}$, constitutes a right inverse of S^*.

Answer 1.2.14 ([10]) Define $T : \ell^2 \to \ell^2$ by

$$T(x_1, x_2, x_3, \cdots) = (x_1, x_2/2, x_3/3, \cdots).$$

Obviously, $T \in B(\ell^2)$, and besides, it is clearly one-to-one. Assume for the sake of contradiction that T has some left inverse $B \in B(\ell^2)$, i.e. $BT = I$. Choosing e_n from the standard orthonormal basis of ℓ^2 yields

$$Be_n = B(nTe_n) = n(BT)e_n = ne_n.$$

In other language, B would be unbounded. Therefore, T cannot be left invertible despite being injective.

Answer 1.2.15 Let

$$A = \begin{pmatrix} 0 & 1 \\ -1 & 0 \end{pmatrix}$$

on $\mathbb{R}^2$ (considered as an $\mathbb{R}$-Hilbert space). Then one immediately sees that for all $x = (x_1, x_2)^t \in \mathbb{R}^2$

$$\langle Ax, x \rangle = x_2 x_1 - x_1 x_2 = 0$$

while $A \neq \begin{pmatrix} 0 & 0 \\ 0 & 0 \end{pmatrix}$.

Answer 1.2.16 ([163]) Define $T : \mathbb{C}^3 \times \mathbb{C}^3 \to \mathbb{C}^4$ as follows:

$$T(x, y) = (x_1 y_1, x_1 y_2, x_1 y_3 + x_3 y_1 + x_2 y_2, x_3 y_2 + x_2 y_1)$$

where $x = (x_1, x_2, x_3)$ and $y = (y_1, y_2, y_3)$. That T is bilinear, and continuous are clear.

Let us turn to surjectivity. To begin with, assume that we are given $z = (z_1, z_2, z_3, z_4) \in \mathbb{C}^4$. If $z_1 = z_2 = 0$, choosing $x = (0, z_4, z_3)$ and $y = (1, 0, 0)$ yields $T(x, y) = z$. If for instance, $z_1 \neq 0$, letting $x = (1, z_4/z_1, 0)$ and $y = (z_1, z_2, z_3 - (z_4 z_2/z_1))$ gives again $T(x, y) = z$. The case $z_2 \neq 0$ may be treated similarly, and hence, we leave it to readers. There only remains to show that T is not open at the origin. To this end, assume that $T(x, y) = z = (a, a, a, 1)$ where a is a small positive number ($a > 0$). By the expression of T, we get

$$y_1 = y_2 = \frac{a}{x_1}, \ x_2 + x_3 = \frac{x_1}{a} \text{ and } y_3 = \frac{a-1}{x_1}$$

thereby

$$(x_1 + x_2 + x_3)(y_1 + y_2 + y_3) = 3a + 2 - \frac{1}{a}.$$

If T were open at the origin, each bounded subset of $\mathbb{C}^4$ could be obtained as the image of a bounded subset of $\mathbb{C}^3 \times \mathbb{C}^3$. This is evidently not the case as seen by sending a to 0. Accordingly, T is not open at the origin.

Chapter 2
Basic Classes of Bounded Linear Operators

2.1 Basics

Definition 2.1.1 Let H be a Hilbert space, and let $A \in B(H)$. Let I be the identity operator on H. We say that A is:

(1) Self-adjoint (or symmetric or Hermitian) if $A = A^*$.
(2) Skew-adjoint (or skew-symmetric or skew Hermitian) when $A^* = -A$.
(3) Normal if $AA^* = A^*A$.
(4) Unitary if $AA^* = A^*A = I$ (hence, $A^{-1} = A^*$).
(5) An isometry if $A^*A = I$.
(6) A co-isometry if $AA^* = I$.
(7) A partial isometry if A restricted to $(\ker A)^{\perp}$ is an isometry, that is, for all $x \in (\ker A)^{\perp}$: $\|Ax\| = \|x\|$.
(8) A projection or idempotent if $A^2 = A$.
(9) An orthogonal projection if $A^2 = A$ and $A^* = A$.
(10) A fundamental symmetry if $A = A^*$ and $A^2 = I$.
(11) Bounded below if there exists an $\alpha > 0$ such that

$$\|Ax\| \geq \alpha \|x\|$$

for all $x \in H$.

Definition 2.1.2 Let $A, B \in B(H)$. We say that A and B are similar if there exists an invertible operator $T \in B(H)$ such that

$$T^{-1}AT = B.$$

We say that A and B are unitarily equivalent if there exists some unitary operator $U \in B(H)$ such that

$$U^* A U = B.$$

In the end, we introduce the Cartesian decomposition of operators. The idea is inspired from the usual Cartesian decomposition of a complex number z that is expressed as $z = x + iy$, where $x, y \in \mathbb{R}$.

Definition 2.1.3 Let H be a complex Hilbert space, and consider $A \in B(H)$. Let "i" denote the usual complex number. Then

$$\operatorname{Re} A = \frac{A + A^*}{2} \text{ and } \operatorname{Im} A = \frac{A - A^*}{2i}$$

are called the real and the imaginary parts of A (respectively).

Lemma 2.1.1 *Let $A \in B(H)$. Then $\operatorname{Re} A$ and $\operatorname{Im} A$ are both self-adjoint.*

Theorem 2.1.2 *Let $T \in B(H)$. Then there exist two self-adjoint operators $A, B \in B(H)$ such that $T = A + iB$. Necessarily, $A = \operatorname{Re} T$ and $B = \operatorname{Im} T$.*

Definition 2.1.4 The decomposition $T = A + iB$ is called the Cartesian decomposition (or Toeplitz decomposition) of the operator $T \in B(H)$.

Remark It is easy to see that if $T = A + iB$ where A and B are self-adjoint, then $T = 0$ iff $A = B = 0$ (the right to left implication does not require self-adjointness).

The following result is easily shown:

Proposition 2.1.3 *Let $T := A + iB \in B(H)$ (where A and B are self-adjoint). Then T is normal iff A commutes with B.*

2.2   Questions

2.2.1   A Non-unitary Isometry

It is clear that a unitary operator is an isometry and a co-isometry simultaneously. It is also easy to see why on finite-dimensional spaces all these three classes coincide.

Question 2.2.1 Provide an example of a non-unitary isometry.

2.2.2 A Nonnormal A Such That ker $A = $ ker A^*

Since a normal $A \in B(H)$ satisfies: $\|Ax\| = \|A^*x\|$ for every $x \in H$, it is easily seen that ker $A = $ ker A^*. What about the converse?

Question 2.2.2 Find an $A \in B(H)$ with ker $A = $ ker A^* yet A is not normal.

2.2.3 Do Normal Operators A and B Satisfy $\|ABx\| = \|BAx\|$ for All x?

If $A, B \in B(H)$ are normal, then it may be shown that $\|AB\| = \|BA\|$ (see e.g. Exercise 4.3.13 in [256]).

Question 2.2.3 Find two self-adjoint matrices A and B satisfying $\|ABu\| \neq \|BAu\|$ for some vector u.

2.2.4 Do Normal Operators A and B Satisfy $\|ABx\| = \|AB^*x\|$ for All x?

If $A, B \in B(H)$, and A is normal, then $\|A^*Bx\| = \|ABx\|$ for all $x \in H$.

Question 2.2.4 Find normal matrices A and B obeying $\|ABu\| \neq \|AB^*u\|$ for a certain vector u.

2.2.5 Two Operators B and V Such That $\|BV\| \neq \|B\|$ Where V Is an Isometry

Let $A, V \in B(H)$ be such that V is an isometry. Then it is known that $\|VA\| = \|A\|$. How about when the isometry is the right factor of the product?

Question 2.2.5 Find a $B \in B(H)$ and an isometry $V \in B(H)$ such that

$$\|BV\| \neq \|B\|.$$

Hint: Avoid e.g. $\dim H < \infty$.

2.2.6   An Invertible Normal Operator That Is not Unitary

A unitary operator is both normal and invertible. Does the converse hold true?

Question 2.2.6 Give a normal and invertible $A \in B(H)$ that is not unitary.

2.2.7   Two Self-Adjoint Operators Whose Product Is Not Even Normal

Let $A, B \in B(H)$ be self-adjoint. Then AB is self-adjoint if and only if $AB = BA$.

Question 2.2.7 Is the product of two self-adjoint operators necessarily normal?

2.2.8   Two Normal Operators A, B Such That AB Is Normal, but $AB \neq BA$

If $A, B \in B(H)$ are normal and $AB = BA$, then AB is normal. On finite-dimensional spaces (at least for complex ones), this may be shown using the so-called simultaneous unitary diagonalization.

In infinite-dimensional spaces, the proof seems to need the Fuglede theorem or the spectral theorem (both not recalled yet). In fact, the same idea may be applied to show that AB is normal whenever $AB = \lambda BA$ where $\lambda \in \mathbb{C}$, and A and B being normal (see e.g. [54]).

> **Question 2.2.8 (Cf. Question 2.2.12)** Find normal $A, B \in B(H)$ such that AB is normal, but $AB \neq BA$.

Further Reading See [2, 236, 288] and [305].

2.2.9 Two Normal Operators Whose Sum Is Not Normal

Clearly, the sum of two self-adjoints $A, B \in B(H)$ is self-adjoint. Does this result have a chance to remain valid for the larger class of normal operators? The answer is affirmative when the assumption $AB = BA$ is also made. A proof may be found in Exercise 4.3.11 in [256].

> **Question 2.2.9** Find two normal operators whose sum fails to be normal.

Further Reading See [31, 251, 253] and [254].

2.2.10 Two Unitary U, V for Which $U + V$ Is Not Unitary

As we already observed in Answer 2.2.8, if $U, V \in B(H)$ are unitary, then UV is always unitary (that is, even when $UV \neq VU$!). What about the sum of unitary operators?

> **Question 2.2.10** Find commuting unitary $U, V \in B(H)$ such that $U + V$ is not unitary.

2.2.11 Two Anti-commuting Normal Operators Whose Sum Is Not Normal

From the discussion above Question 2.2.9, we know that the sum of two commuting normal operators is normal. Since, in particular, the product of two normal operators A and B such that $AB = -BA$ (A and B are then said to anti-commute) is normal, perhaps the sum of two anti-commuting normal operators would be normal.

Question 2.2.11 Is the sum of two anti-commuting normal operators always normal?

2.2.12   Two Unitary Operators A and B Such That AB, BA, and $A + B$ Are All Normal yet $AB \neq BA$

Question 2.2.12 Find two unitary operators A and B such that AB, BA, and $A + B$ are all normal and yet $AB \neq BA$.

2.2.13   A Non-self-adjoint A Such That A^2 Is Self-Adjoint

Question 2.2.13 Find a non-self-adjoint operator $A \in B(H)$ such that A^2 is self-adjoint.

2.2.14   Three Self-Adjoint Operators A, B, and C Such That ABC Is Self-Adjoint, Yet No Two of A, B, and C Need to Commute

Question 2.2.14 Let $A, B, C \in B(H)$ be self-adjoint and such that ABC is self-adjoint. Does it follow that at least two of A, B, and C must commute?

2.2.15   An Orthogonal Projection P and a Normal A Such That PAP Is Not Normal

Question 2.2.15 Find an orthogonal projection $P \in B(H)$ and a normal $A \in B(H)$ such that PAP is not normal.

2.2.16   A Partial Isometry That Is Not an Isometry

Recall a characterization of partial isometries. It reads: $A \in B(H)$ is partial isometry iff $AA^*A = A$ (see e.g. Theorem 3 on Page 54 in [123]).

Question 2.2.16 Give an example of a partial isometry that is not an isometry.

2.2.17 A Non-partial Isometry V Such That V^2 Is a Partial Isometry

Question 2.2.17 Find a $V \in B(H)$ such that V^2 is a partial isometry, whereas V is not one.

2.2.18 A Partial Isometry V Such That V^2 Is a Partial Isometry, but Neither V^3 Nor V^4 Is One

Question 2.2.18 (Cf. Question 2.2.17) Find a partial isometry V such that V^2 is also a partial isometry, while neither V^3 nor V^4 is one.

2.2.19 No Condition of $U = U^*$, $U^2 = I$ and $U^*U = I$ Needs to Imply Any of the Other Two

Question 2.2.19 Let $U \in B(H)$. Does any of the following conditions imply any of the others

$$U = U^*, \ U^2 = I \text{ and } U^*U = I?$$

2.2.20 A B Such That $BB^* + B^*B = I$ and $B^2 = B^{*2} = 0$

As quoted in [295], in the quantization of wave fields for particles satisfying Fermi–Dirac statistics, the particles are described through ("closed") operators B and B^* on a Hilbert space satisfying the anticommutation relations:

$$BB^* + B^*B = I, \ B^2 = B^{*2} = 0.$$

Notice that such B cannot be normal.

Question 2.2.20 Find a $B \in B(H)$ such that

$$BB^* + B^*B = I, \ B^2 = B^{*2} = 0.$$

2.2.21 A Nonnormal Solution of $(A^*A)^2 = A^{*2}A^2$

It was shown in [374] (see also the proof of Theorem 3.1 in [213]) that if $\dim H < \infty$ and $A \in B(H)$ is such that $(A^*A)^2 = A^{*2}A^2$, then A is automatically normal.

Question 2.2.21 (Cf. Question 14.2.24) Show that this is not necessarily the case when $\dim H = \infty$.

2.2.22 An $A \in B(H)$ Such That $A^n = I$, While $A^{n-1} \neq I$, $n \geq 2$

Question 2.2.22 Let $n \in \mathbb{N}$ be given. Find an $n \times n$ matrix A such that $A^n = I$ with $A^{n-1} \neq I$.

2.2.23 A Unitary A Such That $A^n \neq I$ for All $n \in \mathbb{N}$, $n \geq 2$

Question 2.2.23 Find a unitary $A \in B(H)$ such that

$$A^n \neq I \text{ for all } n \in \mathbb{N}, \ n \geq 2.$$

2.2.24 A Normal Non-self-adjoint Operator $A \in B(H)$ Such That $A^*A = A^n$

It was asked in [210] whether $A^*A = A^2$ entails the self-adjointness of $A \in B(H)$. This was first answered in [374], then independently and differently in [224] in both the finite- and infinite-dimensional settings though the infinite-dimensional case was only alluded to superficially in [374]. See [88] for yet new proofs and more general results.

> **Question 2.2.24 (Cf. Questions 21.2.31 and 25.1.7)** Let $n \in \mathbb{N}$ be given with $n \geq 3$. Find a normal but non-self-adjoint $A \in B(H)$ such that
>
> $$A^*A = A^n.$$

2.2.25 A Nonnormal A Satisfying $A^{*p}A^q = A^n$

> **Question 2.2.25** Let $n, p, q \in \mathbb{N}$ be given. Find a nonnormal $A \in B(H)$ such that
>
> $$A^{*p}A^q = A^n.$$

Answers

Answer 2.2.1 As already observed, it is mandatory to work on a H with $\dim H = \infty$. The prominent example of an isometry that is not unitary (that is, it is not surjective in these circumstances) is the shift operator $S \in B(\ell^2)$. Indeed, we do know that $S^*S = I$, that is, S an isometry. However, S is not surjective. One way of seeing this is that if S were surjective, then S^* would be injective (why?), and this is obviously not the case! Therefore, S cannot be unitary.

Let us give another example on a "functional L^2." Consider the linear operator $V : L^2(0, \infty) \to L^2(0, \infty)$ defined by

$$(Vf)(x) = \begin{cases} f(x-1), & x > 1, \\ 0 & x \leq 1. \end{cases}$$

Then it may be shown that V^*, the adjoint of V, is given by

$$(V^*f)(x) = f(x+1), \ x > 0$$

whereby it becomes clear that (for any $f \in L^2(0, \infty)$)

$$\|Vf\|_2^2 = \int_0^\infty |Vf(x)|^2 dx = \int_1^\infty |f(x-1)|^2 dx = \int_0^\infty |f(t)|^2 dt = \|f\|_2^2$$

or equivalently $V^*V = I$, i.e. V is an isometry. Since V^* is not injective, a similar argument as above applies, making V non-unitary.

Remark Readers are asked to extend V to an isometry $\tilde{V}$ defined on all of $L^2(\mathbb{R})$, and such that $\tilde{V}$ is not unitary.

Answer 2.2.2 We give an example of a *nonnormal* A such that $\ker A = \ker A^*$. Let

$$A = \begin{pmatrix} 0 & 1 \\ 2 & 0 \end{pmatrix}.$$

Then A is not normal, and we may easily check that

$$\ker A = \ker A^* \ (= \{0\}).$$

Answer 2.2.3 Consider the self-adjoint matrices

$$A = \begin{pmatrix} 2 & 0 \\ 0 & 1 \end{pmatrix} \text{ and } B = \begin{pmatrix} 0 & 1 \\ 1 & 0 \end{pmatrix}.$$

Then

$$AB = \begin{pmatrix} 0 & 2 \\ 1 & 0 \end{pmatrix} \text{ and } BA = \begin{pmatrix} 0 & 1 \\ 2 & 0 \end{pmatrix},$$

and so for e.g. $e_1 = (1, 0)$, we see that $\|ABe_1\| \neq \|BAe_1\|$ (with some abuse of notation!).

Answer 2.2.4 Consider the matrices

$$A = \begin{pmatrix} 2 & 0 \\ 0 & 1 \end{pmatrix} \text{ and } B = \begin{pmatrix} 1 & 2 \\ -2 & 1 \end{pmatrix}.$$

Clearly, A is self-adjoint and B is normal. Then

$$AB = \begin{pmatrix} 2 & 4 \\ -2 & 1 \end{pmatrix} \text{ and } BA = \begin{pmatrix} 2 & 2 \\ -4 & 1 \end{pmatrix},$$

and by considering $u = (1, 1)$ say, we see that $\|ABu\| \neq \|AB^*u\|$, as needed.

Answer 2.2.5 Let

$$V(x_1, x_2, \cdots) = (x_1, 0, x_2, 0, \cdots)$$

and

$$B(x_1, x_2, \cdots) = (0, x_2, 0, x_4, 0, \cdots)$$

be both defined on ℓ^2. Then V is an isometry (as it may be checked).

Now,

$$BV(x_1, x_2, \cdots) = B(x_1, 0, x_2, 0, \cdots) = (0, 0, 0, \cdots).$$

Hence, $\|BV\| = 0$. It may also easily be checked that $\|B\| = 1$ so that finally

$$\|BV\| = 0 \neq 1 = \|B\|.$$

Answer 2.2.6 Simply consider $A = 2I$. Then A is invertible and even self-adjoint. It is, however, non-unitary because

$$AA^* = A^*A = 4I \neq I.$$

Answer 2.2.7 The answer is negative. For instance, let

$$A = \begin{pmatrix} -1 & 0 \\ 0 & 2 \end{pmatrix} \text{ and } B = \begin{pmatrix} 0 & 1 \\ 1 & 0 \end{pmatrix}.$$

Then A and B are both self-adjoint, but

$$AB = \begin{pmatrix} 0 & -1 \\ 2 & 0 \end{pmatrix}$$

is not normal (hence, it cannot be self-adjoint either).

Answer 2.2.8 One simple counterexample is to take two *non-commuting unitary* operators. Then, their product is always unitary and hence normal. However, by assumption, these two operators do not commute!

Here is a better counterexample: Let

$$A = \begin{pmatrix} 1 & 0 \\ 0 & -1 \end{pmatrix} \text{ and } B = \begin{pmatrix} 0 & 1 \\ 1 & 0 \end{pmatrix}.$$

Then clearly, A and B *are self-adjoint and also unitary*. Their product is

$$AB = \begin{pmatrix} 0 & 1 \\ -1 & 0 \end{pmatrix}$$

which is clearly normal (it is even unitary). Nonetheless, $AB \neq BA$ because AB is not self-adjoint.

Answer 2.2.9 Consider the normal matrices A and B (B is even self-adjoint) defined as

$$A = \begin{pmatrix} 1 & -1 \\ 1 & 1 \end{pmatrix}, \quad B = \begin{pmatrix} -1 & 2 \\ 2 & 1 \end{pmatrix}.$$

Then

$$A + B = \begin{pmatrix} 0 & 1 \\ 3 & 2 \end{pmatrix}$$

is clearly not normal.

Answer 2.2.10 Let $U \in B(H)$ be unitary and set $V = U$. Then

$$U + V = U + U = 2U$$

is not unitary for

$$(2U)(2U)^* = 4I \neq I!$$

In fact, the sum $U + V$ need not even be normal. Indeed, on $\mathbb{C}^2$, let

$$U = \begin{pmatrix} 0 & i \\ -i & 0 \end{pmatrix} \text{ and } V = \begin{pmatrix} i & 0 \\ 0 & -i \end{pmatrix}.$$

Then U and V are unitary (U is also self-adjoint). Set $A = U + V$. Then A is not normal for

$$AA^* = \begin{pmatrix} 2 & -2 \\ -2 & 2 \end{pmatrix} \neq \begin{pmatrix} 2 & 2 \\ 2 & 2 \end{pmatrix} = A^*A.$$

Answer 2.2.11 (Cf. [254]) Let

$$A = \begin{pmatrix} 2 & 0 \\ 0 & -2 \end{pmatrix} \text{ and } B = \begin{pmatrix} 0 & 1 \\ -1 & 0 \end{pmatrix}.$$

Then A and B are both normal. In fact, A is self-adjoint and B is unitary! They anti-commute because

$$AB = -BA = \begin{pmatrix} 0 & 2 \\ 2 & 0 \end{pmatrix}.$$

Observe in the end that

$$A + B = \begin{pmatrix} 2 & 1 \\ -1 & -2 \end{pmatrix}$$

is not normal.

Answer 2.2.12 The example is borrowed from [125]. On $\mathbb{C}^2$, let

$$A = \begin{pmatrix} 1 & 0 \\ 0 & -1 \end{pmatrix} \text{ and } B = \begin{pmatrix} 0 & 1 \\ 1 & 0 \end{pmatrix}.$$

Then A and B are both unitary (and self-adjoint). Hence, AB and BA both remain unitary. Also, $A + B = \begin{pmatrix} 1 & 1 \\ 1 & -1 \end{pmatrix}$ is normal. In addition,

$$AB = -BA = \begin{pmatrix} 0 & 1 \\ -1 & 0 \end{pmatrix}.$$

Thereby, $AB \neq BA$, as otherwise $AB = 0$!

Answer 2.2.13 Let $A \in B(\mathbb{C}^2)$ be given by

$$A = \begin{pmatrix} a & 1 \\ 0 & -a \end{pmatrix}$$

where $a \in \mathbb{R}$. Then A is obviously non-self-adjoint, but

$$A^2 = \begin{pmatrix} a^2 & 0 \\ 0 & a^2 \end{pmatrix},$$

is self-adjoint.

Answer 2.2.14 ([229]) False! Indeed, we can have a self-adjoint product of three self-adjoint operators that do not commute pairwise. On $\mathbb{C}^2$, consider the following self-adjoint matrices:

$$A = \begin{pmatrix} 1 & 1 \\ 1 & 0 \end{pmatrix}, \ B = \begin{pmatrix} 0 & 1 \\ 1 & 2 \end{pmatrix} \text{ and } C = \begin{pmatrix} 0 & 1 \\ 1 & 0 \end{pmatrix}.$$

Then

$$ABC = \begin{pmatrix} 3 & 1 \\ 1 & 0 \end{pmatrix}$$

is self-adjoint. Nevertheless, as may readily be checked, none of the products AB, AC, and BC is self-adjoint, that is,

$$AB \neq BA, \ AC \neq CA \text{ and } BC \neq CB.$$

Answer 2.2.15 ([375]) On $H = \mathbb{C}^3$, consider

$$A = \begin{pmatrix} 0 & 0 & 1 \\ 1 & 0 & 0 \\ 0 & 1 & 0 \end{pmatrix} \text{ and } P = \begin{pmatrix} 0 & 0 & 0 \\ 0 & 1 & 0 \\ 0 & 0 & 1 \end{pmatrix}.$$

Then A is unitary, and P is an orthogonal projection. However,

$$PAP = \begin{pmatrix} 0 & 0 & 0 \\ 0 & 0 & 0 \\ 0 & 1 & 0 \end{pmatrix}$$

is not normal.

Remark Can you find a counterexample when dim $H = 2$?

Answer 2.2.16 To answer the question, define on $\mathbb{C}^2$:

$$A = \begin{pmatrix} 1 & 0 \\ 0 & 0 \end{pmatrix}.$$

Then A is a partial isometry for A is self-adjoint and $A^3 = A$. It is also plain that A is not an isometry merely because A is not invertible.

Answer 2.2.17 ([106]) Let

$$V = \begin{pmatrix} 0 & 0 & 0 \\ 2 & 0 & 0 \\ 0 & 0 & 1 \end{pmatrix}.$$

Since

$$VV^*V = \begin{pmatrix} 0 & 0 & 0 \\ 8 & 0 & 0 \\ 0 & 0 & 1 \end{pmatrix} \neq V,$$

V is not a partial isometry. Since

$$V^2 = \begin{pmatrix} 0 & 0 & 0 \\ 0 & 0 & 0 \\ 0 & 0 & 1 \end{pmatrix},$$

V^2 is clearly an orthogonal projection, and so it is a partial isometry.

Answer 2.2.18 ([106, 121]) Let

$$V = \frac{1}{2} \begin{pmatrix} 0 & \sqrt{3} & -1 \\ 0 & 1 & \sqrt{3} \\ 0 & 0 & 0 \end{pmatrix}.$$

It can easily be checked that both V and V^2 are partial isometries and that neither V^3 nor V^4 is a partial isometry. We leave details to readers...

Answer 2.2.19 (Cf. [149]) None of the given conditions has to imply any of the remaining two! Indeed:

(1) Let

$$A = \begin{pmatrix} 1 & 0 \\ 0 & 3 \end{pmatrix}.$$

Then clearly $A = A^*$, whereas $A^2 = A^*A \neq I$.
(2) Let

$$B = \begin{pmatrix} 0 & -1 \\ 1 & 0 \end{pmatrix}.$$

Then $B^*B = I$, while neither $B = B^*$ nor $B^2 = I$ is satisfied.
(3) In the end, let

$$C = \begin{pmatrix} 1 & 1 \\ 0 & -1 \end{pmatrix}.$$

Then $C^2 = I$, while neither $C = C^*$ nor $C^*C = I$ is satisfied.

Answer 2.2.20 The easiest example is to consider on $\mathbb{C}^2$ the matrix $B = \begin{pmatrix} 0 & 1 \\ 0 & 0 \end{pmatrix}$ and so $B^* = \begin{pmatrix} 0 & 0 \\ 1 & 0 \end{pmatrix}$. Thereupon,

$$B^2 = B^{*2} = \begin{pmatrix} 0 & 0 \\ 0 & 0 \end{pmatrix}$$

and

$$BB^* + B^*B = \begin{pmatrix} 1 & 0 \\ 0 & 1 \end{pmatrix} = I,$$

as required.

Answer 2.2.21 Let $S \in B(\ell^2)$ be the shift operator. Then obviously

$$(S^* S)^2 = I = S^* S^* S S = S^{*2} S^2$$

and, as is known, S is not normal.

Answer 2.2.22 There are many types of counterexamples even when dim $H < \infty$. The simplest one is to take the following permutation $n \times n$ matrix:

$$A_n = \begin{pmatrix} 0 & 1 & 0 & \cdots \cdots & 0 \\ & 0 & 0 & 1 & 0 & & \vdots \\ & \vdots & & 0 & 1 & \ddots & \vdots \\ & \vdots & & & \ddots & \ddots & 0 \\ & 0 & & & & 0 & 1 \\ & 1 & 0 & \cdots \cdots & & 0 & 0 \end{pmatrix}$$

where the corresponding permutation being $p(i) = i + 1$ (modulo n). Then it may be checked that $A_n^n = I_n$ and $A_n^{n-1} \neq I_n$.

For the purpose of Question 19.2.29 below, I asked Professor R. B. Israel whether we can place some parameters inside the previous matrix and still obtain the same conclusion. He then kindly suggested to conjugate with some nonsingular matrix that does not commute with A_n but contains the parameter α. So, a more general form is

$$A_n = \begin{pmatrix} 0 & 1 & \alpha & \cdots \cdots & 0 \\ & 0 & 0 & 1 & 0 & & \vdots \\ & \vdots & & 0 & 1 & \ddots & \vdots \\ & \vdots & & & \ddots & \ddots & 0 \\ & 0 & & & & 0 & 1 \\ & 1 & -\alpha & \cdots \cdots & & 0 & 0 \end{pmatrix},$$

where readers can again check that $A_n^n = I_n$ and $A_n^{n-1} \neq I_n$ (and for any α).

Answer 2.2.23 In fact, there are counterexamples even when dim $H < \infty$. Consider on a finite-dimensional space H, the following operator:

$$A = e^{ie\pi} I$$

where e is the usual transcendental number. Then

$$A^n = e^{ine\pi} I \neq I, \ \forall n \in \mathbb{N}, \ n \geq 2,$$

as needed.

Answer 2.2.24 Let $n \in \mathbb{N}$ be such that $n \geq 3$, and define

$$A = e^{\frac{2i\pi}{n}} I$$

where I is the usual identity operator on some Hilbert space. Then A is unitary and

$$A^* A = e^{-\frac{2i\pi}{n}} e^{\frac{2i\pi}{n}} I = I = (e^{\frac{2i\pi}{n}})^n I = A^n,$$

and yet A is not self-adjoint.

Answer 2.2.25 Let $k = \inf(n, p, q)$, and let A be a nonzero nilpotent matrix of index k, i.e. $A^k = 0$ (whence A cannot be normal). WLOG, consider $k = p$, say. Then $A^{*p} = 0$, $A^q = 0$, and $A^n = 0$. Hence, $A^{*p} A^q = A^n$ is trivially satisfied (yet A is not normal).

Let us also give a non-nilpotent example that is perhaps more interesting to readers. For instance, let p, q, and n be all even and obeying $p + q = n$. Let

$$A = \begin{pmatrix} 0 & 1 \\ 2 & 0 \end{pmatrix}$$

be defined on $\mathbb{C}^2$. Then A is not normal and $A^{2k} = 2^k I = A^{*2k}$ for all k in $\mathbb{N}$. Hence,

$$A^{*p} A^q = 2^{p/2} 2^{q/2} I = 2^{p/2+q/2} I = 2^{n/2} I = A^n.$$

Chapter 3
Operator Topologies

3.1 Questions

3.1.1 Strong Convergence Does Not Imply Convergence in Norm, and Weak Convergence Does Not Entail Strong Convergence

Let H be a Hilbert space, and let (A_n) be a sequence in $B(H)$:

(1) Say that (A_n) converges in norm (or uniformly) to $A \in B(H)$ if

$$\lim_{n \to \infty} \|A_n - A\| = 0.$$

The topology associated with this convergence is called the topology of the operator norm (or the uniform topology).

(2) We say that (A_n) strongly converges to $A \in B(H)$ if

$$\lim_{n \to \infty} \|(A_n - A)x\| = 0$$

for each $x \in H$. We may then write $s - \lim_{n \to \infty} A_n = A$. The topology associated with this convergence is called the strong operator topology (denoted also by S.O.T.).

(3) We say that (A_n) weakly converges to $A \in B(H)$ if

$$\lim_{n \to \infty} \langle A_n x, y \rangle = \langle Ax, y \rangle$$

for each $x, y \in H$. We may write $w - \lim_{n \to \infty} A_n = A$. The topology associated with this convergence is called the weak operator topology (denoted also by W.O.T.).

© The Author(s), under exclusive license to Springer Nature Switzerland AG 2022
M. H. Mortad, *Counterexamples in Operator Theory*,
https://doi.org/10.1007/978-3-030-97814-3_3

It is fairly easy to see that

Convergence in norm $\implies$ Strong convergence $\implies$ Weak convergence.

The reverse implications do not remain true in general as we will shortly see. They, however, do hold good if e.g. dim $H < \infty$. They are also valid if a certain uniformity on the unit sphere is assumed (as in Problem 107 in [148]).

Question 3.1.1

(1) Show that strong convergence does not imply convergence in norm.
(2) Show that weak convergence does not entail strong convergence.

3.1.2 $s - \lim_{n\to\infty} A_n = A \neq s - \lim_{n\to\infty} A_n^* = A^*$

Let (A_n) be a sequence in $B(H)$, and let $A \in B(H)$. Then $A_n \to A$ in norm if and only if $A_n^* \to A^*$ in norm, because

$$\|A_n - A\| = \|A_n^* - A^*\|$$

for all $n \in \mathbb{N}$.

It should also be mentioned that $A_n \to A$ weakly if and only if $A_n^* \to A^*$ weakly.

Question 3.1.2 Provide an example of a sequence (A_n) in $B(H)$ and an $A \in B(H)$ such that

$$s - \lim_{n\to\infty} A_n = A \nLeftrightarrow s - \lim_{n\to\infty} A_n^* = A^*.$$

3.1.3 $(A, B) \mapsto AB$ *Is Not Weakly Continuous*

Let (A_n) and (B_n) be two sequences in $B(H)$ converging in norm (resp. strongly) to A and B, respectively. Then $A_n B_n \to AB$ in norm (resp. strongly), as in e.g. Problem 3.2 in [206]. What about the weak convergence?

Question 3.1.3 Find two sequences (A_n) and (B_n) in $B(H)$ such that:

$$w - \lim_{n \to \infty} A_n = A \text{ and } w - \lim_{n \to \infty} B_n = B \not\Longrightarrow w - \lim_{n \to \infty} A_n B_n = AB$$

where $A, B \in B(H)$.

3.1.4   The Uniform Limit of a Sequence of Invertible Operators

Question 3.1.4 (Cf. Questions 3.1.6 and 17.1.3) Is the uniform limit of a sequence of invertible operators automatically invertible?
Hint: In order to obtain a counterexample, the sequence of inverses must not be (uniformly) bounded.

3.1.5   A Sequence of Self-adjoint Operators Such That None of Its Terms Commutes with the (Uniform) Limit of the Sequence

Question 3.1.5 Supply a sequence of self-adjoint operators (A_n) in $B(H)$ that converges in norm to some self-adjoint $A \in B(H)$, yet none of the terms of (A_n) commutes with A, that is,

$$A_n A \neq A A_n, \ \forall n \in \mathbb{N}.$$

3.1.6   Strong (or Weak) Limit of Sequences of Unitary or Normal Operators

First, recall that the weak limit of a sequence of self-adjoint operators remains self-adjoint, that is, the set of self-adjoint operators is W.O.T.-closed. Hence, the strong (and uniform) limit of a sequence of self-adjoint operators remains self-adjoint.

Also, the set of normal (resp. unitary) operators is closed w.r.t. to the convergence in norm.

Question 3.1.6 Let H be a Hilbert space. Show that the set of unitary operators on H is not closed w.r.t. the strong operator topology. What about the set of normal operators? What about the weak limit of sequences of normal operators?

Answers

Answer 3.1.1

(1) Let S be the shift operator on ℓ^2, and set $A_n = (S^*)^n$. Let $x = (x_1, x_2, \cdots) \in \ell^2$. Then

$$A_n x = (S^*)^n x = (x_{n+1}, x_{n+2}, \cdots)$$

and whence

$$\|A_n x\|_{\ell^2}^2 = \sum_{k=n+1}^{\infty} |x_k|^2,$$

which, being a tail of a convergent series, necessarily goes to zero. Hence, $A_n x \to 0$ for all $x \in \ell^2$, i.e. (A_n) converges strongly to 0.

On the other hand, it is easy to see that $\|A_n\| = 1$, for all $n \in \mathbb{N}$ (for example, by using $\|A_n\|^2 = \|A_n A_n^*\|$ for each n). Therefore, (A_n) cannot converge in norm to zero.

Another simpler example consists of considering

$$C_n(x) = (x_n, 0, 0, \cdots)$$

on ℓ^2. Then clearly $\|C_n x\|^2 = |x_n|^2$ (for all x in ℓ^2), and so (C_n) converges strongly to zero for the series $\sum_{n \geq 0} |x_n|^2$ is convergent. As above, we may show that (C_n) does not converge in norm to $C := 0$.

Remark Perhaps the sequence (A_n) converges to another limit B w.r.t. the convergence in norm? The answer is no! Indeed, assume that $\|A_n - B\| \to 0$ for some $B \in B(H)$. Hence, necessarily (A_n) converges to B *strongly*, forcing $B = 0$ (by uniqueness of the limit).

(2) Let S be the shift operator on ℓ^2, and then set $B_n = S^n$. Then

$$B_n(x_1, x_2, \cdots) = (\underbrace{0, 0, \cdots, 0}_{n \text{ terms}}, x_1, x_2, \cdots).$$

Hence,

$$\forall x \in \ell^2 : \ \|B_n x\| = \|x\|$$

which says that (B_n) does not converge to zero in the strong operator topology. Let us show now that (B_n) converges weakly to zero. Let $x, y \in \ell^2$. Then

$$\langle B_n x, y \rangle = \sum_{k=n+1}^{\infty} x_{k-n} \overline{y_k}$$

so that

$$|\langle B_n x, y \rangle| = \left| \sum_{k=n+1}^{\infty} x_{k-n} \overline{y_k} \right| \le \sum_{k=n+1}^{\infty} |x_{k-n}| |y_k|.$$

By the Cauchy–Schwarz inequality, we obtain

$$|\langle B_n x, y \rangle| \le \sqrt{\sum_{k=n+1}^{\infty} |x_{k-n}|^2} \sqrt{\sum_{k=n+1}^{\infty} |y_k|^2}.$$

Thus

$$0 \le |\langle B_n x, y \rangle| \le \|x\| \sqrt{\sum_{k=n+1}^{\infty} |y_k|^2} \longrightarrow 0$$

as n tends to ∞, proving that (B_n) weakly converges to zero.

Remark As above, one may wonder whether the sequence (B_n) has a limit different from zero w.r.t. S.O.T. This is impossible as a limit $C \in B(H)$ w.r.t. the strong operator topology is necessarily a limit w.r.t. the weak operator topology and so $C = 0$.

Answer 3.1.2 Set again $A_n = (S^*)^n$ where S is the shift operator on ℓ^2. For each $x = (x_1, x_2, \cdots) \in \ell^2$, we have

$$A_n x = (S^*)^n x = (x_{n+1}, x_{n+2}, \cdots).$$

We already know that (A_n) converges strongly to 0. Clearly,

$$A_n^* = [(S^*)^n]^* = S^n$$

and so

$$A_n^* x = (\underbrace{0, 0, \cdots, 0}_{n \text{ terms}}, x_1, x_2, \cdots)$$

which is nothing but the sequence (B_n) that appeared in Answer 3.1.1. As is already known, (B_n) does not converge strongly to the zero operator. Thus, (A_n) converges strongly to 0, but (A_n^*) does not.

Answer 3.1.3 We need again the examples of Answer 3.1.1. Let

$$A_n x = (x_{n+1}, x_{n+2}, \cdots) (= (S^*)^n x)$$

and

$$B_n x = A_n^* x = (\underbrace{0, 0, \cdots, 0}_{n \text{ terms}}, x_1, x_2, \cdots).$$

Then, we know that both (A_n) and (B_n) converge weakly to the zero operator (so $A = B = 0$). However, for all n in $\mathbb{N}$, we have

$$A_n B_n = (S^*)^n S^n = I$$

making it impossible for $(A_n B_n)$ to converge weakly to $AB = 0$.

Answer 3.1.4 On $\mathbb{C}^2$, consider the sequence of invertible matrices (A_n) given by

$$A_n = \begin{pmatrix} \frac{1}{n} & 0 \\ 0 & \frac{1}{n} \end{pmatrix}$$

where $n \in \mathbb{N}$. Then (A_n) converges in norm to $A := \begin{pmatrix} 0 & 0 \\ 0 & 0 \end{pmatrix}$ because for each n

$$\|A_n - A\| = \frac{1}{n},$$

and so $A_n \to A$ uniformly. Finally, observe that A is clearly not invertible.

Answer 3.1.5 On $\mathbb{C}^2$ say, consider the sequence of self-adjoint matrices (A_n) defined by

$$A_n = \begin{pmatrix} \frac{1}{n} & 1 \\ 1 & 0 \end{pmatrix}$$

where $n \in \mathbb{N}$. Then the sequence (A_n) converges in norm to the self-adjoint $A = \begin{pmatrix} 0 & 1 \\ 1 & 0 \end{pmatrix}$ because for each n

$$\|A_n - A\| = \frac{1}{n}$$

and so $A_n \to A$. Nonetheless,

$$A_n A = \begin{pmatrix} 1 & \frac{1}{n} \\ 0 & 1 \end{pmatrix} \neq A A_n = \begin{pmatrix} 1 & 0 \\ \frac{1}{n} & 1 \end{pmatrix}$$

for every n, i.e. none of the A_n commutes with (the limit) A.

Answer 3.1.6

(1) Consider the sequence of operators defined on ℓ^2 as

$$U_n(x_0, x_1, x_2, \cdots) = (x_n, x_0, x_1, \cdots, x_{n-1}, x_{n+1}, \cdots).$$

Then each U_n is unitary as U_n is onto and

$$\|U_n x\| = \|x\|, \ \forall x \in \ell^2.$$

Now, (U_n) converges strongly to the shift operator S (which is non-unitary).

(2) The set of normal operators is not SOT-closed. A way of seeing this is via the previous counterexample. Indeed, the sequence (U_n) is normal and converges to the shift (which is not normal).

(3) By the previous answer, it is now obvious that the weak limit of a sequence of normal operators need not remain normal either. Alternatively, remember that the weak closure of the set of unitary operators on an infinite- dimensional space H is precisely the set of all contractions (see e.g. Problem 224 in [148]). Then take any nonnormal contraction (e.g. the usual shift S on $\ell^2(\mathbb{N})$ or Volterra's operator on $L^2(0, 1)$), and so there is always a sequence of unitary operators (a fortiori normal) that converges weakly to the nonnormal S.

Remark (Needs the Notion of Subnormality) E. Bishop in [29] showed that an operator $S \in B(H)$ is subnormal (see Definition 14.1.3 below) if and only if there is a sequence of normal operators that converges strongly to S. Therefore, any nonnormal subnormal operator is just perfect as a counterexample to the second question. Note in the end that a simplification of Bishop's result appeared in [69] (see also Page 36 in [67]).

Chapter 4
Positive Operators

4.1 Basics

Definition 4.1.1 Let H be a $\mathbb{C}$-Hilbert space, and let $A \in B(H)$. We say that A is:

(1) Positive (or nonnegative) if $\langle Ax, x \rangle \geq 0$ for all $x \in H$. In symbols, $A \geq 0$.
(2) Strictly positive if $\langle Ax, x \rangle > 0$ for all $x \in H$ with $x \neq 0$. Symbolically, $A > 0$.
(3) Negative if $\langle Ax, x \rangle \leq 0$ for all $x \in H$. We then write $A \leq 0$.

Remark It is easy to see that a positive operator $A \in B(H)$ is automatically self-adjoint because $\langle Ax, x \rangle \in \mathbb{R}$ for all $x \in H$. The proof uses the fact that H is a $\mathbb{C}$-Hilbert.

In the case of an $\mathbb{R}$-Hilbert space, we have the following definition:

Definition 4.1.2 Let H be an $\mathbb{R}$-Hilbert space, and let $A \in B(H)$. Call A positive if A is self-adjoint and $\langle Ax, x \rangle \geq 0$ for all $x \in H$.

Examples 4.1.1

(1) Let $A \in B(H)$. Then AA^* and A^*A are always positive.
(2) Let $a, b, c \in \mathbb{R}$, and consider the *self-adjoint* matrix $A = \begin{pmatrix} a & b \\ b & c \end{pmatrix}$. As is known,

A is positive iff $a \geq 0$ and $\det A = ac - b^2 \geq 0$. For example, $B = \begin{pmatrix} 1 & 2 \\ 2 & 2 \end{pmatrix}$ is

not positive, whereas $C = \begin{pmatrix} 1 & 1 \\ 1 & 1 \end{pmatrix}$ is positive.

Remark Recall also that a (finite) self-adjoint matrix is positive if and only if all of its eigenvalues λ_i are positive (i.e. $\lambda_i \geq 0$ for all i).

© The Author(s), under exclusive license to Springer Nature Switzerland AG 2022
M. H. Mortad, *Counterexamples in Operator Theory*,
https://doi.org/10.1007/978-3-030-97814-3_4

Definition 4.1.3 Let $A, B \in B(H)$. We write $A \geq B$ when $A - B \geq 0$ and if both A and B are self-adjoint.

Remark If $A, B \in B(H)$ are such that $A - B \geq 0$, then one cannot a priori write $A \geq B$ unless A and B are both self-adjoint. For example, let H be a complex Hilbert space, and then set $B = iI$ and $A = B + I$ (hence, both of A and B are non-self-adjoint). While

$$A - B = I \geq 0,$$

writing $A \geq B$ would not make sense for the associated quadratic forms are *complex* numbers!

4.2 Questions

4.2.1 Two Positive Operators A, B Such That $AB = 0$

Question 4.2.1 Find two positive nonzero matrices A, B such that $AB = 0$.

4.2.2 Two A, B Such That $A \not\leq 0$, $A \not\geq 0$, $B \not\leq 0$, $B \not\geq 0$, yet $AB \geq 0$

Recall that if $A, B \in B(H)$ are commuting and both are positive (or both negative), then $AB \geq 0$. The simplest proof perhaps is the one based on positive square roots that are yet to be introduced. See e.g. Exercise 5.3.28 in [256]. Another less known proof by W. T. Reid [306], and based upon his inequality may be consulted in say Exercise 5.3.16 in [256].

Question 4.2.2 Give $A, B \in B(H)$ such that $A \not\leq 0$, $A \not\geq 0$, $B \not\leq 0$, $B \not\geq 0$, yet $AB \geq 0$.

4.2.3 $KAK^* \not\leq A$ Where $A \geq 0$ and K Is a Contraction

Let $A \in B(H)$ be positive, and let $K \in B(H)$ be a contraction. If $KA = AK^*$ or $KA = AK$, then it may be shown that:

$$KAK^* \leq A.$$

Proofs (based on Reid's inequality) could be consulted in Exercises 5.3.15 and 5.3.29 in [256]. What about the general case?

Question 4.2.3 Find a positive $A \in B(H)$ and a contraction $K \in B(H)$ such that

$$KAK^* \not\leq A.$$

4.2.4 $KAK^* \leq A \not\Rightarrow AK = KA$ Where $A \geq 0$ and K Is an Isometry

Let $A \in B(H)$ be positive, and let $K \in B(H)$ be a contraction such that $KAK^* \leq A$. Then, it is not true that $AK = KA$. As an example, let

$$K = \begin{pmatrix} 0 & 1/2 \\ 0 & 0 \end{pmatrix} \text{ and } A = \begin{pmatrix} 1 & 0 \\ 0 & 0 \end{pmatrix}$$

be both defined on $\mathbb{C}^2$, say. Then K is a contraction and $A \geq 0$. Besides, $KA = 0$ whereby $KAK^* \leq A$. In the end, $KA \neq AK$.

What if we remain on a finite-dimensional space, but we assume that K is an isometry, i.e. K is unitary? Now, the answer is yes (by only assuming the self-adjointness of A). For a proof, suppose that $KAK^* \leq A$, i.e. $A - KAK^* \geq 0$. If "tr" denotes the usual trace of a matrix, then

$$\text{tr}(A - KAK^*) = \text{tr}\,A - \text{tr}(KAK^*) = \text{tr}\,A - \text{tr}(K^*KA) = 0.$$

Hence, $A = KAK^*$ and so $AK = KA$. What about when $\dim H = \infty$?

Question 4.2.4 Find $A, K \in B(H)$ such that $A \geq 0$ and K is an isometry for which

$$KAK^* \leq A \not\Longrightarrow AK = KA.$$

4.2.5 $KAK^* \leq A \nRightarrow AK^* = KA$ Where $A \geq 0$ and K Is Unitary

Question 4.2.5 Find $A, K \in B(H)$ such that $A \geq 0$ and K is unitary, both obeying

$$KAK^* \leq A \text{ and } AK^* \neq KA.$$

4.2.6   The Operator Norm Is Not Strictly Increasing

Recall that a norm $\| \cdot \|$ is called strictly increasing if for all (self-adjoint) operators A and B we have

$$0 \leq A \leq B \text{ and } \|A\| = \|B\| \Longrightarrow A = B.$$

Question 4.2.6 Show that the usual operator norm is not strictly increasing.

4.2.7 $A \geq B \geq 0 \nRightarrow A^2 \ngeq B^2$

Let $A, B \in B(H)$ be such that $AB = BA$ and $A \geq B \geq 0$. Then $A^2 \geq B^2$. Indeed, since $AB = BA$, we have

$$A^2 - B^2 = (A + B)(A - B).$$

But, $A \geq B \geq 0$ gives $A - B \geq 0$ and $A + B \geq 0$. The condition $AB = BA$ also says that $A - B$ commutes with $A + B$ whereby

$$(A + B)(A - B) = A^2 - B^2 \geq 0.$$

Therefore, $A^2 \geq B^2$ (for A^2 and B^2 are self-adjoint).

Question 4.2.7 Find $A, B \in B(H)$ such that $A \geq B \geq 0$ with $A^2 \ngeq B^2$.

4.2.8 $A, B \geq 0 \nRightarrow AB + BA \ngeq 0$

Let $A, B \in B(H)$ be two self-adjoint operators. Then, $AB + BA$ is obviously self-adjoint. Do we still have an affirmative answer if we replace "self-adjoint" by "positive" in both the assumption and the conclusion?

> **Question 4.2.8** Find positive $A, B \in B(H)$ such that
>
> $$AB + BA \ngeq 0.$$

4.2.9 Two Non-self-adjoint A and B Such That $A^n + B^n \geq 0$ for All n

It was shown in [366] that: If $A, B \in B(H)$ are such that $\operatorname{Re} A \geq \operatorname{Re} B$, then $A^n + B^n \geq 0$ for all n forces A and B to be self-adjoint.

> **Question 4.2.9** Show that the hypothesis $\operatorname{Re} A \geq \operatorname{Re} B$ may not just be dropped.

4.2.10 Two Positive A, B (with $A \neq 0$ and $B \neq 0$) and Such That $AB \geq 0$ but $A^2 + B^2$ Is Not Invertible

If $S, T \in B(H)$ where S is invertible and $T \geq S \geq 0$, then T is invertible (see Exercise 5.3.30 in [256] for a different proof of this well known result).

So, if $A, B \in B(H)$ are self-adjoint (one of them is invertible), then $A^2 + B^2$ is invertible. Indeed, if e.g. A is invertible, then $A^2 + B^2$ is invertible as $A^2 + B^2 \geq A^2$ by remembering that A^2 is invertible.

Nevertheless, and if one of the operators is not self-adjoint (even if it remains normal), the sum $A^2 + B^2$ is not necessarily invertible. Consider any self-adjoint invertible operator A and then set $B = iA$. We therefore see that $A^2 + B^2 = 0$ is not invertible.

Question 4.2.10 Find $A, B \in B(H)$ both positive (with $A \neq 0$ and $B \neq 0$) and such that $AB \geq 0$ but $A^2 + B^2$ is not invertible.
Hint: You must avoid an invertible AB.

4.2.11   *Two A, B Satisfying $\|AB - BA\| = 1/2\|A\|\|B\|$*

Let $A, B \in B(H)$. F. Kittaneh showed in [198] that if *both* A and B are positive, then

$$\|AB - BA\| \leq \frac{1}{2}\|A\|\|B\|.$$

Question 4.2.11 Give an example of a nontrivial pair A and B for which the equality

$$\|AB - BA\| = \frac{1}{2}\|A\|\|B\|$$

holds.

4.2.12   *Two A, B Satisfying $\|AB - BA\| = \|A\|\|B\|$*

Let A, B be both in $B(H)$. F. Kittaneh showed in [199] that if *either A or B* is positive, then

$$\|AB - BA\| \leq \|A\|\|B\|.$$

Question 4.2.12 Give an example of a (nontrivial) pair A and B for which the equality

$$\|AB - BA\| = \|A\|\|B\|$$

holds.

4.2.13   On Normal Solutions of the Equations $AA^* = qA^*A$, $q \in \mathbb{R}$

Question 4.2.13 (Cf. Question 25.1.6) Let $q \in \mathbb{R}$. Can you find a nonnormal operator $A \in B(H)$ such that

$$AA^* = qA^*A?$$

We call such an operator q-normal.

Answers

Answer 4.2.1 On $\mathbb{C}^2$, define

$$A = \begin{pmatrix} 1 & 0 \\ 0 & 0 \end{pmatrix} \text{ and } B = \begin{pmatrix} 0 & 0 \\ 0 & 1 \end{pmatrix}.$$

Then $A, B \geq 0$ and

$$AB = \begin{pmatrix} 0 & 0 \\ 0 & 0 \end{pmatrix}.$$

Answer 4.2.2 This is easy! Let

$$A = B = \begin{pmatrix} -1 & 0 \\ 0 & 1 \end{pmatrix}.$$

Then $A \not\geq 0$ and $A \not> 0$, while $AB = I \geq 0$.

Answer 4.2.3 On $\mathbb{C}^2$, let

$$K = \begin{pmatrix} 0 & 1 \\ 1 & 0 \end{pmatrix} \text{ and } A = \begin{pmatrix} 1 & 0 \\ 0 & 0 \end{pmatrix}.$$

Then $K = K^*$, $K^2 = I$, and $A \geq 0$. Hence,

$$A - KAK = \begin{pmatrix} 1 & 0 \\ 0 & 0 \end{pmatrix} - \begin{pmatrix} 0 & 0 \\ 0 & 1 \end{pmatrix} = \begin{pmatrix} 1 & 0 \\ 0 & -1 \end{pmatrix} \not\geq \begin{pmatrix} 0 & 0 \\ 0 & 0 \end{pmatrix}.$$

Answer 4.2.4 An efficient place to find such an example is ℓ^2. Let $S \in B(\ell^2)$ be the shift operator, i.e.

$$S(x_1, x_2, \cdots) = (0, x_1, x_2, \cdots).$$

Then

$$S^*(x_1, x_2, x_3, \cdots) = (x_2, x_3, \cdots)$$

and so

$$SS^*(x_1, x_2, x_3, \cdots) = (0, x_2, x_3, \cdots).$$

Hence,

$$S^2(x_1, x_2, \cdots) = (0, 0, x_1, x_2, \cdots)$$

and

$$S^{*^2}(x_1, x_2, x_3, \cdots) = (x_3, x_4, \cdots).$$

Therefore,

$$S^2 S^{*^2}(x_1, x_2, \cdots) = (0, 0, x_3, x_4, \cdots).$$

By setting $A = SS^*$ and $K = S$, we see that $A \geq 0$ and that K is an isometry on ℓ^2. Besides, $KAK^* \leq A$ as

$$SSS^*S^* = S^2 S^{*^2} \leq SS^*$$

because

$$(SS^* - S^2 S^{*^2})(x_1, x_2, \cdots) = (0, x_2, 0, 0, \cdots)$$

and

$$\langle (SS^* - S^2 S^{*^2})(x_1, x_2, \cdots), (x_1, x_2, \cdots) \rangle = |x_2|^2 \geq 0$$

for all $(x_1, x_2, \cdots) \in \ell^2$. However, $AK \neq KA$ since $AK = KA$ would lead to $SS^* = I$.

Answer 4.2.5 Let K be a unitary matrix (i.e. $KK^* = I$) that is not self-adjoint (e.g. $K = \begin{pmatrix} 0 & 1 \\ -1 & 0 \end{pmatrix}$ on $\mathbb{C}^2$). Next, take $A = I$. Then obviously $A \geq 0$ and $KAK^* = KK^* = I$ and yet

$$AK^* = K^* \neq K = KA.$$

Answer 4.2.6 We provide a simple counterexample. Let

$$A = \begin{pmatrix} 1 & 0 \\ 0 & 0 \end{pmatrix}, B = \begin{pmatrix} 1 & 0 \\ 0 & 1 \end{pmatrix}.$$

Then $0 \leq A \leq B$, $\|A\| = \|B\| (= 1)$ but $A \neq B$.

Answer 4.2.7 Consider

$$A = \begin{pmatrix} 2 & 1 \\ 1 & 1 \end{pmatrix} \text{ and } B = \begin{pmatrix} 1 & 0 \\ 0 & 0 \end{pmatrix}.$$

First, observe that both A and B are positive. So it only remains to check that $A \geq B$, whereas $A^2 \ngeq B^2$, that is, we need to verify that $A - B \geq 0$ and that $A^2 - B^2 \ngeq 0$. We have

$$A - B = \begin{pmatrix} 1 & 1 \\ 1 & 1 \end{pmatrix} \geq 0$$

while

$$A^2 - B^2 = \begin{pmatrix} 5 & 3 \\ 3 & 2 \end{pmatrix} - \begin{pmatrix} 1 & 0 \\ 0 & 0 \end{pmatrix} = \begin{pmatrix} 4 & 3 \\ 3 & 2 \end{pmatrix} \ngeq 0$$

(as e.g. $\det(A^2 - B^2) < 0$).

One might be tempted to think that under the invertibility of both A and B, the result may become true. This is refuted by the example: Let $T = A + I$ and $S = B + I$, where A and B are as above. Then S and T are positive and invertible and $T \geq S$. Nevertheless,

$$T^2 - S^2 = \begin{pmatrix} 6 & 5 \\ 5 & 4 \end{pmatrix} \ngeq 0$$

for $\det(T^2 - S^2) < 0$.

Answer 4.2.8 Consider the positive matrices

$$A = \begin{pmatrix} 1 & 1 \\ 1 & 1 \end{pmatrix} \text{ and } B = \begin{pmatrix} 0 & 0 \\ 0 & 1 \end{pmatrix}.$$

Then,

$$AB = \begin{pmatrix} 0 & 1 \\ 0 & 1 \end{pmatrix} \text{ and } BA = (AB)^* = \begin{pmatrix} 0 & 0 \\ 1 & 1 \end{pmatrix}.$$

Hence,

$$AB + BA = \begin{pmatrix} 0 & 1 \\ 1 & 2 \end{pmatrix}$$

is not positive (as e.g. $\det(AB + BA) = -1 < 0$).

Answer 4.2.9 Consider

$$A = \begin{pmatrix} 1 & 1 \\ 0 & 0 \end{pmatrix} \text{ and } B = \begin{pmatrix} 0 & 0 \\ 1 & 1 \end{pmatrix}.$$

Both A and B are not self-adjoint. Then for all $n \in \mathbb{N}$,

$$A^n + B^n = \begin{pmatrix} 1 & 1 \\ 1 & 1 \end{pmatrix}$$

which is clearly positive. Observe in the end that Re A and Re B are not comparable.

Answer 4.2.10 Let A be the positive operator defined by

$$Af(x) = xf(x), \ f \in L^2[0, 1].$$

Setting $B = A$, we then see that

$$ABf(x) = x^2 f(x)$$

is clearly positive. However,

$$(A^2 + B^2)f(x) = 2x^2 f(x)$$

is not invertible.

Alternatively, merely consider:

$$A = \begin{pmatrix} 1 & 0 & 0 \\ 0 & 0 & 0 \\ 0 & 0 & 0 \end{pmatrix} \text{ and } B = \begin{pmatrix} 0 & 0 & 0 \\ 0 & 1 & 0 \\ 0 & 0 & 0 \end{pmatrix}.$$

Answer 4.2.11 Consider

$$A = \begin{pmatrix} 1 & 1 \\ 1 & 1 \end{pmatrix} \text{ and } B = \begin{pmatrix} 1 & 0 \\ 0 & 0 \end{pmatrix}.$$

Then both A and B are positive. Also,

$$\|AB - BA\| = 1, \ \|A\| = 2, \ \|B\| = 1,$$

that is,

$$\|AB - BA\| = \frac{1}{2}\|A\|\|B\|,$$

as wished.

Answer 4.2.12 Define on $\mathbb{C}^2$:

$$A = \begin{pmatrix} 0 & 1 \\ 0 & 0 \end{pmatrix} \text{ and } B = \begin{pmatrix} 1 & 0 \\ 0 & 0 \end{pmatrix}.$$

Then B is positive and $\|A\| = \|B\| = 1$. Since $AB = 0$, it follows that $\|AB - BA\| = \|BA\|$. But

$$\|BA\|^2 = \left\| \begin{pmatrix} 0 & 1 \\ 0 & 0 \end{pmatrix} \right\|^2 = \left\| \begin{pmatrix} 0 & 0 \\ 1 & 0 \end{pmatrix} \begin{pmatrix} 0 & 1 \\ 0 & 0 \end{pmatrix} \right\| = \left\| \begin{pmatrix} 0 & 0 \\ 0 & 1 \end{pmatrix} \right\| = 1,$$

and we are done.

Answer 4.2.13 The answer is negative! Observe first that such q cannot be in $\mathbb{R}^- := (-\infty, 0]$. Indeed, if q were in $\mathbb{R}^-$ and given the positivity of both AA^* and A^*A, then $AA^* = qA^*A$ would necessarily force $AA^* = A^*A = 0$ or simply $A = 0$, whereby A is normal.

Assume now that $q > 0$. Clearly,

$$AA^* = qA^*A \implies \|AA^*\| = q\|A^*A\|.$$

Since $\|AA^*\| = \|A^*A\|$ (for any $A \in B(H)$), it follows that

$$(1 - q)\|AA^*\| = 0.$$

Hence, either $q = 1$ or $\|AA^*\| = 0$. In the former case, we obtain $AA^* = A^*A$, and in the latter, we obtain $A = 0$. Accordingly, in all cases, A is normal.

Remark In the theory of quantum groups, one has to work with equations of the type

$$AA^* = qA^*A$$

where q is a positive real number. See Question 25.1.6 for the unbounded case.

Chapter 5
Matrices of Bounded Operators

5.1 Basics

Before giving the formal definition of matrices of operators, we recall the following:

Definition 5.1.1 Let H_1 and H_2 be two Hilbert spaces. The set

$$\{(x_1, x_2) : x_1 \in H_1, \ x_2 \in H_2\}$$

is denoted by $H_1 \oplus H_2$ (or $H_1 \times H_2$), and it is called the direct sum of H_1 and H_2. It is a Hilbert space with respect to the inner product

$$\langle (x_1, x_2), (y_1, y_2) \rangle_{H_1 \oplus H_2} = \langle x_1, y_1 \rangle_{H_1} + \langle x_2, y_2 \rangle_{H_2}.$$

Definition 5.1.2 Let $A_{ij} \in B(H_j, H_i)$, where $i, j = 1, 2$. The matrix

$$A = \begin{pmatrix} A_{11} & A_{12} \\ A_{21} & A_{22} \end{pmatrix}$$

is called an operator matrix, and it defines a linear operator from $H_1 \oplus H_2$ into $H_1 \oplus H_2$ by

$$A \begin{pmatrix} x_1 \\ x_2 \end{pmatrix} = \begin{pmatrix} A_{11}x_1 + A_{12}x_2 \\ A_{21}x_1 + A_{22}x_2 \end{pmatrix}.$$

Remark We tolerate the abuse of notation $A(x_1, x_2)$ in place of $A \begin{pmatrix} x_1 \\ x_2 \end{pmatrix}$ in many cases.

© The Author(s), under exclusive license to Springer Nature Switzerland AG 2022
M. H. Mortad, *Counterexamples in Operator Theory*,
https://doi.org/10.1007/978-3-030-97814-3_5

Example 5.1.1 Let H and K be two Hilbert spaces. Let I_H be the identity operator on H, and let I_K be the identity operator on K. The matrix of operators

$$I_{H \oplus K} = \begin{pmatrix} I_H & 0 \\ 0 & I_K \end{pmatrix}$$

is called the identity operator on $H \oplus K$ (where the upper 0 is the zero operator between K and H and the lower one is the zero operator from H into K).

Theorem 5.1.1 Let $A_{ij} \in B(H_j, H_i)$ and $B_{ij} \in B(H_j, H_i)$, where $i, j = 1, 2$. Set

$$A = \begin{pmatrix} A_{11} & A_{12} \\ A_{21} & A_{22} \end{pmatrix} \text{ and } B = \begin{pmatrix} B_{11} & B_{12} \\ B_{21} & B_{22} \end{pmatrix}.$$

Then:

(1) $A \in B(H_1 \oplus H_2)$ *and*

$$\max_{1 \leq i, j \leq 2} \|A_{ij}\| \leq \|A\| \leq 4 \max_{1 \leq i, j \leq 2} \|A_{ij}\|.$$

(2) *The sum $A + B$ is given by*

$$A + B = \begin{pmatrix} A_{11} + B_{11} & A_{12} + B_{12} \\ A_{21} + B_{21} & A_{22} + B_{22} \end{pmatrix}.$$

(3) *The product AB is given by*

$$AB = \begin{pmatrix} A_{11}B_{11} + A_{12}B_{21} & A_{11}B_{12} + A_{12}B_{22} \\ A_{21}B_{11} + A_{22}B_{21} & A_{21}B_{12} + A_{22}B_{22} \end{pmatrix}.$$

(4) *The adjoint A^* of A is given by*

$$A^* = \begin{pmatrix} A_{11} & A_{12} \\ A_{21} & A_{22} \end{pmatrix}^* = \begin{pmatrix} A_{11}^* & A_{21}^* \\ A_{12}^* & A_{22}^* \end{pmatrix}.$$

As a particular case, we have:

Proposition 5.1.2 *If H and K are two Hilbert spaces, and if $A \in B(H)$ and $B \in B(K)$, then*

$$T(x, y) = (Ax, By)$$

*defines a bounded operator on $H \oplus K$, which may be denoted by $A \oplus B$. We call it
the direct sum of the operators A and B. Furthermore,*

$$\|T\| = \|A \oplus B\| = \max(\|A\|, \|B\|)$$

and

$$(A \oplus B)^* = A^* \oplus B^*.$$

The direct sum of two operators defined previously may easily be generalized to
finite sums. Now, we give the appropriate generalization to infinite (not necessarily
countable) sums. Readers may wish to consult [184] from which we have borrowed
the following lines:

Let I be any set. Let $(H_i)_{i \in I}$ be a collection of Hilbert spaces. Define

$$H := \bigoplus H_i = \left\{ (x_i)_{i \in I}, x_i \in H_i : \sum_{i \in I} \|x_i\|_{H_i}^2 < \infty \right\}.$$

Then it may be shown that H is a Hilbert space w.r.t. the inner product

$$\langle (x_i), (y_i) \rangle = \sum_{i \in I} \langle x_i, y_i \rangle$$

(the induced norm being $\|(x_i)\|_H = \sqrt{\sum_{i \in I} \|x_i\|_{H_i}^2}$).

Example 5.1.2 Clearly,

$$\mathbb{C} \oplus \mathbb{C} \oplus \cdots \mathbb{C} \oplus \mathbb{C} \oplus \cdots = \ell^2.$$

Suppose now that H_i and K_i are Hilbert spaces and $A_i \in B(H_i, K_i)$ for each $i \in I$.
If $\sup_{i \in I} \|A_i\| < \infty$, then $A(x_i) = (A_i x_i)$ defines a bounded linear operator A from
$\oplus H_i$ into $\oplus K_i$. The operator $A := \oplus A_i$ is called the direct sum of the operators
(A_i). It can be shown that

$$\left\| \bigoplus_{i \in I} A_i \right\| = \sup_{i \in I} \|A_i\|,$$

$$\left(\bigoplus_{i \in I} A_i \right)^* = \bigoplus_{i \in I} A_i^*,$$

$$\bigoplus_{i \in I} (\alpha A_i + \beta B_i) = \alpha \bigoplus_{i \in I} A_i + \beta \bigoplus_{i \in I} B_i$$

and

$$\left(\bigoplus_{i \in I} A_i\right)\left(\bigoplus_{i \in I} C_i\right) = \bigoplus_{i \in I} A_i C_i$$

where $B_i \in B(H_i, K_i)$ and $C_i \in B(K_i, H_i)$.

5.2   Questions

5.2.1   A Non-invertible Matrix Whose Formal Determinant Is Invertible

Let $A \in B(H \oplus H)$. Then A is said to be invertible if there exists a $B \in B(H \oplus H)$ such that

$$AB = BA = I_{H \oplus H}.$$

Proposition 5.2.1 *Let H and K be two Hilbert spaces, and consider the operator $T = A \oplus B$ on $H \oplus K$. Then*

$$A \oplus B \text{ is invertible } \Longleftrightarrow A \text{ and } B \text{ are invertible}.$$

In such case,

$$(A \oplus B)^{-1} = A^{-1} \oplus B^{-1}.$$

In other words,

$$\begin{pmatrix} A & 0 \\ 0 & B \end{pmatrix} \text{ is invertible } \Longleftrightarrow A \text{ and } B \text{ are invertible}$$

and in such case

$$\begin{pmatrix} A & 0 \\ 0 & B \end{pmatrix}^{-1} = \begin{pmatrix} A^{-1} & 0 \\ 0 & B^{-1} \end{pmatrix}.$$

More general results on invertibility are the following (see e.g. [148]):

Proposition 5.2.2 *Let $A, B, C, D \in B(H)$ and define*

$$T = \begin{pmatrix} A & B \\ C & D \end{pmatrix}.$$

If all the entries pairwise commute, then T is invertible iff $AD - BC$ (more known as the formal determinant of T, noted $\det T$) is invertible.

Proposition 5.2.3 *Let $A, B, C, D \in B(H)$, and then set*

$$T = \begin{pmatrix} A & B \\ C & D \end{pmatrix}.$$

If C commutes with D, and D is invertible, then T is invertible iff $\det T$ is invertible.

Remark Due to emotional connotations, readers should be careful and must not precipitate when calculating $\det T$. Indeed, one major pitfall is that if $T = \begin{pmatrix} A & B \\ C & D \end{pmatrix}$ is defined on say $H \oplus K$ (with $H \neq K$), then none of the products AD and BC need to be well defined.

Question 5.2.1 Find a matrix of operators A that is not invertible, yet its formal determinant is invertible.

5.2.2 An Invertible Matrix Whose Formal Determinant Is Not Invertible

Question 5.2.2 Find an invertible matrix of operators, yet its formal determinant is not invertible.

5.2.3 Invertible Triangular Matrix vs. Left and Right Invertibility of Its Diagonal Elements

Let $T = \begin{pmatrix} A & C \\ 0 & B \end{pmatrix}$ be defined on $H \oplus H$.

If T is invertible, then A is left invertible and B is right invertible (see e.g. [152] for necessary and sufficient conditions of the invertibility of T).

Question 5.2.3 Find an example of an invertible T, yet A is not right invertible and B is not left invertible.

5.2.4   Non-invertible Triangular Matrix vs. Left and Right Invertibility of Its Diagonal Elements

Question 5.2.4 Find $T = \begin{pmatrix} A & C \\ 0 & B \end{pmatrix}$ on $H \oplus H$ where A is left invertible, B is right invertible, and T is not invertible.

5.2.5   An Invertible Matrix yet None of Its Entries Is Invertible

It is quite conceivable to have a non-invertible matrix where all of its entries are invertible. The simplest example perhaps is to consider on $\mathbb{C}^2$ the matrix $\begin{pmatrix} 1 & 1 \\ 1 & 1 \end{pmatrix}$, or its clone on $H \oplus H$ given by

$$T = \begin{pmatrix} I & I \\ I & I \end{pmatrix}$$

where I is the usual identity operator on H. One reason why T is not invertible is that it is clearly not one-to-one.

Question 5.2.5 Find a matrix of bounded operators

$$T = \begin{pmatrix} A & B \\ C & D \end{pmatrix}$$

such that *none* of A, B, C, and D is invertible yet T is invertible.

5.2.6   A Normal Matrix yet None of Its Entries Is Normal

Let $T = \begin{pmatrix} A & B \\ C & D \end{pmatrix}$ where $A, B, C, D \in B(H)$. If we assume that T is self-adjoint, then it may be shown that A and D must be self-adjoint and that $B^* = C$. In particular, $\begin{pmatrix} 0 & A \\ A^* & 0 \end{pmatrix}$ is always self-adjoint, that is, regardless of what $A \in B(H)$ can be.

Question 5.2.6 Find a matrix of bounded operators

$$T = \begin{pmatrix} A & B \\ C & D \end{pmatrix}$$

such that *all* of A, B, C, and D are *not* normal yet T is normal.

5.2.7 A Unitary Matrix yet None of Its Entries Is Unitary

Question 5.2.7 Find a matrix of bounded operators

$$T = \begin{pmatrix} A & B \\ C & D \end{pmatrix}$$

where *none* of A, B, C, and D is unitary, but T is unitary.

5.2.8 Two Non-comparable Self-Adjoint Matrices yet the Corresponding Entries Are Comparable

Question 5.2.8 Find self-adjoint A, B, C; A', B', $C' \in B(H)$ such that $A \le A'$, $B \le B'$, and $C \le C'$, but

$$T := \begin{pmatrix} A & B \\ B & C \end{pmatrix} \nleq T' := \begin{pmatrix} A' & B' \\ B' & C' \end{pmatrix}.$$

5.2.9 An Isometry S Such That S^2 Is Unitarily Equivalent to $S \oplus S$

Question 5.2.9 Find an isometry S such that S^2 is unitarily equivalent to $S \oplus S$.

5.2.10   An Infinite Direct Sum of Invertible Operators Need Not Be Invertible

Question 5.2.10   Let (T_n) be a sequence of bounded operators on H such that $(\|T_n\|)_n$ is bounded. Set $T = \bigoplus_{n \in \mathbb{N}} T_n$. Show that it may well happen that *all* T_n are invertible, while T is not.
Hint: Recall that (see e.g. Exercise 2.8.16 in [184]): $\bigoplus_{n \in \mathbb{N}} T_n$ *is invertible if and only if each T_n is invertible and* $\sup_n \|T_n^{-1}\| < \infty$.

5.2.11   The Similarity of $A \oplus B$ to $C \oplus D$ Does Not Entail the Similarity of A to C or That of B to D

Question 5.2.11   Let $A, B, C, D \in B(H)$ be such that $A \oplus B$ is similar to $C \oplus D$. Does it follow that A is similar to C or that B is similar to D?

5.2.12   A Matrix of Operators T on H^2 Such That $T^3 = 0$ But $T^2 \neq 0$

As is known to readers, if $\dim H < \infty$ and N is a nilpotent matrix on H, then

$$N^{\dim H} = 0$$

(see e.g. 8.18 in [9]).

In the case of matrices of operators, and given their apparent resemblance to usual matrices with real or complex entries, it seems therefore reasonable to conjecture that: If T is a matrix of bounded operators defined on $H \oplus H \oplus \cdots \oplus H$ (n times), that is, on $H \times H \times \cdots \times H = H^n$, and $T^p = 0$ for some integer $p \geq n$, then necessarily $T^n = 0$.

Question 5.2.12   Find a matrix of operators T on some Hilbert space $H^2 = H \times H$ such that $T^3 = 0$ while $T^2 \neq 0$.

5.2.13   Block Circulant Matrices Are Not Necessarily Circulant

First, recall the following definition: A matrix $A \in \mathcal{M}_n$ of the form

$$
A = \begin{pmatrix}
a_1 & a_2 & \cdots & a_{n-1} & a_n \\
a_n & a_1 & a_2 & \cdots & a_{n-1} \\
a_{n-1} & a_n & \ddots & \ddots & \vdots \\
\vdots & \ddots & \ddots & \ddots & a_2 \\
a_2 & a_3 & \cdots & a_n & a_1
\end{pmatrix}
$$

is called a circulant matrix.

Circulant matrices arise in many areas of applications. For example, when solving some difference and differential equations as well as delay differential equations. They also arise in cryptography and also in applications involving the discrete Fourier transform (see [276]).

Inspired by the above definition, we say (cf. [276]) that T, defined on $H \oplus H \oplus \cdots \oplus H$ (n copies), is a block circulant matrix if it is of the form

$$
A = \begin{pmatrix}
A_1 & A_2 & \cdots & A_{n-1} & A_n \\
A_n & A_1 & A_2 & \cdots & A_{n-1} \\
A_{n-1} & A_n & \ddots & \ddots & \vdots \\
\vdots & \ddots & \ddots & \ddots & A_2 \\
A_2 & A_3 & \cdots & A_n & A_1
\end{pmatrix}
$$

where $A_i \in B(H)$ for $i = 1, \cdots, n$.

Question 5.2.13 Find a block circulant matrix that is not circulant.

Answers

Answer 5.2.1 Let S be the shift operator on ℓ^2, and let S^* be its adjoint. Define A on $\ell^2 \oplus \ell^2$ by

$$
A = \begin{pmatrix}
S^* & 0 \\
0 & S
\end{pmatrix}
$$

where the zero entries represent the zero operator on ℓ^2. Then, the formal determinant of A is $S^*S - 0$, that is, it is just I, which is obviously invertible.

To show that A is not invertible, it suffices (for example) to show that its kernel is nontrivial. Let $e_1 = (1, 0, 0, \cdots)$ from the standard basis of ℓ^2, so $(e_1, 0_{\ell^2}) \neq (0_{\ell^2}, 0_{\ell^2})$. But

$$A \begin{pmatrix} e_1 \\ 0_{\ell^2} \end{pmatrix} = \begin{pmatrix} S^* & 0 \\ 0 & S \end{pmatrix} \begin{pmatrix} e_1 \\ 0_{\ell^2} \end{pmatrix} = \begin{pmatrix} S^* e_1 \\ 0_{\ell^2} \end{pmatrix} = \begin{pmatrix} 0_{\ell^2} \\ 0_{\ell^2} \end{pmatrix}.$$

Answer 5.2.2 ([148]) Let S be the shift operator on ℓ^2, and define T on $\ell^2 \oplus \ell^2$ by

$$T = \begin{pmatrix} S & A \\ 0 & S^* \end{pmatrix},$$

where $A \in B(\ell^2)$ is the self-adjoint operator defined by

$$A(x_1, x_2, \cdots) = (x_1, 0, 0, \cdots).$$

Then T is invertible with

$$T^{-1} = \begin{pmatrix} S^* & 0 \\ A & S \end{pmatrix}$$

(observe that here $T^{-1} = T^*$). Indeed,

$$TT^{-1} = \begin{pmatrix} SS^* + A^2 & AS \\ S^*A & I \end{pmatrix}$$

where $I \in B(\ell^2)$ is the usual identity operator. Clearly, $AS = 0$ and also $S^*A = 0$. It is also easy to see that $SS^* + A^2 = I$. Therefore, $TT^{-1} = I$. Similarly, readers can check that $T^{-1}T = I$. Finally, $\det T = SS^*$ is clearly non-invertible on ℓ^2 (as e.g. non-injective).

Answer 5.2.3 Define T on $\ell^2 \oplus \ell^2$ by

$$T = \begin{pmatrix} S & A \\ 0 & S^* \end{pmatrix},$$

where S is the shift operator and where $A \in B(\ell^2)$ is the self-adjoint operator defined by

$$A(x_1, x_2, \cdots) = (x_1, 0, 0, \cdots).$$

Then T is invertible (see Answer 5.2.2). However, S is not right invertible, and S^* is not left invertible.

Answer 5.2.4 Let $C = 0$, and let $A = S$ and $B = S^*$, where S is the shift operator on ℓ^2. Then A is left invertible, and B is right invertible yet $\begin{pmatrix} S & 0 \\ 0 & S^* \end{pmatrix}$ is not invertible for its adjoint $\begin{pmatrix} S^* & 0 \\ 0 & S \end{pmatrix}$ is not invertible (see Answer 5.2.1).

Answer 5.2.5 On $L^2(\mathbb{R})$, define A by $Af(x) = \cos x f(x)$ and set $D = A$. Also, define on $L^2(\mathbb{R})$ an operator B by $Bf(x) = \sin x f(x)$ and then set $C = -B$. As is known, A, B, C, and D are *all* non-invertible operators. Since these operators commute pairwise, by Proposition 5.2.2 (just above Question 5.2.1), T is invertible on $L^2(\mathbb{R}) \oplus L^2(\mathbb{R})$ if and only if $AD - BC$ is invertible. This is indeed the case as for all $f \in L^2(\mathbb{R})$

$$(AD - BC)f(x) = f(x),$$

i.e. $AD - BC = I$.

Answer 5.2.6 Let $A \in B(H)$ be nonnormal. Hence, A^* is not normal either. The matrix T defined on $H \oplus H$ by

$$T = \begin{pmatrix} A & A^* \\ A^* & A \end{pmatrix}$$

is normal for

$$TT^* = T^*T = \begin{pmatrix} A^*A + AA^* & A^{*2} + A^2 \\ A^{*2} + A^2 & A^*A + AA^* \end{pmatrix}.$$

Answer 5.2.7 Let H be a Hilbert space, and let I be the identity operator on H. Then *all* of

$$A = \begin{pmatrix} I & 0 \\ 0 & 0 \end{pmatrix}, \ B = \begin{pmatrix} 0 & 0 \\ I & 0 \end{pmatrix}, \ C = B^* \text{ and } D = \begin{pmatrix} 0 & 0 \\ 0 & I \end{pmatrix}$$

are *non-unitary*. Nevertheless,

$$T = \begin{pmatrix} A & B \\ C & D \end{pmatrix} = \begin{pmatrix} I & 0 & 0 & 0 \\ 0 & 0 & I & 0 \\ 0 & I & 0 & 0 \\ 0 & 0 & 0 & I \end{pmatrix}$$

is unitary for $T^{-1} = T^* = T$ as also $T^2 = I_{H \oplus H \oplus H \oplus H}$.

Answer 5.2.8 Let $A \in B(H)$ be positive, and let $B = A + I$ where I is the identity operator on H (we also assume that dim $H \geq 2$). Hence, $A \leq B$. Setting

$$T = \begin{pmatrix} 0 & A \\ A & 0 \end{pmatrix} \text{ and } T' = \begin{pmatrix} 0 & B \\ B & 0 \end{pmatrix},$$

we see that all the assumptions of the question are fulfilled. If we had $T \leq T'$, then we would have for all $(x, y) \in H \oplus H$

$$\left\langle T \begin{pmatrix} x \\ y \end{pmatrix}, \begin{pmatrix} x \\ y \end{pmatrix} \right\rangle \leq \left\langle T' \begin{pmatrix} x \\ y \end{pmatrix}, \begin{pmatrix} x \\ y \end{pmatrix} \right\rangle$$

or

$$\langle Ay, x \rangle + \langle Ax, y \rangle \leq \langle By, x \rangle + \langle Bx, y \rangle,$$

i.e. we would have for

$$\langle Ay, x \rangle + \langle Ax, y \rangle \leq \langle Ay, x \rangle + \langle y, x \rangle + \langle Ax, y \rangle + \langle x, y \rangle.$$

Therefore, we would end up having

$$\langle y, x \rangle + \langle x, y \rangle \geq 0$$

for all $x, y \in H$! This is impossible as seen by taking for instance $y = -x \neq 0$.

Answer 5.2.9 Let $S \in B(\ell^2)$ be the unilateral shift operator. For instance, it can be shown that $S \oplus S$ is a shift of "multiplicity" 2 and so is the case with S^2. Then, remember that two shifts having the same "multiplicity," here 2, are automatically unitarily equivalent. Readers are referred to Chapter 2 (Propositions 2.2 to 2.6) in [205] for details about all that.

Remark See also [72] or [141] for some interesting related results.

Let us provide a relatively simple and explicit example of a unitary operator U such that $U S^2 = (S \oplus S) U$. Let (e_n) be the standard orthonormal basis in ℓ^2, and define $U : H \to H \oplus H$ by taking

$$U e_{2n} = (e_n, 0) \text{ and } U e_{2n+1} = (0, e_n)$$

for all $n \geq 0$. Then U is unitary. By considering both odd and even numbers, we may check that for all $n \geq 0$:

$$U S^2 e_n = (S \oplus S) U e_n.$$

For example, when n is even, we have

$$US^2 e_n = USe_{n+1} = Ue_{n+2} = (e_{\frac{n}{2}+1}, 0),$$

whereas

$$(S \oplus S)U e_n = (S \oplus S)(e_{\frac{n}{2}}, 0) = (e_{\frac{n}{2}+1}, 0).$$

The very similar "odd" case is left to readers.

Answer 5.2.10 Consider the sequence A_n of size $n \times n$ (this example should be remembered by readers) defined by

$$A_n = \begin{pmatrix} 0 & 0 & . & . & . & 0 \\ 1 & 0 & 0 & . & & \\ 0 & 1 & 0 & & . & \\ & . & . & . & 0 & 0 & . \\ & . & & & . & . & . \\ 0 & & & 0 & 1 & 0 \end{pmatrix}$$

(for instance, $A_2 = \begin{pmatrix} 0 & 0 \\ 1 & 0 \end{pmatrix}$ and $A_3 = \begin{pmatrix} 0 & 0 & 0 \\ 1 & 0 & 0 \\ 0 & 1 & 0 \end{pmatrix}$, and so on). Set $T_n = I_n - A_n$,

where I_n is the identity matrix of order n. Clearly,

$$\|T_n\| \leq 2, \quad \text{for } n = 1, 2, \cdots .$$

Besides, each T_n is invertible, and since T_n is the sum of a nilpotent matrix and the identity matrix, we know that

$$T_n^{-1} = I_n + A_n + A_n^2 + \cdots + A_n^{n-1} := S_n.$$

Since

$$\|S_n e_1\|^2 = \|e_1 + \cdots + e_n\|^2 = n,$$

it ensures that $\|S_n\| \geq \sqrt{n}$. According to the hint, this makes it impossible for $\bigoplus_{n \in \mathbb{N}} T_n$ to be invertible and yet each T_n is invertible.

Remark In fact, there are simpler counterexamples. For instance, one may consider $T_n = n^{-1} I$. However, the example above will be needed on other occasions.

Answer 5.2.11 The answer is negative (even when $\dim H < \infty$). For example, $I \oplus 0$ is similar to $0 \oplus I$ via the invertible $\begin{pmatrix} 0 & C \\ C & 0 \end{pmatrix}$, where C is invertible. Indeed, we easily see that:

$$\begin{pmatrix} 0 & C \\ C & 0 \end{pmatrix} \begin{pmatrix} I & 0 \\ 0 & 0 \end{pmatrix} = \begin{pmatrix} 0 & 0 \\ 0 & I \end{pmatrix} \begin{pmatrix} 0 & C \\ C & 0 \end{pmatrix},$$

while I is not similar to 0, as needed.

Answer 5.2.12 A counterexample is available on finite- dimensional spaces. Let $H = \mathbb{C}^2$, and let

$$A = \begin{pmatrix} 0 & 1 \\ 0 & 0 \end{pmatrix} \text{ and } B = \begin{pmatrix} 1 & 0 \\ 0 & 0 \end{pmatrix}$$

be both defined on H. Then

$$AB = \begin{pmatrix} 0 & 0 \\ 0 & 0 \end{pmatrix} \text{ and } BA = \begin{pmatrix} 0 & 1 \\ 0 & 0 \end{pmatrix}$$

and so $ABA = BAB = \begin{pmatrix} 0 & 0 \\ 0 & 0 \end{pmatrix}$. Finally, set

$$T = \begin{pmatrix} \mathbf{0} & A \\ B & \mathbf{0} \end{pmatrix}$$

which is defined on $H \times H$ (where $\mathbf{0} \in B(\mathbb{C}^2)$). Thus,

$$T^2 = \begin{pmatrix} \mathbf{0} & \mathbf{0} \\ \mathbf{0} & BA \end{pmatrix} \text{ and } T^3 = \begin{pmatrix} \mathbf{0} & \mathbf{0} \\ \mathbf{0} & \mathbf{0} \end{pmatrix}$$

and so $T^2 \neq 0$, while $T^3 = 0$, marking the end of the proof.

Answer 5.2.13 ([276]) On $\mathbb{C}^2$ say, let

$$A = \begin{pmatrix} 2 & -1 \\ -1 & 2 \end{pmatrix} \text{ and } B = \begin{pmatrix} -1 & 0 \\ 0 & -1 \end{pmatrix}.$$

Let

$$C = \begin{pmatrix} A & B & 0 & B \\ B & A & B & 0 \\ 0 & B & A & B \\ B & 0 & B & A \end{pmatrix}$$

$$= \begin{pmatrix} 2 & -1 & -1 & 0 & 0 & 0 & -1 & 0 \\ -1 & 2 & 0 & -1 & 0 & 0 & 0 & -1 \\ -1 & 0 & 2 & -1 & -1 & 0 & 0 & 0 \\ 0 & -1 & -1 & 2 & 0 & -1 & 0 & 0 \\ 0 & 0 & -1 & 0 & 2 & -1 & -1 & 0 \\ 0 & 0 & 0 & -1 & -1 & 2 & 0 & -1 \\ -1 & 0 & 0 & 0 & -1 & 0 & 2 & -1 \\ 0 & -1 & 0 & 0 & 0 & -1 & -1 & 2 \end{pmatrix}.$$

Then C is a block circulant matrix that is not circulant.

Chapter 6
(Square) Roots of Bounded Operators

6.1 Basics

Definition 6.1.1 Let $A \in B(H)$.

(1) We say that $B \in B(H)$ is a square root of A if $B^2 = A$.
(2) We say that $B \in B(H)$ is a cube root of A if $B^3 = A$.
(3) More generally, we say that $B \in B(H)$ is an nth root of A if $B^n = A$ (where $n \in \mathbb{N}$).

Definition 6.1.2 (Cf. [395]) Let $A \in B(H)$. We say that A is rootless if there is no $B \in B(H)$ and no integer $p \geq 2$ such that $B^p = A$.

Example 6.1.1 A square root of

$$A = \begin{pmatrix} 1 & 0 \\ 0 & 2 \end{pmatrix}$$

is

$$\begin{pmatrix} 1 & 0 \\ 0 & \sqrt{2} \end{pmatrix}.$$

Another one is

$$\begin{pmatrix} -1 & 0 \\ 0 & \sqrt{2} \end{pmatrix}.$$

Remark Are there more? The answer is yes. For example $\begin{pmatrix} \mp 1 & 0 \\ 0 & -\sqrt{2} \end{pmatrix}$ are two other square roots of A. Are there more hidden square roots? In this case, the answer

© The Author(s), under exclusive license to Springer Nature Switzerland AG 2022
M. H. Mortad, *Counterexamples in Operator Theory*,
https://doi.org/10.1007/978-3-030-97814-3_6

is negative. Indeed, it can be shown that an $n \times n$ matrix with n nonzero distinct eigenvalues has precisely 2^n square roots (over $\mathbb{C}$).

Remark There is a nifty way for seeing when a real matrix does not possess a square root. It goes as follows: Let A be a square matrix with $\det A < 0$. Then A does not have any square root with solely *real* entries. Indeed, if B is a real matrix such that $B^2 = A$, then

$$\det A = \det B^2 = (\det B)^2$$

which is not coherent with $\det A < 0$ and $\det B \in \mathbb{R}$.

However, and as in the case of positive numbers: *A positive operator has only one positive square root*:

Theorem 6.1.1 *Let $A \in B(H)$ be positive. Then A possesses a unique positive square root denoted exclusively by $\sqrt{A}$ (or $A^{\frac{1}{2}}$). Moreover, if $B \in B(H)$ is such that $AB = BA$, then $\sqrt{A}B = B\sqrt{A}$.*

Corollary 6.1.2 *If $A, B \in B(H)$ are two commuting and positive operators, then AB is positive.*

Theorem 6.1.3 (See e.g. Exercise 5.3.30 in [256]) *Let A, B be two self-adjoint operators such that $0 \leq A \leq B$. Then $\sqrt{A} \leq \sqrt{B}$.*

6.2   Questions

6.2.1   A Self-Adjoint Operator with an Infinitude of Self-Adjoint Square Roots

A self-adjoint, or even normal, matrix has always a square root. A proof is based on the spectral theorem (the same argument works for infinite-dimensional spaces, even for unbounded operators).

A self-adjoint matrix A may possess infinitely many square roots. For instance, let I be the identity 2×2 matrix, i.e.

$$I = \begin{pmatrix} 1 & 0 \\ 0 & 1 \end{pmatrix}.$$

By solving the system

$$\begin{pmatrix} a & b \\ c & d \end{pmatrix}^2 = \begin{pmatrix} 1 & 0 \\ 0 & 1 \end{pmatrix}$$

we may obtain that

$$\begin{pmatrix} x & 1 \\ 1 - x^2 & -x \end{pmatrix}$$

constitutes, for each $x \in \mathbb{R}$, a square root of I.

Remark Observe that this also says that a self-adjoint matrix may have nonnormal square roots. There are other examples such as $A = \begin{pmatrix} 0 & a \\ 0 & 0 \end{pmatrix}$, $a \in \mathbb{R}$, which is a nonnormal square root of the zero 2×2 matrix.

Question 6.2.1 Provide a self-adjoint $A \in B(H)$ with infinitely many self-adjoint square roots.

6.2.2 An Operator Without Any Square Root

Question 6.2.2 Give an $A \in B(H)$ without any square root.

6.2.3 A Nilpotent Operator with Infinitely Many Square Roots

Question 6.2.3 We saw in Answer 6.2.2 two nilpotent matrices having no square roots. Can you provide a nilpotent matrix with infinitely many square roots?

6.2.4 An Operator Having a Cube Root but Without Any Square Root

Question 6.2.4 (⊛) Provide an operator $A \in B(H)$ without any square root, but A does have a cube root (treat the cases of both real and complex H).

6.2.5 An Operator Having a Square Root but Without Any Cube Root

Question 6.2.5 Give an operator $A \in B(H)$ having a square root and yet A does not have any cube root.

6.2.6 A Non-invertible Operator with Infinitely Many Square Roots

It is known that an invertible operator $A \in B(H)$ where H is a finite-dimensional *complex* Hilbert space has always a square root (see e.g. 8.33 in [9]). That the generalization to infinite-dimensional spaces fails to hold will be seen in Question 14.2.36. Recall also that this result is no longer true when the finite-dimensional space is over the real field. For the sake of completeness, we give a simple counterexample. Any real matrix A with $\det A < 0$ (so A is invertible) will do. For example, let A

$$A = \begin{pmatrix} 1 & 0 \\ 0 & -1 \end{pmatrix}$$

be defined on $\mathbb{R}^2$. As we already saw above, A does not possess any square root with real entries, yet A is invertible.

Question 6.2.6 Find a non-invertible operator that has lots of square roots. **Hint:** You may remain in a finite-dimensional setting.

6.2.7 An Operator A Without Any Square Root, but $A + \alpha I$ Always Has One ($\alpha \in \mathbb{C}^*$)

Question 6.2.7 Find an operator A without any square root, yet $A + \alpha I$ does admit a square root for any $\alpha \in \mathbb{C}^*$.

6.2.8 $A^2 \geq 0 \not\Rightarrow A \geq 0$ *Even When A Is Normal*

In general, A^2 is not always a positive operator (consider e.g. $A = iI$). However, and by analogy to real numbers, if $A \in B(H)$ is *self-adjoint*, then $A^2 \geq 0$. The proof is very simple: Let $x \in H$; then by the self-adjointness of A, it would ensue that

$$\langle A^2 x, x \rangle = \langle Ax, Ax \rangle = \|Ax\|^2 \geq 0.$$

Conversely, if $A \in B(H)$ is such that $A^2 \geq 0$, then what can be said about A? A priori, nothing! (See e.g. Answer 2.2.13). If, however, A is normal and $A^2 \geq 0$, then A is (only) self-adjoint.

Remark The proof of the previous claim is not hard. It uses some standard arguments from the spectral theorem. Readers may wish to consult the proof of Theorem 3.2 in [88] where an even more general result is established.

Question 6.2.8 Give an example of a self-adjoint unitary non-positive operator A for which A^2 is positive.

6.2.9 $A^3 \geq 0 \not\Rightarrow A \geq 0$ *Even When A Is Normal*

Let $A \in B(H)$ be self-adjoint. If $A^3 \geq 0$, then $A \geq 0$. The proof seems to need some notions that we have not discussed yet.

Further Reading In [87], I have shown jointly with S. Dehimi that if A is a normal operator such that A^p and A^q are both positive for some relatively prime numbers p and q, then A must be positive as well (in fact, the result was shown for unbounded hyponormal operators, a notion not yet defined). The importance of having two co-prime numbers is seen by considering $A^3 \geq 0$ (hence $A^6 \geq 0$ too) with $A \not\geq 0$.

Question 6.2.9 Give an example of a unitary $A \in B(H)$ such that $A^3 \geq 0$, but $A \not\geq 0$.

6.2.10   An Operator Having Only Two Square Roots

Question 6.2.10 Can you provide an operator having only two square roots?

6.2.11   Can an Operator Have Only One Square Root?

Can an operator have only one square root? A priori, the answer is positive. One could just consider the operator $A = 0$ on the one-dimensional space.

Question 6.2.11 Can you provide an operator having only one nonzero square root?

6.2.12   Can an Operator Have Only Two Cube Roots?

Question 6.2.12 Can you provide an operator having only two cube roots?

6.2.13   A Rootless Operator

Question 6.2.13 Give an example of a rootless operator.

Further Reading Readers may wish to consult [327] and [395] for more results on rootless matrices.

6.2.14   On Some Result By B. Yood on Rootless Matrices

It was proven in [395] that: If A is an $n \times n$ matrix over the complex field, $n \geq 2$, then A is rootless if $A^{n-1} \neq 0$ and $A^n = 0$.

Question 6.2.14 Give an example that shows that this result is not necessarily valid if $A^{k-1} \neq 0$ and $A^k = 0$ for some positive integer $k < n$.

6.2.15 A Non-nilpotent Rootless Matrix

Question 6.2.15 Provide an example of a rootless matrix that is not nilpotent.

6.2.16 Two (Self-Adjoint) Square Roots of a Self-Adjoint Operator Need Not Commute

Let $T \in B(H)$ be a self-adjoint operator. As is known, T could have infinitely many (self-adjoint) square roots. Do all these square roots pairwise commute?

Question 6.2.16 Show that it can occur that two self-adjoint square roots of some T do not commute.

6.2.17 A $B \in B(H)$ Commuting with A Need Not Commute with an Arbitrary Root of A

From Theorem 6.1.1, we know that if $BA = AB$ and $A \geq 0$, then $B\sqrt{A} = \sqrt{A}B$. Does this result remain valid for arbitrary square roots?

Question 6.2.17 Find $A, B \in B(H)$ such that $BA = AB$ but $BR \neq RB$ where R is a square root of A.

Further Reading Readers might be interested in the following result that was shown in [87] (holding even for unbounded operators):

Proposition 6.2.1 *Let $A \in B(H)$ be invertible. If $T \in B(H)$ commutes with both A^p and A^q, i.e. $TA^p = A^pT$ and $TA^q = A^qT$ for some relatively prime numbers p and q, then $TA = AT$, i.e. T commutes with A.*

6.2.18   A Self-Adjoint Operator Without Any Positive Square Root

Question 6.2.18 Give a self-adjoint operator not possessing any positive (or even self-adjoint) square root.

6.2.19   Three Positive Operators A, B, $C \in B(H)$ Such That $A \geq B \geq 0$ and C Is Invertible Yet $(CA^2C)^{\frac{1}{2}} \not\geq (CB^2C)^{\frac{1}{2}}$

Question 6.2.19 ([51]) Find positive operators $A, B, C \in B(H)$ such that $A \geq B \geq 0$ and C is invertible yet

$$(CA^2C)^{\frac{1}{2}} \not\geq (CB^2C)^{\frac{1}{2}}.$$

6.2.20   Three Positive Operators A, B, $C \in B(H)$ Such That $A \leq C$ and $B \leq C$ Yet $(A^2 + B^2)^{\frac{1}{2}} \not\leq \sqrt{2}C$

It was asked in [51] whether for all positive operators $A, B, C \in B(H)$ such that $A \leq C$ and $B \leq C$:

$$(A^2 + B^2)^{\frac{1}{2}} \leq \sqrt{2}C.$$

A counterexample was then found by T. Furuta in [122].

Question 6.2.20 Find positive operators $A, B, C \in B(H)$ such that $A \leq C$ and $B \leq C$, yet

$$(A^2 + B^2)^{\frac{1}{2}} \not\leq \sqrt{2}C.$$

6.2.21   On Some Result by F. Kittaneh on Normal Square Roots

We have already observed before that normal matrices may have nonnormal square roots. The next result by F. Kittaneh [194] is therefore of some interest:

Theorem 6.2.2 *Let $B, N \in B(H)$ be commuting such that B is normal and $N^2 = 0$. Set $A = B + N$. If A is invertible and A^2 is normal, then A is normal.*

> **Question 6.2.21** Show that the hypothesis A being invertible cannot just be dispensed with.

6.2.22   On the Normality of Roots of Normal Operators Having Co-prime Powers

In [87], the following result was shown among others:

Theorem 6.2.3 *If A is an invertible bounded operator, such that A^p and A^q are normal, where p and q are two relatively prime numbers, then A is normal.*

Proof We have that $ap + bq = 1$ for some integers a and b, by Bézout's theorem in arithmetic. Since A is invertible, so are A^p and A^q. Since A^p and A^q are also normal, A^{ap} and A^{bq} remain normal. Since A^{ap} and A^{bq} commute, their product is normal, i.e.

$$A^{ap} A^{bq} = A^{ap+bq} = A$$

is normal, as needed.   □

> **Question 6.2.22** Show that the condition "p and q being co-primes" may not just be dropped.

6.2.23   An Isometry Without Square or Cube Roots

Recall that an isometry $A \in B(H)$ is unitary if $\dim H < \infty$ (cf. the discussion before Question 1.2.13). Since invertible operators on finite-dimensional (complex) spaces do possess square roots, one might expect isometries (on infinite-dimensional spaces) to be a class of operators that enjoys such a property. Is this true?

Question 6.2.23 Give an isometry $A \in B(H)$ without any square root. Also, give an isometry without any cube root.

6.2.24   Two Operators A and B Without Square Roots, Yet A ⊕ B Has a Square Root

Let $A, B \in B(H)$, and let $T := A \oplus B$, i.e. $T := \begin{pmatrix} A & 0 \\ 0 & B \end{pmatrix}$ be defined on $H \oplus H$. It is easy to see that if A and B both have a *pth* root, then so does T. Indeed, if C and D are *pth* roots of A and B, respectively, then

$$\begin{pmatrix} C & 0 \\ 0 & D \end{pmatrix}^p = \begin{pmatrix} C^p & 0 \\ 0 & D^p \end{pmatrix} = \begin{pmatrix} A & 0 \\ 0 & B \end{pmatrix} = T.$$

Question 6.2.24 Show that it may happen that $A \oplus B$ possesses a square root, yet neither A nor B has a square root. What about the case of *pth* roots?

Answers

Answer 6.2.1 Let I be the identity matrix on $\mathbb{C}^2$. By solving

$$\begin{pmatrix} a & b \\ b & c \end{pmatrix}^2 = \begin{pmatrix} 1 & 0 \\ 0 & 1 \end{pmatrix},$$

we may find that each member of the infinite family of self-adjoint matrices

$$\begin{pmatrix} x & \sqrt{1 - x^2} \\ \sqrt{1 - x^2} & -x \end{pmatrix},$$

where $x \in [-1, 1]$, is a square root of I.

Answer 6.2.2 Let

$$A = \begin{pmatrix} 0 & 1 \\ 0 & 0 \end{pmatrix}.$$

Then A does not have any square root. To see this, let us presume that A has a square root, B say, which is a 2×2 matrix of the form

$$B = \begin{pmatrix} a & b \\ c & d \end{pmatrix}.$$

Then solve the system

$$B^2 = A \Longleftrightarrow \begin{pmatrix} a^2 + bc & ab + bd \\ ac + cd & bc + d^2 \end{pmatrix} = \begin{pmatrix} 0 & 1 \\ 0 & 0 \end{pmatrix}$$

to get a contradiction (see also Answer 6.2.13).

We give another example to use a different strategy (cf. [149]). Let

$$A = \begin{pmatrix} 0 & 1 & 0 \\ 0 & 0 & 1 \\ 0 & 0 & 0 \end{pmatrix}.$$

Then

$$A^2 = \begin{pmatrix} 0 & 0 & 1 \\ 0 & 0 & 0 \\ 0 & 0 & 0 \end{pmatrix} \text{ and } A^3 = \begin{pmatrix} 0 & 0 & 0 \\ 0 & 0 & 0 \\ 0 & 0 & 0 \end{pmatrix}.$$

Assume that A has a square root B, viz., $B^2 = A$. Hence,

$$B^6 = A^3 = 0,$$

i.e. B is nilpotent. Since B is a 3×3 matrix, its index cannot exceed 3. Therefore, $B^3 = 0$, but this is just not consistent with $B^4 = A^2 \neq 0$. Thus A has no square root.

Answer 6.2.3 The answer is in the affirmative! Let

$$A = \begin{pmatrix} 0 & 0 & 1 \\ 0 & 0 & 0 \\ 0 & 0 & 0 \end{pmatrix}.$$

Hence, A is nilpotent ($A^2 = 0$). Readers may then check that each

$$B_x = \begin{pmatrix} 0 & x & 0 \\ 0 & 0 & x^{-1} \\ 0 & 0 & 0 \end{pmatrix}$$

is a square root of A (x being a nonzero scalar).

Answer 6.2.4 An obvious example is to consider on $\mathbb{R}$ the 1×1 matrix $A = (-1)$. Then, A does not have any square root on $\mathbb{R}$, and A does admit a cube root, namely the operator A itself. Another example is to let

$$A = \begin{pmatrix} 1 & 0 \\ 0 & -1 \end{pmatrix}.$$

Since $\det A < 0$, we already know that A cannot possess a real (matrix) square root, yet A does have a cube root that is A itself for $A^3 = A$.

Another example over $\mathbb{C}$ is desirable (it was kindly pointed out to me by Professor Robert B. Israel, cf. [30]). Consider the 7×7 matrix:

$$A = \begin{pmatrix} 0 & 0 & 0 & 1 & 0 & 0 & 0 \\ 0 & 0 & 0 & 0 & 1 & 0 & 0 \\ 0 & 0 & 0 & 0 & 0 & 1 & 0 \\ 0 & 0 & 0 & 0 & 0 & 0 & 1 \\ 0 & 0 & 0 & 0 & 0 & 0 & 0 \\ 0 & 0 & 0 & 0 & 0 & 0 & 0 \\ 0 & 0 & 0 & 0 & 0 & 0 & 0 \end{pmatrix}.$$

Then A has a cube root, e.g.

$$B = \begin{pmatrix} 0 & 1 & 0 & 0 & 0 & 0 & 0 \\ 0 & 0 & 1 & 0 & 0 & 0 & 0 \\ 0 & 0 & 0 & 1 & 0 & 0 & 0 \\ 0 & 0 & 0 & 0 & 1 & 0 & 0 \\ 0 & 0 & 0 & 0 & 0 & 1 & 0 \\ 0 & 0 & 0 & 0 & 0 & 0 & 1 \\ 0 & 0 & 0 & 0 & 0 & 0 & 0 \end{pmatrix}$$

(indeed, $B^3 = A$ as it may be checked).

Its Jordan form consists of three blocks of sizes 3, 2, and 2 for eigenvalue 0, namely

$$J_A = \begin{pmatrix} 0 & 1 & 0 & 0 & 0 & 0 & 0 \\ 0 & 0 & 0 & 0 & 0 & 0 & 0 \\ 0 & 0 & 0 & 1 & 0 & 0 & 0 \\ 0 & 0 & 0 & 0 & 0 & 0 & 0 \\ 0 & 0 & 0 & 0 & 0 & 1 & 0 \\ 0 & 0 & 0 & 0 & 0 & 0 & 1 \\ 0 & 0 & 0 & 0 & 0 & 0 & 0 \end{pmatrix}$$

and so it does not have a square root: In order to have a square root, it must be possible to arrange the Jordan blocks for eigenvalue 0 so that each such block of size > 1 is paired with another block of size differing from it by at most 1.

Answer 6.2.5 Let

$$A = \begin{pmatrix} 0 & 0 & 0 \\ 0 & 0 & 1 \\ 0 & 0 & 0 \end{pmatrix}, \quad B = \begin{pmatrix} 0 & 0 & 1 \\ 0 & 0 & 0 \\ 0 & 0 & 0 \end{pmatrix} \text{ and } U = \begin{pmatrix} 0 & 1 & 0 \\ 1 & 0 & 0 \\ 0 & 0 & 1 \end{pmatrix}.$$

Then U is a *fundamental symmetry*. Besides,

$$A = UBU.$$

We already know from Answer 6.2.3 that B possesses lots of square roots (noted $B_x, x \in \mathbb{R}, x \neq 0$). Hence, each $A_x := U B_x U$ is a square root of A as for all $x \neq 0$:

$$A_x^2 = U B_x U U B_x U = U B_x^2 U = UBU = A$$

which tells us that A has an infinitude of square roots. Now, A does not have any cube root. To see why, suppose that A has a cube root denoted by C, i.e. $C^3 = A$. Since $A^2 = 0$, it follows $C^6 = 0$ and so $C^3 = 0$ in view of the size of C. Hence, we would have $A = 0$, which is impossible! Therefore, A does not have any cube root whatsoever.

Answer 6.2.6 Let

$$A = \begin{pmatrix} 0 & 0 \\ 0 & 0 \end{pmatrix}.$$

Then, as readers may easily check, A has infinitely many square roots given by the family say

$$A_t = t \begin{pmatrix} 1 & -1 \\ 1 & -1 \end{pmatrix}$$

with $t \in \mathbb{R}$.

Answer 6.2.7 This is easy. Indeed, let $A = \begin{pmatrix} 0 & 1 \\ 0 & 0 \end{pmatrix}$ be defined on $\mathbb{C}^2$ say. As we already know, A does not have any square root. However, if we let $\alpha \in \mathbb{C}^*$, then

$$A + \alpha I = \begin{pmatrix} \alpha & 1 \\ 0 & \alpha \end{pmatrix}$$

has always a square root for it is *invertible on a finite-dimensional complex space*.

Answer 6.2.8 On $\mathbb{C}^2$, merely consider the self-adjoint unitary matrix

$$A = \begin{pmatrix} 1 & 0 \\ 0 & -1 \end{pmatrix}.$$

Then A is not positive, while $A^2 = I$ is obviously positive.

Answer 6.2.9 On $\mathbb{C}^2$, consider the unitary matrix

$$A = \begin{pmatrix} e^{\frac{2i\pi}{3}} & 0 \\ 0 & e^{\frac{2i\pi}{3}} \end{pmatrix} = e^{\frac{2i\pi}{3}} I,$$

where I designates the identity matrix. Then A is not positive for it is not self-adjoint, whereas A^3 is obviously positive because

$$A^3 = I.$$

Answer 6.2.10 Let

$$A = \begin{pmatrix} 1 & 0 \\ 0 & 0 \end{pmatrix}.$$

Then A has the following two square roots:

$$B = \begin{pmatrix} 1 & 0 \\ 0 & 0 \end{pmatrix} \text{ and } C = \begin{pmatrix} -1 & 0 \\ 0 & 0 \end{pmatrix}.$$

By using a method similar to that of Answer 6.2.2 say, we may show that B and C are the only square roots of A. Let the interested reader complete the proof.

Answer 6.2.11 This is impossible for a simple reason. If B is a *nonzero* square root of a given operator A, i.e. $B^2 = A$, then $-B$ too becomes another square root of A as

$$(-B)^2 = B^2 = A.$$

Remark Therefore, the number of (nonzero) square roots is always even.

Answer 6.2.12 This is not possible for a similar reason as just above. If an operator A does not have any cube root, then obviously it cannot have two cube roots! Now, as soon as B is some *nonzero* cube root of A, i.e. $B^3 = A$, then it will be clear that $e^{2i\pi/3} B$ and $e^{4i\pi/3} B$ also work as two other cube roots of A as may readily be checked. Therefore, whenever we know a (nonzero) cube root of A, we will construct two others.

Remark Readers should now be capable of answering the following question: Can you find an $A \in B(H)$ having *three* fourth nonzero roots only?

Answer 6.2.13 Let $A = \begin{pmatrix} 0 & 1 \\ 0 & 0 \end{pmatrix}$. Then A is rootless. Suppose for the sake of contradiction that there is a matrix B such that $B^p = A$ for some integer $p \geq 2$. Hence,

$$B^{2p} = A^2 = 0$$

whereby $B^2 = 0$. This would also give $B^p = 0$, which is absurd.

Answer 6.2.14 Take $n = 3$ and $k = 2$. Consider

$$A = \begin{pmatrix} 0 & 0 & 1 \\ 0 & 0 & 0 \\ 0 & 0 & 0 \end{pmatrix}.$$

We have already seen that A has infinitely many square roots, one of them being

$$B = \begin{pmatrix} 0 & 1 & 0 \\ 0 & 0 & 1 \\ 0 & 0 & 0 \end{pmatrix},$$

whereby A is not rootless. Observe in the end that $A^2 = 0$ (and $A \neq 0$).

Answer 6.2.15 Consider over the complex field the matrix

$$A = \begin{pmatrix} 1 & 0 & 0 \\ 0 & 0 & 1 \\ 0 & 0 & 0 \end{pmatrix}.$$

Since $A^3 \neq 0$, A cannot be nilpotent. To show that A is rootless, we use the following practical result (which appeared in [327]): *If an $n \times n$ matrix T is expressed as*

$$T = \begin{pmatrix} R & 0 \\ 0 & N \end{pmatrix}$$

where R and N are square matrices such that R is invertible and N is nilpotent, then T is rootless if and only if N is rootless.

Clearly, we can write our A as $\begin{pmatrix} B & 0 \\ 0 & N \end{pmatrix}$ where B is the 1×1 invertible matrix $B := (1)$ and $N = \begin{pmatrix} 0 & 1 \\ 0 & 0 \end{pmatrix}$. Since N is nilpotent and rootless (by Answer 6.2.13), we infer that A is rootless, as desired.

Answer 6.2.16 Let

$$A = \begin{pmatrix} -1 & 0 \\ 0 & 1 \end{pmatrix} \text{ and } B = \begin{pmatrix} 0 & 1 \\ 1 & 0 \end{pmatrix}.$$

Then, A and B are self-adjoint (and also unitary), and each of them is a square root of the identity matrix I because $A^2 = B^2 = I$. Nevertheless,

$$AB = \begin{pmatrix} 0 & -1 \\ 1 & 0 \end{pmatrix} \neq \begin{pmatrix} 0 & 1 \\ -1 & 0 \end{pmatrix} = BA,$$

as wished.

Answer 6.2.17 The question does not even hold true on finite-dimensional spaces. Just consider:

$$A = \begin{pmatrix} a & 0 \\ 0 & a \end{pmatrix}, \quad B = \begin{pmatrix} 0 & 1 \\ 1 & 0 \end{pmatrix} \text{ and } R = \begin{pmatrix} 0 & a \\ 1 & 0 \end{pmatrix}$$

where $a \in \mathbb{C}$ (and $a \neq 1$). Then $BA = AB$, $R^2 = A$, while $BR \neq RB$, as may be checked by readers.

Answer 6.2.18 On $\mathbb{C}^2$, let

$$A = \begin{pmatrix} 0 & 1 \\ 1 & 0 \end{pmatrix}.$$

Then A is self-adjoint (but non-positive) and has two distinct nonzero eigenvalues and so A ought to have *four and only four* square roots. To find them explicitly, we need to diagonalize A. We find

$$A = TDT^{-1} \text{ where } D = \begin{pmatrix} 1 & 0 \\ 0 & -1 \end{pmatrix} \text{ and } T = \begin{pmatrix} 1 & 0 \\ 0 & -1 \end{pmatrix}.$$

The four square roots of D are given by

$$\begin{pmatrix} -1 & 0 \\ 0 & -i \end{pmatrix}, \quad \begin{pmatrix} -1 & 0 \\ 0 & i \end{pmatrix}, \quad \begin{pmatrix} 1 & 0 \\ 0 & -i \end{pmatrix} \text{ and } \begin{pmatrix} 1 & 0 \\ 0 & i \end{pmatrix}.$$

The square roots of A are therefore

$$\frac{1}{2}\begin{pmatrix} -1-i & -1+i \\ -1+i & -1-i \end{pmatrix}, \ \frac{1}{2}\begin{pmatrix} -1+i & -1-i \\ -1-i & -1+i \end{pmatrix}, \ \frac{1}{2}\begin{pmatrix} 1-i & 1+i \\ 1+i & 1-i \end{pmatrix}$$

and

$$\frac{1}{2}\begin{pmatrix} 1+i & 1-i \\ 1-i & 1+i \end{pmatrix}.$$

As one observes none of the previous matrices is self-adjoint, let alone their positiveness. In other words, *all square roots in this example are non-self-adjoint*. Another simple example is the self-adjoint matrix:

$$A = \begin{pmatrix} -1 & 0 \\ 0 & 1 \end{pmatrix}.$$

Clearly,

$$\begin{pmatrix} \pm i & 0 \\ 0 & \pm 1 \end{pmatrix}$$

are four (and the only four) non-self-adjoint square roots of A.

These examples come as no surprise. More precisely, there is nothing special about them. Indeed, any self-adjoint non-positive operator will do! To see that, let A be any self-adjoint non-positive operator, and let B be a positive or even self-adjoint square root of A, i.e. $B^2 = A$. But, this is absurd because B^2 is always positive here, and A is taken to be non-positive.

Answer 6.2.19 ([51]) Let

$$A = \begin{pmatrix} 5 & 2 \\ 2 & 101 \end{pmatrix}, \ B = \begin{pmatrix} 1 & 0 \\ 0 & 100 \end{pmatrix} \text{ and } C = \begin{pmatrix} 1 & 0 \\ 0 & 1/5 \end{pmatrix}.$$

Then all of A, B, C are positive, $A \geq B$, and C is invertible. Approximately,

$$(CA^2C)^{\frac{1}{2}} - (CB^2C)^{\frac{1}{2}} = \begin{pmatrix} 4.116681 & 1.679158 \\ 1.679158 & 0.134061 \end{pmatrix}.$$

The eigenvalues of the previous self-adjoint matrix are of opposite signs (namely: -0.4794 and 4.730), which is enough for declaring

$$(CA^2C)^{\frac{1}{2}} \not\geq (CB^2C)^{\frac{1}{2}},$$

as desired.

Answer 6.2.20 ([122]) Let

$$A = \begin{pmatrix} 1 & 0 \\ 0 & 0 \end{pmatrix}, \ B = \begin{pmatrix} 1 & -1 \\ -1 & 1 \end{pmatrix} \text{ and } C := A + B = \begin{pmatrix} 2 & -1 \\ -1 & 1 \end{pmatrix}$$

Then obviously $A \leq C$ and $B \leq C$. Nevertheless,

$$\sqrt{2}C - (A^2 + B^2)^{\frac{1}{2}} = \begin{pmatrix} 1.250757\ldots & -0.699400\ldots \\ -0.699400\ldots & 0.193951\ldots \end{pmatrix}$$

is a self-adjoint matrix whose eigenvalues are $-0.154214\ldots$ and $1.598922\ldots$, i.e. it cannot be positive. Accordingly,

$$(A^2 + B^2)^{\frac{1}{2}} \not\leq \sqrt{2}C,$$

as desired.

Answer 6.2.21 ([194]) Let $\dim H = 2$. Then every $A \in B(H)$ may be decomposed as $A = B + N$ where B is normal, $N^2 = 0$, and $BN = NB$ (more known as the Dunford decomposition).

If the result held without the invertibility of A, then it would follow that A is normal iff A^2 is normal. Obviously, this is not always true as nonnormal matrices having normal squares exist (abundantly!).

Answer 6.2.22 ([87]) To see why the condition "p and q being co-prime numbers" may not just dropped, it suffices to take an invertible nonnormal square root A of the identity matrix, and then $A^2 = A^4 = I$. An explicit example would be $A = \begin{pmatrix} 2 & 1 \\ -3 & -2 \end{pmatrix}$.

Answer 6.2.23 The simplest example is the unilateral shift on ℓ^2. First, recall that an operator has a square root if and only if its adjoint does. In the case of the shift, it is slightly simpler to work with S^*. Aiming for a contradiction, assume that S^* does have a square root, i.e. $A^2 = S^*$ for some $A \in B(\ell^2)$. Then, $A^2 S = S^* S = I$, and by the general theory, A is right invertible and so it is surjective. Notice also that A cannot be injective (indeed, this would imply that $A^2 = S^*$ is injective, which is untrue).

Let us now show that $\ker A = \ker S^* = \mathbb{R}e_1$, where $e_1 = (1, 0, 0, \cdots)$. Clearly, $\ker S^* = \mathbb{R}e_1$ and so $\dim \ker S^* = 1$. Now, we obviously have $\ker A \subset \ker S^*$ because $A^2 = S^*$. Since A is not injective, we are forced to have $\ker A = \ker S^*$.

Since A is onto, for $e_1 \in \ell^2$, there is an $x \in \ell^2$ such that $Ax = e_1$ (and so $x \notin \ker A = \ker S^*$). Thus (as $e_1 \in \ker A$)

$$A^2 x = Ae_1 = 0 \neq S^* x.$$

This shows that S^* does not have any square root. Accordingly, S cannot have a square root either!

Before dealing with the case of cube roots, recall a practical result that permits to see when a given operator does not have square roots (this result provides another proof that the shift does not have any square root).

Proposition 6.2.4 (Conway–Morrel, [70]) *Let p be an integer with $p \geq 2$. If $T \in B(H)$ is such that $0 < \dim(\ker T) < p$ and $\ker T \neq \ker T^2$, then T does not have a pth root.*

With this result, S^* above clearly satisfies the assumptions of the theorem. Hence, S^* (or S) does not possess any pth root, i.e. the shift is rootless (in particular, S does not have any cube root!).

Remarks

(1) More generally, no unilateral weighted shift with nonzero weights has a pth root.
(2) Another consequence of the result above is the fact that $S \oplus S^*$, defined on $\ell^2 \oplus \ell^2$, is rootless.

Answer 6.2.24 Let S be the usual unilateral shift on ℓ^2. We already know that S is without any square root. Define on $\ell^2 \oplus \ell^2$ the operator

$$V = \begin{pmatrix} S & 0 \\ 0 & S \end{pmatrix}.$$

Let us show that V has indeed a square root. A square root of V is given by $R = \begin{pmatrix} 0 & S \\ I & 0 \end{pmatrix}$ for

$$R^2 = \begin{pmatrix} 0 & S \\ I & 0 \end{pmatrix}\begin{pmatrix} 0 & S \\ I & 0 \end{pmatrix} = \begin{pmatrix} S & 0 \\ 0 & S \end{pmatrix}.$$

Another way of showing that $S \oplus S$ has a square root is to remember that $S \oplus S$ is *unitarily equivalent* to S^2 (see Question 5.2.9) and that S^2 has an obvious square root (namely itself!).

To answer the case of pth roots, notice that the construction of R above is in fact borrowed from [70] where it is shown that if $T \in B(H)$, then $T \oplus T \oplus \cdots \oplus T$ (p times, $p \geq 2$) defined on $H \oplus H \oplus \cdots \oplus H$ has a pth root given by

$$\begin{pmatrix} 0 & 0 & \cdots & 0 & T \\ I & 0 & \cdots & 0 & 0 \\ 0 & I & \cdots & 0 & 0 \\ \vdots & \ddots & \ddots & \vdots & \vdots \\ 0 & 0 & \cdots & I & 0 \end{pmatrix}.$$

Now, it suffices to replace T by the shift S and H by ℓ^2. For example, one *cube* root

$$\begin{pmatrix} S & 0 & 0 \\ 0 & S & 0 \\ 0 & 0 & S \end{pmatrix}$$

is

$$R = \begin{pmatrix} 0 & 0 & S \\ I & 0 & 0 \\ 0 & I & 0 \end{pmatrix}$$

Remark Related to the main question, it was shown in [70] that, inter alia, if $A \oplus B$ has a *pth* root and if there is *no nonzero* operator X such that $AX = XB$, then both A and B have a *pth* root.

Chapter 7
Absolute Value, Polar Decomposition

7.1 Basics

It is well known that if $z \in \mathbb{C}$, then $|z| = \sqrt{\bar{z}z}$ denotes the modulus of z. It is possible to extend this definition to the case of linear operators.

Definition 7.1.1 Let $A \in B(H)$. The (unique) positive square root of A^*A is called the absolute value (or modulus) of A, and it is denoted by $|A|$. Symbolically,

$$|A| = \sqrt{A^*A}.$$

Proposition 7.1.1 *Let $A \in B(H)$. Then:*

(1) *A is positive if and only if $|A| = A$.*
(2) *A is normal if and only if $|A| = |A^*|$.*
(3) *(See e.g. Exercise 6.3.7 in [256]) If A is self-adjoint, then*

$$-|A| \leq A \leq |A|.$$

Examples 7.1.1

(1) On $\mathbb{C}^2$, let $A = \begin{pmatrix} a & 0 \\ 0 & b \end{pmatrix}$, where $a, b \in \mathbb{R}$ or $\mathbb{C}$. Then clearly

$$|A| = \begin{pmatrix} |a| & 0 \\ 0 & |b| \end{pmatrix}.$$

(2) If A is normal over $\mathbb{C}^n$ say, then $U^*AU = D$, where D is diagonal, and for some unitary U. Hence,

$$|A| = U|D|U^*.$$

© The Author(s), under exclusive license to Springer Nature Switzerland AG 2022
M. H. Mortad, *Counterexamples in Operator Theory*,
https://doi.org/10.1007/978-3-030-97814-3_7

Remark For basic properties about the absolute value of an operator, see [256]. See also [257].

It is also known that each complex z is expressible as $z = e^{i\theta}|z|$. This too may be carried over to $B(H)$.

Definition 7.1.2 (Cf. the Remark Below Theorem 7.1.4) Let $A \in B(H)$. If we can write $A = UP$, where U is unitary and P is positive, then UP is called the polar decomposition of A.

Theorem 7.1.2 (For a Proof, See E.g. Theorem 12.35 in [314]) *Let $A \in B(H)$ be invertible. Then A has a unique polar decomposition $A = U|A|$.*

Theorem 7.1.3 (For a Proof, See E.g. Theorem 12.35 in [314]) *Let $A \in B(H)$ be normal. Then A has a polar decomposition*

$$A = U|A| = |A|U$$

where $U \in B(H)$ is unitary.

In particular, if A is self-adjoint, then U may be taken to be a fundamental symmetry.

In fact, we can always write $A \in B(H)$ as $V|A|$ where V is something weaker than unitary, namely:

Theorem 7.1.4 *Each $A \in B(H)$ may be decomposed as*

$$A = V|A| \text{ and } |A| = V^*A$$

where $V \in B(H)$ is a partial isometry. Moreover, V is uniquely determined by the kernel condition $\ker V = \ker A$.

Remark The foregoing result is also called a polar decomposition. In the present manuscript, we shall indicate, when necessary, which polar decomposition we are referring to.

7.2 Questions

7.2.1 *An A Such That $|\operatorname{Re} A| \not\leq |A|$ and $|\operatorname{Im} A| \not\leq |A|$*

Let $z \in \mathbb{C}$. Then

$$|\operatorname{Re} z| \leq |z| \text{ and } |\operatorname{Im} z| \leq |z|.$$

If $A \in B(H)$ is normal, then it can be shown that

$$|\operatorname{Re} A| \leq |A| \text{ and } |\operatorname{Im} A| \leq |A|.$$

What about general operators?

Question 7.2.1 Find an $A \in B(H)$ such that

$$|\operatorname{Re} A| \not\leq |A| \text{ and } |\operatorname{Im} A| \not\leq |A|.$$

Hint: You may remain in a finite-dimensional setting.

7.2.2 A Weakly Normal T Such That T^2 Is Not Normal

First, we give a definition: Say that $T \in B(H)$ is weakly normal (symbolically, T belongs to the class (WN)) if $(\operatorname{Re} T)^2 \leq |T|^2$.

It is interesting, perhaps unexpected though, to see that number theory could be called on to prove some results in operator theory! It was shown in [194] that if T is in (WN) such that T^p and T^q are normal for some relatively prime natural numbers p and q, then T is normal.

Question 7.2.2 Give an example of a T in (WN) such that T^2 is normal, but T is not normal.

7.2.3 Two Self-Adjoints A, B Such That $|A + B| \not\leq |A| + |B|$

Let $A, B \in B(H)$ be self-adjoint and such that $AB = BA$. Then

$$|A + B| \leq |A| + |B|.$$

The proof is simple: Since $AB = BA$, it follows that

$$|A + B|^2 = (A + B)^2 = A^2 + 2AB + B^2$$

and

$$(|A| + |B|)^2 = A^2 + 2|A||B| + B^2.$$

But $|A||B| = |AB| \geq AB$ and so

$$(|A + B|)^2 \leq (|A| + |B|)^2.$$

Theorem 6.1.3 finally gives the desired inequality.

Question 7.2.3 Find self-adjoints $A, B \in B(H)$ such that

$$|A + B| \nleq |A| + |B|.$$

Hint: Work on a two-dimensional space.

7.2.4 Two Self-Adjoint Operators A, B That Do Not Satisfy $|A||B| + |B||A| \geq AB + BA$

Question 7.2.4 Find self-adjoint operators A, B such that

$$|A||B| + |B||A| \ngeq AB + BA.$$

Hint: Obviously, one has to avoid commuting A and B.

7.2.5 Two Self-Adjoint Operators A and B Such That $|||A| - |B||| \nleq ||A - B||$

Question 7.2.5 Find two self-adjoint matrices A and B such that it is not true that

$$||\, |A| - |B|\, || \leq ||A - B||.$$

Hint: To get a counterexample, one has two avoid commuting normal operators (see e.g. Exercise 6.3.15 in [256]). Two 2×2 off-diagonal matrices will not do either as may easily be checked.

7.2.6 Two Non-commuting Operators A and B That Are Not Normal and Yet $|A + B| = |A| + |B|$

Question 7.2.6 Find nonnormal $A, B \in B(H)$ with $AB \neq BA$ yet

$$|A + B| = |A| + |B|.$$

7.2.7 Two Positive Operators A and B with $|A - B| \not\leq A + B$

Question 7.2.7 Find two positive matrices A and B such that

$$|A - B| \not\leq A + B.$$

7.2.8 Two Self-adjoint Operators A and B Such That $I + |AB - I| \not\leq (I + |A - I|)(I + |B - I|)$

Recall that for any complex numbers z, z':

$$1 + |zz' - 1| \leq (1 + |z - 1|)(1 + |z' - 1|)$$

as in say Exercise 1.1.17 in [4].

We leave it to readers to check that one can establish the inequality

$$I + |AB - I| \leq (I + |A - I|)(I + |B - I|)$$

if A and B are self-adjoint (or even normal) and such that $AB = BA$.

Question 7.2.8 Find self-adjoint matrices A, B such that

$$I + |AB - I| \not\leq (I + |A - I|)(I + |B - I|).$$

7.2.9   Two Self-Adjoints A, $B \in B(H)$ Such That $|AB| \neq |A||B|$

As mentioned above if A and B are self-adjoint operators such that $AB = BA$, then

$$|AB| = |A||B|.$$

In fact, just the normality of A suffices for the result to be valid (using the Fuglede theorem yet to be recalled). See e.g. [257].

Question 7.2.9 Provide self-adjoints (matrices) A and B such that:

$$|AB| \neq |A|\,|B|.$$

7.2.10   Two Operators A and B Such That $AB = BA$, However, $|A||B| \neq |B||A|$

Question 7.2.10 Find $A, B \in B(H)$ such that $AB = BA$, but

$$|A||B| \neq |B||A|.$$

7.2.11   A Pair of Operators A and B Such That $A|B| = |B|A$ and $B|A| = |A|B$, But $AB \neq BA$ and $AB^* \neq B^*A$

Let $A, B \in B(H)$ be such that $AB = BA$ and $AB^* = B^*A$. Then

$$AB^*B = B^*AB = B^*BA$$

and so $A|B| = |B|A$. Similarly, it is seen that $B|A| = |A|B$.

Question 7.2.11 Give two operators A and B with $A|B| = |B|A$ and $B|A| = |A|B$, but neither $AB = BA$ nor $AB^* = B^*A$.

7.2.12 An Operator A Such That $A|A| \neq |A|A$

Let A be normal, i.e. $AA^* = A^*A$. Hence,

$$AA^*A = A^*AA$$

thereby $A|A| = |A|A$, by Theorem 6.1.1.

Question 7.2.12 (See Also Question 7.2.23, cf. Question 14.2.23) Find an $A \in B(H)$ such that $A|A| \neq |A|A$.

7.2.13 An A Such That $|A||A^*| = |A^*||A|$ But $AA^* \neq A^*A$

Question 7.2.13 Find an $A \in B(H)$ such that $|A||A^*| = |A^*||A|$ but $AA^* \neq A^*A$.

7.2.14 An Operator A Such That $|A^2| \neq |A|^2$

If $A \in B(H)$ is normal, then

$$|A^2| = \sqrt{A^{*2}A^2} = \sqrt{A^*AA^*A} = \sqrt{|A|^4} = |A|^2.$$

What about general operators?

Question 7.2.14 (Cf. Question 14.2.22) Find an $A \in B(H)$ such that

$$|A^2| \neq |A|^2.$$

7.2.15   A Non-surjective A Such That |A| Is Surjective

Let $A \in B(H)$ be invertible. Then A^*A is clearly invertible, and hence so is $|A|$. What about right or left invertibility?

When $\dim H = \infty$, right invertibility is just not sufficient. For example, take the shift's adjoint on ℓ^2. What about left invertibility still when $\dim H = \infty$? The answer is affirmative in this case, i.e. the left invertibility of A does yield the (full) invertibility of A^*A, hence that of $|A|$. Two proofs appeared in [277]. One of them reads: Let $A = V|A|$ be the polar decomposition of A, where $V \in B(H)$ is a partial isometry. Since $A \in B(H)$ is left invertible, $BA = I$ for some $B \in B(H)$. Hence,

$$I = BA = BV|A|$$

and so $|A|$ is left invertible. Since $|A|$ is self-adjoint, it becomes (bilaterally) invertible. Thus, A^*A is invertible.

Question 7.2.15 Show that it may happen that $|A|$ is invertible, while A is not. What about ran $A = $ ran $|A|$?

7.2.16   Two Self-Adjoint Operators A, B with B ≥ 0 Such That −B ≤ A ≤ B but |A| ≰ B

If $A, B \in B(H)$ are self-adjoint, then

$$|A| \leq B \Longrightarrow -B \leq A \leq B$$

(see e.g. Exercise 6.3.8 in [256]).

Question 7.2.16 Let $A, B \in B(H)$ be self-adjoint with $B \geq 0$. Show that it may well happen that $-B \leq A \leq B$ without having $|A| \leq B$.

7.2.17 The Failure of the Inequality $|\langle Ax, x \rangle| \leq \langle |A|x, x \rangle$

Let $A \in B(H)$ be self-adjoint (or even a weaker class, see [85]). Then for each $x \in H$:

$$|\langle Ax, x \rangle| \leq \langle |A|x, x \rangle$$

(as in say Exercise 6.3.7 in [256]).

Question 7.2.17 Let $A \in B(H)$. Show that in general

$$|\langle Ax, x \rangle| \leq \langle |A|x, x \rangle,$$

for all $x \in H$, need not hold.

7.2.18 On the Generalized Cauchy–Schwarz Inequality

First, recall the following generalization of the Cauchy–Schwarz inequality:

Theorem 7.2.1 (Generalized Cauchy–Schwarz Inequality) *Let $A \in B(H)$ be positive. Then*

$$|\langle Ax, y \rangle|^2 \leq \langle Ax, x \rangle \langle Ay, y \rangle$$

for all $x, y \in H$.

The previous inequality still has a generalization, namely (cf. [123] or [148]): If $A \in B(H)$, then the following generalized Cauchy–Schwarz inequality

$$|\langle Ax, y \rangle|^2 \leq \langle |A|x, x \rangle \langle |A^*|y, y \rangle$$

holds for all $x, y \in H$. It is also known that the equality holds in the previous inequality iff $|A|x$ and A^*y are linearly dependent iff Ax and $|A^*|y$ are linearly dependent.

Question 7.2.18 Give an example when the inequality above is strict when $|A^*|y$ and $|A|x$ are linearly dependent.

7.2.19   On the Failure of Some Variants of the Generalized Cauchy–Schwarz Inequality

Question 7.2.19 ([148]) Let $A \in B(H)$. Do we always have for all $x, y \in H$:

$$|\langle Ax, y \rangle|^2 \leq \langle |A|x, x \rangle \langle |A|y, y \rangle?$$

How about

$$|\langle Ax, y \rangle|^2 \leq \langle |A^*|x, x \rangle \langle |A^*|y, y \rangle?$$

Or

$$|\langle Ax, y \rangle|^2 \leq \langle |A^*|x, x \rangle \langle |A|y, y \rangle?$$

7.2.20   A Sequence of Self-Adjoint Operators (A_n) Such That $\||A_n| - |A|\| \to 0$ But $\|A_n - A\| \not\to 0$

Let (A_n) be a sequence in $B(H)$, and let $A \in B(H)$. Recall that

$$\|A_n - A\| \longrightarrow 0 \Longrightarrow \||A_n| - |A|\| \longrightarrow 0.$$

Let us include a simple proof of the previous result. First, we show that if $B, C \in B(H)$ are positive, then

$$\|\sqrt{B} - \sqrt{C}\| \leq \sqrt{\|B - C\|}$$

(mutatis mutandis, the same result holds by replacing each square root by the nth root): Indeed, $B - C$ is self-adjoint and so (cf. Exercise 5.3.11 in [256])

$$B - C \leq \|B - C\|I$$

or $B \leq C + \|B - C\|I$. Upon passing to the positive square root, we obtain

$$\sqrt{B} \leq \sqrt{C + \|B - C\|I} \leq \sqrt{C} + \sqrt{\|B - C\|I}$$

(also, by a glance at Exercise 5.3.31 in [256]). Therefore,

$$\sqrt{B} - \sqrt{C} \leq \sqrt{\|B - C\|}I.$$

By inverting the roles of B and C, we obtain

$$\sqrt{C} - \sqrt{B} \le \sqrt{\|B - C\|} I,$$

and hence we get the desired inequality (using again Exercise 5.3.11 in [256]).

To show the very first implication, remember that as $A_n \to A$ w.r.t. $\|\cdot\|$, we also have $A_n^* \to A^*$ w.r.t. $\|\cdot\|$. Thus, $A_n^* A_n \to A^* A$. Accordingly,

$$\||A_n| - |A|\| = \|\sqrt{A_n^* A_n} - \sqrt{A^* A}\| \le \sqrt{\|A_n^* A_n - A^* A\|} \longrightarrow 0,$$

as wished.

Question 7.2.20 Find a sequence of self-adjoint operators (A_n) in $B(H)$ and a self-adjoint $A \in B(H)$ such that $\||A_n| - |A|\| \to 0$ but $\|A_n - A\| \not\to 0$. What about other types of convergence?
Hint: Look for diagonal matrices.

7.2.21 The Non-weakly Continuity of $A \mapsto |A|$

Question 7.2.21 Show that $A \mapsto |A|$ is not weakly continuous on $B(H)$, that is, find a sequence (A_n) in $B(H)$ that converges weakly to some $A \in B(H)$, but $|A_n|$ does not converge weakly to $|A|$.

7.2.22 A Sequence of Operators (A_n) Converging Strongly to A, but $(|A_n|)$ Does Not Converge Strongly to $|A|$

Let (A_n) be a sequence that converges strongly to A in $B(H)$ such that A_n^* also converges strongly to A^*. Then $\sqrt{A_n^* A_n}$ converges strongly to $\sqrt{A^* A}$, i.e. $|A_n|$ converges strongly to $|A|$.

Question 7.2.22 Find a sequence of (non-self-adjoint) operators (A_n) in $B(H)$ that converges strongly to A, yet $(|A_n|)$ does not converge strongly to $|A|$.
Hint: A counterexample is impossible when dim $H < \infty$.

7.2.23   An Invertible $A = U|A|$ with $U|A| \neq |A|U$,
$UA \neq AU$, and $A|A| \neq |A|A$

Question 7.2.23 Let $A = U|A|$ be the polar decomposition of an invertible operator A. Do we have $U|A| = |A|U$? What about $UA = AU$? Or $A|A| = |A|A$?

7.2.24   Left or Right Invertible Operators Do Not Enjoy a
("Unitary") Polar Decomposition

Question 7.2.24 Give an example of a left invertible operator $A \in B(H)$ that does not have a polar decomposition $U|A|$ where U is unitary. What about right invertible operators?

7.2.25   A Normal Operator Whose Polar Decomposition Is Not
Unique

Question 7.2.25 Give an example of a normal $A \in B(H)$ whose polar decomposition is not unique.

7.2.26   On a Result of the Uniqueness of the Polar
Decomposition By Ichinose–Iwashita

In [167], W. Ichinose and K. Iwashita established a simple and a quite interesting result characterizing the uniqueness of the polar decomposition (the "partial isometry version") of a bounded operator. It reads:

Theorem 7.2.2 *Let H and K be two complex Hilbert spaces, and let $A \in B(H, K)$. Then the polar decomposition $A = V|A|$ (where V is a partial isometry) is unique if and only if either* $\ker A = \{0\}$ *or* $[\operatorname{ran} A]^{\perp} = \{0\}$.

It is just amazing that such a simple and elegant result was discovered recently (2013), that is, after several decades of the first proof of the polar decomposition.

The authors of the above result pointed out that the statement in Section 2.6.3 of the reference "A. Arai, *mathematical structure I for quantum mechanics*, Asakura Publ., Tokyo 1999" is wrong. Unfortunately, I do not know any Japanese; however, the authors said that A. Arai mistakenly showed that if ker $A \neq \{0\}$, then the polar decomposition is not unique. They then provided a counterexample.

Question 7.2.26 Find an operator A such that ker $A \neq \{0\}$, yet its polar decomposition is unique.

7.2.27 An Operator A Expressed as $A = V|A|$ with $A^3 = 0$ but $V^3 \neq 0$

Let $A \in B(H)$ be decomposed as $A = V|A|$ where V is a partial isometry. It is shown in [123] that

$$A^2 = 0 \Longleftrightarrow V^2 = 0.$$

Question 7.2.27 Can the previous be generalized to higher powers?

7.2.28 An Invertible Operator A Expressed as $A = U|A|$ with $A^3 = I$ but $U^3 \neq I$

Let $A \in B(H)$ be decomposed as $A = U|A|$ where $A^2 = I$, and so A is invertible and U is unitary. It is shown in [266] that if $A^2 = I$, then $U^2 = I$.

Question 7.2.28 Can the previous be generalized to higher powers?

Answers

Answer 7.2.1 Let

$$A = \begin{pmatrix} 0 & 2 \\ 0 & 0 \end{pmatrix}.$$

Hence,

$$\text{Re}\, A = \begin{pmatrix} 0 & 1 \\ 1 & 0 \end{pmatrix} \quad \text{and} \quad \text{Im}\, A = \begin{pmatrix} 0 & -i \\ i & 0 \end{pmatrix}.$$

Then, we easily get

$$|A| = \begin{pmatrix} 0 & 0 \\ 0 & 2 \end{pmatrix} \quad \text{and} \quad |\text{Re}\, A| = |\text{Im}\, A| = I.$$

Hence,

$$|\text{Re}\, A| = |\text{Im}\, A| \nleq |A|$$

for the simple reason that the matrix $\begin{pmatrix} -1 & 0 \\ 0 & 1 \end{pmatrix}$ is not positive.

Answer 7.2.2 ([194], and Also [111]) On $\mathbb{C}^2$, let

$$T = \begin{pmatrix} i & 1 \\ 0 & -i \end{pmatrix}.$$

Since

$$|T|^2 - (\text{Re}\, T)^2 = \begin{pmatrix} 3/4 & -i \\ i & 7/4 \end{pmatrix} \geq 0,$$

T is in (WN). In addition, $T^2 = -I$ and so T^2 is normal. However, T is not normal.

Answer 7.2.3 (An Example Due to E. Nelson) Let

$$A = \begin{pmatrix} -1 & 1 \\ 1 & -1 \end{pmatrix} \quad \text{and} \quad B = \begin{pmatrix} 2 & 0 \\ 0 & 0 \end{pmatrix}.$$

Since A is negative and B is positive, we have

$$|A| = -A = \begin{pmatrix} 1 & -1 \\ -1 & 1 \end{pmatrix} \quad \text{and} \quad |B| = \begin{pmatrix} 2 & 0 \\ 0 & 0 \end{pmatrix}.$$

Doing some arithmetic yields

$$|A + B| = \begin{pmatrix} \sqrt{2} & 0 \\ 0 & \sqrt{2} \end{pmatrix}.$$

If $|A+B| \leq |A|+|B|$ were true, then we would have for any vector, and in particular for $e_2 = \begin{pmatrix} 0 \\ 1 \end{pmatrix}$

$$\langle |A+B|e_2, e_2 \rangle \leq \langle |A|e_2, e_2 \rangle + \langle |B|e_2, e_2 \rangle.$$

But

$$\langle |A+B|e_2, e_2 \rangle = \sqrt{2} \nleq 1 = \langle |A|e_2, e_2 \rangle + \langle |B|e_2, e_2 \rangle.$$

Answer 7.2.4 Observe first that since A and B are required to be self-adjoint, $AB + BA$ is clearly self-adjoint. Since $|A||B| + |B||A|$ is always self-adjoint, comparing it with $|A||B| + |B||A|$ therefore makes sense. For a counterexample, let

$$A = \begin{pmatrix} 0 & 1 \\ 1 & 0 \end{pmatrix} \text{ and } B = \begin{pmatrix} 2 & 0 \\ 0 & 1 \end{pmatrix}.$$

Then both A and B are self-adjoint (B is even positive). It is easily seen that

$$|A| = \begin{pmatrix} 1 & 0 \\ 0 & 1 \end{pmatrix} \text{ and } |B| = B = \begin{pmatrix} 2 & 0 \\ 0 & 1 \end{pmatrix}.$$

Hence,

$$AB + BA = \begin{pmatrix} 0 & 3 \\ 3 & 0 \end{pmatrix}$$

and (since $|A| = I$)

$$|A||B| + |B||A| = 2B = \begin{pmatrix} 4 & 0 \\ 0 & 2 \end{pmatrix}.$$

Therefore,

$$|A||B| + |B||A| - (AB + BA) = \begin{pmatrix} 4 & -3 \\ -3 & 2 \end{pmatrix}$$

which is clearly not positive, whereby

$$|A||B| + |B||A| \ngeq AB + BA,$$

as wished.

Answer 7.2.5 Take on $\mathbb{C}^2$

$$A = \begin{pmatrix} 2 & 1 \\ 1 & 0 \end{pmatrix} \text{ and } B = \begin{pmatrix} 1 & 1 \\ 1 & 1 \end{pmatrix}.$$

Then A and B are both self-adjoint and

$$A - B = \begin{pmatrix} 1 & 0 \\ 0 & -1 \end{pmatrix}$$

and so $\|A - B\| = 1$. Since B is positive, $|B| = B$. Since A is self-adjoint, $|A| = \sqrt{A^2}$. But

$$A^2 = \begin{pmatrix} 5 & 2 \\ 2 & 1 \end{pmatrix}.$$

By diagonalizing A^2 first, we then obtain

$$|A| =$$

$$\frac{1}{4} \begin{pmatrix} 1 - \sqrt{2} & 1 + \sqrt{2} \\ 1 & 1 \end{pmatrix} \begin{pmatrix} \sqrt{3 - 2\sqrt{2}} & 0 \\ 0 & \sqrt{3 + 2\sqrt{2}} \end{pmatrix} \begin{pmatrix} -\sqrt{2} & 2 + \sqrt{2} \\ \sqrt{2} & 2 - \sqrt{2} \end{pmatrix}.$$

So

$$|A| = \frac{\sqrt{2}}{2} \begin{pmatrix} 3 & 1 \\ 1 & 1 \end{pmatrix}.$$

Therefore and approximately,

$$\| \, |A| - |B| \, \| \approx 1.179.$$

Accordingly, the inequality

$$\| \, |A| - |B| \, \| \leq \|A - B\|$$

is violated.

Further Reading Some relevant papers that are in relation with this inequality are: [25, 47, 189, 196] and [201].

Answer 7.2.6 Such pairs exist on finite-dimensional spaces. Consider $A = \begin{pmatrix} 0 & a \\ b & 0 \end{pmatrix}$ and $B = \begin{pmatrix} 0 & c \\ d & 0 \end{pmatrix}$ where $a, b, c,$ and d are the positive constants such that $ad \neq bc$.

Choose the constants such that A and B are nonnormal. Clearly, $AB \neq BA$. It is straightforward to check in the end that

$$|A| + |B| = \begin{pmatrix} b & 0 \\ 0 & a \end{pmatrix} + \begin{pmatrix} d & 0 \\ 0 & c \end{pmatrix} = \begin{pmatrix} b+d & 0 \\ 0 & a+c \end{pmatrix} = |A+B|,$$

as wished.

Answer 7.2.7 ([26]) Let

$$A = \begin{pmatrix} 4 & -2 \\ -2 & 1 \end{pmatrix} \text{ and } B = \begin{pmatrix} 1 & -2 \\ -2 & 4 \end{pmatrix}.$$

Then $A \geq 0$ and $B \geq 0$. In addition,

$$|A - B| = \begin{pmatrix} 3 & 0 \\ 0 & 3 \end{pmatrix} \text{ and } A + B = \begin{pmatrix} 5 & -4 \\ -4 & 5 \end{pmatrix}.$$

Finally, $|A - B| \not\leq A + B$ for

$$A + B - |A - B| = \begin{pmatrix} 2 & -4 \\ -4 & 2 \end{pmatrix} \not\geq 0.$$

Answer 7.2.8 First, observe that $(I + |A - I|)(I + |B - I|)$ may well be non-self-adjoint, in which case comparing it with $I + |AB - I|$ would not make sense. For the answer to be exhaustive, it is desirable to give a counterexample when $(I + |A - I|)(I + |B - I|)$ is self-adjoint, that is, when $I + |A - I|$ and $I + |B - I|$ (or $|A - I|$ and $|B - I|$) commute. Let us, therefore, provide two self-adjoint matrices A and B such that:

$$|A - I| + |B - I| + |A - I||B - I| \not\geq |AB - I|$$

and $|A - I||B - I| = |B - I||A - I|$.
   Let

$$A = \begin{pmatrix} 2 & 0 \\ 0 & 0 \end{pmatrix} \text{ and } B = \begin{pmatrix} 2 & 1 \\ 1 & 2 \end{pmatrix}.$$

Hence,

$$A - I = \begin{pmatrix} 1 & 0 \\ 0 & -1 \end{pmatrix}, \ B - I = \begin{pmatrix} 1 & 1 \\ 1 & 1 \end{pmatrix} \text{ and } AB - I = \begin{pmatrix} 3 & 2 \\ 0 & -1 \end{pmatrix}.$$

Obviously, $|A - I| = I$ and $|B - I| = B - I$. Using diagonalization, readers may check that approximately:

$$|AB - I| \approx \begin{pmatrix} 2.6833 & 1.3416 \\ 1.3416 & 1.7889 \end{pmatrix}.$$

Since

$$|A - I| + |B - I| + |A - I||B - I| = \begin{pmatrix} 3 & 2 \\ 2 & 3 \end{pmatrix},$$

it follows that

$$|A - I| + |B - I| + |A - I||B - I| - |AB - I| = \begin{pmatrix} 0.3167 & 0.6584 \\ 0.6584 & 1.2111 \end{pmatrix} \not\geq 0$$

as (for example) $0.3167 \geq 0$, and the main determinant is $-0.0498 < 0$.

Answer 7.2.9 First, observe that $|AB|$ is always self-adjoint, whereas $|A||B|$ may fail to be so. For the product $|A||B|$ to be self-adjoint, recall that it is necessary and sufficient to have $|A||B| = |B||A|$, given that $|A|$ and $|B|$ are self-adjoint. Once that is known, a wide choice of counterexamples becomes available to us. Indeed, *any two non-commuting positive* operators A and B will do. To give an explicit pair, let

$$A = \begin{pmatrix} 1 & 1 \\ 1 & 1 \end{pmatrix} \text{ and } B = \begin{pmatrix} 2 & 0 \\ 0 & 0 \end{pmatrix}$$

be defined on $\mathbb{C}^2$. Then $|A| = A$ and $|B| = B$. Besides,

$$AB = \begin{pmatrix} 2 & 0 \\ 2 & 0 \end{pmatrix} \text{ and } (AB)^* AB = \begin{pmatrix} 8 & 0 \\ 0 & 0 \end{pmatrix}.$$

Therefore,

$$|AB| = \begin{pmatrix} 2\sqrt{2} & 0 \\ 0 & 0 \end{pmatrix} \neq AB = |A||B|,$$

as wished.

Answer 7.2.10 First of all, neither A nor B should be taken normal. For instance, let

$$A = \begin{pmatrix} 1 & 1 \\ 0 & 1 \end{pmatrix} \text{ and } B = \begin{pmatrix} 0 & 1 \\ 0 & 0 \end{pmatrix}.$$

We can easily check that

$$AB = \begin{pmatrix} 0 & 1 \\ 0 & 0 \end{pmatrix} = BA.$$

Now,

$$A^*A = \begin{pmatrix} 1 & 1 \\ 1 & 2 \end{pmatrix} \text{ and } B^*B = \begin{pmatrix} 0 & 0 \\ 0 & 1 \end{pmatrix}.$$

It is clear that

$$|B| = \sqrt{B^*B} = \begin{pmatrix} 0 & 0 \\ 0 & 1 \end{pmatrix}.$$

We could also compute $\sqrt{A^*A}$ to check that $|A||B| \neq |B||A|$. Alternatively, we could just say that if $|A||B| = |B||A|$ held, so would do $A^*A|B| = |B|A^*A$. This also would mean that $A^*A|B|$ is self-adjoint. This is not the case as

$$A^*A|B| = \begin{pmatrix} 1 & 1 \\ 1 & 2 \end{pmatrix} \begin{pmatrix} 0 & 0 \\ 0 & 1 \end{pmatrix} = \begin{pmatrix} 0 & 1 \\ 0 & 2 \end{pmatrix}$$

is not self-adjoint.

Answer 7.2.11 The answer is simple. Consider any two unitary matrices A and B. Hence, $|A| = |B| = I$ and so $A|B| = |B|A$ and $B|A| = |A|B$. Remembering that $AB = BA$ if and only if $A^*B = BA^*$ (when A or B say is unitary), to get the desired counterexample, merely choose two non-commuting unitary operators! An explicit example would be

$$A = \begin{pmatrix} 1 & 0 \\ 0 & -1 \end{pmatrix} \text{ and } B = \begin{pmatrix} 0 & -1 \\ 1 & 0 \end{pmatrix}.$$

Then A and B are unitary (A is also self-adjoint). Besides, $AB = -BA$ and $AB \neq 0$. Accordingly, $AB \neq BA$ and $AB^* \neq B^*A$.

Answer 7.2.12 On $\mathbb{C}^2$, define

$$A = \begin{pmatrix} 0 & 1 \\ 0 & 0 \end{pmatrix} \text{ so that } A^* = \begin{pmatrix} 0 & 0 \\ 1 & 0 \end{pmatrix}.$$

Hence,

$$|A| = \begin{pmatrix} 0 & 0 \\ 0 & 1 \end{pmatrix}.$$

Therefore,

$$A|A| = \begin{pmatrix} 0 & 1 \\ 0 & 0 \end{pmatrix} \neq \begin{pmatrix} 0 & 0 \\ 0 & 0 \end{pmatrix} = |A|A,$$

as wished.

Answer 7.2.13 Let

$$A = \begin{pmatrix} 0 & 1 \\ 0 & 0 \end{pmatrix} \text{ and so } A^* = \begin{pmatrix} 0 & 0 \\ 1 & 0 \end{pmatrix}.$$

Hence,

$$AA^* = \begin{pmatrix} 1 & 0 \\ 0 & 0 \end{pmatrix} \neq \begin{pmatrix} 0 & 0 \\ 0 & 1 \end{pmatrix} = A^*A.$$

Since

$$|A^*| = \begin{pmatrix} 1 & 0 \\ 0 & 0 \end{pmatrix} \text{ and } |A| = \begin{pmatrix} 0 & 0 \\ 0 & 1 \end{pmatrix},$$

it results that

$$|A||A^*| = |A^*||A| = \begin{pmatrix} 0 & 0 \\ 0 & 0 \end{pmatrix}.$$

Answer 7.2.14 For instance, let

$$A = \begin{pmatrix} 0 & 2 \\ 1 & 0 \end{pmatrix}.$$

Then

$$|A^2| = \begin{pmatrix} 2 & 0 \\ 0 & 2 \end{pmatrix} \neq |A|^2 = \begin{pmatrix} 1 & 0 \\ 0 & 4 \end{pmatrix}.$$

Alternatively, we may consider any nonzero A such that $A^2 = 0$ (such an operator cannot be normal). Then, clearly $|A^2| \neq |A|^2$.

Answer 7.2.15 The first thing to think of is the shift operator S on ℓ^2. Then

$$|S| = \sqrt{S^*S} = I$$

is obviously invertible, whereas we all know that S is not invertible. Hence, ran $|S| = \ell^2$. But ran $S = \overline{\text{span}\{e_2, e_3, \cdots\}}$, that is,

$$\text{ran } |S| \neq \text{ran } S.$$

Remark It should be noted that if $A \in B(H)$, then

$$\ker A = \ker |A|.$$

Answer 7.2.16 Consider the self-adjoint matrices

$$A = \begin{pmatrix} 1 & 1 \\ 1 & -1 \end{pmatrix} \text{ and } B = \begin{pmatrix} 3 & -1 \\ -1 & 1 \end{pmatrix}.$$

Observe that B is also positive. Then

$$B - A = \begin{pmatrix} 2 & -2 \\ -2 & 2 \end{pmatrix} \text{ and } A + B = \begin{pmatrix} 4 & 0 \\ 0 & 0 \end{pmatrix}.$$

Both $B - A$ and $A + B$ are positive, i.e. $-B \leq A \leq B$. However,

$$|A| = \begin{pmatrix} \sqrt{2} & 0 \\ 0 & \sqrt{2} \end{pmatrix} \nleq B = \begin{pmatrix} 3 & -1 \\ -1 & 1 \end{pmatrix}$$

as may be checked by interested readers.

Answer 7.2.17 Let

$$A = \begin{pmatrix} 0 & 1 \\ 0 & 0 \end{pmatrix} \text{ and so } |A| = \begin{pmatrix} 0 & 0 \\ 0 & 1 \end{pmatrix}.$$

Then for $x = (2, 1)^t$, we have

$$|\langle Ax, x \rangle| = 2 \nleq 1 = \langle |A|x, x \rangle.$$

Answer 7.2.18 Let

$$A = \begin{pmatrix} 0 & 2 \\ 1 & 0 \end{pmatrix}, \quad x = \begin{pmatrix} 1 \\ 1 \end{pmatrix} \text{ and } y = \begin{pmatrix} 1 \\ 4 \end{pmatrix}.$$

Then clearly

$$|A| = \begin{pmatrix} 1 & 0 \\ 0 & 2 \end{pmatrix} \text{ and } |A^*| = \begin{pmatrix} 2 & 0 \\ 0 & 1 \end{pmatrix}.$$

Therefore,

$$2|A|x = |A^*|y,$$

i.e. $|A^*|y$ and $|A|x$ are linearly dependent. Finally,

$$|\langle Ax, y\rangle|^2 = 36 < 54 = \langle |A|x, x\rangle\langle |A^*|y, y\rangle,$$

as needed.

Answer 7.2.19 Let

$$A = \begin{pmatrix} 0 & 0 \\ 1 & 0 \end{pmatrix}, \quad x = \begin{pmatrix} 1 \\ 0 \end{pmatrix} \text{ and } y = \begin{pmatrix} 0 \\ 1 \end{pmatrix}.$$

Then clearly $\langle Ax, y\rangle = 1$, and one may readily check that

$$\begin{aligned}
\langle |A|x, x\rangle\langle |A|y, y\rangle &= \langle |A^*|x, x\rangle\langle |A^*|y, y\rangle \\
&= \langle |A^*|x, x\rangle\langle |A|y, y\rangle \\
&= 0.
\end{aligned}$$

In other words, all "Cauchy-Schwarz inequalities" that appeared in the question are violated.

Answer 7.2.20 On $\mathbb{C}^2$, let

$$A_n = \begin{pmatrix} 1 & 0 \\ 0 & (-1)^n \end{pmatrix}$$

where $n \in \mathbb{N}$. Then each A_n is self-adjoint, and it is plain that for all n

$$|A_n| = \begin{pmatrix} 1 & 0 \\ 0 & 1 \end{pmatrix} = I.$$

Setting $A = I$, we clearly see that $\||A_n| - |A|\| \to 0$, while $\|A_n - A\| \not\to 0$ for

$$\lim_{n \to \infty} \|A_{2n+1} - A\| \neq 0.$$

Since this counterexample is given on a finite-dimensional Hilbert space (where weak, strong, and uniform convergences all coincide), this also says that $|A_n| \to |A|$ strongly does not necessarily entail $A_n \to A$ strongly, and that $|A_n| \to |A|$ weakly does not necessarily imply that $A_n \to A$ weakly either!

Answer 7.2.21 Let $A_n = S^n$ where $S \in B(\ell^2)$ is the shift operator. Then (cf. Answer 3.1.1), we easily see that $A_n \to A$ weakly, where A is the zero operator. Besides, $A_n^* A_n = I$ for all n, and so $|A_n| = I$ for all n. Thus, $|A_n|$ cannot converge weakly to $|A|$.

Answer 7.2.22 Let S be the shift on ℓ^2, and set $A_n = (S^*)^n + I$, where I is the usual identity operator on ℓ^2. Since we already know that $(S^*)^n$ converges strongly to 0, clearly (A_n) converges strongly to I. Now assume that $(|A_n|)$ were convergent strongly to $|I| = I$. Obviously, it would follow that

$$|A_n|^2 = A_n^* A_n \xrightarrow{s} I^2 = I$$

or that

$$S^n (S^*)^n + S^n + (S^*)^n + I \xrightarrow{s} I.$$

Hence, $S^n (S^*)^n + S^n + (S^*)^n \xrightarrow{s} 0$. Since it can be shown that $S^n (S^*)^n \xrightarrow{s} 0$, and by remembering that $(S^*)^n \xrightarrow{s} 0$, it would follow that $S^n \xrightarrow{s} 0$. This is the contradiction for we already know that (S^n) does not converge strongly to 0 (in fact, it does not converge to any $B \in B(\ell^2)$). Consequently, (A_n) converges strongly to I, and yet $(|A_n|)$ does not converge strongly to I.

Answer 7.2.23 None of the equalities holds. Consider

$$A = \begin{pmatrix} 0 & 1 \\ 3 & 0 \end{pmatrix} = \begin{pmatrix} 0 & 1 \\ 1 & 0 \end{pmatrix} \begin{pmatrix} 3 & 0 \\ 0 & 1 \end{pmatrix} = U|A|$$

which is the *only polar decomposition* of A. Readers can then easily check that

$$U|A| \neq |A|U, \ \ UA \neq AU, \ \text{and} \ A|A| \neq |A|A.$$

Answer 7.2.24 The answer is no for both types of invertibility. Let S be the usual shift operator on ℓ^2. Obviously, S is left invertible. To see why S does not have a ("unitary") polar decomposition, assume that $S = UP$, with U unitary and $P \geq 0$. Then

$$S^* = (UP)^* = PU^*.$$

From here

$$S^* S = PU^* UP = P^2.$$

Since $S^* S = I$, $P^2 = I$. But P is positive and so is I. Thus $P = I$, which would imply that $S = U$, i.e. S would be unitary, which is impossible.

For the case of right invertibility, consider S^* (the shift's adjoint) that is right invertible. Assume that $S^* = U|S^*|$ where U is unitary. Since $SS^*(x_1, x_2, \cdots) = (0, x_2, \cdots)$ for all $(x_1, x_2, \cdots) \in \ell^2$, clearly

$$|S^*| = \sqrt{SS^*} = SS^*.$$

We would then obtain that $S^* = USS^*$ and so $US = I$ or merely $S = U^*$, i.e. S would again be unitary, the same contradiction as before!

Answer 7.2.25 Let $A = 0$ be the zero operator on H. Then A is clearly normal yet for *any unitary* $U \in B(H)$:

$$0 = U|0| = |0|U.$$

Answer 7.2.26 ([167]) Consider the matrix $A = \begin{pmatrix} a & 0 \end{pmatrix}$ where $a \in \mathbb{C}^*$ (A is in $B(\mathbb{C}^2, \mathbb{C})$). Clearly, $\ker A \neq \{0\}$. We show that the polar decomposition of A is unique. Since

$$A^*A = \begin{pmatrix} \bar{a} \\ 0 \end{pmatrix} \begin{pmatrix} a & 0 \end{pmatrix} = \begin{pmatrix} |a|^2 & 0 \\ 0 & 0 \end{pmatrix},$$

it follows that $|A| = \begin{pmatrix} |a| & 0 \\ 0 & 0 \end{pmatrix}$. Let $V = \begin{pmatrix} u_1 & u_2 \end{pmatrix} \in B(\mathbb{C}^2, \mathbb{C})$ be a partial isometry such that $A = V|A|$. A little calculation shows that

$$V = \begin{pmatrix} a/|a| & u_2 \end{pmatrix}.$$

It may also be shown that $(\ker V)^\perp = \{\beta(1, |a|\bar{u}_2/\bar{a})^t : \beta \in \mathbb{C}\}$. Since V is a partial isometry,

$$\|Vw\| = \|w\| \text{ for } w = \beta(1, |a|\bar{u}_2/\bar{a})^t.$$

But

$$\|w\|^2 = |\beta|^2(1 + |u_2|^2) \text{ and } \|Vw\|^2 = |\beta|^2(1 + |u_2|^2)^2$$

and so necessarily $u_2 = 0$. Thus,

$$V = \begin{pmatrix} a/|a| & 0 \end{pmatrix}$$

and so V is clearly uniquely determined.

Answer 7.2.27 ([123], See Also [124] for Related Results) The answer is no! For this purpose, we need to find an $A \in B(H)$ in some Hilbert space H such that $A = V|A|$ with $A^3 = 0$ but $V^3 \neq 0$.

On $\mathbb{C}^3$, let

$$A = \begin{pmatrix} 0\,0\,0 \\ 1\,0\,0 \\ 1\,1\,0 \end{pmatrix}.$$

Clearly, $A^3 = 0$ (the zero matrix). To find V, we could first find $|A|$, which turns out to be

$$|A| = \sqrt{A^*A} = \frac{1}{\sqrt{5}} \begin{pmatrix} 3\,1\,0 \\ 1\,2\,0 \\ 0\,0\,0 \end{pmatrix}.$$

Then, we deduce that

$$V = \frac{1}{\sqrt{5}} \begin{pmatrix} 0\;0\;0 \\ 2\,{-1}\,0 \\ 1\;\;2\;\;0 \end{pmatrix}.$$

Hence,

$$V^3 = \frac{1}{5\sqrt{5}} \begin{pmatrix} 0\;\;0\;0 \\ 2\;{-1}\,0 \\ {-4}\;\;2\;\;0 \end{pmatrix},$$

that is, $V^3 \neq 0$, as wished.

Answer 7.2.28 ([266]) Let

$$A = \begin{pmatrix} 0 & 1 \\ -1 & -1 \end{pmatrix}.$$

Then $|A| = \begin{pmatrix} 2\sqrt{5}/5 & \sqrt{5}/5 \\ \sqrt{5}/5 & 3\sqrt{5}/5 \end{pmatrix}$. By the proof of the polar decomposition of invertible operators, we know that $U = A|A|^{-1}$. Hence, $U = \begin{pmatrix} -\sqrt{5}/5 & 2\sqrt{5}/5 \\ -2\sqrt{5}/5 & -\sqrt{5}/5 \end{pmatrix}$.

In the end, we leave it to readers to check that $A^3 = I$ and that $U^3 \neq I$, as needed.

Chapter 8
Spectrum

8.1 Basics

Definition 8.1.1 Let $A \in B(H)$ where H is a complex Hilbert space. The set

$$\sigma(A) = \{\lambda \in \mathbb{C} : \lambda I - A \text{ is not invertible}\}$$

is called the spectrum of A.

The resolvent set of A, denoted by $\rho(A)$, is the complement of $\sigma(A)$, i.e.

$$\rho(A) = \mathbb{C} \setminus \sigma(A).$$

Remark On a *real* Hilbert space H, the spectrum of $A \in B(H)$ is defined as

$$\sigma(A) = \{\lambda \in \mathbb{R} : \lambda I - A \text{ is not invertible}\},$$

and so

$$\rho(A) = \mathbb{R} \setminus \sigma(A).$$

Remark When $\dim H < \infty$, then $\sigma(A)$ coincides with the set of the eigenvalues of A.

Examples 8.1.1

(1) Let

$$A = \begin{pmatrix} 1 & 2 \\ 0 & -3 \end{pmatrix}.$$

© The Author(s), under exclusive license to Springer Nature Switzerland AG 2022
M. H. Mortad, *Counterexamples in Operator Theory*,
https://doi.org/10.1007/978-3-030-97814-3_8

Then

$$\sigma(A) = \{-3, 1\}.$$

(2) (See e.g. Exercise 7.3.14 in [256]) Let S be the shift operator on ℓ^2. Then

$$\sigma(S) = \{\lambda \in \mathbb{C} : |\lambda| \le 1\}.$$

(3) (See e.g. Problem 84 in [148]) Let R be the bilateral shift on $\ell^2(\mathbb{Z})$. Then

$$\sigma(R) = \{\lambda \in \mathbb{C} : |\lambda| = 1\}.$$

(4) (See e.g. Example 4 on Page 112 in [136]) Let A be the multiplication operator on $L^2[a, b]$ by the continuous function φ on $[a, b]$. Then

$$\sigma(A) = \{\varphi(x) : x \in [a, b]\} = \operatorname{ran} \varphi.$$

(5) Let $A \in B(H)$ be nilpotent. Then

$$\sigma(A) = \{0\}.$$

Next, we give some of the important subsets of $\sigma(A)$.

Definition 8.1.2 Let H be a complex Hilbert space, and consider $A \in B(H)$.

(1) The set of the eigenvalues of A, denoted by $\sigma_p(A)$, is called the point spectrum of A and is given by

$$\sigma_p(A) = \{\lambda \in \mathbb{C} : \exists x \in H, x \ne 0 : Ax = \lambda x\},$$

that is, the set of complex λ for which $\lambda I - A$ is not injective. The nonzero vector x is called an eigenvector, whereas the space $\ker(\lambda I - A)$ is called an eigenspace.

(2) The continuous spectrum is defined by

$$\sigma_c(A) = \{\lambda \in \mathbb{C} : \lambda I - A \text{ is injective, has a dense range, but non-surjective}\}.$$

(3) The residual spectrum is defined by

$$\sigma_r(A) = \{\lambda \in \mathbb{C} : \lambda I - A \text{ is injective, but it does not have a dense range}\}.$$

(4) The approximate point spectrum is the set of approximate eigenvalues. An approximate eigenvalue of A is a $\lambda \in \mathbb{C}$ such that for any $\varepsilon > 0$, there exists $x \in H$, with $\|x\| = 1$, such that

$$\|(\lambda - A)x\| < \varepsilon.$$

The set of approximate eigenvalues is denoted by $\sigma_a(A)$.

Remark The same definition is valid for real Hilbert spaces with the obvious changes.

Proposition 8.1.1 *The residual spectrum of a normal operator is empty.*

8.2 Questions

8.2.1 An $A \in B(H)$ Such That $\sigma(A) = \varnothing$

Recall the following elementary result:

Theorem 8.2.1 *Let $H \neq \{0\}$ be a complex Hilbert space, and let $A \in B(H)$. Then:*

(1) $\sigma(A)$ *is never empty.*
(2) $\sigma(A)$ *is a bounded closed subset of $\mathbb{C}$, i.e. $\sigma(A)$ is compact. Furthermore, $\sigma(A) \subset B_c(0, \|A\|)$, i.e. $\sigma(A)$ is contained in the closed disc of radius $\|A\|$ centered at the origin.*

Question 8.2.1 Find an $A \in B(H)$ such that $\sigma(A) = \varnothing$ (where H is necessarily a real Hilbert space).

8.2.2 An A Such That $\sigma_p(A^*) \neq \{\bar{\lambda} : \lambda \in \sigma_p(A)\}$

If we know the spectrum of an operator A, then we may find the spectrum of its adjoint as well as that of its inverse (in case the invertibility is available). That is, if $A \in B(H)$, then

$$\sigma(A^*) = \{\bar{\lambda} : \lambda \in \sigma(A)\};$$

and if A is invertible, then

$$\sigma(A^{-1}) = \{\lambda^{-1} : \lambda \in \sigma(A)\}.$$

Question 8.2.2 Provide an $A \in B(H)$ for which

$$\sigma_p(A^*) \neq \{\bar{\lambda} : \lambda \in \sigma_p(A)\}.$$

Hint: Avoid (finite-dimensional spaces and) normal operators.

8.2.3   An $A \in B(H)$ with $p[\sigma(A)] \neq \sigma[p(A)]$ Where p Is a Polynomial

Recall the well known result that states that if we know the spectrum of an operator A, then we should be able to find the spectrum of polynomials of A:

Theorem 8.2.2 (Spectral Mapping Theorem)   *Let $p \in \mathbb{C}[X]$ be a polynomial, and let $A \in B(H)$ where H is a complex Hilbert space. Then*

$$\sigma[p(A)] = p[\sigma(A)].$$

Next, we provide a simple example that shows that it is important to work on complex Hilbert spaces for the previous theorem to be valid.

Question 8.2.3 Give an example of an $A \in B(H)$ and a polynomial p such that

$$p[\sigma(A)] \neq \sigma[p(A)].$$

8.2.4   A Non-self-adjoint Operator Having a Real Spectrum

We have substantial information about the spectrum of some classes of operators. Indeed, if $A \in B(H)$ where H is a complex Hilbert space, then:

(1) $\sigma(A)$ is real whenever A is self-adjoint.
(2) $\sigma(A) \subset \{\lambda \in \mathbb{C} : |\lambda| = 1\}$ if A is unitary.
(3) $\sigma(A) = \{0, 1\}$ if A is a nontrivial orthogonal projection.

Question 8.2.4 (Cf. Question 11.2.2) Give an example of a non-self-adjoint operator $A \in B(H)$ having a real spectrum.

8.2.5 A Non-positive Operator with a Positive Spectrum

Let $A \in B(H)$. If A is positive, then $\sigma(A) \subset \mathbb{R}^+$. There is a simple proof that appeared in [260] (most probably known to specialists). It goes as follows:

If $\lambda < 0$, then $-\lambda I > 0$ and so $A - \lambda I \geq -\lambda I$ because $A \geq 0$. Hence, $A - \lambda I$ is invertible as $-\lambda I$ is (see the discussion before Question 4.2.10), i.e. $\lambda \notin \sigma(A)$.

If $\sigma(A) \subset \mathbb{R}^+$ and A is normal say, then it may be shown that A is a positive operator.

Question 8.2.5 Find a non-positive operator having a positive spectrum.

8.2.6 A Non-orthogonal Projection A with $\sigma(A) = \{0, 1\}$

Question 8.2.6 Give an example of a projection (non-orthogonal) A such that $\sigma(A) = \{0, 1\}$.

8.2.7 A Self-Adjoint Operator Without Any Eigenvalue

Let $H \neq \{0\}$ be a finite-dimensional vector space. Every self-adjoint matrix on H has an eigenvalue. See e.g. 7.27 in [9]. Can this feature be carried over to infinite-dimensional spaces?

Question 8.2.7 Give an example of a self-adjoint operator in $B(H)$ not possessing any eigenvalue.

8.2.8 A Self-Adjoint A Such That $\sigma(A) \cap \sigma(-A) = \varnothing$, Yet A Is Neither Positive Nor Negative

If $A \in B(H)$ is positive, then $\sigma(A) \subset \mathbb{R}^+$, whereby

$$\sigma(A) \cap \sigma(-A) \subseteq \{0\}.$$

As we shall see below, the previous condition does not imply the positiveness of A nor it implies its negativeness. It is interesting, however, to notice that the condition $\sigma(A) \cap \sigma(-A) \subseteq \{0\}$ could replace the condition $A \geq 0$ in some situations and still get similar conclusions. For example, it could be used to define the square root of A^2 (A being self-adjoint) as carried out in say [2].

Question 8.2.8 Find a self-adjoint $A \in B(H)$ such that

$$\sigma(A) \cap \sigma(-A) = \varnothing$$

and A is neither positive nor negative.

8.2.9 A Self-Adjoint A Such That $\sigma(A) \cap \sigma(-A) = \{0\}$, Yet A Is Neither Positive Nor Negative

Question 8.2.9 Find a self-adjoint $A \in B(H)$ such that

$$\sigma(A) \cap \sigma(-A) = \{0\},$$

yet A is neither positive nor negative.

8.2.10 A Self-Adjoint Matrix A Such That $\sigma(A) \cap \sigma(-A) = \{0\}$ and tr $A = 0$, Yet A Is Neither Positive Nor Negative

Question 8.2.10 Let H be a finite-dimensional Hilbert space. Find a (nonzero) self-adjoint $A \in B(H)$ such that tr $A = 0$ and

$$\sigma(A) \cap \sigma(-A) = \{0\},$$

yet A is neither positive nor negative.

8.2.11 A Nilpotent $T = A + iB$ Such That $\sigma(A) \cap \sigma(-A) = \{0\}$

Question 8.2.11 (Cf. [113]) Find a $T = A + iB \in B(H)$ that is nilpotent and such that $\sigma(A) \cap \sigma(-A) = \{0\}$ (where A and B are self-adjoint).

8.2.12 A Nilpotent $T = A + iB$ Such That $\sigma(A) \cap \sigma(-A) = \varnothing$

Question 8.2.12 (Cf. [113]) Find a nilpotent $T = A + iB \in B(H)$ such that $\sigma(A) \cap \sigma(-A) = \varnothing$ (where A and B are self-adjoint).

8.2.13 Two Operators A, B Such That $\sigma(AB) \neq \sigma(BA)$

It is well known that if A and B are two square matrices, then they have the same eigenvalues, i.e. in this context $\sigma(AB) = \sigma(BA)$. What about the infinite-dimensional case? In other words, if $A, B \in B(H)$ with dim $H = \infty$, then is it true that $\sigma(AB) = \sigma(BA)$?

What we are always sure of is the following result, more commonly known as the Jacobson lemma:

Theorem 8.2.3 Let $A, B \in B(H)$. Then

$$\sigma(AB) \cup \{0\} = \sigma(BA) \cup \{0\}$$

(or $\sigma(AB) - \{0\} = \sigma(BA) - \{0\}$).

There are quite a few cases when $\sigma(AB) = \sigma(BA)$ holds. For example, when A is invertible. We may also add the compactness of A (to be defined later). Less obviously, notice that there are papers dealing with when $\sigma(AB) = \sigma(BA)$ holds (e.g. [14, 139, 158] and [383]). The ultimate generalization so far is the following (first appeared in [84] and works for unbounded operators just as good):

Theorem 8.2.4 ([84]) *Let $A, B \in B(H)$. If* $\ker(A^*) = \ker(A)$, *then*

$$\sigma(BA) = \sigma(AB).$$

Question 8.2.13 Find $A, B \in B(H)$ such that

$$\sigma(AB) \neq \sigma(BA).$$

8.2.14   Two Self-Adjoints A, B with $\sigma(A) + \sigma(B) \not\supset \sigma(A + B)$

Recall the following result that probably does not have any elementary proof in the literature yet. A proof based on Gelfand's transform may be found in Theorem 11.23 in [314].

Proposition 8.2.5 *Let $A, B \in B(H)$ be commuting. Then*

$$\sigma(AB) \subset \sigma(A)\sigma(B) \text{ and } \sigma(A + B) \subset \sigma(A) + \sigma(B),$$

where

$$\sigma(A)\sigma(B) = \{\lambda\mu : \lambda \in \sigma(A), \mu \in \sigma(B)\}$$

and

$$\sigma(A) + \sigma(B) = \{\lambda + \mu : \lambda \in \sigma(A), \mu \in \sigma(B)\}.$$

Question 8.2.14 Give an example of self-adjoint matrices A and B such that

$$\sigma(A) + \sigma(B) \not\supset \sigma(A + B).$$

8.2.15 Two Commuting A, B with $\sigma(A) + \sigma(B) \not\subset \sigma(A + B)$

Question 8.2.15 Provide an example of commuting self-adjoint matrices A and B such that

$$\sigma(A) + \sigma(B) \not\subset \sigma(A + B).$$

8.2.16 Two Self-Adjoints A and B with $\sigma(A)\sigma(B) \not\supset \sigma(AB)$

Question 8.2.16 Give an example of self-adjoint matrices A and B such that

$$\sigma(A)\sigma(B) \not\supset \sigma(AB).$$

8.2.17 Two Commuting A, B Such That $\sigma(A)\sigma(B) \not\subset \sigma(AB)$

Question 8.2.17 Provide an example of commuting self-adjoint matrices A and B such that

$$\sigma(A)\sigma(B) \not\subset \sigma(AB).$$

8.2.18 Two Commuting Self-Adjoint Operators A and B Such That $\sigma_p(AB) \not\subset \sigma_p(A)\sigma_p(B)$ and $\sigma_p(A + B) \not\subset \sigma_p(A) + \sigma_p(B)$

The objective of this question is to investigate the validity of Proposition 8.2.5 (just before Question 8.2.14) in the case of the point spectrum σ_p. Recall that the point spectrum coincides with the usual spectrum when $\dim H < \infty$. Hence, Proposition 8.2.5 is obviously true for the point spectrum, but on a finite-dimensional *complex H*.

Let us give counterexamples in the case of finite-dimensional real Hilbert spaces. On $\mathbb{R}^2$, consider

$$A = \begin{pmatrix} 0 & 1 \\ -1 & 0 \end{pmatrix},$$

and so $A^2 = -I$. Then, $\sigma_p(A) = \varnothing$. Hence,

$$\sigma_p(A^2) = \{-1\} \not\subset \varnothing = \sigma_p(A)\sigma_p(A).$$

As for the sum, keep the same A, and set $B = -A$. It is then seen that

$$\sigma_p(A - A) = \{0\} \not\subset \varnothing = \sigma_p(A) + \sigma_p(-A).$$

Observe in the end that the commutativity is patent in both cases.

Question 8.2.18 Find two commuting self-adjoint operators $A, B \in B(H)$ (where H is a complex infinite-dimensional Hilbert space) such that $\sigma_p(AB) \not\subset \sigma_p(A)\sigma_p(B)$ and $\sigma_p(A + B) \not\subset \sigma_p(A) + \sigma_p(B)$ (not necessarily the same pair for both inclusions).

8.2.19 A Unitary Operator A Such That $|\sigma(A)|^2 \neq \sigma(A^*)\sigma(A)$ and $\sigma(A^*)\sigma(A) \not\subset \mathbb{R}$

Let $A \in B(H)$. Define

$$\sigma(A^*)\sigma(A) := \{\lambda\mu : \lambda \in \sigma(A^*), \mu \in \sigma(A)\}$$

and

$$|\sigma(A)|^2 := \{|\lambda|^2 : \lambda \in \sigma(A)\}.$$

Question 8.2.19 Find a unitary $A \in B(H)$ such that

$$|\sigma(A)|^2 \neq \sigma(A^*)\sigma(A) \text{ and } \sigma(A^*)\sigma(A) \not\subset \mathbb{R}.$$

8.2.20 Two Operators A and B Such That $\sigma(A) \cap \sigma(-B) = \varnothing$, but A + B Is Not Invertible

If $A, B \in B(H)$ are commuting and such that $\sigma(A) \cap \sigma(-B) = \varnothing$, then $A + B$ is invertible. For if $0 \in \sigma(A + B)$, then it would follow that $0 \in \sigma(A) + \sigma(B)$ (by

Proposition 8.2.5 just above Question 8.2.14). That is, $0 = \alpha + \beta$ for some $\alpha \in \sigma(A)$ and some $\beta \in \sigma(B)$. Hence,

$$\sigma(A) \ni \alpha = -\beta \in \sigma(-B).$$

But, this is not coherent with the assumption $\sigma(A) \cap \sigma(-B) = \varnothing$. Consequently, $A + B$ must be invertible.

Question 8.2.20 Find $A, B \in B(H)$ such that $\sigma(A) \cap \sigma(-B) = \varnothing$, but $A+B$ is not invertible (showing that the commutativity condition in the result above may not just dropped).

8.2.21 "Bigger" Operators Defined on Larger Hilbert Spaces Do Not Necessarily Have "Bigger" Spectra

Question 8.2.21 (Cf. Question 8.2.26) Let $A \in B(H)$, and let $B \in B(K)$ where H and K are two Hilbert spaces such that $H \subset K$, with the conditions $BH \subset H$ and $A = B|H$ (we may say that B is an extension of A).
Is it necessarily true that $\sigma(A) \subset \sigma(B)$?

8.2.22 An Operator A Such That $A \neq 0$ and $\sigma(A) = \{0\}$

We will see in the next chapter that if a normal operator $A \in B(H)$ is such that $\sigma(A) = \{\alpha\}, \alpha \in \mathbb{C}$, then necessarily $A = \alpha I$.

Question 8.2.22 Give an example of an operator $A \in B(H)$ such that

$$A \neq 0 \text{ and } \sigma(A) = \{0\}.$$

8.2.23 A Self-Adjoint Operator A Such That $\sigma_p(A) = \{0, 1\}$, $\sigma_c(A) = (0, 1)$, and $\sigma(A) = [0, 1]$

Question 8.2.23 Give an example of a self-adjoint operator A such that

$$\sigma_p(A) = \{0, 1\},\ \sigma_c(A) = (0, 1) \text{ and } \sigma(A) = [0, 1].$$

8.2.24 An Operator A with $\sigma(A) = \{1\}$, But A Is Neither Self-Adjoint Nor Unitary

Question 8.2.24 (Cf. Question 9.2.4) Give an example of an operator $A \in B(H)$ with $\sigma(A) = \{1\}$, and A is neither self-adjoint nor unitary.

8.2.25 Two Self-Adjoint Operators A, B Such That $\sigma(AB)$ Is Purely Imaginary

Recall that if A, $B \in B(H)$ and B is positive, then

$$\sigma(AB) = \sigma(BA) = \sigma(\sqrt{B}A\sqrt{B})$$

(see e.g. Exercise 7.3.19 in [256]).
  If A is also self-adjoint, then $\sqrt{B}A\sqrt{A}$ remains self-adjoint. Therefore: If A, $B \in B(H)$ are self-adjoint, one of them is positive, then $\sigma(AB) \subset \mathbb{R}$.

Question 8.2.25 Give an example of two self-adjoint operators $A, B \in B(H)$ such that $\sigma(AB)$ is purely imaginary.

8.2.26 $A \leq B$ Does Not Yield $\sigma(A) \subset \sigma(B)$

It may be shown that if A and B are self-adjoint operators obeying $\sigma(A) \subset \sigma(B)$, then $A \leq B$ (see e.g. the "True or False" Section 8.2 in [256]). What about the converse?

Question 8.2.26 Find self-adjoints $A, B \in B(H)$ with $A \leq B$ and $\sigma(A) \not\subset \sigma(B)$.

8.2.27 An Operator A with $\overline{\sigma_p(A)} \subsetneq \sigma_a(A) \subsetneq \sigma(A)$

Let A be in $B(H)$. Then it may be shown that (see e.g. Exercise 7.3.10 in [256]):

$$\overline{\sigma_p(A)} \subset \sigma_a(A) \subset \sigma(A).$$

Question 8.2.27 Let H be a Hilbert space, and let $A \in B(H)$. Give an example showing that the inclusions

$$\overline{\sigma_p(A)} \subset \sigma_a(A) \subset \sigma(A)$$

may be strict.

8.2.28 A $T \in B(H)$ Such That $|\lambda| \in \sigma(|T|)$, but $\lambda \notin \sigma(T)$

Question 8.2.28 Find a $T \in B(H)$ such that $|\lambda| \in \sigma(|T|)$, but $\lambda \notin \sigma(T)$.

8.2.29 A $T \in B(H)$ Such That $\lambda \in \sigma(T)$, yet $|\lambda| \notin \sigma(|T|)$

Question 8.2.29 (Cf. Question 14.2.5) Find a $T \in B(H)$ such that $\lambda \in \sigma(T)$, yet $|\lambda| \notin \sigma(|T|)$.

Hint: Avoid (for instance) normal operators.

8.2.30   A Sequence of Normal Operators A_m Converging Strongly to A, but $\sigma(A) \neq \lim_{m \to \infty} \sigma(A_m)$

Define

$$\lim_{m \to \infty} \sigma(A_m) := \{z \in \mathbb{C} : \exists z_m \in \mathbb{C}, \ z_m \in \sigma(A_m) \text{ and } z_m \to z\}.$$

Question 8.2.30 Find a sequence of normal operators A_m in $B(H)$ and an $A \in B(H)$ such that $\|A_m x - Ax\| \to 0$ as $m \to \infty$ (for all $x \in H$) but

$$\sigma(A) \neq \lim_{m \to \infty} \sigma(A_m).$$

Remark A counterexample is not available anymore if the strong convergence is replaced by the convergence in norm.

8.2.31 $T = \bigoplus_{n \in \mathbb{N}} T_n$, but $\overline{\bigcup_{n \in \mathbb{N}} \sigma(T_n)} \neq \sigma(T)$

Question 8.2.31 Let (T_n) be a sequence of bounded operators on H. Set $T = \bigoplus_{n \in \mathbb{N}} T_n$. Show that it could happen that

$$\overline{\bigcup_{n \in \mathbb{N}} \sigma(T_n)} \neq \sigma(T).$$

Hint: Notice that (see e.g. Exercise 3.5.26 in [184]) we always have

$$\overline{\bigcup_{n \in \mathbb{N}} \sigma(T_n)} \subset \sigma(T)$$

and that full equality holds if e.g. *each T_n is normal*.

8.2.32 $A = \bigoplus_{n \in \mathbb{N}} A_n$ Where All (A_n) Are Nilpotent, but A Is Not Nilpotent

Question 8.2.32 Let (A_n) be a sequence of bounded operators on H such that $(\|A_n\|)_n$ is bounded. Set $A = \bigoplus_{n \in \mathbb{N}} A_n$. Give a sequence of (all) nilpotent operators (A_n) such that A is not nilpotent.

Answers

Answer 8.2.1 On $\mathbb{R}^2$, let

$$A = \begin{pmatrix} 0 & 1 \\ -1 & 0 \end{pmatrix}.$$

Then A does not possess any real eigenvalue, that is to say that

$$\sigma(A) = \sigma_p(A) = \varnothing.$$

Answer 8.2.2 Let S be the shift operator on ℓ^2. Recall that $\sigma_p(S) = \varnothing$ and $\sigma_p(S^*) = \{\lambda \in \mathbb{C} : |\lambda| < 1\}$ (see e.g. Exercise 7.3.14 in [256]). So

$$\sigma_p(S^*) \neq \varnothing = \{\bar{\lambda} : \lambda \in \sigma_p(S)\}.$$

Answer 8.2.3 On $\mathbb{R}^2$, let

$$A = \begin{pmatrix} 0 & 1 \\ -1 & 0 \end{pmatrix}.$$

Then A has no eigenvalues in $\mathbb{R}$, i.e.

$$\sigma(A) = \varnothing.$$

Let $p(x) = x^2$. Then

$$p(A) = A^2 = \begin{pmatrix} -1 & 0 \\ 0 & -1 \end{pmatrix}$$

whose spectrum is reduced to

$$\sigma(p(A)) = \{-1\}.$$

Thus

$$\sigma[p(A)] = \{-1\} \neq p[\sigma(A)] = p(\varnothing) = \varnothing.$$

Answer 8.2.4 This is easy. On $\mathbb{C}^2$, take

$$A = \begin{pmatrix} 1 & 2 \\ 0 & -1 \end{pmatrix}.$$

Then A is clearly not self-adjoint and yet

$$\sigma(A) = \{1, -1\} \subset \mathbb{R}.$$

Answer 8.2.5 Let

$$A = \begin{pmatrix} 1 & 1 \\ 0 & 2 \end{pmatrix}.$$

Then A is not positive since it is not self-adjoint, and yet

$$\sigma(A) = \{1, 2\} \subset \mathbb{R}^+.$$

Answer 8.2.6 Just take

$$A = \begin{pmatrix} 0 & 0 \\ 2021 & 1 \end{pmatrix}.$$

Then $A \neq A^*$ and $A^2 = A$, while

$$\sigma(A) = \{0, 1\}.$$

Answer 8.2.7 Define $A \in B(L^2(0, 1))$ by

$$Af(x) = xf(x).$$

Then A is self-adjoint without any eigenvalue. Indeed, if λ were an eigenvalue, we would have $(\lambda - x)f(x) = 0$, which clearly would mean that $f(x) = 0$ almost everywhere. That is, $f = 0$ on $L^2(0, 1)$. Thus,

$$\sigma_p(A) = \varnothing.$$

Answer 8.2.8 The counterexample is simple. Let

$$A = \begin{pmatrix} 2 & 0 \\ 0 & -1 \end{pmatrix}.$$

Then $\sigma(A) = \{2, -1\}$ and so $\sigma(-A) = \{-2, 1\}$. Thus,

$$\sigma(A) \cap \sigma(-A) = \varnothing,$$

yet A is neither positive nor negative.

Answer 8.2.9 Just let

$$A = \begin{pmatrix} 2 & 0 & 0 \\ 0 & -1 & 0 \\ 0 & 0 & 0 \end{pmatrix}.$$

Hence,

$$\sigma(A) = \{2, -1, 0\} \text{ and so } \sigma(-A) = \{-2, 1, 0\}.$$

Accordingly,

$$\sigma(A) \cap \sigma(-A) = \{0\},$$

yet A is neither positive nor negative.

Answer 8.2.10 Take

$$A = \begin{pmatrix} 0 & 0 & 0 & 0 \\ 0 & 3 & 0 & 0 \\ 0 & 0 & -2 & 0 \\ 0 & 0 & 0 & -1 \end{pmatrix}.$$

Clearly, $\operatorname{tr} A = 0$ and

$$\sigma(A) \cap \sigma(-A) = \{0\},$$

yet A is neither positive nor negative.

Answer 8.2.11 Let

$$T = \begin{pmatrix} 2 & 2 & -2 & 0 \\ 5 & 1 & -3 & 0 \\ 1 & 5 & -3 & 0 \\ 0 & 0 & 0 & 0 \end{pmatrix}$$

and so

$$A = \begin{pmatrix} 2 & 7/2 & -1/2 & 0 \\ 7/2 & 1 & 1 & 0 \\ -1/2 & 1 & -3 & 0 \\ 0 & 0 & 0 & 0 \end{pmatrix}.$$

Hence (approximately),

$$\sigma(A) = \{0, -3.71, -1.33, 5.04\},$$

and so

$$\sigma(A) \cap \sigma(-A) = \{0\}$$

is satisfied. Observe finally that $T \neq 0$, whereas $T^3 = 0$, i.e. T is nilpotent.

Answer 8.2.12 Let

$$T = \begin{pmatrix} 0 & 1 & 2 & 4 \\ 0 & 0 & 2 & 1 \\ 0 & 0 & 0 & 5 \\ 0 & 0 & 0 & 0 \end{pmatrix}.$$

Hence,

$$A = \begin{pmatrix} 0 & 1/2 & 1 & 2 \\ 1/2 & 0 & 1 & 1/2 \\ 1 & 1 & 0 & 5/2 \\ 2 & 1/2 & 5/2 & 0 \end{pmatrix}.$$

Then, it may be checked that approximately

$$\sigma(A) = \{+4.058, -1.043, -2.811, -0.205\}.$$

Hence, the requirement $\sigma(A) \cap \sigma(-A) = \varnothing$ is clearly fulfilled. In the end, $T^4 = 0$ (with $T^3 \neq 0$).

Answer 8.2.13 An efficient example is the shift operator S on ℓ^2. As is known $S^*S = I$, and so $\sigma(S^*S) = \{1\}$. Since SS^* is an orthogonal projection, we know that $\sigma(SS^*) \subset \{0, 1\}$. Since SS^* and S^*S share the same nonzero elements, we see that $1 \in \sigma(SS^*)$. Since it may also be shown that 0 is an eigenvalue for SS^*, it ensures that:

$$\sigma(SS^*) = \{0, 1\} \neq \{1\} = \sigma(S^*S).$$

Remark The discussion just before Question 8.2.22 below, even though using an argument not yet introduced, provides another way of showing that $\sigma(SS^*) = \{0, 1\}$. Indeed, we already know that $\sigma(SS^*) \subset \{0, 1\}$. If $\sigma(SS^*) = \{0\}$ or $\sigma(SS^*) = \{1\}$, the normality of SS^* would yield $SS^* = 0$ or $SS^* = I$ (respectively), both being impossible. Therefore, the only possible outcome is

$$\sigma(SS^*) = \{0, 1\}.$$

Answer 8.2.14 Consider the self-adjoint matrices:

$$A = \begin{pmatrix} 0 & 1 \\ 1 & 0 \end{pmatrix} \text{ and } B = \begin{pmatrix} 1 & 0 \\ 0 & 2 \end{pmatrix}.$$

Then $\sigma(A) = \{1, -1\}$ and $\sigma(B) = \{1, 2\}$, and so $\sigma(A + B) = \{0, 1, 2, 3\}$. Since $\sigma(A + B) = \{(3 - \sqrt{5})/2, (3 + \sqrt{5})/2\}$, it follows that

$$\sigma(A + B) \not\subset \sigma(A) + \sigma(B)$$

(observe in the end that $AB \neq BA$).

Answer 8.2.15 Let

$$A = \begin{pmatrix} 1 & 0 \\ 0 & 2 \end{pmatrix}.$$

Then $\sigma(A) = \{1, 2\}$. Set $B = -A$. Then A commutes with B and $\sigma(B) = \{-1, -2\}$. To conclude, it is seen that

$$\sigma(A) + \sigma(B) \not\subset \sigma(A + B) = \{0\}$$

because e.g. $1 \in \sigma(A) + \sigma(B)$.

Answer 8.2.16 Reuse the counterexamples of Answer 8.2.14! Then

$$\sigma(AB) = \{\sqrt{2}, -\sqrt{2}\} \not\subset \sigma(A)\sigma(B).$$

Answer 8.2.17 Let

$$A = \begin{pmatrix} 1 & 0 \\ 0 & 2 \end{pmatrix}.$$

Then $\sigma(A) = \{1, 2\}$. Setting $B = A^{-1}$, we see that $\sigma(B) = \{1, 1/2\}$. Then

$$\sigma(A)\sigma(B) \not\subset \{1\} = \sigma(AB)$$

as e.g. $2 \in \sigma(A)\sigma(B)$.

Answer 8.2.18 Let A be a self-adjoint operator defined on $L^2(0, 1)$ and such that $\sigma_p(A) = \varnothing$ (see Answer 8.2.7), and then take $B = 0$ (hence $BA = AB$). Thence, $\sigma_p(B) = \{0\}$. Therefore,

$$\sigma_p(AB) = \{0\} \not\subset \varnothing = \sigma_p(A)\sigma_p(B).$$

As for the sum, keep the same A and take $B = -A$ (so $AB = BA$). Then

$$\sigma_p(A - A) = \{0\} \not\subset \varnothing = \sigma_p(A) + \sigma_p(-A)$$

Answer 8.2.19 ([207]) On $\mathbb{C}^3$, consider the unitary matrix:

$$A = \begin{pmatrix} -i & 0 & 0 \\ 0 & 1 & 0 \\ 0 & 0 & i \end{pmatrix}.$$

Clearly, $\sigma(A) = \sigma_p(A) = \{-i, 1, i\}$, and so $\sigma(A^*) = \{i, 1, -i\}$ and $|\sigma(A)|^2 = \{1\}$. Hence,

$$\sigma(A^*)\sigma(A) = \{1, i, -1, -i\} \not\subset \mathbb{R}$$

and

$$|\sigma(A)|^2 = \{1\} \neq \sigma(A^*)\sigma(A),$$

as suggested.

Answer 8.2.20 ([125]) Consider

$$A = \begin{pmatrix} 1 & 0 \\ 2 & 1 \end{pmatrix} \text{ and } B = \begin{pmatrix} 1 & 2 \\ 0 & 1 \end{pmatrix}.$$

Then $\sigma(A) = \sigma(B) = \{1\}$ and so

$$\sigma(A) \cap \sigma(-B) = \varnothing.$$

However,

$$A + B = \begin{pmatrix} 2 & 2 \\ 2 & 2 \end{pmatrix}$$

is obviously not invertible (observe that $AB \neq BA$).

Answer 8.2.21 The answer is no! Let S be the unilateral shift on $\ell^2(\mathbb{N})$, and let R be the bilateral shift on $\ell^2(\mathbb{Z})$. Then $\ell^2(\mathbb{N}) \subset \ell^2(\mathbb{Z})$, but

$$\sigma(S) = \{\lambda \in \mathbb{C} : |\lambda| \leq 1\} \not\subset \{\lambda \in \mathbb{C} : |\lambda| = 1\} = \sigma(R).$$

Answer 8.2.22 Simply, consider

$$A = \begin{pmatrix} 0 & 1 \\ 0 & 0 \end{pmatrix}.$$

Then $\sigma(A) = \{0\}$ yet $A \neq 0$.

Answer 8.2.23 Let A be the multiplication operator by the real-valued function:

$$\varphi(x) = \begin{cases} 0, & 0 < x < \frac{1}{3}, \\ 3(x - \frac{1}{3}), & \frac{1}{3} \leq x \leq \frac{2}{3}, \\ 1, & \frac{2}{3} < x < 1, \end{cases}$$

defined on $L^2(0, 1)$. That is, $Af(x) = \varphi(x)f(x)$ where $f \in L^2(0, 1)$. Then A is obviously self-adjoint (and so $\sigma_r(A) = \varnothing$). By considering $f(x) = \mathbb{1}_{(0,1/3)}(x)$ and also $g(x) = \mathbb{1}_{(2/3,1)}(x)$, we see that

$$\sigma_p(A) = \{0, 1\}.$$

Therefore,

$$\sigma_c(A) = (0, 1)$$

for $\sigma(A) = [0, 1]$.

Answer 8.2.24 Merely consider a nilpotent matrix B, for instance, let $B = \begin{pmatrix} 0 & 1 \\ 0 & 0 \end{pmatrix}$, and then set $A = B + I$. Hence, $\sigma(A) = \{1\}$. However,

$$A = \begin{pmatrix} 1 & 1 \\ 0 & 1 \end{pmatrix}$$

is not even normal, and so it cannot be self-adjoint nor it can be unitary.

Answer 8.2.25 On $\mathbb{C}^2$, let

$$A = \begin{pmatrix} 0 & 1 \\ 1 & 0 \end{pmatrix} \text{ and } B = \begin{pmatrix} 1 & 0 \\ 0 & -1 \end{pmatrix}.$$

Then A and B are self-adjoint. Also

$$AB = \begin{pmatrix} 0 & -1 \\ 1 & 0 \end{pmatrix}.$$

Thus

$$\sigma(AB) = \{i, -i\},$$

as wished.

Answer 8.2.26 Even in a finite-dimensional setting, the result is untrue as may readily be checked by considering the simple matrices

$$A = \begin{pmatrix} 1 & 0 \\ 0 & 2 \end{pmatrix} \text{ and } B = \begin{pmatrix} 1 & 0 \\ 0 & 3 \end{pmatrix}.$$

Answer 8.2.27 Let $S \in B(\ell^2)$ be the shift operator. Then

$$\overline{\sigma_p(S)} = \varnothing \subsetneq \sigma_a(S) = \{\lambda \in \mathbb{C} : |\lambda| = 1\} \subsetneq \sigma(S) = \{\lambda \in \mathbb{C} : |\lambda| \leq 1\}.$$

Answer 8.2.28 For example, consider

$$T = \begin{pmatrix} 1 & 0 \\ 0 & 0 \end{pmatrix}$$

and so $\sigma(T) = \{0, 1\}$. Then

$$1 = |-1| \in \sigma(T) = \sigma(|T|)$$

and yet $-1 \notin \sigma(T)$.

Answer 8.2.29 On $\mathbb{C}^2$, take

$$T = \begin{pmatrix} 0 & 2 \\ 1/2 & 0 \end{pmatrix}.$$

Then $\sigma(T) = \{1, -1\}$. In addition,

$$T^*T = \begin{pmatrix} 1/4 & 0 \\ 0 & 4 \end{pmatrix}, \text{ and so } |T| = \begin{pmatrix} 1/2 & 0 \\ 0 & 2 \end{pmatrix}.$$

Thus e.g. $1 \in \sigma(T)$, but $1 \notin \sigma(|T|)$.

Remark This example is telling us that, in general:

$$\sigma(|T|) \neq \{|\lambda| : \lambda \in \sigma(T)\}.$$

Answer 8.2.30 Let $H = \ell^2$. Define a sequence (A_m) by

$$A_m(x_n)_n = (x_1, x_2, \cdots, x_m, 0, 0, \cdots).$$

Since each A_m is a nontrivial orthogonal projection, it is clear that $\sigma(A_m) = \{0, 1\}$. Now, let $x \in \ell^2$. Then

$$\|A_m x - Ix\|^2 = \sum_{k=m+1}^{\infty} |x_k|^2 \longrightarrow 0$$

as $m \to \infty$ (for this is a tail of a convergent series). Thus, $A_m \to I$ strongly and obviously I is normal. Finally, remember that $\sigma(I) = \{1\}$

Answer 8.2.31 As suggested, we need to avoid normal operators. Consider again the sequence of matrices A_n (which appeared in Answer 5.2.10) of order n defined by

$$A_n = \begin{pmatrix} 0 & 0 & . & . & . & 0 \\ 1 & 0 & 0 & . & & \\ 0 & 1 & 0 & & . & \\ & . & . & . & 0 & 0 & . \\ & . & & & . & . & . \\ 0 & & & 0 & 1 & 0 \end{pmatrix}.$$

Setting $T_n = I_n - A_n$, we easily see that $\sigma(T_n) = \{1\}$ for each n, leading to

$$\overline{\bigcup_{n \in \mathbb{N}} \sigma(T_n)} = \overline{\{1\}} = \{1\}.$$

Since we already know from Answer 5.2.10 that T is not invertible, we have $0 \in \sigma(T)$. Therefore, this *violates* the equality

$$\overline{\bigcup_{n \in \mathbb{N}} \sigma(T_n)} = \sigma(T),$$

as desired.

Answer 8.2.32 Consider again the sequence of nilpotent matrices A_n given in Answer 8.2.31. Then A_n is nilpotent and also $(\|A_n\|)_n$ is bounded, both statements holding for each n.

Nonetheless, $A = \bigoplus_{n \in \mathbb{N}} A_n$ is not nilpotent. One reason preventing A from being nilpotent is that we shall see in Answer 15.1.13 that $\sigma(A) = \{\lambda \in \mathbb{C} : |\lambda| \leq 1\}$, which is not consistent with the known fact that the spectrum of any nilpotent operator is always equal to $\{0\}$.

Chapter 9
Spectral Radius, Numerical Range

9.1 Basics

Definition 9.1.1 Let A be in $B(H)$. The spectral radius of A is defined as

$$r(A) = \sup\{|\lambda| : \lambda \in \sigma(A)\}.$$

Proposition 9.1.1 (Gelfand-Beurling Formula) *Let $A \in B(H)$. Then the limit* $\lim_{n\to\infty} \|A^n\|^{\frac{1}{n}}$ *exists, and we have*

$$\lim_{n\to\infty} \|A^n\|^{\frac{1}{n}} = \inf_{n\in\mathbb{N}} \|A^n\|^{\frac{1}{n}} = r(A).$$

Remark It is straightforward to see that $r(A) \leq \|A\|$.

Theorem 9.1.2 (Spectral Radius Theorem) *Let $A \in B(H)$ be normal. Then*

$$r(A) = \|A\|.$$

Corollary 9.1.3 *Let $A \in B(H)$ be normal and let $\alpha \in \mathbb{C}$. If $\sigma(A) = \{\alpha\}$, then $A = \alpha I$.*

We finish by introducing an important subset of $\mathbb{C}$.

Definition 9.1.2 Let $A \in B(H)$. The numerical range of A is defined by

$$W(A) = \{\langle Ax, x \rangle : x \in H, \|x\| = 1\}.$$

Theorem 9.1.4 (Toeplitz-Hausdorff) *Let $A \in B(H)$. Then $W(A)$ is a convex set in the complex plane.*

In the end, we give the numerical range of a multiplication operator.

M. H. Mortad, *Counterexamples in Operator Theory*,
https://doi.org/10.1007/978-3-030-97814-3_9

Proposition 9.1.5 *Let H be a separable Hilbert space and let $A \in B(H)$ be the diagonal operator defined for all n by*

$$Ae_n = \lambda_n e_n,$$

where $(\lambda_n)_n$ is a bounded sequence of complex numbers and (e_n) is an orthonormal basis of H. Then

$$W(A) = \mathrm{conv}(\{\lambda_n\}_n)$$

(where "conv" designates the convex hull).

Remark The previous result is important in the sense that it may be used to find interesting counterexamples. For example, let A be a self-adjoint operator on ℓ^2 such that $\sigma(A) = [0, 1]$. Then depending on whether 0 or 1 or both or none of them appear on the diagonal, it is clear that the numerical range of A can be any of the following intervals: $(0, 1]$ or $[0, 1)$ or $(0, 1)$ or $[0, 1]$.

9.2   Questions

9.2.1   An $A \in B(H)$ Such That $r(A) < \|A\|$

Question 9.2.1 Give an example of an $A \in B(H)$ such that $r(A) < \|A\|$.

9.2.2   A Quasinilpotent Operator Which Is Not Nilpotent

The next notion is a generalization of nilpotence: Let $A \in B(H)$. We say that A is quasinilpotent if

$$\lim_{n \to \infty} \|A^n\|^{\frac{1}{n}} = 0.$$

Equivalently, this amounts to say that $\sigma(A) = \{0\}$.

It is not hard to see that when $\dim H < \infty$, then $T \in B(H)$ is nilpotent if and only if T is quasinilpotent.

Question 9.2.2 Find a quasinilpotent operator which is not nilpotent.

9.2.3 A Non-normaloid Operator T Such That Re T ≥ 0

Many classes of operators (e.g. normal operators) have a spectral radius equal to their norms. So, we give the following definition: An operator $A \in B(H)$ satisfying

$$r(A) = \|A\|$$

is called normaloid . It may also be shown that $A \in B(H)$ is normaloid iff $\|A^n\| = \|A\|^n$ for all n.

It was established in [261] that if $T \in B(H)$ is such that $T^2 = 0$ and Re $T \geq 0$ (or Im $T \geq 0$), then T is normal, whereby $T = 0$. This result was generalized to the case $T^n = 0$ in say [113] or [114].

Remark Notice that the results above might be known to some specialists, however, different approaches were adopted to them.

Question 9.2.3 Find a $T \in B(H)$ with Re $T \geq 0$ (or Im $T \geq 0$) yet T is not normaloid.

9.2.4 An Operator A with A ≠ I, σ(A) = {1} and ‖A‖ = 1

Let A be a matrix such that $\sigma(A) = \{1\}$. Obviously, A need not be equal to I. In fact, such a matrix need not even be normal, e.g.

$$A = \begin{pmatrix} 1 & 0 \\ 1 & 1 \end{pmatrix}.$$

What if we assume that A has all its eigenvalues equal to 1 and that A diagonalizable? The answer now is yes. Indeed, in such a case

$$T^{-1}AT = I$$

for some invertible matrix T. Therefore, $A = I$.

The norm of the matrix above is equal to 1.64. This is not special to this matrix. In fact, each matrix A is unitarily equivalent to an upper triangular matrix (Schur decomposition), denoted here by R. So if $\sigma(A) = \{1\}$, then R has only "ones" on the main diagonal, i.e. $\sigma(A) = \sigma(R)$ and so

$$1 = r(A) \leq \|A\| = \|R\|.$$

Clearly, the case $\|A\| = 1$ can only occur in the event $A = I$. Is this a purely finite-dimensional phenomenon?

Question 9.2.4 Using Volterra's operator or else, find an example of a bounded operator A such that $A \neq I$, $\sigma(A) = \{1\}$ and $\|A\| = 1$.

9.2.5   No Inequality Between $r(AB)$ and $r(A)r(B)$ Need to Hold

Let $r(\cdot)$ denote the spectral radius defined as a function from $B(H)$ into $\mathbb{R}^+$. Let us state some basic properties of the spectral radius as regards products and sums.

Proposition 9.2.1 *Let A and B be two operators on a Hilbert space. Then*

$$r(AB) = r(BA).$$

Proposition 9.2.2 *Let $A, B \in B(H)$ be commuting. Then*

$$r(A + B) \leq r(A) + r(B).$$

Proposition 9.2.3 *Let $A, B \in B(H)$ be commuting. Then*

$$r(AB) \leq r(A)r(B).$$

Question 9.2.5 Let $A, B \in B(H)$. Show that, in general, no inequality between $r(AB)$ and $r(A)r(B)$ is possible.

9.2.6   The Discontinuity of $A \mapsto r(A)$ with Respect to the S.O.T.

Question 9.2.6 Let $r(\cdot)$ denote the spectral radius. Show that r is not continuous w.r.t. the strong operator topology, i.e. find a sequence (A_n) in $B(H)$ such that $A_n \to A$ strongly but $r(A_n) \nrightarrow r(A)$.

9.2.7   Two Commuting Normaloid Operators with a Non-normaloid Product

Question 9.2.7 Show that the product of two normaloid operators need not remain normaloid, even when they commute.

9.2.8   Two Commuting Self-Adjoint Operators A and B Such That $W(A + B) \not\supset W(A) + W(B)$

If $A, B \in B(H)$, then it may be shown that

$$W(A + B) \subset W(A) + W(B).$$

Question 9.2.8 Find two commuting self-adjoint matrices A and B such that

$$W(A + B) \not\supset W(A) + W(B)$$

(the usual addition of sets).

9.2.9   A Bounded Operator Whose Numerical Range Is Not a Closed Set in $\mathbb{C}$

Let $A \in B(H)$. We know that on finite-dimensional spaces, $W(A)$ is always compact in $\mathbb{C}$, hence it is closed (see e.g. Proposition 8.1.18 in [256]). This, however, may not be carried over to infinite-dimensional spaces.

Question 9.2.9 Find a bounded operator whose numerical range is not a closed set in $\mathbb{C}$.

9.2.10   Neither $\sigma(A) \subset W(A)$ Nor $W(A) \subset \sigma(A)$ Need to Hold, Even for Self-Adjoint Operators

A well known result states that

$$\sigma(A) \subset \overline{W(A)}$$

whenever $A \in B(H)$. By an elementary topological observation, we know that $\sigma(A) \subset W(A)$ for any finite matrix A.

Question 9.2.10 Provide a self-adjoint operator A such that $\sigma(A) \not\subset W(A)$, and another self-adjoint A with $W(A) \not\subset \sigma(A)$.

9.2.11   Two Self-Adjoint $A, B \in B(H)$ with Equal Spectra but Different Numerical Ranges

Question 9.2.11 Give two self-adjoint $A, B \in B(H)$ with equal spectra and different numerical ranges.

9.2.12   On the Failure of $\sigma(AB) \subset \overline{W(A)} \cdot \overline{W(B)}$

Let A and B be in $B(H)$. In view of $\sigma(A) \subset \overline{W(A)}$, it seems reasonable to conjecture that

$$\sigma(AB) \subset \overline{W(A)} \cdot \overline{W(B)}$$

where $\overline{W(A)} \cdot \overline{W(B)} = \{\lambda\mu : \lambda \in \overline{W(A)}, \mu \in \overline{W(B)}\}$? Does it help to further assume that $0 \notin \overline{W(A)} \cup \overline{W(B)}$?

Question 9.2.12 Show that this is not true even when dim $H = 2$.

9.2.13 On the Failure of $\sigma(A^{-1}B) \subset \overline{W(B)}/\overline{W(A)}$ for An Invertible A

In [383], J. P. Williams showed that if $A, B \in B(H)$ and if $0 \notin \overline{W(A)}$, then

$$\sigma(A^{-1}B) \subset \overline{W(B)}/\overline{W(A)}$$

where the set on the right is by definition the set of quotients b/a where $b \in \overline{W(B)}$ and $a \in \overline{W(A)}$.

Observe that the condition $0 \notin \overline{W(A)}$ is stronger than the invertibility of A.

Question 9.2.13 Show that the result above may fail to hold if A is only assumed to be invertible (i.e. without assuming $0 \notin \overline{W(A)}$ anymore).

9.2.14 A Nonnormal Operator A with $\overline{W(A)} = \text{conv}(\sigma(A))$

It is a well known result that if $A \in B(H)$ is normal, then

$$\overline{W(A)} = \text{conv}(\sigma(A))$$

where $\text{conv}(\sigma(A))$ is the convex hull of $\sigma(A)$. See e.g. Problem 216 in [148]. What about the converse? It is shown in [267] that the converse also holds, but only when $\dim H \leq 4$.

Question 9.2.14 (⊛) Find a nonnormal 5×5 complex matrix A such that $W(A) = \text{conv}(\sigma(A))$.

Answers

Answer 9.2.1 Let

$$A = \begin{pmatrix} 0 & 1 \\ 0 & 0 \end{pmatrix}.$$

Then

$$\|A\|^2 = \|A^*A\| = 1.$$

The eigenvalues of A are just zero and so

$$r(A) = 0 < 1 = \|A\|.$$

Answer 9.2.2 Let V be the Volterra operator defined on $L^2[0, 1]$ by

$$(Vf)(x) = \int_0^x f(t)dt.$$

First, recall that $\sigma(V) = \{0\}$ (see e.g. Exercise 9.3.21 in [256]), that is, V is quasinilpotent. It may also be shown that

$$V^n f(x) = \frac{1}{(n-1)!} \int_0^x (x-t)^{n-1} f(t)dt, \ n \in \mathbb{N}.$$

(we leave the proof by induction to interested readers.) By considering $f(t) = 1$ say, it is seen that the Volterra operator is not nilpotent.

Remark Another example is to consider the weighted shift with weights n^{-1}.

Answer 9.2.3 If V is the Volterra operator on $L^2[0, 1]$, then

$$\|V\| = \frac{2}{\pi}.$$

Hence V is not normaloid as

$$r(V) = 0 \neq \frac{2}{\pi} = \|V\|$$

where $r(V)$ denotes the spectral radius of V. To conclude, we must show that $(V + V^*)/2$ is positive. Let $f \in L^2[0, 1]$ (and let $x \in [0, 1]$). We have

$$(V + V^*)f(x) = Vf(x) + V^*f(x) = \int_0^x f(t)dt + \int_x^1 f(t)dt = \int_0^1 f(t)dt$$

so that

$$\langle (V + V^*)f, f \rangle = \int_0^1 \left(\int_0^1 f(t)dt \right) \overline{f(x)}dx$$

$$= \int_0^1 \int_0^1 f(t)\overline{f(x)}dtdx$$

$$= \int_0^1 f(t)dt \int_0^1 \overline{f(x)}dx$$

$$= \int_0^1 f(t)dt \overline{\int_0^1 f(x)dx}$$

$$= \left| \int_0^1 f(t)dt \right|^2 \geq 0.$$

Accordingly, Re $V \geq 0$, as needed.

Answer 9.2.4 (Cf. [148]) Consider again the Volterra operator V defined on $L^2[0, 1]$ by

$$(Vf)(x) = \int_0^x f(t)dt$$

(remember that $\sigma(V) = \{0\}$). Set

$$A = (I + V)^{-1}.$$

Then $I + V$ is invertible, and it is evident that $A \neq I$. Since $\sigma(V + I) = \{1\}$, we may write

$$\sigma(A) = \left\{ \lambda^{-1} : \lambda \in \sigma(V + I) \right\} = \{1\}.$$

It only remains to show that $\|A\| = 1$. One inequality is immediate as

$$r(A) = \sup\{|\lambda| : \lambda \in \sigma(A)\} = 1 \leq \|A\|.$$

To show the reverse inequality, we may show that $\|Af\| \leq \|f\|$ for all f, or equivalently,

$$\|A^{-1}f\| \geq \|f\|, \ \forall f \in L^2[0, 1].$$

Let $f \in L^2[0, 1]$. We then have

$$\begin{aligned}
\|A^{-1}f\|^2 &= \|(V + I)f\|^2 \\
&= \langle (V + I)f, (V + I)f \rangle \\
&= \|Vf\|^2 + \|f\|^2 + \langle Vf, f \rangle + \langle V^*f, f \rangle \\
&= \|Vf\|^2 + \|f\|^2 + \langle (V + V^*)f, f \rangle \\
&\geq 0
\end{aligned}$$

since $V + V^*$ is positive (see Answer 9.2.3).

Remark Another example is to consider the weighted shift with weights $(n-1)/n$.

Answer 9.2.5 ([135]) There are counterexamples in spaces as simple as $\mathbb{C}^2$. Take

$$A = \begin{pmatrix} 1 & 0 \\ 0 & 0 \end{pmatrix} \text{ and } B = \begin{pmatrix} 0 & 0 \\ 0 & 1 \end{pmatrix}.$$

Then A and B are even orthogonal projections and so $r(A) = r(B) = 1$. Since $AB = \begin{pmatrix} 0 & 0 \\ 0 & 0 \end{pmatrix}$, $r(AB) = 0$ whereby

$$r(A)r(B) \not\leq r(AB).$$

To deal with the other inequality, let

$$A = \begin{pmatrix} 0 & 1 \\ 0 & 0 \end{pmatrix} \text{ and } B = \begin{pmatrix} 0 & 0 \\ 1 & 0 \end{pmatrix}$$

and so $r(A) = r(B) = 0$. Since $AB = \begin{pmatrix} 1 & 0 \\ 0 & 0 \end{pmatrix}$, it results that $r(AB) = 1$, thereby

$$r(A)r(B) \not\geq r(AB).$$

Answer 9.2.6 (Cf. [148]) Let S be the usual shift operator on ℓ^2 and take $A_n = (S^*)^n$. Then $(S^*)^n$ converges strongly to $A := 0$. But $r(A_n) = 1$ for all n, and so

$$r(A_n) \not\to 0 = r(A),$$

as needed.

Remark It is worth mentioning that the spectral radius is discontinuous even with respect to the norm topology. A famous example was given by S. Kakutani (see e.g. [148]).

Answer 9.2.7 ([135]) Consider on a two-dimensional space:

$$T = \begin{pmatrix} 0 & 0 \\ 1 & 0 \end{pmatrix}.$$

Hence, $\|T\| = 1$ and also $\sigma(T) = \{0\}$. Now, set

$$A = \begin{pmatrix} I_2 & 0 \\ 0 & T \end{pmatrix} \text{ and } B = \begin{pmatrix} T & 0 \\ 0 & I_2 \end{pmatrix},$$

where the zeros now are the zero matrices over $\mathbb{C}^2$. It is clear that

$$r(A) = \|A\| = 1 \text{ and } r(B) = \|B\| = 1,$$

meaning that both A and B are normaloid.

The product AB, however, is not normaloid. To see that, first observe that

$$AB = \begin{pmatrix} I_2 & 0 \\ 0 & T \end{pmatrix} \begin{pmatrix} T & 0 \\ 0 & I_2 \end{pmatrix} = \begin{pmatrix} T & 0 \\ 0 & T \end{pmatrix}.$$

Therefore

$$r(AB) = 0 \neq 1 = \|AB\|,$$

and yet A commutes with B.

Answer 9.2.8

$$A = \begin{pmatrix} 1 & 0 \\ 0 & 0 \end{pmatrix} \text{ and } B = \begin{pmatrix} 0 & 0 \\ 0 & 1 \end{pmatrix}.$$

Both A and B are nontrivial orthogonal projections. Hence,

$$W(A) = W(B) = [0, 1].$$

But

$$A + B = I \implies W(A + B) = \{1\}.$$

Therefore,

$$W(A + B) \not\supset W(A) + W(B).$$

Answer 9.2.9 Let S be the unilateral shift operator on ℓ^2. Then (see Exercise 8.3.14 in [256])

$$W(S) = B(0, 1) = \{\lambda \in \mathbb{C} : |\lambda| < 1\}.$$

Obviously, $W(S)$ is not closed.

Answer 9.2.10 By the remark below Proposition 9.1.5, we may have a self-adjoint A with spectrum $[0, 1]$, and with numerical range e.g. $(0, 1]$. This shows that

$$\sigma(A) \not\subset W(A).$$

A similar argument applies to show the failure of the other inclusion. Indeed, if A is a self-adjoint multiplication operator on ℓ^2 (multiplication by $(\frac{1}{n})$, say), then

$$\sigma(A) = \overline{\left\{\frac{1}{n} : n \in \mathbb{N}\right\}} = \{0\} \cup \left\{\frac{1}{n} : n \in \mathbb{N}\right\} \not\supseteq W(A) = (0, 1].$$

Answer 9.2.11 ([214]) On ℓ^2, define

$$A(x_1, x_2, x_3, \cdots) = (x_1, \frac{1}{2}x_2, \frac{1}{3}x_3, \cdots)$$

and

$$B(x_1, x_2, x_3, \cdots) = (0, x_2, \frac{1}{2}x_3, \frac{1}{3}x_4, \cdots).$$

Then A and B are clearly self-adjoint,

$$\sigma(A) = \sigma(B) = \overline{\left\{\frac{1}{n} : n \in \mathbb{N}\right\}} = \left\{\frac{1}{n} : n \in \mathbb{N}\right\} \cup \{0\}$$

while

$$W(A) = (0, 1] \neq W(B) = [0, 1].$$

Answer 9.2.12 ([383]) On $\mathbb{C}^2$, let

$$A = \begin{pmatrix} 1 & 0 \\ 1 & 1 \end{pmatrix}.$$

It is fairly easy to show that

$$W(A) = W(A^*) = \{z \in \mathbb{C} : |z - 1| \leq 1/2\}$$

(this is made even easier by consulting Proposition 8.1.16 and Exercise 8.3.11 in [256]). Observe that $0 \notin W(A)$ and $0 \notin W(A^*)$ (remember that in $\mathbb{C}^2$ both $W(A)$ and $W(A^*)$ are evidently closed subsets). The set $W(A) \cdot W(A^*)$ lies to the left of $\operatorname{Re} z = 9/4$. It can also be checked that $(3 + \sqrt{5})/2$ is an eigenvalue for AA^* but $9/4 < (3 + \sqrt{5})/2$, that is

$$\sigma(AA^*) \not\subset W(A) \cdot W(A^*),$$

as wished.

Answer 9.2.13 On $\mathbb{C}^2$ say, requiring $\sigma(A^{-1}B) \subset \overline{W(B)}/\overline{W(A)}$ is just equivalent to requiring $\sigma(A^{-1}B) \subset W(B)/W(A)$. We are done if we have a pair of two self-adjoint matrices whose product (the invertible matrix being on the left) has only complex eigenvalues. More explicitly, consider the two self-adjoint matrices

$$A = \begin{pmatrix} 0 & i \\ -i & 0 \end{pmatrix} \text{ and } B = \begin{pmatrix} 0 & 1 \\ 1 & 0 \end{pmatrix}$$

(where A is also invertible and $A^{-1} = A$). Then

$$A^{-1}B = \begin{pmatrix} 0 & i \\ -i & 0 \end{pmatrix}\begin{pmatrix} 0 & 1 \\ 1 & 0 \end{pmatrix} = \begin{pmatrix} i & 0 \\ 0 & -i \end{pmatrix}.$$

Thus,

$$\sigma(A^{-1}B) \not\subset W(B)/W(A)$$

as $W(B)/W(A)$ is a subset of $\mathbb{R}$.

Answer 9.2.14 ([178]) Let

$$A = \begin{pmatrix} B & 0 \\ 0 & C \end{pmatrix}$$

where B is a normal 3×3 matrix and C is a nonnormal 2×2 matrix. Choose B having three non-collinear eigenvalues and choose $C = \begin{pmatrix} 0 & 1 \\ 0 & 0 \end{pmatrix}$ and such that $W(C) \subset W(B)$. Since we know that $W(C) = \{z \in \mathbb{C} : |z| \leq 1/2\}$ (see e.g. Exercise 8.3.11 in [256]), we may choose, for instance, the eigenvalues of B to be $2i, 2 - 2i$, and $-2 - 2i$.

Hence

$$W(A) = W(B \oplus C) = \text{conv}(W(B) \cup W(C)) = \text{conv}(W(B)) = W(B).$$

But

$$W(B) = \text{conv}(\sigma(B) \cup \sigma(C)) = \text{conv}(\sigma(B \oplus C)) = \text{conv}(\sigma(A)).$$

Thus, $W(A) = \text{conv}(\sigma(A))$ but A is not normal for C is not normal.

Chapter 10
Compact Operators

10.1 Basics

Definition 10.1.1 Let H and H' be two Hilbert spaces and let $A \in L(H, H')$. We say that A is compact if for any sequence (x_n) in H with $\|x_n\| = 1$, (Ax_n) has a convergent subsequence in H'.

The set of compact operators from H into H' will be denoted by $K(H, H')$. In particular, $K(H) := K(H, H)$.

Examples 10.1.1

(1) Let $A \in L(H)$ be such that $Ax = 0$ for all $x \in H$. Obviously, $A \in K(H)$.
(2) When $\dim H < \infty$, any $A \in B(H)$ is in $K(H)$.
(3) The identity operator on H is not compact unless $\dim H < \infty$.

Proposition 10.1.1 *Let H be a Hilbert space and let $A \in K(H)$. Then $A \in B(H)$. That is,*

$$K(H) \subset B(H).$$

Proposition 10.1.2

(1) *The sum of two compact operators is compact.*
(2) *The products AB and BA are compact if A is compact and B is bounded.*
(3) *A is compact iff A^* is compact.*

Corollary 10.1.3 *Let $A \in K(H)$ where H is infinite-dimensional. Then A is not invertible.*

Definition 10.1.2 Let A be a linear operator. We say that A has finite rank if $\dim[\mathrm{ran}(A)]$ is finite.

Proposition 10.1.4 *If $A \in B(H)$ is of a finite rank, then A is compact.*

© The Author(s), under exclusive license to Springer Nature Switzerland AG 2022
M. H. Mortad, *Counterexamples in Operator Theory*,
https://doi.org/10.1007/978-3-030-97814-3_10

Corollary 10.1.5 *If (A_n) is a sequence of bounded and finite rank operators such that $\|A_n - A\| \to 0$ when n goes to ∞, where $A \in B(H)$, then $A \in K(H)$.*

Conversely, if $A \in K(H)$, then there exists a sequence of finite rank operators (A_n) which converges to $A \in B(H)$.

10.2   Questions

10.2.1   A Non-compact A Such That A^2 Is Compact

Recall that if $A \in B(H)$ is compact, then AB is always compact for each $B \in B(H)$. In particular, if A is compact, so are its powers A^n, $n \in \mathbb{N}$. What about the converse?

Question 10.2.1 Give an example of a non-compact $A \in B(H)$ such that A^2 is compact.

10.2.2   A Non-compact Operator A Such That $\|Ae_n\| \to 0$ as $n \to \infty$ Where (e_n) Is An Orthonormal Basis

It is well known to readers that if A is a compact operator on an *infinite*-dimensional separable Hilbert space H with an orthonormal basis (e_n), then

$$\lim_{n \to \infty} \|Ae_n\| = 0.$$

See e.g. Exercise 6.7 in [315]. The converse does not always hold as we shall shortly see. It is worth noticing that the converse holds in the following form: If $\lim_{n \to \infty} \|Ae_n\| = 0$ for *any* orthonormal sequence in H, then A is compact. See [107] (cf. [12]).

Question 10.2.2 Show a non-compact operator $A \in B(H)$ such that $\|Ae_n\| \to 0$ as $n \to \infty$ where (e_n) is some orthonormal basis in H.

10.2.3 Does the Shift Operator Commute with a Nonzero Compact Operator?

Let S be the shift operator on ℓ^2. Obviously, S commutes with the zero operator (which is compact). Is there any compact nonzero operator which commutes with S?

Question 10.2.3 Show that the shift operator does not commute with any nonzero compact operator.

10.2.4 A Compact Operator Which Is Not Hilbert-Schmidt

Let H be an infinite-dimensional Hilbert space and let (e_n) be an orthonormal basis in H. If $A \in B(H)$ is such that

$$\sum_{n=1}^{\infty} \|Ae_n\|^2 < \infty,$$

then A is called a Hilbert-Schmidt operator.

Theorem 10.2.1 *Hilbert-Schmidt operators are compact.*

The following result says that the definition does not depend on the choice of the orthonormal basis, which is quite practical.

Theorem 10.2.2 *Let H be an infinite-dimensional Hilbert space and let (e_n) and (f_n) be two orthonormal bases in H. Let $A \in B(H)$. Then*

$$\sum_{n=1}^{\infty} \|Af_n\|^2 = \sum_{n=1}^{\infty} \|Ae_n\|^2 = \sum_{n=1}^{\infty} \|A^*e_n\|^2$$

(in the sense that all three quantities can be infinite simultaneously or finite simultaneously).

The primary application of the Hilbert-Schmidt theory is that an integral operator on an adequate setting is Hilbert-Schmidt.

Theorem 10.2.3 *Let $K \in L^2((a, b) \times (c, d))$ and define an operator $A : L^2(a, b) \to L^2(c, d)$ by*

$$(Af)(x) = \int_c^d K(x, y) f(y) dy, \ a < x < b.$$

Then A is Hilbert-Schmidt (and so compact).

Example 10.2.1 The Volterra operator V defined on $L^2(0, 1)$ by

$$(Vf)(x) = \int_0^x f(t)dt$$

(with $0 < x < 1$) is therefore compact.

Question 10.2.4 Give an example of a compact operator which is not Hilbert-Schmidt.

10.2.5 A Compact Operator Without Eigenvalues

The next theorem gives some information about the spectrum of compact operators.

Theorem 10.2.4 *Let $A \in B(H)$ be compact. Then*

(1) *If $\dim H = \infty$, then $0 \in \sigma(A)$.*
(2) *Also,*

$$\sigma(A) = \sigma_p(A) \cup \{0\}.$$

(3) *$\sigma(A)$ is countable. In case $\sigma(A)$ is infinite, then the eigenvalues (λ_n) may be arranged so that for all n*

$$|\lambda_{n+1}| \leq |\lambda_n| \text{ and } \lim_{n\to\infty} \lambda_n = 0.$$

Example 10.2.2 Let V be the Volterra operator defined on $L^2[0, 1]$. Then $\sigma(V) = \{0\}$ (see e.g. Exercise 9.3.21 in [256]).

Question 10.2.5 Give a compact operator having no eigenvalues.

10.2.6 A Non-positive Integral Operator Having a Positive Kernel

In general, an integral operator with a positive "self-adjoint" kernel K (i.e. $K(x, y) = \overline{K(y, x)}$ and $K(x, y) \geq 0$ for all x, y) does not have to be positive as shall be seen below.

It is worth recalling, however, a known sufficient condition for an integral operator to be positive. For instance, assume that $K \in C([0, 1] \times [0, 1])$ is such that $K(x, y) = \overline{K(y, x)}$ for all $x, y \in [0, 1]$. If K also satisfies the so-called Bochner condition, i.e.

$$\sum_{1 \leq i, j \leq n} K(x_i, x_j) z_i \overline{z_j} \geq 0, \ \forall x_1, \cdots, x_n \in [0, 1]; \forall z_1, \cdots, z_n \in \mathbb{C},$$

then $T : L^2[0, 1] \rightarrow L^2[0, 1]$ defined by $Tf(x) = \int_0^1 K(x, y) f(y) dy$ is a positive operator. See e.g. Exercise VIII.5 in [52].

Question 10.2.6 Find an integral self-adjoint operator $A \in B(H)$ with a positive kernel, yet $A \not\geq 0$.

10.2.7 On λ-Commutativity Related to Compact and Finite Rank Operators

We have already defined commuting and anti-commuting operators. These two definitions can be put into one more general definition, namely: Let λ be some scalar. Say that $S, T \in B(H)$ are λ-commuting if $ST = \lambda TS$.

Now, set (cf. [71]):

$$\mathcal{C}_1 = \{T \in B(H) : \text{there is a finite rank } S \in B(H), \ S \neq 0 : TS = \lambda ST\},$$

$$\mathcal{C}_2 = \{T \in B(H) : \text{there is a compact } S \in B(H), \ S \neq 0 : TS = \lambda ST\}$$

and

$$\mathcal{C}_3 = \{T \in B(H) : \text{there is an } S \in B(H), \ S \neq 0 : TS = \lambda ST\}.$$

It is relatively easy to see that if T is in $\mathcal{C}_1$, then $\sigma_p(T) \neq \varnothing$. It is also plain that $\mathcal{C}_1 \subset \mathcal{C}_2 \subset \mathcal{C}_3$.

Question 10.2.7 Show that $\mathcal{C}_2 \not\subset \mathcal{C}_1$ and that $\mathcal{C}_3 \not\subset \mathcal{C}_2$.

10.2.8   The Infinite Direct Sum of Compact Operators Need Not Be Compact

Question 10.2.8 Let (A_n) be a sequence of bounded operators on H such that $(\|A_n\|)_n$ is bounded. Set $A = \bigoplus_{n\in\mathbb{N}} A_n$. Show that even when *all* A_n are compact, A need not remain compact.

  Hint: Remember that (see e.g. Exercise 13 on Page 46 in [66]) $\bigoplus_{n\in\mathbb{N}} A_n$ is compact if and only if each A_n is compact *and* $\|A_n\| \to 0$.

10.2.9   A Compact Quasinilpotent Operator A with $0 \notin W(A)$

Recall that if A is a quasinilpotent operator, then $\sigma(A) = \{0\}$ and so $0 \in \overline{W(A)}$.

Question 10.2.9 Find a compact quasinilpotent $A \in B(H)$ with $0 \notin W(A)$.

Answers

Answer 10.2.1 Let $A \in B(\ell^2)$ be defined by

$$A(x_1, x_2, x_3, \cdots) = (0, x_1, 0, x_3, \cdots).$$

  Then A^2 is compact for $A^2 = 0$. But, A cannot be compact. Indeed, if it were, so would be A^*A and AA^* and hence so would be $A^*A + AA^*$ too. But as one may easily check, $A^*A + AA^* = I$, i.e. I would be compact. Since dim $\ell^2 = \infty$, it results that A is not compact, as wished.
  Let us give another example. Before that, recall that if $A_{ij} \in B(H)$, where $i, j = 1, 2$, then $A = \begin{pmatrix} A_{11} & A_{12} \\ A_{21} & A_{22} \end{pmatrix}$ *is compact on $B(H \oplus H)$ iff each of the four operators A_{ij} is compact.* Again, and for the purpose of the counterexample, it is primordial to work on an infinite-dimensional space. Let $I \in B(\ell^2)$ be the identity operator (which is not compact), then consider

$$A = \begin{pmatrix} 0 & I \\ 0 & 0 \end{pmatrix}$$

(defined on $\ell^2 \oplus \ell^2$). Then A is not compact while

$$A^2 = \begin{pmatrix} 0 & 0 \\ 0 & 0 \end{pmatrix}$$

is clearly compact.

Remark As baffling as it may sound, there are *unbounded* operators having compact squares! See Question 22.2.1 for details.

Answer 10.2.2 ([148]) On ℓ^2, let

$$A = \begin{pmatrix} 1 & 0 & 0 & 0 & 0 & 0 & \cdots \\ 0 & 1/2 & 1/2 & 0 & 0 & 0 & \cdots \\ 0 & 1/2 & 1/2 & 0 & 0 & 0 & \cdots \\ 0 & 0 & 0 & 1/3 & 1/3 & 1/3 & \cdots \\ 0 & 0 & 0 & 1/3 & 1/3 & 1/3 & \cdots \\ 0 & 0 & 0 & 1/3 & 1/3 & 1/3 & \cdots \\ \vdots & \vdots & \vdots & & & & \ddots \end{pmatrix}.$$

Then A is bounded and non-compact (it is a direct sum of an infinite sequence of projections). In the end, we leave it to readers to check that

$$\lim_{n \to \infty} \|Ae_n\| = 0.$$

Answer 10.2.3 Let S be the shift operator on ℓ^2, i.e. $Se_n = e_{n+1}$ where (e_n) is some basis in ℓ^2. Let $K \in B(\ell^2)$ be any compact operator which commutes with S, that is, $SK = KS$. We show that the only possible outcome is $K = 0$.

Since S is an isometry, it follows that

$$\|Ke_n\| = \|SKe_n\| = \|KSe_n\| = \|Ke_{n+1}\|, \ n = 1, 2, 3, \cdots$$

Hence

$$\|Ke_1\| = \|Ke_2\| = \cdots = \|Ke_n\| = \cdots$$

In other words, the real sequence $n \mapsto \|Ke_n\|$ is constant. By the compactness of K, we know that $\lim_{n \to \infty} \|Ke_n\| = 0$. Therefore, each of the terms of the sequence is null. Accordingly,

$$Ke_n = 0, \ n = 1, 2, \cdots$$

and so $K = 0$, as needed.

Remark Does the shift operator anti-commute with a nonzero compact operator? By using a similar method as above, and assuming that $SK = -KS$, it may again be shown that the only possible conclusion is $K = 0$.

Answer 10.2.4 Take the operator A defined on ℓ^2 by:

$$A(x_1, x_2, \cdots, x_n, \cdots) = \left(x_1, \frac{1}{\sqrt{2}} x_2, \cdots, \frac{1}{\sqrt{n}} x_n, \cdots \right).$$

Since $\lim_{n \to \infty} 1/\sqrt{n} = 0$, A is compact (see e.g. Exercise 9.3.5 in [256]).

It remains to check that A is not Hilbert-Schmidt. Let (e_n) be the standard orthonormal basis of ℓ^2. It is seen that

$$\sum_{n=1}^{\infty} \|Ae_n\|^2 = \sum_{n=1}^{\infty} \frac{1}{n} = \infty,$$

showing that A is not Hilbert-Schmidt

Answer 10.2.5 Let V be the Volterra operator defined on $L^2[0, 1]$ by

$$(Vf)(x) = \int_0^x f(t)dt$$

(with $0 < x < 1$). Then V is compact and $\sigma(V) = \{0\}$. Hence if λ is an eigenvalue for V, then it is necessarily equal to 0. We claim that V is one-to-one. Indeed, observe that V is absolutely continuous (see Definition B.5 below for a definition) and hence its derivative exists almost everywhere. Then $Vf = 0$ leads to $f = 0$ on L^2, i.e. V is injective. Thus, $\sigma_p(V) = \varnothing$.

Answer 10.2.6 ([52]) Define $A \in B(L^2[0, 1])$ by

$$Af(x) = \int_0^1 K(x, y)f(y)dy,$$

where $x \in [0, 1]$ and

$$K(x, y) = \begin{cases} y(x+1), & 0 \le x \le y \le 1, \\ x(y+1), & 0 \le y \le x \le 1. \end{cases}$$

Then clearly

$$\forall x, y \in [0, 1] : K(x, y) = \overline{K(y, x)} \text{ and } K(x, y) \ge 0.$$

If A were positive, the fact that A is self-adjoint would then mean that necessarily $\sigma(A) \subset \mathbb{R}^+$. But, $\sigma(A)$ does contain negative values (for instance, $\lambda = -(n^2\pi^2)^{-1}$, $n \in \mathbb{N}$). Therefore, A is not positive.

Answer 10.2.7 ([71]) Let S be the unilateral shift on ℓ^2, i.e. $Se_n = e_{n+1}$ where (e_n) is some orthonormal basis in ℓ^2.

For $|\lambda| > 1$, define $K_\lambda e_n = \lambda^{-n} e_n$ where $n \geq 1$. That K_λ is compact is clear. Moreover, $SK_\lambda = \lambda K_\lambda S$ and so S is in $\mathcal{C}_2$. However, S cannot be in $\mathcal{C}_1$ for the simple reason that S does not possess any eigenvalues.

Now, we deal with the other inclusion. Let T be the operator defined by $Te_n = (-1)^n e_n$ where $n \geq 1$. Then S anti-commutes with T, i.e. $ST = -TS$. Hence $S \in \mathcal{C}_3$. But by the remark below Answer 10.2.3, S does not anti-commute with any nonzero compact operator K, whereby $S \notin \mathcal{C}_2$.

Answer 10.2.8 Consider once more the sequence of matrices A_n given in Answer 5.2.10. Obviously, each A_n is compact. Nonetheless, $A = \bigoplus_{n \in \mathbb{N}} A_n$ cannot be compact. One reason is that $\|A_n\| \nrightarrow 0$, and according to the hint this yields the non-compactness of A (another reason being $\sigma(A) = \{\lambda \in \mathbb{C} : |\lambda| \leq 1\}$, as will be seen in Answer 15.1.13 below).

Answer 10.2.9 ([148]) Obviously, a counterexample cannot be found unless $\dim H = \infty$. Let V be the Volterra operator on $L^2(0, 1)$. Hence $I + V$ is invertible. Now, set

$$A = I - (I + V)^{-1}.$$

Since V is compact and we can write

$$A = (I + V)(I + V)^{-1} - (I + V)^{-1} = V(I + V)^{-1},$$

it follows that A is compact. By Answer 9.2.4, we already know that $\sigma(A) = \{0\}$ and so A is quasinilpotent. It only remains to show that $0 \notin W(A)$. It suffices to show that if f is such that $\langle Af, f \rangle = 0$, then $f = 0$. If $\langle Af, f \rangle = 0$, then $\langle f, f \rangle = \langle (I + V)^{-1} f, f \rangle$. By the Cauchy-Schwarz inequality,

$$\|f\|^2 = \langle (I + V)^{-1} f, f \rangle \leq \|(I + V)^{-1}\| \|f\|^2 = \|f\|^2$$

for $\|(I + V)^{-1}\| = 1$ (still by Answer 9.2.4).

Since the Cauchy-Schwarz inequality has become a full equality here, we know that $(I + V)^{-1} f = \alpha f$ for some α, that is, α is an eigenvalue for $(I + V)^{-1}$. But, $\sigma[(I + V)^{-1}] = \{1\}$ and so $\alpha = 1$. Therefore, $(I + V)^{-1} f = f$ and so $f = (I + V)f$. Thus, $Vf = 0$ whereby $f = 0$. Accordingly, $0 \notin W(A)$, as wished.

Chapter 11
Functional Calculi

11.1 Basics

The functional calculus aims to define $f(A)$ where A is a fixed operator, and f belongs to some classes of functions defined in domains containing $\sigma(A)$, say. We already know that this is possible for any polynomial p. We also know how to define the exponential of A at an undergraduate level (this will be recalled in Chap. 13).

Now, we give a first version of the spectral theorem for self-adjoint operators (for proof, see e.g. Exercise 11.3.1 in [256]):

Theorem 11.1.1 (Spectral Theorem for Self-Adjoint Operators: Continuous Functional Calculus Form) *Let $A \in B(H)$ be a self-adjoint operator. Then there is a unique map $\varphi : C(\sigma(A)) \to B(H)$, defined by: $\varphi(f) = f(A)$, such that:*

(1) *$(f+g)(A) = f(A)+g(A)$, $(\alpha f)(A) = \alpha f(A)$ for any complex α, $f(A)g(A) = fg(A)$ (where fg is the pointwise product), $[f(A)]^* = \overline{f}(A)$, $1(A) = I$ and $id(A) = A$ (where $g \in C(\sigma(A))$ and $id(z) = z$).*
(2) *$\sigma(f(A)) = f(\sigma(A))$ (spectral mapping theorem).*
(3) *$\|f(A)\|_{B(H)} = \|f\|_\infty$.*

Corollary 11.1.2 *Let $A \in B(H)$ be self-adjoint and let $f : \sigma(A) \to \mathbb{C}$ be continuous. Then*

(1) *$f(A)$ is self-adjoint iff f is real-valued.*
(2) *$f(A) \geq 0$ iff $f \geq 0$.*

Before stating a more advanced form of the spectral theorem, we first introduce the concept of a spectral measure:

Definition 11.1.1 Let X be a non-vacuous set and let Σ be a sigma algebra of subsets of X. Let H be a Hilbert space. A spectral measure is a function $E : \Sigma \to$

© The Author(s), under exclusive license to Springer Nature Switzerland AG 2022
M. H. Mortad, *Counterexamples in Operator Theory*,
https://doi.org/10.1007/978-3-030-97814-3_11

$B(H)$ such that:

(1) $E(\varnothing) = 0$ and $E(X) = I$ (I is the identity operator).
(2) For every $\Delta \in \Sigma$, $E(\Delta)$ is an orthogonal projection.
(3) For any $\Delta_1, \Delta_2 \in \Sigma$: $E(\Delta_1 \cap \Delta_2) = E(\Delta_1)E(\Delta_2)$.
(4) If $(\Delta_n)_{n \in \mathbb{N}} \in \Sigma$ are pairwise disjoint, then

$$E\left(\bigcup_{n=1}^{\infty} \Delta_n\right) = \sum_{n=1}^{\infty} E(\Delta_n).$$

Theorem 11.1.3 (Spectral Theorem for Self-Adjoint Operators: Integral Form)
Let $A \in B(H)$ be self-adjoint. Then there exists a unique spectral measure E on the Borel subsets of $\sigma(A)$ such that:

$$A = \int_{\sigma(A)} \lambda \, dE,$$

also written as (for all $x, y \in H$)

$$\langle Ax, y \rangle = \int_{\sigma(A)} \lambda \, d\mu_{x,y},$$

where $\mu_{x,y}$, defined by

$$\mu_{x,y}(\Delta) = \langle E(\Delta)x, y \rangle$$

($x, y \in H$), is a countably additive measure.
   Finally, if f is bounded and Borel on $\sigma(A)$, then

$$f(A) = \int_{\sigma(A)} f(\lambda) \, dE.$$

Remark There are other forms of the spectral theorem, see e.g. [66] or [302].

Remark There is also a functional calculus for normal operators. For example, if A is normal and f is a bounded Borel function on $\sigma(A)$, then $f(A)$ is a well defined normal operator. In particular, $[f(A)]^* = \overline{f}(A)$. See e.g. [326] for more details about all these topics.

11.2 Questions

11.2.1 A Bounded Borel Measurable Function f and a Self-Adjoint $A \in B(H)$ Such That $\sigma(f(A)) \neq f(\sigma(A))$

Question 11.2.1 Show that the spectral mapping theorem fails to hold true for bounded measurable functions, i.e. find a Borel measurable bounded function f and a self-adjoint $A \in B(H)$ such that

$$\sigma[f(A)] \neq f[\sigma(A)].$$

11.2.2 A Normal Operator A Having Only Real Eigenvalues But A Is Not Self-Adjoint

It is known that a normal operator with a real spectrum is automatically self-adjoint. The proof is well known (see e.g. Theorem 12.26 in [314]) and quite easy if we know the spectral theorem of normal operators (whose proof is very much involved).

Readers might be interested in knowing that there is a quite different and interesting proof of this result bypassing the spectral theorem. This appeared in [260] (see also Exercise 8.3.9 in [256]). We include this original proof here for readers' convenience:

Theorem 11.2.1 If $T \in B(H)$ is normal with $\sigma(T) \subset \mathbb{R}$, then T is self-adjoint.

Proof Write $T = A + iB$ where A and B are self-adjoint. The normality of T is then equivalent to $AB = BA$. Hence also $TA = AT$. Since $T - A = iB$, we have by Proposition 8.2.5 (just above Question 8.2.14)

$$i\sigma(B) = \sigma(iB) = \sigma(T - A) \subset \sigma(T) + \sigma(-A) \subset \mathbb{R}.$$

Since $\sigma(B) \subset \mathbb{R}$, we necessarily have $\sigma(B) = \{0\}$. Accordingly, the spectral radius theorem gives $B = 0$ and so $T = A$. In other words, we have established the self-adjointness of T. □

Question 11.2.2 Let H be an *infinite*-dimensional space. Give an example of a normal $A \in B(H)$ *having only real eigenvalues*, and yet A is not self-adjoint.

11.2.3   A Non-self-adjoint Operator Whose Square Root Is Its Adjoint

Question 11.2.3 (See [88] for Related Results) Find a non-self-adjoint operator $A \in B(H)$ whose square root is its adjoint.

11.2.4 $A, B, C \in B(H)$, A Being Self-Adjoint, $AB = CA$ but $f(A)B \neq Cf(A)$ for Some Continuous Function f

It is known that: If $A, B, C \in B(H)$ are such that A and C are self-adjoint, then

$$AB = BC \implies f(A)B = Bf(C)$$

for any continuous function f on both $\sigma(A)$ and $\sigma(C)$ (see e.g. Corollary 11.1.8 in [256]).

Question 11.2.4 Find $A, B, C \in B(H)$ such that A is self-adjoint and $AB = CA$, but $f(A)B \neq Cf(A)$ for some continuous function f on $\sigma(A)$.

11.2.5   A Positive A Such That $BA = AB^*$ and $B\sqrt{A} \neq \sqrt{A}B^*$

Let $A, B \in B(H)$ be such that A is positive and $BA = AB^*$. Is it true that $B\sqrt{A} = \sqrt{A}B^*$? The answer is no in general.

Related to this, notice an interesting result by Z. Sebestyén who proved in [330] that if $A, B \in B(H)$ with $A \geq 0$, then

$$AB^* = BA \implies \sqrt{A}B^* = C\sqrt{A}$$

for some self-adjoint $C \in B(H)$. See also [155] for more related results.

Question 11.2.5 Find $A, B \in B(H)$ such that

$$BA = AB^* \text{ and } B\sqrt{A} \neq \sqrt{A}B^*$$

(even when B is invertible).

11.2.6 An $A \in B(H)$ Such That $\|A\| \neq \sup_{\|x\| \leq 1} |\langle Ax, x \rangle|$

If $A \in B(H)$ is a normal operator where H is a complex Hilbert space, then it may be shown as a consequence of the spectral theorem that (see e.g. Theorem 12.25 in [314]):

$$\|A\| = \sup\{|\langle Ax, x \rangle| : x \in H, \|x\| \leq 1\}.$$

A direct proof may be consulted in [24].

Question 11.2.6 Find an $A \in B(H)$ such that

$$\|A\| \neq \sup\{|\langle Ax, x \rangle| : x \in H, \|x\| \leq 1\}$$

11.2.7 A Normal $A \in B(H)$ (Where H Is Over the Field $\mathbb{R}$) Such That $\|A\| \neq \sup_{\|x\| \leq 1} |\langle Ax, x \rangle|$

Question 11.2.7 Find a normal $A \in B(H)$ such that

$$\|A\| \neq \sup\{|\langle Ax, x \rangle| : x \in H, \|x\| \leq 1\}$$

11.2.8 An A Such That $p(A) \neq A^*$ for Any Polynomial p

Let $A \in B(H)$. If $p(A) = A^*$, then A is clearly normal for A commutes with $p(A)$ (for any polynomial p).

Question 11.2.8 Find an operator (nonnormal) A such that $p(A) \neq A^*$ for any polynomial p.

Answers

Answer 11.2.1 Denote the identity operator by I as usual. Consider the self-adjoint multiplication operator A on $L^2[0, 1]$ defined by

$$A\varphi(x) = x\varphi(x).$$

Then consider the Borel measurable function f defined from $[0, 1]$ into $\mathbb{R}$ by

$$f(x) = \begin{cases} 1, \ x \neq 0 \\ 0 \ x = 0 \end{cases}$$

It is seen that

$$f(A)\varphi(x) = \varphi(x) = I\varphi(x)$$

for almost every x, i.e. we have equality on $L^2[0, 1]$. So

$$\sigma[f(A)] = \sigma(I) = \{1\}.$$

Thus,

$$\sigma[f(A)] = \{1\} \neq f[\sigma(A)] = \{0, 1\}.$$

Answer 11.2.2 Consider on $L^2[0, 1]$ the multiplication operator A by the function:

$$\varphi(x) = \begin{cases} 0, \ 0 < x < \frac{1}{3}, \\ ix, \ \frac{1}{3} \leq x \leq \frac{2}{3}, \\ 1, \ \frac{2}{3} < x < 1. \end{cases}$$

Then A is clearly bounded and normal. The only eigenvalues of A are 0 and 1. Nonetheless, A cannot be self-adjoint as $\langle Af, f \rangle$ is not real for every $f \in L^2[0, 1]$. For instance, if $f(x) = 1$, then

$$\langle Af, f \rangle = \int_0^1 \varphi(x)|f(x)|^2 dx = \int_0^1 \varphi(x)dx = \frac{1}{3} + i\frac{1}{6} \notin \mathbb{R}.$$

Answer 11.2.3 ([88]) We are asked to find a non-self-adjoint $A \in B(H)$ such that $A^* = A^2$. Observe that such operator is necessarily normal as obviously $AA^* = A^*A \, (= A^3)$. Let us give the characterization of this class of operators. The complex function given by

$$\varphi(z) = \bar{z} - z^2, \ z \in \mathbb{C}$$

vanishes on $\sigma(A)$. From that it is readily seen that if $\lambda \in \sigma(A)$, then either $\lambda = 0$ or λ is a solution of $\lambda^3 = 1$. Whence, we conclude that

$$\sigma(A) \subseteq \{0\} \cup \{e^{\frac{2k\pi i}{3}}, \ k = 0, 1, 2\}.$$

From the spectral theorem, it follows that for some orthogonal projections P_0, P_1, P_2 with pairwise orthogonal ranges, we may write

$$A = P_0 + e^{\frac{2\pi i}{3}} P_1 + e^{\frac{4\pi i}{3}} P_2.$$

Answer 11.2.4 There are counterexamples even on $\mathbb{C}^2$. Let

$$A = \begin{pmatrix} 1 & 0 \\ 0 & -1 \end{pmatrix} \text{ and } B = \begin{pmatrix} 0 & 0 \\ 1 & 0 \end{pmatrix}.$$

Then A is a fundamental symmetry, that is, A is both self-adjoint and unitary. In addition, it may readily be checked that $AB = -BA$. However, even for a function as elementary as $x \mapsto f(x) = x^2$ on $\mathbb{R}$,

$$f(A)B \neq -Bf(A).$$

Indeed, for this particular f, the equation $f(A)B = -Bf(A)$ yields $B = -B$, i.e. $B = 0$ which is impossible!

Let us give another counterexample. Consider

$$B = \begin{pmatrix} 0 & 2 \\ 1 & 0 \end{pmatrix} \text{ and } A = \begin{pmatrix} 2 & 0 \\ 0 & 1 \end{pmatrix}.$$

Ergo $BA = AB^*$ but

$$BA^2 = \begin{pmatrix} 0 & 2 \\ 4 & 0 \end{pmatrix} \neq A^2 B^* = \begin{pmatrix} 0 & 4 \\ 2 & 0 \end{pmatrix},$$

as desired.

Answer 11.2.5 Consider again:

$$B = \begin{pmatrix} 0 & 2 \\ 1 & 0 \end{pmatrix} \text{ and } A = \begin{pmatrix} 2 & 0 \\ 0 & 1 \end{pmatrix}.$$

Then B is invertible and A is positive (also invertible). Clearly

$$\sqrt{A} = \begin{pmatrix} \sqrt{2} & 0 \\ 0 & 1 \end{pmatrix}.$$

It is easy to see that

$$BA = AB^* = \begin{pmatrix} 0 & 2 \\ 2 & 0 \end{pmatrix}$$

whereas

$$B\sqrt{A} = \begin{pmatrix} 0 & 2 \\ \sqrt{2} & 0 \end{pmatrix} \neq \begin{pmatrix} 0 & \sqrt{2} \\ 2 & 0 \end{pmatrix} = \sqrt{A}B^*.$$

Answer 11.2.6 Let

$$A = \begin{pmatrix} 0 & 1 \\ 0 & 0 \end{pmatrix}.$$

Then

$$\|A\|^2 = \|A^*A\| = 1$$

and so $\|A\| = 1$. Now, if $x = (x_1, x_2)$, then

$$|\langle Ax, x \rangle| = |x_2 x_1| \leq \frac{1}{2}$$

if $\|x\| \leq 1$. Therefore,

$$\|A\| \neq \sup\{|\langle Ax, x \rangle| : x \in H, \ \|x\| \leq 1\},$$

as required.

Answer 11.2.7 Consider the normal (in fact unitary!) matrix on $\mathbb{R}^2$:

$$A = \begin{pmatrix} 0 & 1 \\ -1 & 0 \end{pmatrix}.$$

Then $\|A\| = 1$ but for *all* $x \in \mathbb{R}^2$

$$\langle Ax, x \rangle = 0.$$

Consequently,

$$\|A\| \neq \sup\{|\langle Ax, x \rangle| : x \in H, \|x\| \le 1\},$$

as wished.

Answer 11.2.8 ([149]) The difficulty in this sort of questions lies in showing that *there is no* polynomial p such that $p(A) \neq A^*$. Thankfully, an example of such an A is available on $\mathbb{C}^2$ and so things are easier to handle. Let

$$A = \begin{pmatrix} 0 & 0 \\ 1 & 0 \end{pmatrix}$$

(and so A is not normal). Then $A^2 = 0$, and so *any* polynomial in A is necessarily of the form

$$p(A) = \begin{pmatrix} a & 0 \\ b & a \end{pmatrix}$$

where $a, b \in \mathbb{C}$. Since $A^* = \begin{pmatrix} 0 & 1 \\ 0 & 0 \end{pmatrix}$, clearly $p(A) \neq A^*$, as suggested.

Chapter 12
Fuglede-Putnam Theorems and Intertwining Relations

12.1 Basics

One of the most powerful tools in the theory of normal operators is the following Fuglede theorem.

Theorem 12.1.1 ([117]) *Let $A, B \in B(H)$ be such that $AB = BA$. If A (or B) is normal, then $A^*B = BA^*$.*

Remark See Chap. 27 for some historical notes.

Corollary 12.1.2 *Let $A, B \in B(H)$ be such that A is normal. Then*

$$BA = AB \Longleftrightarrow BA^* = A^*B \Longleftrightarrow B^*A = AB^* \Longleftrightarrow B^*A^* = A^*B^*.$$

As immediate applications of the Fuglede theorem, we have:

Theorem 12.1.3 *Let $A, B \in B(H)$ be both normal and such that $AB = BA$. Then*

(1) AB, A^*B^*, AB^*, A^*B, BA, BA^*, B^*A and B^*A^* are all normal.
(2) $A + B$, $A^* + B^*$, $A^* + B$ and $A + B^*$ are all normal.

An important generalization of the Fuglede theorem (established by C. R. Putnam, see [294]) commonly known as the Fuglede-Putnam theorem, and presented in a basic form, is:

Theorem 12.1.4 *Let $A, B, T \in B(H)$ and assume that A and B are normal. Then*

$$TA = BT \Longleftrightarrow TA^* = B^*T.$$

There is a particular terminology for the transformation which occurs in the Fuglede-Putnam theorem.

© The Author(s), under exclusive license to Springer Nature Switzerland AG 2022   179
M. H. Mortad, *Counterexamples in Operator Theory*,
https://doi.org/10.1007/978-3-030-97814-3_12

Definition 12.1.1 Say that an operator $T \in B(H)$ intertwines two operators $A, B \in B(H)$ when $TA = BT$.

We finish with a definition in relation with the results above:

Definition 12.1.2 We say that $A, B \in B(H)$ are double commuting if

$$AB = BA \text{ and } AB^* = B^*A.$$

12.2 Questions

12.2.1 Does $TA = AT$ Give $TA^* = A^*T$ When $\ker A = \ker A^*$?

We already know that a normal $A \in B(H)$ satisfies $\ker A = \ker A^*$, and that the converse is not true. Nevertheless, the condition $\ker A = \ker A^*$ could replace the normality of A in some situations. For example, see ([84], Theorem 1.2 and Corollary 3.4). Does this apply to the Fuglede theorem?

Question 12.2.1 Does $TA = AT$ imply that $TA^* = A^*T$ when $A, T \in B(H)$ and $\ker A = \ker A^*$?

12.2.2 If $T, N, M \in B(H)$ Are Such That N and M Are Normal, Do We Have $\|TN - MT\| = \|TN^* - M^*T\|$? How About $|TN - MT| = |TN^* - M^*T|$?

Obviously, if $T, N, M \in B(H)$ are such that N and M are normal, and $\|TN - MT\| = \|TN^* - M^*T\|$ (or $|TN - MT| = |TN^* - M^*T|$), then Fuglede-Putnam theorem would follow.

Question 12.2.2 Find $T, N, M \in B(H)$ such that N and M are normal and $\|TN - MT\| \neq \|TN^* - M^*T\|$, and $|TN - MT| \neq |TN^* - M^*T|$.

Further Reading There are many papers (as e.g. [376], cf. [187]) dealing with versions of the Fuglede-Putnam theorem such that say $\|AX - XB\|_2 = \|A^*X - XB^*\|_2$, where A and B are normal and X is arbitrary, and $\| \cdot \|_2$ is the Hilbert-Schmidt norm (see e.g. [302] for its definition). Interested readers are referred to

[120, 138], and [193] for similar versions involving other classes of operators. See also [265].

12.2.3   A Self-Adjoint A and a Normal B Such That $AB = BA$ but $B^*A \neq AB$

Question 12.2.3 Assume that $B \in B(H)$ is normal and that $A \in B(H)$ is self-adjoint. Do we have

$$BA = AB \Longrightarrow B^*A = AB? \text{ Or } BA = AB^*?$$

12.2.4   A Unitary $U \in B(H)$ and a Self-Adjoint $A \in B(H)$ Such That $AU^* = UA$ and $AU \neq UA$

Question 12.2.4 Find a unitary $U \in B(H)$ and a self-adjoint $A \in B(H)$ such that

$$AU^* = UA \text{ and } AU \neq UA.$$

12.2.5   Two Nonnormal Double Commuting Matrices

Question 12.2.5 (Cf. Question 14.2.19) Let $\dim H < \infty$. Can you find two matrices $A, B \in B(H)$, none of them is normal and yet they double commute, i.e.

$$AB = BA \text{ and } AB^* = B^*A?$$

12.2.6 $AN = MB \nRightarrow AN^* = M^*B$ Even When All of M, N, A and B Are Unitary

Question 12.2.6 Let M, N, A, and B be all *unitary* operators such that A and B are also self-adjoint such that $AN = MB$. Does it follow that $AN^* = M^*B$?

12.2.7 $AN = MB \nRightarrow BN^* = M^*A$ Where M Is an Isometry or N Is a Co-isometry

It is easy to see that if M is an isometry and N is a co-isometry, then

$$AN = MB \Longrightarrow BN^* = M^*A$$

where A and B are arbitrary.

Question 12.2.7 Let M be an isometry *or* let N be a co-isometry. Let A and B be two bounded linear operators such that $AN = MB$. Does it follow that $BN^* = M^*A$?

12.2.8 Positive Invertible Operators A, B, N, M Such That $AN = MB$ and $AB = BA$ Yet $BN^* \neq M^*A$

In the examples met in Answer 12.2.6, we may check that $BN^* = M^*A$. So, one could be tempted to believe that $AN = MB$ does entail $BN^* = M^*A$.

Question 12.2.8 Provide self-adjoint, positive, and invertible A, B, N, $M \in B(H)$ such that $AN = MB$ and $AB = BA$ yet $BN^* \neq M^*A$.

12.2.9 On Some Generalization of the Fuglede-Putnam Theorem Involving Contractions

Assume that N is unitary and that A and B are two bounded operators. It is straightforward to see that

$$NA = BN \Longrightarrow NA^* = B^*N.$$

To drop the "unitarity" assumption on N, one may use a trick of matrix operators via the so-called Julia operator (see e.g. [256]). The following result was then obtained in [241]:

Theorem 12.2.1 *Let $A, B \in B(H)$. Suppose $N \in B(H)$ is a contraction such that*

$$(1 - N^*N)^{\frac{1}{2}}A = B(1 - NN^*)^{\frac{1}{2}} = (1 - N^*N)^{\frac{1}{2}}A^* = B^*(1 - NN^*)^{\frac{1}{2}} = 0.$$

Then

$$NA = BN \Longrightarrow NA^* = B^*N.$$

Corollary 12.2.2 *Let $A, B \in B(H)$. If $N \in B(H)$ is an isometry such that*

$$B(1 - NN^*)^{\frac{1}{2}} = B^*(1 - NN^*)^{\frac{1}{2}} = 0,$$

then

$$BN = NA \Longrightarrow B^*N = NA^*.$$

Question 12.2.9 Show that the two conditions $B(1 - NN^*)^{\frac{1}{2}} = B^*(1 - NN^*)^{\frac{1}{2}} = 0$ cannot be completely eliminated in the previous corollary.

12.2.10 $A, B, C \in B(H)$ with A Being Self-Adjoint Such That $AB = \lambda CA$ and λ Is Arbitrary

Question 12.2.10 (Cf. [54] and [242], and the references therein) Find $A, B, C \in B(H)$ with A being self-adjoint such that $AB = \lambda CA$ and λ is arbitrary (i.e. it does not have to be real unlike when $B = C$, as in say [35] or [394]).

12.2.11 Are There Two Normal Operators A and B Such That $AB = 2BA$?

Let

$$A = \begin{pmatrix} 0 & 1 \\ 0 & 0 \end{pmatrix} \text{ and } B = \begin{pmatrix} 1 & 0 \\ 0 & 2 \end{pmatrix}.$$

Then A and B 2-commute, i.e. $AB = 2BA$, as may be checked. Observe that A is not normal, whereas B is self-adjoint. Can we find a similar example in the case of two normal matrices?

> **Question 12.2.11 (Cf. Question 32.2.7)** Show that it is impossible for two bounded normal operators A and B to satisfy $AB = \lambda BA \neq 0$, where $\lambda \in \mathbb{C}$ with $|\lambda| \neq 1$.

12.2.12 A Normal A and a Self-Adjoint B Such That AB Is Normal but BA Is Not Normal

It has been known for quite some time that if A, B, and AB are normal (finite) matrices, then BA is also normal. See e.g. Theorem 12.3.4 in [125] or Problem 151 in [149]. Does this result have a chance to hold good in infinite-dimensional spaces? The answer is no as will be seen below. However, I. Kaplansky showed the following result (whose proof seems to need the Fuglede-Putnam theorem):

Theorem 12.2.3 ([188], cf. Question 14.2.30) *Let $A, B \in B(H)$ be such that AB and A are normal. Then B commutes with A^*A iff BA is normal.*

> **Question 12.2.12 (⊛⊛)** Provide $A, B \in B(H)$ such that A is normal, B is self-adjoint, AB is normal but BA is not normal.
>
> **Hint:** As mentioned, remember the impossibility of a counterexample when $\dim H < \infty$, or if $A, B \in K(H)$.

Further Reading The counterexample to be given was obtained by I. Kaplansky in [188], and it was reproduced in [132] with slightly further details. Some related papers are [17, 132] and [197]. Readers may also wish to consult the seemingly pioneering papers on this topic which are [380] and [381].

12.2.13 About Some Embry's Theorem

Recall the following result by M. R. Embry in [101]:

Theorem 12.2.4 (Embry Theorem) *Let $A, B \in B(H)$ be two commuting normal operators. Let $T \in B(H)$ be such that $0 \notin W(T)$. If $TA = BT$, then $A = B$.*

Remark Readers might be interested in knowing that a simple proof in the finite-dimensional context is available in [168].

Question 12.2.13 Let $A, B, T \in B(H)$. Provide examples of normal A and normal B such that $TA = BT$, and

(1) A and B commute, but $A \neq B$;
(2) $0 \notin W(T)$ but $A \neq B$.

Answers

Answer 12.2.1 The answer is negative. Let A be an invertible operator, and so $\ker A = \ker A^* = \{0\}$. So, if the claim were true, we would have $AA = AA \Rightarrow AA^* = A^*A$. This is obviously untrue as this would mean that an invertible operator is necessarily normal. So, for a counterexample, it suffices to consider any invertible operator which is not normal.

Answer 12.2.2 We answer both questions together. The counterexample was obtained by M. J. Crabb and P. G. Spain in [76], but it had a different aim in their paper. On $\mathbb{C}^2$, let

$$T = \begin{pmatrix} 1+i & 2 \\ -1+i & -2 \end{pmatrix}, \quad N = \begin{pmatrix} i & 0 \\ 0 & 0 \end{pmatrix} \quad \text{and} \quad M = \begin{pmatrix} 1 & 0 \\ 0 & -1 \end{pmatrix}.$$

Then N and M are both normal. Besides

$$TN - MT = \begin{pmatrix} -2 & -2 \\ -2 & -2 \end{pmatrix} \quad \text{and} \quad TN^* - M^*T = \begin{pmatrix} -2i & -2 \\ 2i & -2 \end{pmatrix}.$$

Since $TN - MT$ is unitarily equivalent to the diagonal 2×2 matrix with 0 and -4 on its diagonal, it follows that $\|TN - MT\| = 4$. Also, $\|TN^* - M^*T\| = 2\sqrt{2}$ as $(TN^* - M^*T)^*(TN^* - M^*T) = 8I$, say. Thus

$$\|TN - MT\| \neq \|TN^* - M^*T\|.$$

For the case of the absolute value, just remember that $\||S|\| = \|S\|$ provided $S \in B(H)$.

Remark This counterexample kills all hope of other similar generalizations. For example, now neither $\|TN - MT\| \leq \|TN^* - M^*T\|$ nor $\|TN - MT\| \geq \|TN^* - M^*T\|$ needs to hold. The former is seen by the aforementioned example while the latter is obtained by replacing N by N^* (M already being self-adjoint) in the example above.

Answer 12.2.3 Both implications are false. The same counterexample works for either of them. Consider a normal but non-self-adjoint operator $B \in B(H)$ and take $A = I$. Then plainly $BA = AB$, but neither $B^*A = AB$ nor $BA = AB^*$ holds because $B \neq B^*$.

Answer 12.2.4 (Cf. [16]) Let

$$A = \begin{pmatrix} 1 & 0 \\ 0 & -1 \end{pmatrix} \text{ and } U = \begin{pmatrix} 0 & -1 \\ 1 & 0 \end{pmatrix}.$$

Then A is self-adjoint and U is unitary. Hence,

$$UA = \begin{pmatrix} 0 & 1 \\ 1 & 0 \end{pmatrix}, \ AU^* = \begin{pmatrix} 0 & 1 \\ 1 & 0 \end{pmatrix} \text{ and } AU = \begin{pmatrix} 0 & -1 \\ -1 & 0 \end{pmatrix}$$

and so

$$AU^* = UA \text{ and } AU \neq UA,$$

as needed.

Answer 12.2.5 Define on $\mathbb{C}^4$:

$$A = \begin{pmatrix} 0 & 1 & 0 & 0 \\ 0 & 0 & 0 & 0 \\ 0 & 0 & 0 & 0 \\ 0 & 0 & 0 & 0 \end{pmatrix} \text{ and } B = \begin{pmatrix} 0 & 0 & 0 & 0 \\ 0 & 0 & 0 & 0 \\ 0 & 0 & 0 & 0 \\ 0 & 0 & 1 & 0 \end{pmatrix}.$$

Then it is seen that neither A nor B is normal. But $AB = BA$ and $AB^* = B^*A$. In fact, we even have

$$AB = BA = AB^* = B^*A = 0_{\mathcal{M}_4(\mathbb{C})}.$$

Answer 12.2.6 ([241]) No, it does not! On $\mathbb{C}^2$, let

$$M = \begin{pmatrix} 0 & 1 \\ -1 & 0 \end{pmatrix}; \ N = \begin{pmatrix} 1 & 0 \\ 0 & -1 \end{pmatrix}; \ A = \begin{pmatrix} 0 & 1 \\ 1 & 0 \end{pmatrix} \text{ and } B = \begin{pmatrix} -1 & 0 \\ 0 & -1 \end{pmatrix}.$$

As it is easily seen, all operators involved are unitary where, in addition, A and B are self-adjoint. Readers may also check that

$$AN = MB \text{ while } AN^* \neq M^*B.$$

Answer 12.2.7 ([241]) If one takes S to be the unilateral shift defined on ℓ^2, then by setting

$$M = N = A = B = S \text{ (hence } N \text{ is not a co-isometry)},$$

one sees that $AN = MB$ while $BN^* \neq M^*A$.

Similarly, if one sets

$$M = N = A = B = S^* \text{ (hence } M \text{ is not an isometry)},$$

then $AN = MB$ whereas $BN^* \neq M^*A$.

Answer 12.2.8 Let

$$A = \begin{pmatrix} 2 & 0 \\ 0 & 1 \end{pmatrix} \text{ and } B = \begin{pmatrix} 3 & 0 \\ 0 & 1 \end{pmatrix}$$

and put

$$N = A^{-1} = \begin{pmatrix} \frac{1}{2} & 0 \\ 0 & 1 \end{pmatrix} \text{ and } M = B^{-1} = \begin{pmatrix} \frac{1}{3} & 0 \\ 0 & 1 \end{pmatrix}.$$

Then A, B, N, and M are all positive and invertible. Besides, $AB = BA$ and

$$AN = MB = I.$$

Finally,

$$BN^* = \begin{pmatrix} \frac{3}{2} & * \\ * & * \end{pmatrix} \neq \begin{pmatrix} \frac{2}{3} & * \\ * & * \end{pmatrix} = M^*A.$$

Answer 12.2.9 ([241]) For instance, if one takes again the unilateral shift S on ℓ^2 and sets $N = B = S$, then N is an isometry and one can check that it does not verify $B(1 - NN^*)^{\frac{1}{2}} = B^*(1 - NN^*)^{\frac{1}{2}} = 0$. If we also set $A = S$, then $BN = S^2 = NA$ while

$$B^*N = S^*S \neq SS^* = NA^*.$$

Answer 12.2.10 Take $\lambda \in \mathbb{C}^*$ and consider

$$A = I = \begin{pmatrix} 1 & 0 \\ 0 & 1 \end{pmatrix}, \ B = \begin{pmatrix} 0 & 0 \\ \lambda & 0 \end{pmatrix} \text{ and } C = \begin{pmatrix} 0 & 0 \\ 1 & 0 \end{pmatrix}.$$

Then A is self-adjoint and $AB = \lambda CA \, (\neq 0)$, but λ is arbitrary.

Answer 12.2.11 ([394] or [59]) Recall that when $A, B \in B(H)$ are normal, then $\|AB\| = \|BA\|$. So if $AB = \lambda BA \neq 0$, then

$$\|AB\| = |\lambda| \|BA\| = |\lambda| \|AB\|,$$

whereby $|\lambda| = 1$, which answers the question.

Remark If $AB = \lambda BA$ and A and B are normal, then AB is always normal (see Question 27.2.10 for a more general result). Accordingly, if $AB = \lambda BA \neq 0$ and A and B are normal, then

$$AB \text{ is normal} \iff |\lambda| = 1.$$

Answer 12.2.12 ([188]) The counterexample is by no means trivial and there is, alas, nothing simpler yet. Let P be the diagonal matrix with the numbers 1, 2, and $1/2$ repeated infinitely often down the diagonal, i.e. P may be the operator $2I_{\ell^2} \oplus I_{\ell^2} \oplus \frac{1}{2}I_{\ell^2}$. Now, introduce

$$Q = \begin{pmatrix} P & 0 \\ 0 & 0 \end{pmatrix} \text{ and } R = \begin{pmatrix} 0 & 0 \\ 0 & I \end{pmatrix}$$

which are defined on $\ell^2 \oplus \ell^2 \oplus \ell^2 \oplus \ell^2$. It is straightforward to see that

$$QR = RQ = 0.$$

Then, and as observed in [132] by a canonical unitary identification of $\ell^2 \oplus \ell^2$ with ℓ^2 by matching odd and even indices, $4Q^2 + 4R^2$ is unitarily equivalent to $4Q^2 + R^2$, that is, for some unitary $U \in B(\ell^2 \oplus \ell^2 \oplus \ell^2 \oplus \ell^2)$

$$U(4Q^2 + 4R^2)U^* = 4Q^2 + R^2.$$

In a similar way, there is a unitary $V \in B(\ell^2 \oplus \ell^2 \oplus \ell^2 \oplus \ell^2)$ such that

$$V(Q^2 + R^2)V^* = Q^2 + 4R^2.$$

Now, let

$$A = \begin{pmatrix} 2U & 0 \\ 0 & V \end{pmatrix} \text{ and } B = \begin{pmatrix} Q & R \\ R & Q \end{pmatrix}.$$

Then A is clearly normal and B is even self-adjoint. Since

$$AB(AB)^* = (AB)(BA^*) = \begin{pmatrix} 4Q^2 + R^2 & 0 \\ 0 & Q^2 + 4R^2 \end{pmatrix} = (AB)^*(AB),$$

it follows that AB is normal. Finally, readers may easily check that BA is not normal.

Answer 12.2.13 ([101])

(1) Let

$$A = \begin{pmatrix} 2 & 0 \\ 0 & 1 \end{pmatrix}, \quad B = \begin{pmatrix} 1 & 0 \\ 0 & 2 \end{pmatrix} \text{ and } T = \begin{pmatrix} 0 & 1 \\ 1 & 0 \end{pmatrix}.$$

Then A and B are self-adjoint and $A \neq B$. We may also easily check that A commutes with B and that $TA = BT$. But $0 \in W(T)$ because if $X = \begin{pmatrix} 0 \\ 1 \end{pmatrix}$, then $\|X\| = 1$ and

$$\langle TX, X \rangle = 1 \times 0 + 0 \times 1 = 0.$$

(2) Let

$$A = \begin{pmatrix} 1 & i \\ -i & 2 \end{pmatrix}, \quad B = \begin{pmatrix} 1 & 1 \\ 1 & 2 \end{pmatrix} \text{ and } T = \begin{pmatrix} 1 & 0 \\ 0 & i \end{pmatrix}$$

(i being the usual complex number). Then observe that A and B are self-adjoint. Readers may then check that $TA = BT$ and that A does not commute with B. If 0 were in $W(T)$, then we would have for any $(x, y) \in \mathbb{C}^2$ such that $\|(x, y)\| = 1$

$$\left\langle T \begin{pmatrix} x \\ y \end{pmatrix}, \begin{pmatrix} x \\ y \end{pmatrix} \right\rangle = |x|^2 + i|y|^2 = 0,$$

forcing $x = y = 0$, which is just not consistent with $\|(x, y)\| = 1$!

.

Chapter 13
Operator Exponentials

13.1 Basics

Let $A \in B(H)$. It is known that the series $\sum_{n=0}^{\infty} A^n/n!$ converges absolutely in $B(H)$, and hence it converges. This allows us to define e^A (where $A \in B(H)$) without using all the theory of the functional calculus.

Definition 13.1.1 The sum $\sum_{n=0}^{\infty} A^n/n!$ is denoted by e^A, and it is called the exponential of A.

Recall some other well established facts.

Proposition 13.1.1 *Let $A, B \in B(H)$ be such that $AB = BA$. Then*

$$e^{A+B} = e^A e^B = e^B e^A.$$

Corollary 13.1.2 *Let $A \in B(H)$. Then e^A is always invertible and*

$$(e^A)^{-1} = e^{-A}.$$

13.2 Questions

13.2.1 An Invertible Matrix Which is Not the Exponential of Any Other Matrix

It is known that each invertible matrix A over $\mathbb{C}$ is the exponential of some other matrix B (of the same size).

© The Author(s), under exclusive license to Springer Nature Switzerland AG 2022
M. H. Mortad, *Counterexamples in Operator Theory*,
https://doi.org/10.1007/978-3-030-97814-3_13

Question 13.2.1 (Cf. Question 14.2.37) Show that this conclusion does not need to remain true over the field $\mathbb{R}$.

13.2.2   A Compact $A \in B(H)$ for Which e^A is Not Compact

Question 13.2.2 Find a compact $A \in B(H)$ for which e^A is not compact.

13.2.3 $e^A = e^B \nRightarrow A = B$

The function $A \mapsto e^A$ (defined from $B(H)$ into $B(H)$) is one-to-one when restricted to the set of self-adjoint operators (for a proof, see e.g. Example 11.1.22 in [256]).

Question 13.2.3 Let $A, B \in B(H)$. Do we have

$$e^A = e^B \Longrightarrow A = B?$$

13.2.4   A Non-Self-Adjoint $A \in B(H)$ Such That e^{iA} Is Unitary

Recall that A is self-adjoint, then e^{iA} is unitary. Does the converse hold?

Question 13.2.4 Find a non-self-adjoint $A \in B(H)$ such that e^{iA} is unitary.

13.2.5   A Normal A Such That $e^A = e^B$ But $AB \neq BA$

It was shown in [49] that if $A, B \in B(H)$ where A is normal and such that $\sigma(\mathrm{Im}A) \subset (0, \pi)$, then

$$e^A = e^B \Longrightarrow AB = BA.$$

Question 13.2.5 Find $A, B \in B(H)$ such that A is normal and $e^A = e^B$ but $AB \neq BA$.

13.2.6   Two Self-Adjoint A, B Such That $A \geq B$ and $e^A \ngeq e^B$

Let $A, B \in B(H)$ be self-adjoint such that $e^A \geq e^B$. Since e^B is positive and invertible, it follows that

$$A = \log(e^A) \geq \log(e^B) = B$$

(by say Proposition 11.1.25 and Example 11.1.22 in [256]). Does the converse hold?

Question 13.2.6 Produce two self-adjoint matrices A and B such that:

$$A \geq B \text{ and } e^A \ngeq e^B.$$

Hint: You must avoid commuting A and B.

13.2.7   A Self-Adjoint $A \in B(H)$ Such That $\|e^A\| \neq e^{\|A\|}$

We know that $\|e^A\| \leq e^{\|A\|}$ holds for any $A \in B(H)$. If A is positive, then $\|A\| \in \sigma(A)$. Hence

$$\|e^A\| = \sup\{|\lambda| : \lambda \in \sigma(e^A)\} = \sup\{e^\mu : \mu \in \sigma(A)\} \geq e^{\|A\|}.$$

Thus, $\|e^A\| = e^{\|A\|}$ whenever $A \in B(H)$ is positive. Are things that nice when A is supposed to be self-adjoint only?

Question 13.2.7 Find a self-adjoint $A \in B(H)$ such that

$$\|e^A\| \neq e^{\|A\|}.$$

13.2.8   A Nonnormal Operator A Such That e^A Is Normal

It is easy to see that if A is normal, then so is e^A. What about the converse?

Question 13.2.8 Find a nonnormal operator A such that e^A is normal.

13.2.9 $A, B \in B(H)$, $AB \neq BA$ and $e^{A+B} \neq e^A e^B \neq e^B e^A$

Question 13.2.9 Find $A, B \in B(H)$ such that $AB \neq BA$ and

$$e^{A+B} \neq e^A e^B \neq e^B e^A.$$

13.2.10 $A, B \in B(H)$, $AB \neq BA$ and $e^{A+B} = e^A e^B = e^B e^A$

Question 13.2.10 Find $A, B \in B(H)$ such that $AB \neq BA$ and

$$e^{A+B} = e^A e^B = e^B e^A.$$

13.2.11 $A, B \in B(H)$, $AB \neq BA$ and $e^{A+B} \neq e^A e^B = e^B e^A$

Question 13.2.11 Find $A, B \in B(H)$ such that $AB \neq BA$ and

$$e^{A+B} \neq e^A e^B = e^B e^A.$$

13.2.12 $A, B \in B(H)$, $AB \neq BA$ and $e^{A+B} = e^A e^B$ $\neq e^B e^A$

Question 13.2.12 (⊛) Find $A, B \in B(H)$ such that $AB \neq BA$ and

$$e^{A+B} = e^A e^B \neq e^B e^A.$$

13.2.13   Real Matrices A and B Such That $AB \neq BA$ and $e^{A+B} = e^A e^B \neq e^B e^A$

Question 13.2.13 Since in the previous counterexample, we needed results from complex analysis, we ask readers now to provide *real matrices* A, B such that $AB \neq BA$ and

$$e^{A+B} = e^A e^B \neq e^B e^A.$$

13.2.14   An Operator A with $e^A e^{A^*} = e^{A^*} e^A \neq e^{A+A^*}$

Question 13.2.14 Find an operator A such that

$$e^A e^{A^*} = e^{A^*} e^A \neq e^{A+A^*}.$$

Hint: By a result in [49] say, we know that

$$e^A e^{A^*} = e^{A^*} e^A = e^{A+A^*} \Longleftrightarrow A \text{ is normal}.$$

13.2.15   An Operator T Such That $|e^T| \neq e^{\operatorname{Re} T}$ and $|e^T| \not\leq e^{|T|}$

It is known that

$$|e^z| = e^{\operatorname{Re} z} \leq e^{|z|}.$$

for any $z \in \mathbb{C}$ (see e.g. Exercise 1.2.1 in [4]). It is seen via the spectral theorem say that $|e^T| \leq e^{|T|}$ for any normal $T \in B(H)$. Using the Cartesian decomposition we may also show that $|e^T| = e^{\operatorname{Re} T}$ when $T \in B(H)$ is normal.

Question 13.2.15 Find a $T \in B(H)$ such that

$$|e^T| \neq e^{\operatorname{Re} T} \text{ and } |e^T| \not\leq e^{|T|}.$$

Hint: Work on $\mathbb{C}^2$.

Answers

Answer 13.2.1 On $\mathbb{R}^2$, let

$$A = \begin{pmatrix} -1 & 0 \\ 0 & 1 \end{pmatrix}.$$

Then A is unitary. If there were a matrix B (with real entries only) such that $e^B = A$, then it would follow that

$$-1 = \det A = \det(e^B) = e^{\operatorname{tr} B}.$$

This, however, is impossible since $\operatorname{tr} B \in \mathbb{R}$! Therefore, no such a matrix B exists.

Answer 13.2.2 A simple counterexample is to take the compact $A := 0$ on an *infinite* dimensional Hilbert space. In this case, $e^0 = I$ is not compact. In fact, if $A \in B(H)$, we may say that e^A is rarely compact in the sense that when $\dim H = \infty$, then e^A is not compact for it is invertible. In the end, readers may show that $e^A - I$ is compact whenever A is so, even when $\dim H = \infty$ (this is easily seen by considering $\sum_{k=1}^{n} A^k/k!$ and then passing to the uniform limit).

Answer 13.2.3 The answer is no in general. Take

$$A = 2\pi i I \neq 4\pi i I = B$$

where i is the complex number and I is the identity operator on H. Then

$$e^A = e^{2\pi i I} = I + (2\pi i)I + \frac{(2\pi i)^2 I^2}{2!} + \cdots$$

$$= \left(1 + (2\pi i) + \frac{(2\pi i)^2}{2!} + \cdots\right) I$$

$$= e^{2\pi i} I = I.$$

Similarly, we obtain

$$e^B = e^{4\pi i I} = I.$$

Consequently, $e^A = e^B$ whereas $A \neq B$.

Answer 13.2.4 ([256]) To get a counterexample, we must avoid normality. On $\mathbb{C}^2$, consider

$$A = \begin{pmatrix} -i\pi & 2i\pi \\ -i\pi & i\pi \end{pmatrix} \text{ so that } i A = \begin{pmatrix} \pi & -2\pi \\ \pi & -\pi \end{pmatrix} = \pi \underbrace{\begin{pmatrix} 1 & -2 \\ 1 & -1 \end{pmatrix}}_{\text{call it } B}.$$

Then it is seen that A is not self-adjoint.

Next, we claim that $e^{iA} = e^{\pi B} = -I$ (hence it is unitary). We need to compute integer powers of πB. We have

$$\pi^2 B^2 = \begin{pmatrix} 1 & -2 \\ 1 & -1 \end{pmatrix}\begin{pmatrix} 1 & -2 \\ 1 & -1 \end{pmatrix} = \pi^2 \begin{pmatrix} -1 & 0 \\ 0 & -1 \end{pmatrix} = -\pi^2 I.$$

Whence

$$\pi^3 B^3 = \pi B(-\pi^2 I) = -\pi^3 B, \ \pi^4 B^4 = \pi^4 I, \text{ and so on...}$$

Therefore,

$$e^{iA} = e^{\pi B}$$

$$= I + \pi B + \frac{\pi^2}{2!} B^2 + \frac{\pi^3}{3!} B^3 + \frac{\pi^4}{4!} B^4 + \frac{\pi^5}{5!} B^5 + \cdots$$

$$= \underbrace{\left(1 - \frac{\pi^2}{2!} + \frac{\pi^4}{4!} + \cdots\right)}_{\cos \pi = -1} I + \underbrace{\left(\pi - \frac{\pi^3}{3!} + \frac{\pi^5}{5!} + \cdots\right)}_{\sin \pi = 0} B$$

$$= -I.$$

Consequently, e^{iA} is obviously unitary while A is not self-adjoint.

Answer 13.2.5 Let

$$A = \begin{pmatrix} 0 & \pi \\ -\pi & 0 \end{pmatrix} \text{ and } B = \begin{pmatrix} \pi & -2\pi \\ \pi & -\pi \end{pmatrix}.$$

Then A is normal. Moreover, it is easy to see that

$$AB = \pi^2 \begin{pmatrix} 1 & -1 \\ -1 & 2 \end{pmatrix} \neq \pi^2 \begin{pmatrix} 2 & 1 \\ 1 & 1 \end{pmatrix} = BA.$$

A short calculation reveals that

$$e^A = \begin{pmatrix} \cos\pi & \sin\pi \\ -\sin\pi & \cos\pi \end{pmatrix} = \begin{pmatrix} -1 & 0 \\ 0 & -1 \end{pmatrix} = -I.$$

Similarly, we find that $e^B = -I$ and so $e^A = e^B$.

Answer 13.2.6 Let

$$A = \begin{pmatrix} 2 & 1 \\ 1 & 1 \end{pmatrix} \text{ and } B = \begin{pmatrix} 1 & 0 \\ 0 & 0 \end{pmatrix}.$$

Then $A \geq B$ as A and B are self-adjoint, and

$$A - B = \begin{pmatrix} 1 & 1 \\ 1 & 1 \end{pmatrix} \geq 0.$$

While the exact form of e^B is obvious the exact form of e^A is slightly messy. So, it is preferable to carry on the calculations of e^A and e^B in approximate forms. We find that

$$e^A = \begin{pmatrix} 10.3247... & 5.4755... \\ 5.4755... & 4.8492... \end{pmatrix}$$

and

$$e^B = \begin{pmatrix} 2.7182... & 0 \\ 0 & 1 \end{pmatrix}.$$

Hence

$$e^A - e^B = \begin{pmatrix} 7.6065... & 5.4755... \\ 5.4755... & 3.8492... \end{pmatrix}.$$

Therefore, $e^A \ngeq e^B$ for the self-adjoint matrix $e^A - e^B$ satisfies

$$\det(e^A - e^B) = -0.7020... < 0.$$

Answer 13.2.7 Simply consider

$$A = \begin{pmatrix} -2 & 0 \\ 0 & 1 \end{pmatrix}.$$

Then $\|A\| = 2$. Moreover,

$$e^A = \begin{pmatrix} e^{-2} & 0 \\ 0 & e \end{pmatrix}.$$

Therefore,

$$\|e^A\| = e \neq e^2 = e^{\|A\|}.$$

Answer 13.2.8 The example is inspired by one from [343] though its aim there differs from ours. On $\mathbb{C}^2$, take

$$A = \begin{pmatrix} i\pi & 1 \\ 0 & -i\pi \end{pmatrix}$$

which is obviously not normal. It is easy to see that

$$e^A = e^{A^*} = \begin{pmatrix} -1 & 0 \\ 0 & -1 \end{pmatrix} = -I.$$

Therefore, e^A is even a fundamental symmetry.

Answer 13.2.9 ([256]) Consider the following matrices

$$A = \begin{pmatrix} 0 & 1 \\ 0 & 0 \end{pmatrix} \text{ and } B = \begin{pmatrix} 0 & 0 \\ 1 & 0 \end{pmatrix}.$$

Clearly, A and B do not commute. Since for all $n \geq 2$, $A^n = 0$ and $B^n = 0$ (the null matrices!), one has

$$e^A = I + A = \begin{pmatrix} 1 & 1 \\ 0 & 1 \end{pmatrix} \text{ and } e^B = I + B = \begin{pmatrix} 1 & 0 \\ 1 & 1 \end{pmatrix}.$$

Then

$$e^A e^B = \begin{pmatrix} 2 & * \\ * & * \end{pmatrix} \neq \begin{pmatrix} 1 & * \\ * & * \end{pmatrix} = e^B e^A.$$

Now, set

$$A + B := C = \begin{pmatrix} 0 & 1 \\ 1 & 0 \end{pmatrix}.$$

Then

$$C^2 = I, \ C^3 = C, \ C^4 = I, \ ... \text{ etc.}$$

Thus

$$\begin{aligned} e^{A+B} &= I + C + \frac{I}{2!} + \frac{C}{3!} + \frac{I}{4!} + \cdots \\ &= \left(1 + \frac{1}{2!} + \frac{1}{4!} + \cdots \right) I + \left(1 + \frac{1}{3!} + \frac{1}{5!} + \cdots \right) C \\ &= \begin{pmatrix} \cosh 1 & \sinh 1 \\ \sinh 1 & \cosh 1 \end{pmatrix}. \end{aligned}$$

Therefore,

$$e^{A+B} \neq e^A e^B \neq e^B e^A,$$

as required.

Answer 13.2.10 ([377]) On $\mathbb{C}^2$, let

$$A = \begin{pmatrix} \pi i & 0 \\ 0 & -\pi i \end{pmatrix} \text{ and } B = \begin{pmatrix} 0 & 1 \\ 0 & -2\pi i \end{pmatrix}.$$

Then

$$AB = \begin{pmatrix} 0 & \pi i \\ 0 & -2\pi^2 \end{pmatrix} \neq \begin{pmatrix} 0 & -\pi i \\ 0 & -2\pi^2 \end{pmatrix} = BA.$$

By doing some arithmetic, we find that

$$e^A = -I, \ e^B = I \text{ and } e^{A+B} = -I.$$

Thus,

$$e^{A+B} = e^A e^B = e^B e^A \ (= -I).$$

Answer 13.2.11 ([377]) Let

$$A = \begin{pmatrix} \pi i & 0 \\ 0 & -\pi i \end{pmatrix} \text{ and } B = \begin{pmatrix} 0 & 1 \\ 0 & 0 \end{pmatrix}.$$

We have already computed e^A and e^B above. More precisely,

$$e^A = -I, \ e^B = \begin{pmatrix} 1 & 1 \\ 0 & 1 \end{pmatrix} \text{ and } e^{A+B} = -I.$$

Accordingly,

$$e^{A+B} = -I \neq \begin{pmatrix} -1 & -1 \\ 0 & -1 \end{pmatrix} = e^A e^B = e^B e^A$$

Answer 13.2.12 ([377]) Consider

$$A = \begin{pmatrix} a & 0 \\ 0 & b \end{pmatrix} \text{ and } B = \begin{pmatrix} 0 & 1 \\ 0 & 0 \end{pmatrix}$$

with a and b being complex and such that $a \neq b$. Hence

$$e^A e^B = \begin{pmatrix} e^a & e^a \\ 0 & e^b \end{pmatrix} \neq \begin{pmatrix} e^a & e^b \\ 0 & e^b \end{pmatrix} = e^B e^A.$$

To see why $e^{A+B} = e^A e^B$ we call for some complex analysis, in particular Picard's theorem ([313]: *Every entire function which is not a polynomial attains each value infinitely many times, with one possible exception*).

We have

$$e^{A+B} = \begin{pmatrix} e^a & \frac{e^a - e^b}{a-b} \\ 0 & e^b \end{pmatrix}.$$

It is clear that for $e^{A+B} = e^A e^B$ to be true, we need to see why

$$\frac{e^a - e^b}{a - b} = e^a.$$

for at least some value of a and b. Setting $z = b - a$ (and so $z \in \mathbb{C}^*$), the last displayed equation becomes equivalent to $e^z - z = 1$. By Picard's theorem, there are lots of values of a and b such that

$$e^{A+B} = e^A e^B.$$

Answer 13.2.13 ([377]) First, recall that there is an isomorphism $f : \mathbb{C} \to X$ defined by $f(x + iy) = xI + yJ$ where

$$I = \begin{pmatrix} 1 & 0 \\ 0 & 1 \end{pmatrix}, \quad J = \begin{pmatrix} 0 & -1 \\ 1 & 0 \end{pmatrix} \text{ and } X = \{xI + yJ : x, y \in \mathbb{R}\} \ (J^2 = -I).$$

Hence, we can identify each complex number $x + iy$ with the real matrix

$$\begin{pmatrix} x & -y \\ y & x \end{pmatrix}.$$

Going back to the counterexample, let $z = a + ib$ be a solution of $e^z - z = 1$ (there are lots of them but, for example, $a = 2.088843...$ and $b = 7.461489...$). Set

$$A = \begin{pmatrix} 0 & 0 & 0 & 0 \\ 0 & 0 & 0 & 0 \\ 0 & 0 & a & -b \\ 0 & 0 & b & a \end{pmatrix} \text{ and } B = \begin{pmatrix} 0 & 0 & 1 & 0 \\ 0 & 0 & 0 & 1 \\ 0 & 0 & 0 & 0 \\ 0 & 0 & 0 & 0 \end{pmatrix}.$$

Noting

$$\tilde{A} = \begin{pmatrix} a & -b \\ b & a \end{pmatrix}, \quad \tilde{B} = \begin{pmatrix} 0 & 1 \\ 1 & 0 \end{pmatrix} \text{ and } \mathbf{0} = \begin{pmatrix} 0 & 0 \\ 0 & 0 \end{pmatrix},$$

we can rewrite A and B as

$$A = \begin{pmatrix} \mathbf{0} & \mathbf{0} \\ \mathbf{0} & \tilde{A} \end{pmatrix} \text{ and } B = \begin{pmatrix} \mathbf{0} & \tilde{B} \\ \mathbf{0} & \mathbf{0} \end{pmatrix}.$$

Clearly

$$e^B = B + I = \begin{pmatrix} 1 & 0 & 1 & 0 \\ 0 & 1 & 0 & 1 \\ 0 & 0 & 1 & 0 \\ 0 & 0 & 0 & 1 \end{pmatrix}.$$

Also,

$$e^A = \begin{pmatrix} e^0 & \mathbf{0} \\ \mathbf{0} & e^{\tilde{A}} \end{pmatrix} = \begin{pmatrix} I_2 & \mathbf{0} \\ \mathbf{0} & e^{\tilde{A}} \end{pmatrix}.$$

To compute $e^{\tilde{A}}$, we can proceed as usual or we can instead exploit the fact that we have identified $\tilde{A}$ with $a + ib$ which is solution of $e^z = z + 1$. Consequently,

$$e^{\tilde{A}} = I + \tilde{A} = \begin{pmatrix} 1 + a & -b \\ b & 1 + a \end{pmatrix}.$$

Remark Before continuing, we say that thanks to this neat idea by E. M. E. Wermuth, we have avoided some long calculations. Indeed, we could have found

$$e^{\tilde{A}} = \begin{pmatrix} e^a \cos b & -e^a \sin b \\ e^a \sin b & e^a \cos b \end{pmatrix}.$$

Then, we would have to divine that $e^a \cos b = 1 + a$ and $e^a \sin b = b$ for the following reason

$$e^a \cos b + i e^a \sin b = 1 + a + ib \text{ or } e^z = z + 1.$$

Now, we finish the solution. We then find that:

$$e^A = \begin{pmatrix} 1 & 0 & 0 & 0 \\ 0 & 1 & 0 & 0 \\ 0 & 0 & a+1 & -b \\ 0 & 0 & b & a+1 \end{pmatrix}.$$

Also,

$$e^{A+B} = \begin{pmatrix} 1 & 0 & 1 & 0 \\ 0 & 1 & 0 & 1 \\ 0 & 0 & a+1 & -b \\ 0 & 0 & b & a+1 \end{pmatrix}.$$

Finally,

$$e^{A+B} = e^A e^B = \begin{pmatrix} 1 & 0 & 1 & 0 \\ 0 & 1 & 0 & 1 \\ 0 & 0 & a+1 & -b \\ 0 & 0 & b & a+1 \end{pmatrix} \neq \begin{pmatrix} 1 & 0 & a+1 & -b \\ 0 & 1 & b & a+1 \\ 0 & 0 & a+1 & -b \\ 0 & 0 & b & a+1 \end{pmatrix} = e^B e^A.$$

Further Reading For related results to similar results on exponentials, readers may wish to consult [32, 49, 166, 234, 235, 249, 256, 319, 377, 378].

Answer 13.2.14 (See [390] for yet another example) Reconsider the example given in Answer 13.2.8, i.e.

$$A = \begin{pmatrix} i\pi & 1 \\ 0 & -i\pi \end{pmatrix}.$$

Then, we already know that $e^A = e^{A^*} = -I$. Since $A + A^* = \begin{pmatrix} 0 & 1 \\ 1 & 0 \end{pmatrix}$, it can be checked that

$$e^{A+A^*} = \frac{1}{2} \begin{pmatrix} e + e^{-1} & e - e^{-1} \\ e - e^{-1} & e + e^{-1} \end{pmatrix},$$

thereby

$$e^A e^{A^*} = e^{A^*} e^A \neq e^{A+A^*},$$

as required.

Answer 13.2.15 On $\mathbb{C}^2$, consider

$$T = \begin{pmatrix} 0 & 1/2 \\ 0 & 0 \end{pmatrix}.$$

Since T is nilpotent, it results that $e^T = I + T$, i.e.

$$e^T = \begin{pmatrix} 1 & 1/2 \\ 0 & 1 \end{pmatrix}.$$

Clearly, $|T| = \begin{pmatrix} 0 & 0 \\ 0 & 1/2 \end{pmatrix}$ and so

$$e^{|T|} = \begin{pmatrix} 1 & 0 \\ 0 & \sqrt{e} \end{pmatrix} \approx \begin{pmatrix} 1 & 0 \\ 0 & 1.6487 \end{pmatrix}.$$

Clearly Re $T = \begin{pmatrix} 0 & 1/4 \\ 1/4 & 0 \end{pmatrix}$. So it can be shown that approximately

$$e^{\mathrm{Re}\, T} \approx \begin{pmatrix} 1.0314 & 0.2526 \\ 0.2526 & 1.0314 \end{pmatrix}.$$

So it only remains to compute $|e^T|$. Clearly

$$|e^T|^2 = \begin{pmatrix} 1 & 1/2 \\ 1/2 & 5/4 \end{pmatrix}$$

which has two eigenvalues: $(\sqrt{17}+9)/8$ and $(-\sqrt{17}+9)/8$. Upon diagonalizing $|e^T|^2$ we may show that approximately (up to four decimal places)

$$|e^T| \approx \begin{pmatrix} 0.9701 & 0.2425 \\ 0.2425 & 1.0914 \end{pmatrix}.$$

Accordingly, $|e^T| \neq e^{\operatorname{Re} T}$. That $e^{|T|} \not\geq |e^T|$ is because e.g. the self-adjoint matrix

$$e^{|T|} - |e^T| \approx \begin{pmatrix} 0.0299 & -0.2425 \\ -0.2425 & 0.5573 \end{pmatrix}$$

has two eigenvalues of opposite signs, namely 0.6519 and -0.0647.

Chapter 14
Nonnormal Operators

14.1 Basics

This chapter is entirely devoted to nonnormal operators. By a nonnormal operator here, we obviously mean an operator which is not normal but not only that, that is, we mean in particular those operators which are in some sense connected or bear a certain resemblance to normal ones. Moreover, most of these classes coincide with normal operators on finite-dimensional spaces.

The first class to be recalled is that of hyponormal operators. It was introduced by P. R. Halmos in [146].

Definition 14.1.1 Let $A \in B(H)$. Say that A is hyponormal if

$$\forall x \in H : \|A^*x\| \leq \|Ax\|.$$

Examples 14.1.1

(1) Obviously, every normal operator is hyponormal.
(2) The shift operator is hyponormal. The shift is the prominent example of a hyponormal operator which is not normal.
(3) More generally, every isometry is hyponormal.

Theorem 14.1.1 Let $A \in B(H)$. Then

$$A \text{ is hyponormal} \iff AA^* \leq A^*A.$$

A practical characterization of hyponormality is the following (a proof may be found in Exercise 12.3.2 in [256]):

Proposition 14.1.2 Let $A \in B(H)$. Then A is hyponormal iff there exists a contraction $K \in B(H)$ such that $A^* = KA$.

M. H. Mortad, *Counterexamples in Operator Theory*,
https://doi.org/10.1007/978-3-030-97814-3_14

Let us introduce more classes of nonnormal operators.

Definition 14.1.2 Let $A \in B(H)$. We say that the operator A is co-hyponormal if A^* is hyponormal, that is

$$\forall x \in H : \|Ax\| \leq \|A^*x\|.$$

Equivalently, this amounts to say that $AA^* \geq A^*A$.

Remark Clearly $A \in B(H)$ is normal *if and only if* A is both hyponormal *and* co-hyponormal.

Remark Following C. R. Putnam [295], an $A \in B(H)$ such that either $AA^* \leq A^*A$ or $AA^* \geq A^*A$ is called semi-normal.

The next class too was introduced by P. R. Halmos in [146].

Definition 14.1.3 Let $A \in B(H)$. We say that A is subnormal if it possesses a normal extension $N \in B(K)$ (with $N(H) \subset H$) where K is a Hilbert space larger than H, i.e. $H \subset K$. In other words, $A \in B(H)$ is subnormal if $K = H \oplus H^\perp$ and $N \in B(K)$, defined by

$$N = \begin{pmatrix} A & B \\ 0 & C \end{pmatrix},$$

is normal for some $B \in B(H^\perp, H)$ and $C \in B(H^\perp)$.

The next class was first introduced by A. Brown in [37].

Definition 14.1.4 Let $A \in B(H)$. Say that A is quasinormal if

$$A(A^*A) = (A^*A)A.$$

Remark Clearly,

$$A \text{ is quasinormal} \iff A|A| = |A|A.$$

The following result is an equivalent definition of quasinormality.

Proposition 14.1.3 *If $A = U|A|$ is the polar decomposition of A (in terms of partial isometries), then*

$$A \text{ is quasinormal} \iff U|A| = |A|U.$$

Remark Readers may wish to consult a quick and interesting review on the above classes (and other ones) by C. R. Putnam [296].

The last class to be recalled here was introduced in [169]:

Definition 14.1.5 Let $A \in B(H)$. We say that A is paranormal if

$$\|Ax\|^2 \leq \|A^2x\|$$

for any *unit* vector $x \in H$. Equivalently, $\|Ax\|^2 \leq \|A^2x\|\|x\|$ for any $x \in H$.

The following condition, equivalent to paranormality, was established by T. Ando in [5]:

Theorem 14.1.4 *Let $A \in B(H)$. Then A is paranormal if and only if*

$$A^{*2}A^2 - 2\lambda A^*A + \lambda^2 I \geq 0$$

for all $\lambda > 0$.

Remark Clearly, A is paranormal iff $A^{*2}A^2 - 2\lambda A^*A + \lambda^2 I \geq 0$ for *all real* λ (why?).

Examples 14.1.2

(1) Each normal operator is quasinormal.
(2) The shift operator $S \in B(\ell^2)$ is subnormal. One normal extension being the bilateral shift $R \in B(\ell^2(\mathbb{Z}))$.
(3) If A is subnormal, then so is A^2.
(4) The shift operator is also quasinormal. In fact, each isometry is quasinormal.
(5) The shift is a quasinormal operator which is not normal.
(6) If A is quasinormal, so are its powers A^n, $n \in \mathbb{N}$.
(7) If S is the shift operator, then $S + I$ is not quasinormal.
(8) A compact paranormal operator is normal. That was shown in [169], then differently in [93]. Hence, a paranormal (finite) matrix is normal. Readers might be interested in seeing a direct proof in the finite-dimensional setting, and this appeared in [168].

Theorem 14.1.5 *We have the following inclusions among those classes:*

$$Quasinormal \subset Subnormal \subset Hyponormal \subset Paranormal \subset Normaloid.$$

Remark When $\dim H < \infty$, subnormality, paranormality, quasinormality, and hyponormality all coincide with normality.

Using Proposition 14.1.3, we may easily prove:

Proposition 14.1.6 *If $A \in B(H)$ is an invertible quasinormal operator, then A is normal.*

A left invertible hyponormal operator need not be invertible. Simply consider the shift operator S on ℓ^2, which is hyponormal and left invertible, but it is not invertible. Nonetheless, since a right invertible hyponormal operator is automatically invertible (see e.g. Exercise 12.3.16 in [256]), an immediate consequence is:

Corollary 14.1.7 *A right invertible quasinormal operator is normal.*

14.2   Questions

14.2.1   A Hyponormal Operator Which Is Not Normal

Question 14.2.1 Give an example of a hyponormal operator which is not normal (differing from the shift).

14.2.2   An Invertible Hyponormal Operator Which Is Not Normal

Question 14.2.2 (Cf. Proposition 14.1.6) Give an example of an invertible hyponormal operator which is not normal.

14.2.3   A Hyponormal Operator with a Non-Hyponormal Square

Let $A \in B(H)$ be hyponormal and let f be some bounded Borel function, then $f(A)$ need not be hyponormal. For example, if $f(z) = \overline{z}$ and S is the shift operator, then $f(S) = S^*$ is not hyponormal. What if f is a real polynomial?

Question 14.2.3 Give an example of a hyponormal operator A such that A^2 is not hyponormal.

14.2.4   A Hyponormal Operator $A \in B(H)$ Such That $A + \lambda A^*$ is Not Hyponormal for Some $\lambda \in \mathbb{C}$

Question 14.2.4 Give a hyponormal operator $A \in B(H)$ such that $A + \lambda A^*$ is not hyponormal for some $\lambda \in \mathbb{C}$.

14.2.5   On the Failure of Some Property on the Spectrum for Hyponormal Operators

C. R. Putnam showed in [297] that: If T is hyponormal and $\lambda \in \partial\sigma(T)$ (the boundary of $\sigma(T)$), then $|\lambda| \in \sigma(|T|) \cap \sigma(|T^*|)$. In the case of the normality of T, we may replace $\lambda \in \partial\sigma(T)$ by $\lambda \in \sigma(T)$.

Question 14.2.5 Show that this slight improvement in the class of normal operators is not extendable to the class of hyponormal ones.

14.2.6   On the Failure of Some Friedland's Conjecture

In [115], S. Friedland showed that if $A \in B(H)$, then A is normal if and only if

$$(\alpha I + A + A^*)^2 \geq AA^* - A^*A \geq -(\alpha I + A + A^*)^2$$

holds for all $\alpha \in \mathbb{R}$.

He then conjectured that if $A \in B(H)$ and the inequality

$$(\alpha I + A + A^*)^2 \geq AA^* - A^*A$$

holds for all $\alpha \in \mathbb{R}$, then A is hyponormal.

Question 14.2.6 (⊛⊛) Show that this conjecture is untrue.

14.2.7   On the Failure of the ("Unitary") Polar Decomposition for Quasinormal Operators

Question 14.2.7 Show that the polar decomposition of normal operators cannot be extended to quasinormal operators.

14.2.8   The Inclusions Among the Classes of Quasinormals, Subnormals, Hyponormals, Paranormals, and Normaloids are Proper

Recall that:

Quasinormal $\subset$ Subnormal $\subset$ Hyponormal $\subset$ Paranormal $\subset$ Normaloid.

Question 14.2.8 Show that none of the reverse inclusions holds in general.

14.2.9   A Subnormal (Or Quasinormal) Operator Whose Adjoint Is Not Subnormal

Question 14.2.9 Give an example of a subnormal (resp. quasinormal) operator A such that A^* is not subnormal (resp. non-quasinormal).

14.2.10   A Paranormal Operator Whose Adjoint Is Not Paranormal

We know that if $T \in B(H)$ is hyponormal such that T^* too is hyponormal, then T is normal. Hence this applies to larger classes such as quasinormal and subnormal operators. What about the same question for the weaker class of paranormal operators? A priori, and according to Ando's paper, "the answer should be negative". Indeed, he showed in [5] that $T \in B(H)$ is normal if and only if both T and T^* are paranormal such that $\ker T = \ker T^*$. So, for a counterexample, it would suffice to choose a $T \in B(H)$ such that T and T^* are paranormal but $\ker T \neq \ker T^*$ (whereby T would not be normal). Is this possible? Curiously, T. Ando did not provide any counterexample showing the importance of the hypothesis "$\ker T = \ker T^*$". In other words, is there any explicit operator T for which T and T^* are paranormal without being normal? What has complicated this situation a little more is that no counterexample has appeared since Ando's paper. An answer finally was obtained when I asked Professor Kotaro Tanahashi about a (possible) counterexample. He then suggested that an answer should be known to either Professor Takeaki Yamazaki or Professor Masatoshi Ito. Both of them sent me the reference [393] in which T. Yamazaki and M. Yanagida showed (Corollary 3) that the paranormality of both T and T^* does make T normal. In other words, the condition $\ker T = \ker T^*$ is guaranteed by the paranormality of both T and T^*. In the end, readers who know unbounded operators might be interested in knowing that this result is not true for unbounded paranormal operators even when $\ker T = \ker T^* = \{0\}$. See Question 25.1.5.

Question 14.2.10 Find a paranormal operator whose adjoint is not paranormal.

14.2.11 A Semi-Hyponormal Operator Which Is Not Hyponormal

We start with a definition: An operator $A \in B(H)$ such that $|A^*| \leq |A|$, that is, $(AA^*)^{\frac{1}{2}} \leq (A^*A)^{\frac{1}{2}}$ is called semi-hyponormal. More generally, an $A \in B(H)$ such that

$$(AA^*)^p \leq (A^*A)^p,$$

for some positive p, is called p-hyponormal.

Remark A hyponormal operator is necessarily p-hyponormal for any $p \in (0, 1]$ thanks to the Löwner-Heinz inequality (see e.g. Page 127 in [123]).

The following result is of interest.

Proposition 14.2.1 (Page 183 in [123]) *Let $p \in (0, 1]$ and let A be p-hyponormal. Then A^n is $\frac{p}{n}$-hyponormal for any $n \in \mathbb{N}$.*

Question 14.2.11 Provide an example of a semi-hyponormal operator which is not hyponormal.

14.2.12 Embry's Criterion for Subnormality: $A = A^*A^2$ Does Not Even Yield the Hyponormality of $A \in B(H)$

Among the practical characterizations of subnormality, we have: $A \in B(H)$ is subnormal and the minimal normal extension of A is a partial isometry if and only if $\|A\| \leq 1$ and $A = A^*A^2$. This appeared in [103]. In particular, a simple *sufficient* condition for subnormality is:

$$\|A\| \leq 1 \text{ and } A = A^*A^2.$$

Question 14.2.12 Show that assuming $A = A^*A^2$ only does not force $A \in B(H)$ to even be hyponormal.

14.2.13   An Invertible Hyponormal Operator Which Is Not Subnormal

We already know that any invertible quasinormal operator is normal. We also saw above that an invertible hyponormal need not be normal (see Answer 14.2.2). By looking closely at that counterexample, we see that it is actually subnormal. Therefore, an invertible subnormal operator cannot be normal either.

Remark It is also easy to see that an invertible subnormal operator need not be quasinormal (why?). See also the last remark below Answer 14.2.35.

Question 14.2.13 Find an invertible hyponormal operator which is not subnormal.

14.2.14   Two Commuting Hyponormal Operators A and B Such That $A + B$ Is Not Hyponormal

Is the product of two commuting hyponormal operators necessarily hyponormal? The answer is negative. In fact, we already have an even stronger counterexample. Indeed, in Answer 14.2.3 we have a hyponormal operator A such that A^2 is not one. What about the sum of two commuting hyponormal operators?

It is noteworthy that the product and the sum of two hyponormal operators A and B such that $BA^* = A^*B$ (hence $AB^* = B^*A$) are both hyponormal. Indeed,

$$(A + B)(A + B)^* = AA^* + BA^* + AB^* + BB^* \leq A^*A + BA^* + AB^* + B^*B.$$

But $BA^* + AB^* = A^*B + B^*A$ whereby

$$A^*A + BA^* + AB^* + B^*B = A^*A + A^*B + B^*A + B^*B = (A + B)^*(A + B),$$

i.e. $A + B$ is hyponormal.

Let us now turn to the product. Clearly

$$AB(AB)^* = ABB^*A^* \leq AB^*BA^* = B^*ABA^* = B^*AA^*B \leq B^*A^*AB,$$

that is, $AB(AB)^* \leq (AB)^*AB$, i.e. AB is hyponormal.

Question 14.2.14 (⊛⊛) Let $A, B \in B(H)$ be two commuting hyponormal operators. Show that $A + B$ does not have to be hyponormal.

14.2.15 Two Quasinormal Operators $A, B \in B(H)$ Such That $AB = BA = 0$ Yet $A + B$ Is Not Even Hyponormal

As we saw just above, we have a pair of a subnormal operator and a quasinormal one which commute, but their sum is not hyponormal. What about the case of two commuting quasinormal operators? Well, their sum may even fail to be hyponormal.

Question 14.2.15 Find two quasinormal operators $A, B \in B(H)$ such that $AB = BA = 0$ yet their sum $A + B$ is not hyponormal.

14.2.16 Two Commuting Subnormal Operators S and T Such That Neither ST nor $S + T$ Is Subnormal

Let S and T be two commuting subnormal operators with normal extensions N and M, respectively. It is natural to ask whether N and M necessarily commute. If that is the case, then $N + M$ and NM would be normal, whereby ST and $S + T$ would be subnormal.

Question 14.2.16 Find two commuting subnormal operators S and T such that neither ST nor $S + T$ is subnormal.

14.2.17 The Failure of the Fuglede Theorem for Quasinormals et al.

Question 14.2.17 Does the Fuglede theorem hold for quasinormal operators? That is, if $A, T \in B(H)$ are such that A is quasinormal and $TA = AT$, then is it true that $TA^* = A^*T$?

14.2.18 A Unitary A and a Quasinormal B with $TB = AT$ but $TB^* \neq A^*T$

Readers were asked in Question 14.2.17 to show the failure of the Fuglede theorem for the class of quasinormal operators. Hence the Fuglede-Putnam theorem would not hold for quasinormal operators either. However, there is still a case which is worth looking at in Fuglede-Putnam's version.

> **Question 14.2.18** Find $A, B, T \in B(H)$ where A is unitary and B is quasinormal such that $TB = AT$, but $TB^* \neq A^*T$.

14.2.19 Two Double Commuting Nonnormal Hyponormal Operators

Many results which hold for normal operators A and B with $AB = BA$, do not necessarily have to remain valid if A and B are assumed to be in some class of nonnormal operators. The main reason is that in the event of the normality we also have $AB^* = B^*A$ thanks to the Fuglede theorem. So, many authors, when trying to generalize results already holding for normal operators to nonnormal ones, assume both conditions $AB = BA$ and $AB^* = B^*A$. Are there nontrivial quasinormal (for instance) operators obeying these two conditions together?

> **Question 14.2.19 (Cf. Question 12.2.5)** Let $\dim H = \infty$. Find two hyponormal (or subnormal or quasinormal) operators A and B both in $B(H)$, none of them is normal such that
>
> $$AB = BA \text{ and } AB^* = B^*A.$$

14.2.20 Two Hyponormal (Or Quasinormal) Operators A and B Such That $AB^* = B^*A$ but $AB \neq BA$

We already know that if A and B are two hyponormal (or even quasinormal) operators, then $AB = BA$ does not necessarily yield $AB^* = B^*A$. The counterexample was pretty simple. Readers, however, should not pass over the converse question "$AB^* = B^*A \Rightarrow AB = BA$?", thinking that the answer should be as easy as in the first case. This is in fact a highly non-obvious question at all.

One of the examples to be given needs the notion of tensor products of bounded linear operators, so we recall some useful properties by *assuming that readers are already familiar with tensor products of Hilbert spaces* (a good reference for all that is [184]). We restrict ourselves to the case of two operators: If H_1, H_2, K_1, K_2 are Hilbert spaces and $A_1 \in B(H_1, K_1)$ and $A_2 \in B(H_2, K_2)$, then there is a unique bounded linear operator A from $H_1 \otimes H_2$ into $K_1 \otimes K_2$ such that

$$A(x_1 \otimes x_2) = A_1 x_1 \otimes A_2 x_2$$

where $x_1 \in H_1$ and $x_2 \in H_2$. We call the operator A the tensor product of the operators A_1 and A_2. Furthermore,

$$(A_1 \otimes A_2)(B_1 \otimes B_2) = A_1 B_1 \otimes A_2 B_2$$

(with $B_1 \in B(K_1, H_1)$ and $B_2 \in B(K_2, H_2)$). Also,

$$(A_1 \otimes A_2)^* = A_1^* \otimes A_2^*.$$

It is also easy to see that if A_1, $A_2 \geq 0$, then $A_1 \otimes A_2 \geq 0$.

Remark Other properties regarding self-adjointness, unitarity, normality, quasinormality, hyponormality, and subnormality may be consulted in: [164, 165, 349]. For example, it is shown among others e.g. two nonzero operators A and B are normal (resp. quasinormal, subnormal, hyponormal) if and only if $A \otimes B$ is normal (resp. quasinormal, subnormal, hyponormal).

See also [39] for the spectrum of tensor products of operators.

Question 14.2.20 (⊛⊛) Find two hyponormal operators A and B such that $AB^* = B^*A$ but $AB \neq BA$. What about quasinormal operators?

14.2.21 Two Commuting (Resp. Double Commuting) Paranormal Operators Whose Tensor (Resp. Usual) Product Fails to Remain Paranormal

We are usually used to see that the operation of taking the tensor product $A \otimes B$ of two (bounded) nonzero linear operators A and B preserves many of the properties of A and B. For example, as noticed above, it preserves quasinormality, subnormality and hyponormality. Does this extend to paranormal operators?

Also, what about the paranormality of the product of two commuting (or even double commuting) paranormal operators?

Question 14.2.21 (⊛) Find two paranormal operators whose tensor product fails to remain paranormal. What about the usual product of two double commuting paranormal operators?

14.2.22   A Subnormal Operator Such That $|A^2| \neq |A|^2$

It is readily seen that if $A \in B(H)$ is normal, then $|A^n| = |A|^n$ for all $n \in \mathbb{N}$. The same result remains true for quasinormal operators as shown in e.g. [365] (see also Question 14.2.24 for more details). What about hyponormal or subnormal operators?

Question 14.2.22 Find a subnormal operator $A \in B(H)$ such that:

$$|A^2| \neq |A|^2.$$

14.2.23   A Subnormal A Such That $A|A| \neq |A|A$

Question 14.2.23 Find a subnormal operator $A \in B(H)$ such that:

$$A|A| \neq |A|A.$$

14.2.24   A Non-Quasinormal T Such That $T^{*2}T^2 = (T^*T)^2$

As indicated by Jabłoński et al. in [172], since late 1980s specialists already knew that if $T \in B(H)$, then

$$T \text{ is quasinormal} \Longleftrightarrow T^{*n}T^n = (T^*T)^n, \ n = 2, 3$$

(cf. [176]). The authors in [172] even extended this result to unbounded closed operators. They also gave many interesting results and examples. Unfortunately, the authors apparently missed a closely related paper to theirs by M. Uchiyama

[365]. The latter contains many interesting characterizations in both the bounded and the unbounded cases. Since these Uchiyama's results are apparently not very well known to readers, and since they are closely related to the topics of this chapter, we recall some of them:

Theorem 14.2.2 ([365]) *Let H be a separable Hilbert space and let $T \in B(H)$. Then the following statements are equivalent:*

(1) *T is quasinormal.*
(2) *$|T^n| = |T|^n$ for some $n \in \mathbb{N}$.*
(3) *There are integers i and k such that $|T^n| = |T|^n$ for $n = i, i+1, k$ and $k+1$ where $1 \leq i < k$.*

Proposition 14.2.3 ([365]) *If $T \in B(H)$ is subnormal and such that $|T^n| = |T|^n$ for some $n \geq 2$, then T is quasinormal.*

In fact, the proof of the preceding proposition contains some interesting inequalities which could be of some interest elsewhere, namely: If T is subnormal, then

$$|T| \leq |T^2|^{1/2} \leq |T^3|^{1/3} \leq \cdots \leq |T^n|^{1/n} \leq \cdots$$

for all $n \in \mathbb{N}$.

Proposition 14.2.4 ([365]) *If $T \in B(H)$ is hyponormal and such that $|T^n| = |T|^n$ and $|T^{n+1}| = |T|^{n+1}$ for some $n \geq 3$, then T is quasinormal.*

Question 14.2.24 Find an operator $T \in B(H)$ such that $T^{*2}T^2 = (T^*T)^2$ but T is not quasinormal.

14.2.25 A Non-Subnormal Operator Whose Powers Are All Hyponormal

Question 14.2.25 (⊛) [345] We have already seen an example of a hyponormal operator whose square is not hyponormal. Readers are asked whether an operator must be subnormal if all its powers are hyponormal (this question was asked by S. Berberian).

Remark A more difficult version of the previous question reads: If $T \in B(H)$ is such that $p(T)$ is hyponormal for every polynomial $p \in \mathbb{C}[z]$, then must T be subnormal? This was answered negatively and was announced in [78].

14.2.26   A Hyponormal Operator $T \in B(H)$ Such That All T^n, $n \geq 2$ Are Subnormal but T Is Not

Let $T \in B(H)$. It is clear that T^n can be normal while T is not. A simple instance is to take any nonzero T such that $T^2 = 0$. Then T^2 is normal while T is not normal (in fact, it cannot even be hyponormal). What if T^n is assumed to be normal for some n and we suppose that T is hyponormal? In this case, T is necessarily normal as shown in [344], say.

Now what if we assume that T^n is subnormal for some n and that T is hyponormal, then could it be true that T must be subnormal? The answer this time is no. Readers will be asked below to manufacture a counterexample. Following J. G. Stampfli [346]: Let H be a separable Hilbert space and let (e_n) be some orthonormal basis in H. Consider the weighted shift, i.e. $T e_n = a_n e_{n+1}$ for $n = 1, 2, \cdots$ and (a_n) is a bounded sequence. Say that T is a monotone shift if

$$|a_n| \leq |a_{n+1}| \leq M$$

for some positive M and all n. Clearly every monotone shift is hyponormal.

Question 14.2.26 (⊛) Find a hyponormal operator $T \in B(H)$ such that all $T^n, n \geq 2$ are subnormal but T is not.

Hint: Choose an appropriate monotone shift. Then use the following result which was established in [346]: If T is a monotone shift as defined above, such that $a_m \neq 0$ for all m, T is subnormal and $|a_k| = |a_{k+1}|$ for some k, then $|a_j| = |a_{j+1}|$ for $j = 2, 3, \cdots$ (and a_1 is arbitrary).

Remark See Question 14.2.38 for related questions.

14.2.27   Two Quasinormal Operators A and B Such That $AB = BA$ and $|AB| \neq |A||B|$

As already observed, one of the virtues of quasinormality is the fact that

$$|A^n| = |A|^n$$

for all $n \in \mathbb{N}$ whenever A is a quasinormal operator (and vice versa). One can perhaps be tempted to conjecture that if we are given a pair of quasinormal operators A and B such that $AB = BA$, then $|AB| = |A||B|$? What supports this conjecture in part is the fact that it holds true when $\dim H < \infty$, where in this setting "quasinormal=normal". Hence the Fuglede theorem becomes available and gives $AB^* = B^*A$, which allows to readily obtain $|AB| = |A||B|$.

Question 14.2.27 Find two quasinormal operators A and B such that $AB = BA$ yet

$$|AB| \neq |A||B|.$$

14.2.28 Two Quasinormal Operators A and B Such That $AB = BA$ Yet $A|B| \neq |B|A$

Question 14.2.28 Find two quasinormal operators A and B such that $AB = BA$ yet $A|B| \neq |B|A$.

14.2.29 Two Quasinormal Operators A and B Such That $AB^* = B^*A$ and $B|A| = |A|B$ Yet $A|B| \neq |B|A$

Question 14.2.29 Find two quasinormal operators A and B such that $AB^* = B^*A$ and $B|A| = |A|B$ yet $A|B| \neq |B|A$.

14.2.30 The Failure of Some Kaplansky's Theorem for Hyponormal Operators

The following result was shown in [17]: Let A and B be two bounded operators on a Hilbert space such that A is normal and AB is hyponormal. Then

$$AA^*B = BAA^* \implies BA \text{ is hyponormal.}$$

Question 14.2.30 Show that the reverse implication does not hold even when A is taken to be self-adjoint.

14.2.31 The Inequality $|\langle Ax, x\rangle| \leq \langle |A|x, x\rangle$ for All x Does Not Yield Hyponormality

It was shown in [179] (cf. [99]) that when dim $H < \infty$, then

$$|\langle Ax, x\rangle| \leq \langle |A|x, x\rangle$$

for all $x \in H$, holds if and only if A is normal.

Question 14.2.31 Show that this is not necessarily the case in the event dim $H = \infty$.

14.2.32 The Failure of Reid's Inequality for Hyponormal Operators

Recall Reid's inequality: Let $A, K \in B(H)$ be such that A is positive and AK is self-adjoint (or even co-hyponormal as in [85]). Then

$$|\langle AKx, x\rangle| \leq ||K||\langle Ax, x\rangle$$

for all $x \in H$.

Question 14.2.32 Does Reid's inequality hold when AK is hyponormal and by keeping A positive?

14.2.33 The Weak Limit of Sequences of Hyponormal Operators

Let (A_n) be a sequence of hyponormal operators in $B(H)$. That is,

$$0 \leq A_n A_n^* \leq A_n^* A_n, \ \forall n \in \mathbb{N}.$$

If (A_n) converges uniformly to A, then A is necessarily hyponormal. This is fairly easy to see. Notice that the same result is true in the case of strong convergence (see e.g. Exercise 12.3.5 in [256]).

Question 14.2.33 Is the weak limit of a sequence of hyponormal operators necessarily hyponormal?

14.2.34 Strong (and Weak) Limit of Sequences of Quasinormal Operators

We have already observed just before Question 14.2.33 that strong limits of sequences of hyponormal operators remain hyponormal. It is also known that strong limits of subnormal operators are subnormal (see e.g. [68]). It should be added that every (nonnormal) subnormal operator is a strong limit of a sequence of hyponormal (non-subnormal) operators. That appeared in [62].

Let us now turn to quasinormal operators. It is easy to see that the uniform limit of a sequence of quasinormal operators is quasinormal.

Indeed, if (A_n) is a sequence of quasinormal operators such that $\|A_n - A\| \to 0$, then A is quasinormal. To see this, observe that the quasinormality of (A_n) signifies that:

$$A_n|A_n| = |A_n|A_n \text{ for } n = 1, 2, \cdots .$$

Then since $\|A_n - A\| \to 0$ and $\||A_n| - |A|\| \to 0$, we have

$$A|A| \longleftarrow A_n|A_n| = |A_n|A_n \longrightarrow |A|A$$

(uniformly). By the uniqueness of the limit, we get $A|A| = |A|A$, i.e. A is quasinormal.

Question 14.2.34 Is the strong limit of a sequence of quasinormal operators necessarily quasinormal? What about their weak limit?

14.2.35 A Quasinormal Operator $A \in B(H)$ Such That e^A Is Not Quasinormal

Question 14.2.35 Find a quasinormal operator $A \in B(H)$ such that e^A is not quasinormal.

14.2.36 An Invertible Subnormal Operator A Without Any Bounded Square Root

As we already know, every normal operator (on any Hilbert space) has a square root, and so does every invertible operator on a finite-dimensional complex Hilbert space. We are also aware that quasinormal operators need not have square roots (e.g. the unilateral shift on ℓ^2). One is therefore tempted to believe that invertible hyponormal or subnormal operators could have square roots? In some sense, this is the best one can hope for in terms of invertible operators which could potentially have square roots. Indeed, the next stronger candidate is the class of quasinormal operators, but they are excluded here for they become normal as soon as we impose invertibility on them, as alluded to on many occasions.

> **Question 14.2.36** (⊛⊛) Find an invertible subnormal operator A not possessing any bounded square root.

14.2.37 An Invertible Operator Which Is Not the Exponential of Any Operator

Each invertible normal operator is the exponential of some normal operator (see e.g. Exercise 10, Page 342 in [314]). Another proof may be found in [320].

> **Question 14.2.37** Find an invertible operator which is not the exponential of any operator (cf. Question 13.2.1). What about if we further assume hyponormality or subnormality (besides invertibility) in both the assumption and the conclusion?

14.2.38 A Positive Answer to the Curto-Lee-Yoon Conjecture About Subnormal and Quasinormal Operators, and a Related Question

Let T be a subnormal operator and suppose that T^2 is quasinormal. Must T be quasinormal? This question was asked in [79]. The authors of the same paper provided a partial (positive) answer by further assuming that T is left invertible.

Then, I too obtained a couple of simple contributions to their conjecture. Recently, after consulting some operator theorists, two different proofs were communicated to me independently, thereby answering the above conjecture positively. The proofs are by Professors Paweł Pietrzycki and Jan Stochel. Notice that Pietrzycki's proof uses [290, 291], whereas Stochel's uses [103, 182].

To summarize, Pietrzycki-Stochel established the following stronger version:

Theorem 14.2.5 ([292]) *If $A \in B(H)$ is subnormal such that A^n is quasinormal for some integer $n \geq 2$, then A is quasinormal.*

Before asking a related and natural question, it is worth noticing (this was actually one of the simple contributions I had before the complete answer came to light) that if T is subnormal, then T^2 is quasinormal iff T^n is quasinormal for $n = 2, 3, \cdots$. In other words, the initial version by Curto et al. and the Pietrzycki a priori stronger version are in fact equivalent.

Let me give a proof for the quasinormality of T^3 and T^4 say.

Since T is subnormal, it follows

$$|T| \leq |T^2|^{1/2} \leq |T^3|^{1/3} \leq \cdots \leq |T^n|^{1/n} \leq \cdots$$

for all $n \in \mathbb{N}$ (see the preamble to Question 14.2.24). Since T^2 is quasinormal, it ensures that $|T^4| = |T^2|^2$. So, by the above inequalities, we have in particular that:

$$|T^2|^{1/2} \leq |T^3|^{1/3} \leq |T^4|^{1/4} = |T^2|^{1/2}.$$

Hence

$$|T^3|^{1/3} = |T^2|^{1/2} \text{ or merely } |T^3|^2 = |T^2|^3.$$

Since T^2 is quasinormal, it follows that $|T^2|^3 = |T^6|$, i.e. $|T^3|^2 = |T^6|$ and so T^3 becomes quasinormal (for T^3 is already subnormal hence hyponormal, cf. [103]). Hence $|T^3|^3 = |T^9|$ and so $|T^3|^{1/3} = |T^9|^{1/9}$. Therefore, by the above inequalities,

$$|T^3|^{1/3} = |T^4|^{1/4} = |T^5|^{1/5} = |T^6|^{1/6} = |T^7|^{1/7} = |T^8|^{1/8}$$

whereby $|T^4|^2 = |T^8|$, i.e. T^4 is quasinormal (since T^4 is subnormal).

Question 14.2.38 Find two commuting subnormal operators A and B whose product AB is (even) unitary yet neither A nor B is quasinormal.

14.2.39 A Subnormal Operator T Such That p(T) Is Quasinormal for Some Non-Constant Polynomial p yet T Is Not Quasinormal

Question 14.2.39 Find a subnormal operator T which is not quasinormal such that $p(T)$ is quasinormal for some non-constant polynomial p.

14.2.40 A Binormal Operator Which Is Not Normal

Following [41], say that $T \in B(H)$ is binormal if TT^* commutes with T^*T. That is

$$(TT^*)(T^*T) = (T^*T)(TT^*).$$

With the aid of the positiveness of both TT^* and T^*T, it is seen that

$$T \text{ is binormal} \iff |T||T^*| = |T^*||T|.$$

The set of binormal operators from H into H is denoted by (BN).

Remark See [168] for a survey of properties of binormal *matrices*.

Question 14.2.40 Give an example of a binormal operator which is not normal. What about when invertibility is also assumed?

14.2.41 An Invertible Binormal Operator T Such That T^2 Is Not Binormal

One may wonder whether the product of two commuting binormal operators must be binormal. The answer is negative even under some extra conditions.

Question 14.2.41 Give an example of an invertible binormal operator T such that T^2 fails to be binormal.

14.2.42 Two Double Commuting Binormal Operators Whose Sum Is Not Binormal

Question 14.2.42 Find two double commuting binormal operators T and S such that $T + S$ is not binormal.

14.2.43 A Non-Binormal Operator T Such That $T^n = 0$ for Some Integer $n \geq 3$

Obviously, if $T \in B(H)$ is such that $T^2 = 0$, then T is binormal. What about higher powers?

Question 14.2.43 Show, by giving a counterexample, that if $T^n = 0$ for some integer $n \geq 3$, then T need not be binormal.

14.2.44 A Subnormal Operator Which Is Not Binormal

It is easy to see that any quasinormal is automatically binormal. What about other classes of nonnormal operators?

Question 14.2.44 Find a subnormal operator which is not binormal.

14.2.45 A Hyponormal Binormal Operator Whose Square Is Not Binormal

Denote by $(BN)^+$ the class of (BN) operators which are also hyponormal.

Question 14.2.45 (⊛) Find an example of an $A \in (BN)^+$ such that $A^2 \notin (BN)$.

14.2.46 Binormal Operators and the Polar Decomposition

In [171], the authors show that if $T = V|T|$ is the polar decomposition (where V is a partial isometry) such that both T and T^2 are both binormal, then

$$T^2 = V^2|T^2| \text{ and } T^4 = V^4|T^4|$$

(this is obvious for a normal T and for all powers).

Question 14.2.46 Find a binormal $T \in B(H)$ whose polar decomposition is $T = V|T|$ (in terms of a partial isometry V) such that T^2 too is binormal and yet

$$T^3 \neq V^3|T^3|.$$

14.2.47 Does the θ-Class Contain Subnormals? Hyponormals?

In [44], S. L. Campbell introduced the θ-class of operators as being the set of all linear bounded operators T acting on a separable Hilbert space H for which T^*T and $T + T^*$ commute. Two related papers followed up [45, 46]. Obviously, each isometry is in the θ-class, and so is any normal operator.

As observed by S. L. Campbell in [45], operators in the θ-class have many hyponormal-like properties. For example, he showed in the previous reference that if T is in the θ-class such that T^* is hyponormal, then T is normal. It is worth noting that compact operators in the θ-class are normal (Corollary 1 in [44]). Thus, when dim $H < \infty$, the θ-class coincides with the class of normal operators.

It is simple to show that any quasinormal operator is in the θ-class. Indeed, let T be quasinormal, i.e. $T|T| = |T|T$. By the self-adjointness of $|T|$, it follows that $T^*|T| = |T|T^*$ and so $(T + T^*)|T| = |T|(T + T^*)$. Whence $(T + T^*)|T|^2 = |T|^2(T + T^*)$, i.e. T^*T and $T + T^*$ commute.

Observe finally that if T is both binormal and in the θ-class, then T is necessarily quasinormal (Theorem 3 in [100]).

Question 14.2.47 Does the θ-class contain subnormals? hyponormals?

14.2.48 An Operator in the θ-Class Which Is Not Even Paranormal

Operators in the θ-class do not have to be quasinormal, e.g. consider the non-quasinormal operator $T = S + I$ where S is the standard shift and I is the identity operator (both defined on ℓ^2). The fact that

$$T^*T = T + T^* = 2I + S^* + S$$

tells us that T is in θ-class.

More generally, we know from [43] that if T is quasinormal and A is self-adjoint such that $AT = TA$, then $T + A$ is in the θ-class. Now, Campbell-Gellar [46] gave an example of an operator in the θ-class which is not hyponormal. So, can operators in the θ-class be paranormal?

> **Question 14.2.48** (⊛) Give an example of an operator in the θ-class which is not even paranormal.

14.2.49 Is the θ-Class Closed Under Addition?

> **Question 14.2.49** Find two commuting operators in the θ-class whose sum is not in the θ-class. What about the sum of two commuting quasinormal operators? Is it in the θ-class?

14.2.50 Is the θ-Class Closed Under the Usual Product of Operators?

> **Question 14.2.50** Find an operator T in the θ-class such that T^2 is not in the θ-class.

14.2.51 Posinormal Operators

Here, we turn to another related class introduced by H. C. Rhaly, Jr [307]:

Definition 14.2.1 Let $A \in B(H)$. We say that A is posinormal if there is a positive $P \in B(H)$ (called the interrupter) such that $AA^* = A^*PA$. If A^* is posinormal, then A is called co-posinormal.

Remark The authors in [173] showed that $AA^* = A^*PA$ is in fact equivalent to $AA^* \leq A^*PA$.

Theorem 14.2.6 ([307]) *Let $A \in B(H)$. Then*

$$A \text{ is posinormal} \iff AA^* \leq \lambda^2 A^* A$$

for some $\lambda \geq 0$.

Corollary 14.2.7 *Hyponormal operators are posinormal.*

Proposition 14.2.8 ([307]) *Each invertible operator is posinormal.*

Corollary 14.2.9 *Every invertible operator is co-posinormal.*

Proposition 14.2.10 ([173]) *If $A \in B(H)$ is posinormal, then*

$$\ker A = \ker A^2.$$

Example 14.2.1 If S is the usual shift operator, then S^* is not posinormal for

$$\ker(S^*) \neq \ker(S^{*2}).$$

Remark More examples will be seen below, in particular, the fact that posinormal operators *do not* coincide with the normal ones in a finite-dimensional setting.

Question 14.2.51

(1) Give an example of a posinormal operator which is not normal.
(2) Find a posinormal operator A such that $A + \lambda I$ is not posinormal (where $\lambda \in \mathbb{C}^*$).
(3) Find $A \in B(H)$ which is both posinormal and co-posinormal yet A is not normal.
(4) Find a posinormal operator $A \in B(H)$ with a real spectrum yet A is not self-adjoint.
(5) Find a posinormal operator which is not normaloid.

14.2.52 (α, β)-*Normal Operators*

Following [94], we say that $A \in B(H)$ is (α, β)-normal $(0 \leq \alpha \leq 1 \leq \beta)$ if:

$$\alpha^2 A^* A \leq A A^* \leq \beta^2 A^* A.$$

Observe that this class constitutes a generalization of the class of posinormal operators.

Question 14.2.52 Find an (α, β)-normal operator which is not normal.

Answers

Answer 14.2.1 We give an example based on the shift operator which will also be used below for a different aim. Let S be the usual shift, and set $A = S^* + 2S$. Then

$$A^* A - A A^* = 3I - 3SS^* = 3(I - SS^*).$$

Therefore, A is not normal while it is hyponormal.

Answer 14.2.2 The example is simple. First, recall that if A is hyponormal, then $A - \lambda I$ stays hyponormal for any scalar λ. So, let $A \in B(H)$ be a hyponormal operator which is not normal. Then $A - \lambda I$ is hyponormal but nonnormal. So to get the desired example, just choose $\lambda \notin \sigma(A)$. More explicitly, let S be the usual shift operator on ℓ^2. Then e.g. $S + 2I$ is hyponormal and invertible without being normal.

Answer 14.2.3 ([170]) Take the operator $A = S^* + 2S$ where S is the shift operator. Then we have already seen that A is hyponormal. So, we need only check that A^2 is not hyponormal. It suffices therefore to find an $x_0 \in \ell^2(\mathbb{N})$ such that

$$\|(A^2)^* x_0\| = \|(A^*)^2 x_0\| \nleq \|A^2 x_0\|,$$

that is, $\|(A^*)^2 x_0\| > \|A^2 x_0\|$. Readers can then that one possible choice is $x_0 = (1, 0, -2, 0, \cdots)$, which finally gives

$$\|(A^*)^2 x_0\| = \sqrt{89} > \sqrt{80} = \|A^2 x_0\|.$$

Answer 14.2.4 Let $\lambda = 3$ and set $A = S + 3S^*$ where S is the shift operator. Then we have

$$AA^* = (S + 3S^*)(S^* + 3S) = SS^* + 3S^{*2} + 3S^2 + 9S^* S$$

and

$$A^* A = (S^* + 3S)(S + 3S^*) = S^* S + 3S^2 + 3S^{*2} + 9SS^*.$$

Remembering that $S^*S = I$, we obtain after simplification

$$AA^* - A^*A = 8I - 8SS^* \not\leq 0.$$

Accordingly, A is not hyponormal (it is, however, co-hyponormal).

Answer 14.2.5 We are asked to find a hyponormal operator T such that $\lambda \in \sigma(T)$ with $|\lambda| \notin \sigma(|T|) \cap \sigma(|T^*|)$.

Let S be the standard shift operator on ℓ^2. Then it is well known to readers that $\sigma(S) = \{z \in \mathbb{C} : |z| \leq 1\}$, $\sigma(S^*S) = \{1\}$ and that $\sigma(SS^*) = \{0, 1\}$. Since SS^* is positive, it results that

$$\sigma(|S^*|) = \sigma(\sqrt{SS^*}) = \{0, 1\}$$

too. Therefore, $\sigma(|S|) \cap \sigma(|S^*|) = \{1\}$, whereby e.g. $1/2 \in \sigma(S)$ whereas $1/2 \notin \sigma(|S|) \cap \sigma(|S^*|)$, as needed.

Answer 14.2.6 ([396]) The conjecture had resisted solutions for almost 10 years. It was then resolved negatively by T. Yoshino in [396]. Without further ado, let us give the counterexample. Let $(e_n)_{n \geq 0}$ be an orthonormal basis of H, and let $Ae_0 = ae_1$ and $Ae_n = e_{n+1}$ for $n = 1, 2, \cdots$ where

$$1 < a \leq \sqrt{\frac{5 - 2\sqrt{2}}{2}}.$$

Hence

$$A^*e_0 = 0, \ A^*e_1 = ae_0, \ A^*e_n = e_{n-1}, \ n = 2, 3, \cdots.$$

In other words, if $x = (x_0, x_1, x_2, \cdots) \in \ell^2$, then

$$A(x_0, x_1, x_2, \cdots) = (0, ax_0, x_1, x_2, \cdots)$$

and

$$A^*(x_0, x_1, x_2, \cdots) = (ax_1, x_2, x_3, \cdots).$$

Note also that $\|A\| = a$.

That A is not hyponormal follows easily from the observation

$$\|A^*e_1\| = \|ae_0\| = a > 1 = \|e_2\| = \|Ae_1\|.$$

It "only" remains to show that $(\alpha I + A + A^*)^2 \geq AA^* - A^*A$ holds for every $\alpha \in \mathbb{R}$. Let $x \in \mathbb{R}$ and let $\alpha \in \mathbb{R}$. It can easily be checked that

$$\langle (AA^* - A^*A)x, x \rangle = \|A^*x\|^2 - \|Ax\|^2 = -a^2|x_0|^2 + (a^2 - 1)|x_1|^2.$$

It is seen that

$$(\alpha I + A + A^*)(x_0, x_1, x_2, \cdots) = (\alpha x_0 + ax_1, ax_0 + \alpha x_1 + x_2, x_1 + \alpha x_2 + x_3, \cdots)$$

and so

$$\|(\alpha I + A + A^*)x\|^2 = |\alpha x_0 + ax_1|^2 + |ax_0 + \alpha x_1 + x_2|^2 + |x_1 + \alpha x_2 + x_3|^2 + \cdots$$

Since $\alpha I + A + A^*$ is self-adjoint, we have

$$\|(\alpha I + A + A^*)x\|^2 = \langle(\alpha I + A + A^*)^2 x, x\rangle.$$

Hence clearly

$$\langle(\alpha I + A + A^*)^2 x, x\rangle - \langle(AA^* - A^*A)x, x\rangle$$
$$\geq |\alpha x_0 + ax_1|^2 + a^2|x_0|^2 - (a^2 - 1)|x_1|^2$$
$$= (\alpha^2 + a^2)|x_0|^2 + \alpha a(x_0\overline{x_1} + \overline{x_0}x_1) + |x_1|^2.$$

If $|\alpha| \leq a/\sqrt{a^2 - 1}$, then $\alpha^2 a^2 \leq \alpha^2 + a^2$. Therefore,

$$\langle(\alpha I + A + A^*)^2 x, x\rangle - \langle(AA^* - A^*A)x, x\rangle$$
$$\geq \alpha^2 a^2 |x_0|^2 + \alpha a(x_0\overline{x_1} + \overline{x_0}x_1) + |x_1|^2$$
$$= |\alpha a x_0 + x_1|^2 \geq 0.$$

Thus, the inequality

$$(\alpha I + A + A^*)^2 \geq AA^* - A^*A$$

holds for all α such that $|\alpha| \leq a/\sqrt{a^2 - 1}$.
  On the other hand, as $\|A\| = a$ we may write

$$\|(\alpha I + A + A^*)x\| \geq |\alpha|\|x\| - \|A + A^*\|\|x\| \geq (|\alpha| - 2a)\|x\|.$$

Since $1 < a \leq \sqrt{(5 - 2\sqrt{2})/2}$, it follows that

$$0 < a^2 - 1 \leq \left(1 - \frac{\sqrt{2}}{2}\right)^2 \text{ or } 0 < \sqrt{a^2 - 1} \leq 1 - \frac{\sqrt{2}}{2}.$$

Consequently,

$$\frac{a}{\sqrt{a^2 - 1}} \geq \frac{a}{1 - \frac{\sqrt{2}}{2}} = (2 + \sqrt{2})a.$$

So, if α is now such that $|\alpha| > a/\sqrt{a^2 - 1}$, then

$$\|(\alpha I + A + A^*)x\| \geq \left(\frac{a}{\sqrt{a^2 - 1}} - 2a \right) \|x\| = \sqrt{2}a\|x\|.$$

Using $\|A\| = a$ once more yields

$$\langle (\alpha I + A + A^*)^2 x, x \rangle = \|(\alpha I + A + A^*)x\|^2 \geq 2a^2\|x\|^2 \geq \|AA^* - A^*A\|\|x\|^2.$$

But

$$\langle (AA^* - A^*A)x, x \rangle \leq \|AA^* - A^*A\|\|x\|^2.$$

We have therefore shown that

$$\langle (\alpha I + A + A^*)^2 x, x \rangle \geq \langle (AA^* - A^*A)x, x \rangle$$

holds for $|\alpha| > a/\sqrt{a^2 - 1}$.

Thus, the inequality

$$(\alpha I + A + A^*)^2 \geq AA^* - A^*A$$

holds for all $\alpha \in \mathbb{R}$. So much for the proof.

Answer 14.2.7 Let S be the unilateral shift operator which is obviously quasi-normal (hence subnormal and also hyponormal) but it is *not normal*. If $S = U|S| = |S|U$ held, then S would be unitary which is untrue. Therefore, no such decomposition exists for quasinormal operators.

Answer 14.2.8 As usual, the shift operator plays a prominent role! Indeed, let $S \in B(\ell^2)$ be the usual shift.

(1) Quasinormal $\not\supset$ Subnormal: Let S be the shift operator. Then it is subnormal as already observed. Hence $T := I + S$ remains subnormal. But, T is not quasinormal because

$$T^*TT - TT^*T = T^*TS - ST^*T = S^*S - SS^* \neq 0.$$

(2) Subnormal $\not\supset$ Hyponormal: Let $A = S^* + 2S$ where S is the shift operator, then A is hyponormal but A^2 is not (see Answer 14.2.3). However, A is not subnormal. Indeed, if A were subnormal, then so would be A^2 and so A^2 would too be hyponormal!

(3) Hyponormal $\not\supset$ Paranormal: The same example as before works perfectly by remembering that if A is paranormal, then so is A^2.

(4) Paranormal $\not\supset$ Normaloid: In fact, there are counterexamples even on finite-dimensional Hilbert spaces. One instance (borrowed from [206]) is to let

$$T = \begin{pmatrix} 1 & 0 & 0 \\ 0 & 0 & 0 \\ 0 & 1 & 0 \end{pmatrix},$$

defined on $\mathbb{C}^3$. Clearly T is not normal and so it is not paranormal (for dim $\mathbb{C}^3 < \infty$). However, T is normaloid because for each natural integer n

$$\|T^n\| = \|T\|^n \ (= 1),$$

as needed.

Answer 14.2.9 Let S be the shift operator. Then S is subnormal. If S^* were subnormal, then it would be hyponormal. Hence the shift operator would be normal given that S is already hyponormal, and this is impossible. Therefore, S^* is not subnormal, as needed.

Similarly, the shift S is quasinormal. However, S^* is not quasinormal as a similar argument as above tells us.

Answer 14.2.10 Recall that a counterexample is to be found in an infinite-dimensional setting. Let S be the usual shift on ℓ^2. Then S is obviously paranormal for it is e.g. hyponormal. However, the shift's adjoint S^* is not paranormal as

$$\|S^* e_2\|^2 = 1 \nleq 0 = \|S^{*2} e_2\|$$

for $e_2 = (0, 1, 0, 0, \cdots)$ say.

Answer 14.2.11 Such counterexamples are not easy to find. We refer readers to [389], where a counterexample is in the form of some "singular integral operator".

We can do it differently here, but first we have already given a hyponormal operator A whose square A^2 *is not hyponormal* (see Answer 14.2.3). Now, since this A is hyponormal, that is, A is 1-hyponormal, by Proposition 14.2.1 (just before Question 14.2.11), A^2 is always $\frac{1}{2}$-hyponormal, i.e. A^2 is always semi-hyponormal. So the sought counterexample is just $B = A^2$. Then B is semi-hyponormal but it is not hyponormal!

Answer 14.2.12 ([103]) Consider the weighted shift on ℓ^2 defined by $A(x_1, x_2, \cdots) = (0, 2x_1, x_2, \cdots)$. Then $A^*(x_1, x_2, \cdots) = (2x_2, x_3, \cdots)$. It is straightforward to check that

$$A^* A^2 (x_1, x_2, \cdots) = (0, 2x_1, x_2, \cdots) = A(x_1, x_2, \cdots)$$

for all $(x_1, x_2, \cdots) \in \ell^2$, i.e. $A = A^* A^2$. However, A is not hyponormal (and so it cannot be subnormal either). Indeed, for instance, if e_2 is picked from the standard orthonormal basis, then

$$\|A^* e_2\| = 2 > \|A e_2\| = 1.$$

Answer 14.2.13 Let A be a hyponormal operator which is not subnormal. Then, $A - \lambda I$ remains hyponormal, and at the same time, it stays non-subnormal. Indeed, if $A - \lambda I$ were subnormal, then $A - \lambda I + \lambda I$ ($= A$) would be subnormal, and this is absurd! Now, if $\lambda \notin \sigma(A)$, then $A - \lambda I$ is an invertible hyponormal operator which is not subnormal.

Answer 14.2.14 ([1]) Let (e_k) be the usual orthonormal basis of $\ell^2(\mathbb{N})$. On $\ell^2(\mathbb{N})$ define a sequence of operators (T_n) by

$$T_n(e_0) = 2^{-n} e_1, \ T_n(e_k) = e_{k+1} \text{ for } k \geq 1.$$

Define also M on $\ell^2(\mathbb{N})$ by

$$M(e_0) = 2e_0 \text{ and } M(e_k) = e_k, \ k \geq 1$$

(and so M is *self-adjoint*). Next, set $H = \ell^2(\mathbb{N}) \oplus \ell^2(\mathbb{N}) \oplus \ell^2(\mathbb{N}) \oplus \cdots$, then define A and B on H by

$$A = \begin{pmatrix} 0 & 0 & 0 & \cdots & & \cdots \\ M & 0 & 0 & 0 & & \\ 0 & M & 0 & 0 & & \ddots \\ & \ddots & M & 0 & 0 & \ddots \\ & & & M & \ddots & \ddots \\ \vdots & & & & \ddots & \ddots \end{pmatrix} \text{ and } B = \begin{pmatrix} T_0 & 0 & & & \cdots & \\ 0 & T_1 & 0 & & \cdots & \\ 0 & 0 & T_2 & \ddots & & \ddots \\ & \ddots & \ddots & T_3 & & \ddots \\ & & & & \ddots & \ddots \\ \vdots & & & & & \ddots & \ddots \end{pmatrix}.$$

It is easy to see that A is hyponormal (it is in fact quasinormal). Indeed,

$$A^*A = \begin{pmatrix} M^2 & 0 & & & \cdots & \\ 0 & M^2 & 0 & & \cdots & \\ 0 & 0 & M^2 & \ddots & & \ddots \\ & \ddots & \ddots & M^2 & & \ddots \\ & & & & \ddots & \ddots \\ \vdots & & & & & \ddots & \ddots \end{pmatrix}.$$

and

$$AA^* = \begin{pmatrix} 0 & 0 & & & & \cdots & \\ 0 & M^2 & 0 & & \cdots & & \\ & 0 & 0 & M^2 & \ddots & & \ddots \\ & & \ddots & \ddots & M^2 & & \ddots \\ & & & & & \ddots & \ddots \\ \vdots & & & & & & \ddots & \ddots \end{pmatrix}$$

and so $AA^* \leq A^*A$. It is not complicated to see that each T_n is hyponormal (in fact subnormal, see again [1]). Therefore, B too is hyponormal. Moreover,

$$AB = \begin{pmatrix} 0 & 0 & 0 & \cdots & & \cdots & \\ MT_0 & 0 & 0 & 0 & & & \\ 0 & MT_1 & 0 & 0 & & \ddots & \\ & \ddots & MT_2 & 0 & 0 & & \ddots \\ & & & MT_3 & \ddots & \ddots & \\ \vdots & & & & & \ddots & \ddots \end{pmatrix}$$

and

$$BA = \begin{pmatrix} 0 & 0 & 0 & \cdots & & \cdots & \\ T_1M & 0 & 0 & 0 & & & \\ 0 & T_2M & 0 & 0 & & \ddots & \\ & \ddots & T_3M & 0 & 0 & & \ddots \\ & & & T_4M & \ddots & \ddots & \\ \vdots & & & & & \ddots & \ddots \end{pmatrix}$$

So, in order that A and B be commuting, it suffices and it necessitates to have $MT_n = T_{n+1}M$ for any $n \geq 0$. This is in effect the case as

$$MT(e_0) = M(e_1) = e_1 = T_1(2e_0) = T_1M(e_0)$$

and for all k (and all n)

$$MT_n(e_k) = M(e_{k+1}) = e_{k+1} = T_{n+1}M(e_k).$$

The last point is to check that $A + B$ is not hyponormal. This contains some calculations. Clearly

$$A + B = \begin{pmatrix} T_0 & 0 & 0 & \cdots & & \cdots \\ M & T_1 & 0 & 0 & & \\ 0 & M & T_2 & 0 & & \ddots \\ & \ddots & M & T_3 & 0 & \ddots \\ & & & M & T_4 & \ddots \\ \vdots & & & & \ddots & \ddots \end{pmatrix}.$$

Hence

$$(A + B)^*(A + B) = \begin{pmatrix} T_0^* T_0 + M^2 & M T_1 & 0 & \cdots & \cdots \\ T_1^* M & T_1^* T_1 + M^2 & M T_2 & 0 & \\ 0 & T_2^* M & T_2^* T_2 + M^2 & M T_3 & \ddots \\ & \ddots & & \ddots & \ddots & \ddots \\ \vdots & & & & \ddots & \ddots \end{pmatrix}$$

and

$$(A + B)(A + B)^* = \begin{pmatrix} T_0 T_0^* & T_0 M & 0 & \cdots & \cdots \\ M T_0^* & T_1 T_1^* + M^2 & T_1 M & 0 & \\ 0 & M T_1^* & T_2 T_2^* + M^2 & & \ddots \\ & \ddots & & \ddots & \ddots & \ddots \\ \vdots & & & & \ddots & \ddots \end{pmatrix}.$$

By setting $x = (e_1, e_0, 0, \cdots) \in H$, we may check that

$$\langle [(A + B)^*(A + B) - (A + B)(A + B)^*]x, x \rangle = -\frac{1}{4}$$

and so

$$(A + B)^*(A + B) \not\geq (A + B)(A + B)^*,$$

that is, $A + B$ is not hyponormal as needed.

Remark Another example may be found in [77].

Remark As observed above, A and B are subnormal. Thus, we also have two commuting subnormal operators whose sum is not subnormal.

Answer 14.2.15 ([217]) Consider a separable Hilbert space H having an orthonormal basis $\{e_n, c_0, f_n : n = 1, 2, \cdots\}$ (for instance, readers may wish to consider $\ell^2(\mathbb{Z})$, in such case e_n plays the role of f_{-n}) and define $A, B \in B(H)$ by:

$$A^*(e_n) = e_{n+1}, A(c_0) = e_1, A(f_n) = 0$$

and

$$B(e_n) = 0, B(c_0) = f_1, B(f_n) = f_{n+1}.$$

As infinite matrices, they are defined by

$$A = \begin{pmatrix} \ddots & \ddots & \ddots & \cdots & & \cdots \\ & \ddots & 0 & 1 & 0 & \\ & & 0 & 0 & 1 & \ddots \\ & & & 0 & \mathbf{0} & 0 & \ddots \\ & & & & 0 & \ddots & \ddots \\ & & & & & \ddots & \ddots \\ & \vdots & & & & \ddots & \ddots \end{pmatrix}$$

and

$$B = \begin{pmatrix} \ddots & \ddots & \ddots & \cdots & & \cdots \\ & \ddots & 0 & 0 & \ddots & \\ & & 0 & 0 & 0 & \ddots \\ & & & 0 & \mathbf{0} & 0 & \ddots \\ & & & 1 & 0 & \ddots \\ & & & & 1 & \ddots \\ & \vdots & & & & \ddots & \ddots \end{pmatrix}$$

where $\mathbf{0}$ is in the position $(0, 0)$. It is easy to see that $AB = BA = 0$. It is also fairly easy to see that both A and B are quasinormal (the same proof works for both of them). First, observe that

$$A^*(e_n) = e_{n-1}, A^*(e_1) = c_0, A^*(f_n) = 0$$

and

$$B^*(e_n) = 0, B^*(f_1) = c_0, B^*(f_n) = f_{n-1}.$$

To see why e.g. A is quasinormal, observe that for all n

$$AA^*Ae_n = AA^*(e_{n+1}) = Ae_n = e_{n+1} = A^*AAe_n.$$

Similarly, $AA^*Af_n = 0 = A^*AAf_n$ for each n. Finally,

$$AA^*Ac_0 = AA^*(e_1) = Ac_0 = e_1 = A^*AAc_0.$$

Let us now proceed to show that $T := A + B$ is not hyponormal. A simple calculation yields

$$(T^*T - TT^*)(e_1 + f_1) = -f_1 - e_1.$$

Hence

$$\langle (T^*T - TT^*)(e_1 + f_1), (e_1 + f_1) \rangle = \langle -f_1 - e_1, e_1 + f_1 \rangle = -2 < 0$$

as e_1 and f_1 are elements of an orthonormal basis. Thus, $A + B$ is not even hyponormal.

Remark It is worth noting that if A and B are two commuting quasinormal operators, then it is not necessary that $A + B$ be quasinormal even when one of them is normal (hence also $AB^* = B^*A$). Indeed, let S be the usual shift operator on ℓ^2 and let I be the identity operator on ℓ^2. Then S is quasinormal, I is normal and they plainly commute yet $I + S$ is not quasinormal.

Answer 14.2.16 The case of the non-subnormality of the sum of two commuting subnormal has already been treated above (in Answer 14.2.14 and also in Answer 14.2.15).

Let us therefore turn to the case of ST. Let A and B be two quasinormal operators such that $AB = BA = 0$ as in Answer 14.2.15. Set $T = A + I$ and $S = B + I$ where I is the identity operator. Since A and B are quasinormal, they are subnormal and hence so are S and T. Besides,

$$ST = A + B + I = TS$$

for $AB = BA = 0$. However, ST is not even hyponormal for if it were, $A + B = (A + B + I) - I$ would be hyponormal. But we already know from Answer 14.2.15 that $A + B$ is not hyponormal. Accordingly, ST is not hyponormal (and so it cannot be subnormal either).

Remark It is noteworthy that A. Lubin in [216] provided a pair of two commuting subnormal operators S and T such that neither ST nor $S + T$ is subnormal.

Answer 14.2.17 An efficient place to find a counterexample here is ℓ^2. Let S be the shift operator on ℓ^2. Then S is quasinormal and obviously $SS = SS$. If the Fuglede-Putnam theorem were to hold for quasinormal operators, then we would have $SS^* = S^*S$, i.e. we would obtain $SS^* = I$ which is definitely a contradiction!

Remark It is worth telling readers that if $A, B^* \in B(H)$ are both hyponormal, then $TB = AT$ implies that $TB^* = A^*T$, where $T \in B(H)$. This was shown in [233].

Answer 14.2.18 ([344]) Let A be the bilateral shift defined on $\ell^2(\mathbb{Z})$, i.e. a linear operator whose action on an orthonormal basis of $\ell^2(\mathbb{Z})$ is given by $Ae_n = e_{n+1}$, $n \in \mathbb{Z}$. Define B as

$$Be_n = \begin{cases} e_{n+1}, & n \geq 0, \\ 0, & n < 0 \end{cases}$$

and take $T = B$. Clearly A is unitary and B is a quasinormal partial isometry (and not just semi-normal as noted by J. G. Stampfli in his paper). Indeed, for all $n \in \mathbb{Z}$

$$B^*BBe_n = BB^*Be_n = \begin{cases} e_{n+1}, & n \geq 0, \\ 0, & n < 0. \end{cases}$$

Now, for all $n \in \mathbb{Z}$, we have

$$TBe_n = ATe_n = \begin{cases} e_{n+2}, & n \geq 0, \\ 0, & n < 0. \end{cases}$$

In other words, $TB = AT$, however,

$$TB^*e_0 = 0 \neq e_0 = A^*Te_0,$$

and so $TB^* \neq A^*T$, as needed.

Answer 14.2.19 Let $H = \ell^2$ and let $S \in B(\ell^2)$ be the usual shift operator, then define

$$A = \begin{pmatrix} S & 0 \\ 0 & 0 \end{pmatrix} \quad \text{and} \quad B = \begin{pmatrix} 0 & 0 \\ 0 & S \end{pmatrix}$$

on $\ell^2 \oplus \ell^2$. Since $S^*S = I$, clearly, $AA^*A = A^*AA$. Similarly, $BB^*B = B^*BB$. Hence A and B are both quasinormal (hence subnormal, and a fortiori hyponormal). Plainly neither A nor B, however, is normal. Observe in the end that

$$AB = BA = AB^* = B^*A$$

(in fact, all products are equal to the zero matrix of operators on $\ell^2 \oplus \ell^2$).

Remark Readers might be interested in consulting [73], in which some situations guaranteeing that either A or B be normal, given that A and B are hyponormal and obeying $AB = BA$ and $AB^* = B^*A$, are established.

Answer 14.2.20 The first example to be given appeared in [73]. Below it, I will give a simpler and stronger example.

Let S be the usual shift operator and let K be an infinite dimensional Hilbert space. Define $A = I_K \otimes S$ and hence $A^* = I_K \otimes S^*$. This plainly yields the hyponormality of A. Let $B = X \otimes I + Y^* \otimes S^*$ where X and Y are yet to be chosen. Hence $B^* = X^* \otimes I + Y \otimes S$. Therefore,

$$AB^* = X^* \otimes S + Y \otimes S^2 = B^*A$$

but

$$AB = X \otimes S + Y^* \otimes SS^* \neq X \otimes S + Y^* \otimes I = BA.$$

So, it only remains to find a hyponormal B with the required properties. First, straightforward computations give

$$B^*B = X^*X \otimes I + X^*Y^* \otimes S^* + YX \otimes S + YY^* \otimes SS^*$$

and

$$BB^* = XX^* \otimes I + XY \otimes S + Y^*X^* \otimes S^* + Y^*Y \otimes I.$$

Assuming that X and Y could be found such that $XY = YX$, it would then follow that

$$B^*B - BB^* = (X^*X - XX^* - Y^*Y) \otimes I + YY^* \otimes SS^*.$$

The hyponormality of B therefore reduces to finding X and Y such that $XY = YX$, $Y \neq 0$ and $X^*X - XX^* - Y^*Y \geq 0$.

In that regard, define the infinite matrices

$$X = \begin{pmatrix} 0 & 0 & 0 \cdots & & \cdots \\ P & 0 & 0 & 0 \\ 0 & I & 0 & 0 \\ & \ddots & I & 0 & 0 & \ddots \\ & & I & \ddots & \ddots \\ \vdots & & & \ddots & \ddots \end{pmatrix} \quad \text{and} \quad Y = \begin{pmatrix} 0 & I - P & 0 \cdots & & \cdots \\ 0 & 0 & 0 & 0 \\ 0 & 0 & 0 & 0 & & \ddots \\ & \ddots & 0 & 0 & 0 & \ddots \\ & & 0 & \ddots & \ddots \\ \vdots & & & \ddots & \ddots \end{pmatrix},$$

whose entries are operators over Hilbert spaces of dimension at least 2 and P is a nontrivial orthogonal projection, i.e. $P^2 = P$, $P^* = P$, $P \neq 0$ and $P \neq I$. Hence

$Y \neq 0$ and $XY = YX \, (= 0)$. Moreover,

$$X^*X = \begin{pmatrix} P & 0 & 0 \cdots & & \cdots \\ 0 & I & 0 & 0 \\ & 0 & 0 & I & 0 \\ & & \ddots & 0 & I & 0 & \ddots \\ & & & 0 & & \ddots & \ddots \\ \vdots & & & & & \ddots & \ddots \end{pmatrix}, \quad XX^* = \begin{pmatrix} 0 & 0 & 0 \cdots & & \cdots \\ 0 & P & 0 & 0 \\ & 0 & 0 & I & 0 \\ & & \ddots & 0 & I & 0 & \ddots \\ & & & 0 & & \ddots & \ddots \\ \vdots & & & & & \ddots & \ddots \end{pmatrix}$$

and

$$Y^*Y = \begin{pmatrix} 0 & 0 & 0 \cdots & & \cdots \\ 0 & I-P & 0 & 0 \\ & 0 & 0 & 0 & 0 \\ & & \ddots & 0 & 0 & 0 & \ddots \\ & & & 0 & & \ddots & \ddots \\ \vdots & & & & & \ddots & \ddots \end{pmatrix},$$

whereby

$$X^*X - XX^* - Y^*Y = \begin{pmatrix} P & 0 & 0 \cdots & & \cdots \\ 0 & 0 & 0 & 0 \\ & 0 & 0 & 0 & 0 \\ & & \ddots & 0 & 0 & 0 & \ddots \\ & & & 0 & & \ddots & \ddots \\ \vdots & & & & & \ddots & \ddots \end{pmatrix}$$

which is clearly positive. So much for the proof (in the first example).

As alluded to above, I now give another example (both simpler and stronger). While reading the paper [288], in which the authors gave several examples, I came across the following examples of A and B:

$$Ae_n = \begin{cases} e_{n-1}, & n \leq 1, \\ 0, & n > 1. \end{cases} \quad \text{and} \quad Be_n = \begin{cases} e_{n+1}, & n < 1, \\ 0, & n \geq 1, \end{cases}$$

both defined on $\ell^2(\mathbb{Z})$, and where (e_n) is an orthonormal basis. This pair of operators was used in [288] for a different purpose as we shall see in Answer 14.2.29. The authors in [288], however, took $n \leq 1$ instead of $n < 1$, and $n > 1$ in place of $n \geq 1$. As adjusted above, the example is just perfect for our case.

There are different ways of finding the adjoints A^* and B^*. Readers should comfortably find that

$$A^*e_n = \begin{cases} e_{n+1}, & n \leq 0, \\ 0, & n > 0. \end{cases} \text{ and } B^*e_n = \begin{cases} e_{n-1}, & n \leq 1, \\ 0, & n > 1. \end{cases}$$

Hence

$$A^*Ae_n = \begin{cases} e_n, & n \leq 1, \\ 0, & n > 1. \end{cases} \text{ and } B^*Be_n = \begin{cases} e_n, & n \leq 0, \\ 0, & n > 0. \end{cases}$$

Therefore, it is seen that

$$AA^*Ae_n = A^*AAe_n \text{ and } BB^*Be_n = B^*BBe_n,$$

both equations holding for all $n \in \mathbb{Z}$. This means that $AA^*A = A^*AA$ and $BB^*B = B^*BB$, but these equations are characteristic of quasinormality. Now, readers can also check that

$$AB^*e_n = e_{n-2} = B^*Ae_n, \ \forall n \leq 1$$

and

$$AB^*e_n = 0 = B^*Ae_n, \ \forall n > 1.$$

Thus, $AB^* = B^*A$. Finally, since e.g.

$$ABe_1 = 0 \neq e_1 = BA(e_1),$$

we see that $AB \neq BA$, as wished.

Answer 14.2.21 ([5, 211]) We answer both questions together. In fact, there is a paranormal operator $T \in B(H)$ such that $T \otimes T$ is not paranormal. Since obviously

$$T \otimes T = (T \otimes I)(I \otimes T),$$

the same counterexample works for the case of the usual product if we say why $T \otimes I$ and $I \otimes T$ are paranormal (observe also that $T \otimes I$ and $I \otimes T$ double commute). For example, $T \otimes I$ is paranormal (using Theorem 14.1.4) since

$$(T \otimes I)^{*2}(T \otimes I)^2 - 2\lambda(T \otimes I)^*(T \otimes I) + \lambda^2(I \otimes I) = (T^{*2}T^2 - 2\lambda T^*T + \lambda^2) \otimes I \geq 0$$

for the tensor product of two positive operator is again positive. The paranormality of $I \otimes T$ follows by a similar argument. So, it "only" remains to find a paranormal operator T such that $T \otimes T$ is not paranormal.

Let $H = \cdots \oplus \mathbb{C}^2 \oplus \mathbb{C}^2 \oplus \mathbb{C}^2 \oplus \cdots$, then define T on H by

$$T = \begin{pmatrix} \ddots & & \cdots & & \cdots \\ & \ddots & 0 \ 0 \ 0 & & \\ & & A \ 0 \ 0 & \ddots & \\ & & 0 \ A \ \mathbf{0} & \ddots & \\ & & 0 \ 0 \ B \ 0 & \ddots & \\ & & \ \ 0 \ B & \ddots & \\ & & & \ddots & \ddots \end{pmatrix}$$

where $\mathbf{0}$ indicates the position $(0, 0)$ in the matrix, and where $A, B \in B(\mathbb{C}^2)$ are both positive. Then, it may be shown that T is paranormal if and only if

$$AB^2A + 2\lambda A^2 + \lambda^2 I \geq 0$$

for any real λ. Next, choose the positive matrices

$$A = \begin{pmatrix} 1 & 1 \\ 1 & 2 \end{pmatrix}^{1/2} \quad \text{and} \quad B = \left[A^{-1/2} \begin{pmatrix} 1 & 2 \\ 2 & 8 \end{pmatrix} A^{-1/2} \right]^{1/2}.$$

Then, for all λ

$$AB^2A + 2\lambda A^2 + \lambda^2 I = \begin{pmatrix} (1+\lambda)^2 & 2(1+\lambda) \\ 2(1+\lambda) & (2+\lambda)^2 + 4 \end{pmatrix}$$

is a positive matrix, making T paranormal. If $T \otimes T$ *were* paranormal, it would ensure that e.g.

$$(T \otimes T)^{*2}(T \otimes T)^2 - 2(T \otimes T)^*(T \otimes T) + I \otimes I \geq 0$$

which would lead to

$$AB^2A \otimes AB^2A - 2A^2 \otimes A^2 + I \otimes I \geq 0$$

and this absurd because

$$AB^2A \otimes AB^2A - 2A^2 \otimes A^2 + I \otimes I$$

$$= \begin{pmatrix} 1 & 2 \\ 2 & 8 \end{pmatrix} \otimes \begin{pmatrix} 1 & 2 \\ 2 & 8 \end{pmatrix} - 2 \begin{pmatrix} 1 & 1 \\ 1 & 2 \end{pmatrix} \otimes \begin{pmatrix} 1 & 1 \\ 1 & 2 \end{pmatrix} + \begin{pmatrix} 1 & 0 \\ 0 & 1 \end{pmatrix} \otimes \begin{pmatrix} 1 & 0 \\ 0 & 1 \end{pmatrix}$$

$$
= \begin{pmatrix} 1 & 2 & 2 & 4 \\ 2 & 8 & 4 & 16 \\ 2 & 4 & 8 & 16 \\ 4 & 16 & 16 & 64 \end{pmatrix} - 2 \begin{pmatrix} 1 & 1 & 1 & 1 \\ 1 & 2 & 1 & 2 \\ 1 & 1 & 2 & 2 \\ 1 & 2 & 2 & 4 \end{pmatrix} + \begin{pmatrix} 1 & 0 & 0 & 0 \\ 0 & 1 & 0 & 0 \\ 0 & 0 & 1 & 0 \\ 0 & 0 & 0 & 1 \end{pmatrix}
$$

$$
= \begin{pmatrix} 0 & 0 & 0 & 2 \\ 0 & 5 & 2 & 12 \\ 0 & 2 & 5 & 12 \\ 2 & 12 & 12 & 57 \end{pmatrix}
$$

is (self-adjoint) not positive as e.g. the main determinant equals $-84 < 0$.

Remarks

(1) The product of two commuting hyponormal operators may even fail to be paranormal. The example to be mentioned was obtained by T. Ando in [5]. This example is also twofold as shall be seen shortly. T. Ando constructed a hyponormal T such that $T^2 - \lambda$ is not paranormal for some complex λ. Next, by letting $\lambda^{1/2}$ to be one of the complex square roots of λ, we see that $T - \lambda^{1/2}$ and $T + \lambda^{1/2}$ become two commuting hyponormal operators having a non-paranormal product. This is not trivial, and details may be consulted in [5].

(2) As mentioned just above, another consequence of the foregoing example is that it also shows that the sum of a paranormal operator and a scalar is not necessarily paranormal.

(3) It is worth noticing that if a *finite* tensor product of operators is paranormal (and nonzero), then each operator in the (tensor) product has to be paranormal. See [352].

Answer 14.2.22 ([31]) Let $S \in B(\ell^2)$ be the usual (unilateral) shift, then take $A = S + I$ so that A is subnormal (observe in passing that A is not quasinormal). Then

$$
|A|^2 = |S + I|^2 = (S^* + I)(S + I) = 2I + S + S^*.
$$

On the other hand,

$$
|A^2| = \sqrt{(S^* + I)^2(S + I)^2} = \sqrt{6I + 4S^* + 4S + S^{*2} + S^2}.
$$

So if $|A^2| = |A|^2$ were true, then we would obtain $|A^2|^2 = |A|^4$. Working out details would then yield $SS^* = I$, which is impossible.

Answer 14.2.23 Recall that A is quasinormal if and only if $A|A| = |A|A$. So, to get a counterexample in the case of subnormality, it then suffices to take any subnormal operator which is not quasinormal. For a concrete example, let $A = S + I$ be defined on ℓ^2 where S is the usual shift. Then A is subnormal (but non-quasinormal), that is,

$$
A|A| \neq |A|A.
$$

Answer 14.2.24 There are at least two examples available in the literature. The first
one I came across is Example 5.1 in [172]. However, the example to be elaborated
could be viewed as a slightly simpler (it appeared in [365] some while before the
one in [172]). It reads: On ℓ^2, let

$$
T = \begin{pmatrix}
0 & 0 & 0 \cdots & & \cdots & \\
1 & 1 & 0 & 0 & & \\
0 & 1 & 1 & 0 & & \ddots \\
& \ddots & 1 & 1 & 0 & \ddots \\
& & & 1 & 1 & \ddots \\
\vdots & & & & \ddots & \ddots
\end{pmatrix}.
$$

The action of the operator T on some orthonormal basis (e_n) is clearly given by:

$$ T e_1 = e_2, \ T e_n = e_n + e_{n+1}, \ n = 2, 3, \cdots $$

(in fact, $T = S + SS^*$ where S is the standard unilateral shift). Hence

$$
T^* = \begin{pmatrix}
0 & 1 & 0 \cdots & & \cdots & \\
0 & 1 & 1 & 0 & & \\
0 & 0 & 1 & 1 & & \ddots \\
& \ddots & 0 & 1 & 1 & \ddots \\
& & & & \ddots & \ddots & \ddots \\
\vdots & & & & & \ddots & \ddots
\end{pmatrix},
$$

that is,

$$ T^* e_1 = 0, \ T^* e_n = e_n + e_{n-1}, \ n = 2, 3, \cdots $$

Readers can then check that

$$
T^{*2} T^2 = (T^* T)^2 = \begin{pmatrix}
2 & 3 & 1 & 0 & \cdots & \cdots \\
3 & 6 & 4 & 1 & \cdots & \\
1 & 4 & 6 & 4 & 1 & \ddots \\
0 & 1 & 4 & 6 & 4 & \ddots \\
& & & \ddots & \ddots & \ddots \\
\vdots & & & & \ddots & \ddots
\end{pmatrix}.
$$

Finally, T is not quasinormal as e.g.

$$T^*TTe_1 = e_1 + 2e_2 + e_3 \neq 2e_2 + e_3 = TT^*Te_1.$$

Answer 14.2.25 The first one to answer this question negatively was J. G. Stampfli (in [345]) by giving the following counterexample: Let $(e_n)_{n \in \mathbb{Z}}$ be an orthonormal basis for $\ell^2(\mathbb{Z})$, then define

$$Te_n = \begin{cases} e_{n+1}, & n \leq 0, \\ 2e_{n+1}, & n > 0. \end{cases}$$

Then $T^k e_n = \alpha_{n,k} e_{n+k}$ where $|\alpha_{n,k}| \leq |\alpha_{n+1,k}|$. Hence each T^k is hyponormal (for $k = 1, 2, \cdots$). Since $\|Te_0\| = \|T^*e_0\|$ but $\|T^2e_0\| = \|T^*Te_0\|$, T cannot be subnormal. Indeed, if T is subnormal, then $\{x \in H : \|Tx\| = \|T^*x\|\}$ is a closed invariant subspace of T (Theorem 5 in [345]).

Answer 14.2.26 ([346]) However, we show the subnormality of the powers T^n, $n \geq 2$ with the aid of "Embry's criterion" given in Question 14.2.12, i.e. differently from Stampfli's original proof.

Let H be a separable Hilbert space and let (e_k) be an orthonormal basis in H. Define

$$Te_1 = \frac{1}{4}e_2, \ Te_2 = \frac{1}{2}e_3, \ Te_k = e_{k+1}, \ k = 2, 3, \cdots$$

(the weight is obviously given by $a_1 = 1/4, a_2 = 1/2$ and $a_k = 1$ otherwise). Hence

$$T^*e_1 = 0, \ T^*e_2 = \frac{1}{4}e_1, \ T^*e_3 = \frac{1}{2}e_2, \ T^*e_k = e_{k-1}, \ k = 4, 5, \cdots.$$

The hyponormality of T is clear. The *non-subnormality* of T easily follows from the result recalled in the hint.

Let us show now that T^2 is subnormal. By "Embry's criterion", we need only check that $\|T^2\| \leq 1$ and $T^2 = T^{*2}T^4$. The former is clear. As for the latter, readers may check that

$$T^2e_1 = T^{*2}T^4e_1 = \frac{1}{8}e_3, \ T^2e_2 = T^{*2}T^4e_2 = \frac{1}{2}e_4, \ T^2e_k = T^{*2}T^4e_k = e_{k+2}$$

for $k = 3, 4, \cdots$. That is, $T^2 = T^{*2}T^4$ and hence T^2 is subnormal. Readers may reason similarly to show that T^n is subnormal for $n \geq 3$.

Answer 14.2.27 Consider the operators A and B which appeared in Answer 14.2.15, i.e.

$$A(e_n) = e_{n+1}, A(c_0) = e_1, A(f_n) = 0$$

and

$$B(e_n) = 0, B(c_0) = f_1, B(f_n) = f_{n+1}$$

defined on a separable Hilbert space H with an orthonormal basis $\{e_n, c_0, f_n : n \geq 1\}$. We also know that $AB = BA = 0$. Next, it is seen that

$$|A|(e_n) = e_n, |A|(c_0) = c_0, |A|(f_n) = 0$$

and

$$|B|(e_n) = 0, |B|(c_0) = c_0, |B|(f_n) = f_n.$$

Since $AB = 0$, plainly $|AB| = 0$. However, $|AB| \neq |A||B|$ because e.g.

$$|AB|c_0 = 0 \neq c_0 = |A||B|c_0,$$

as wished.

Answer 14.2.28 Consider yet again the *commuting* operators A and B which appeared in Answer 14.2.15, that is,

$$A(e_n) = e_{n+1}, A(c_0) = e_1, A(f_n) = 0$$

and

$$B(e_n) = 0, B(c_0) = f_1, B(f_n) = f_{n+1}.$$

From Answer 14.2.27, we know that

$$|B|(e_n) = 0, |B|(c_0) = c_0, |B|(f_n) = f_n.$$

Thus, $A|B| \neq |B|A$ for say

$$A|B|c_0 = e_1 \neq 0 = |B|Ac_0,$$

as is easily checked.

Answer 14.2.29 ([288]) Let

$$Ae_n = \begin{cases} e_{n-1}, & n \leq 1, \\ 0, & n > 1. \end{cases} \quad \text{and} \quad Be_n = \begin{cases} e_{n+1}, & n < 1, \\ 0, & n \geq 1 \end{cases}$$

both defined on $\ell^2(\mathbb{Z})$, and where (e_n) is some orthonormal basis. As originally defined by Patel et al. in [288], the expression of B contained a small typo where it was written $n \leq 1$ instead of $n < 1$, and $n > 1$ in lieu of $n \geq 1$. As adjusted above, the example now works perfectly. Indeed, we already know from Answer 14.2.20 that

$$A^* e_n = \begin{cases} e_{n+1}, & n \leq 0, \\ 0, & n > 0. \end{cases} \quad \text{and } B^* e_n = \begin{cases} e_{n-1}, & n \leq 1, \\ 0, & n > 1, \end{cases}$$

and that both A and B are quasinormal. We also know that $AB^* = B^*A$ still from Answer 14.2.20.

If $A|B| = |B|A$ were true, so would be $AB^*B = B^*BA$. This is clearly not the case for e.g.

$$AB^* B e_1 = AB^*(0) \neq e_0 = B^* e_1 = B^* B e_0 = B^* B A e_1.$$

Finally, since it is clear that

$$|A| e_n = \begin{cases} e_n, & n \leq 1, \\ 0, & n > 1, \end{cases}$$

it results that $B|A| e_n = |A|B e_n$ for each n and so $B|A| = |A|B$.

Answer 14.2.30 ([17]) Let A and B be acting on the standard basis (e_n) of $\ell^2(\mathbb{N})$ by:

$$A e_n = \alpha_n e_n \text{ and } B e_n = e_{n+1}, \ \forall n \geq 1$$

, respectively. Assume further that α_n is bounded, *real-valued*, and *positive*, for all n. Hence A is self-adjoint (in fact positive). Then

$$A B e_n = \alpha_{n+1} e_{n+1} \text{ and } B A e_n = \alpha_n e_{n+1}, \ \forall n \geq 1,$$

meaning that both AB and BA are weighted shifts with weights $\{\alpha_n\}_{n=2}^{\infty}$ and $\{\alpha_n\}_{n=1}^{\infty}$, respectively.

Now recall that (see e.g., [148]): a weighted shift with weight $\{\alpha_n\}_{n=1}^{\infty}$ is hyponormal if and only if $\{|\alpha_n|\}_{n=1}^{\infty}$ is monotonically increasing.

Hence, if $\{\alpha_n\}_{n=1}^{\infty}$ is increasing, then AB and BA are both hyponormal. Moreover, $AB = BA$ (equivalently, $A^2 B = BA^2$) if and only if the sequence $\{\alpha_n\}_{n=1}^{\infty}$ is constant. Taking any non-constant monotonically increasing sequence $\{\alpha_n\}_{n=1}^{\infty}$ does the job.

Answer 14.2.31 By an appeal to a result by F. Kittaneh in [195], we know that if $A \in B(H)$ is hyponormal, then

$$|\langle Ax, x \rangle| \leq \langle |A|x, x \rangle$$

for all $x \in H$. Thus, it becomes apparent what to choose for a counterexample! Take any hyponormal operator which is not normal.

Answer 14.2.32 Let S be the shift operator on ℓ^2. Setting $A = SS^*$, we see that $A \geq 0$. Now, take $K = S$ (and so $\|K\| = 1$). It is clear that $AK = SS^*S = S$ is hyponormal. If Reid's inequality were to hold, then we would have

$$|\langle Sx, x \rangle| \leq \langle SS^*x, x \rangle = \|S^*x\|^2$$

for each $x \in \ell^2$. The previous inequality is clearly violated for say $x = (2, 1, 0, 0, \cdots)$ because

$$|\langle Sx, x \rangle| = 2 \leq \|S^*x\|^2 = 1$$

is impossible.

Answer 14.2.33 The answer is no. We have already observed that the weak closure of unitary operators on H (with dim $H = \infty$) is precisely the set of all contractions (see Answer 3.1.6).

So, take a non-hyponormal contraction (for example, the usual shift's adjoint S^* on $\ell^2(\mathbb{N})$), then there is always a sequence of unitary operators (a fortiori hyponormal) which converges weakly to the non-hyponormal S^*.

Answer 14.2.34 The answer is no! The counterexample is pretty simple. First, remember that an operator is subnormal if and only if it is the strong limit of a sequence of normal operators (Bishop's theorem, see the remark below Answer 3.1.6).

Next, take any subnormal operator which is not quasinormal, e.g. $I + S$ where S is the unilateral shift on ℓ^2. Then there is a sequence of normal (hence quasinormal) operators which converges strongly to $I + S$.

Consequently, the weak limit of a sequence of quasinormal operators need not be quasinormal either!

Answer 14.2.35 Let S be the shift operator on ℓ^2. Remember that S is quasinormal. To find e^S we need to find (integer) powers of S. These are better visualized using their matrix representations. We obtain

$$S = \begin{pmatrix} 0 & 0 & 0 \cdots & & \cdots \\ 1 & 0 & 0 & 0 \\ 0 & 1 & 0 & 0 & \\ & \ddots & 1 & 0 & 0 & \ddots \\ & & & 1 & \ddots & \ddots \\ \vdots & & & & \ddots & \ddots \end{pmatrix}, \quad S^2 = \begin{pmatrix} 0 & 0 & 0 \cdots \cdots \cdots \\ 0 & 0 & 0 & 0 \\ 1 & 0 & 0 & 0 & 0 & \ddots \\ & 1 & 0 & 0 & \ddots \\ & & 1 & \ddots \\ \vdots & & & \ddots & \ddots \end{pmatrix}, \quad \cdots$$

Hence

$$e^S = I + S + \frac{1}{2!}S^2 + \frac{1}{3!}S^3 + \cdots + \frac{1}{n!}S^n + \cdots$$

$$= \begin{pmatrix} 1 & 0 & 0 & \cdots & & \cdots \\ 1 & 1 & 0 & 0 & & \\ 1/2! & 1 & 1 & 0 & & \ddots \\ 1/3! & 1/2! & 1 & 1 & 0 & \ddots \\ 1/4! & 1/3! & 1/2! & 1 & 1 & \ddots \\ \vdots & & \ddots & \ddots & \ddots & \ddots \end{pmatrix}$$

It is readily checked that e^S is not normal. For instance, pick e_1 from the standard orthonormal basis of ℓ^2, then

$$\|(e^S)^* e_1\|^2 = 1 \neq \|e^S e_1\|^2 = 2 + \frac{1}{2!^2} + \frac{1}{3!^2} + \cdots$$

Finally, if e^S were quasinormal, its invertibility would then make it normal. However, we have just seen that this is impossible! Consequently, e^S cannot be quasinormal.

Remarks

(1) Is e^S hyponormal? The answer is yes. There are different ways of seeing that. One way is to show that the sequence $T_n := I + S + \frac{1}{2!}S^2 + \frac{1}{3!}S^3 + \cdots + \frac{1}{n!}S^n$ is hyponormal. Then, pass to the limit (w.r.t. to the uniform topology) to obtain e^S. Hence, e^S is also a strong limit of T_n. Since strong limits of hyponormals are hyponormal, we deduce that e^S must be hyponormal. The same conclusion could have been obtained by using the sequence $(I + S/n)^n$ whose (uniform) limit is also e^S (see e.g. Exercise 2.3.34 in [256]).

(2) Is e^S subnormal? The answer is affirmative and this is maybe simpler to see than the hyponormality of e^S. It was shown in Problem 198 in [148] that if S is the usual shift, then $p(S)$ is always *subnormal*, where p is a polynomial. Then pass to the strong limit to obtain the subnormality of e^S.

Even simpler, if R is the bilateral shift on $\ell^2(\mathbb{Z})$ (which is a normal extension of S), then e^R is a normal extension of e^S, i.e. e^S is subnormal (see [33] for more related results).

(3) This example also tells us that an invertible subnormal operator is not automatically quasinormal. Indeed, e^S is subnormal and invertible, and yet it is not quasinormal!

Answer 14.2.36 ([151]) This is utterly nontrivial! Let D be the annulus $\{z \in \mathbb{C} : r < |z| < R\}$ where $r, R > 0$. Let μ be a planar Lebesgue measure in D. Let $L^2(D)$ be the collection of all complex-valued functions which are analytic throughout D

and square-integrable w.r.t. μ (the Bergman space). That is, $f \in L^2(D)$ if f is analytic in D and $\int_D |f(z)|^2 d\mu(z) < \infty$. Then $L^2(D)$ is a Hilbert space w.r.t. the inner product

$$\langle f, g \rangle = \int_D f(z)\overline{g(z)}d\mu(z).$$

Define now an analytic position operator $A : L^2(D) \to L^2(D)$ by $Af(z) = zf(z)$. Then we claim that A is *bounded, subnormal, invertible, and without any (bounded) square root*. Observe first that A is bounded and that it is clearly invertible. It is also clear that A is subnormal.

Let us proceed now to show that the operator A defined above has no square root, and there seems to be no other way but to follow the idea of the proof in [151]. Recall first the definition of the compression spectrum of an operator $A \in B(H)$, that is, the set

$$\sigma_{comp}(A) = \{\lambda \in \mathbb{C} : \overline{\mathrm{ran}(A - \lambda)} \subsetneq H\}.$$

For the proof, we must fall back on the following auxiliary result.

Lemma 14.2.11 ([151]) *If A is the analytic position operator associated with a domain D, and if $\lambda \in D$, then λ is not an approximate eigenvalue of A, but $\overline{\lambda}$ is a simple eigenvalue of A^*. In particular,*

$$D \subset \sigma_{comp}(A) - \sigma_a(A)$$

where $\sigma_a(A)$ denotes the approximate point spectrum of A.

Let us introduce now the set

$$\sqrt{D} = \{\lambda \in \mathbb{C} : \lambda^2 \in D\} = \{\lambda \in \mathbb{C} : \sqrt{r} < |\lambda| < \sqrt{R}\}.$$

Then $\sqrt{D}$ is connected. Assume now that there is a $B \in L^2(D)$ such that $B^2 = A$. Since $D \subset \sigma_{comp}(A) - \sigma_a(A)$, it results that no point of $\sqrt{D}$ can be in $\sigma_a(B)$ (indeed, by the spectral mapping theorem, if $\lambda^2 \in D$, then $\lambda^2 \notin \sigma_a(A)$ and so $\lambda \notin \sigma_a(B)$). But

$$\sigma(A) = \sigma(B^2) = [\sigma(B)]^2.$$

Hence, if $\lambda \in D$, then at least one of the two square roots of λ is in $\sigma(B)$, and so in $\sigma_{comp}(B)$ too. Now, if both the square roots of a nonzero λ in D were in $\sigma_{comp}(B)$, then their conjugates would be in $\sigma_p(B^*)$ (see e.g. Page 84 of [123]). This, however, would contradict the fact that $\overline{\lambda}$ is a simple eigenvalue of A^*! Recapitulating, we have shown that if $\lambda \in \sqrt{D}$, then only λ or $-\lambda$ is in $\sigma(B)$. Furthermore, the set $(\sqrt{D}) \cap \sigma(B)$ is closed in $\sqrt{D}$ w.r.t. the subspace topology. Therefore, the homomorphism $\lambda \mapsto -\lambda$ carries it onto its relative complement in $\sqrt{D}$ (which is thus closed!). This cannot occur as $\sqrt{D}$ is already connected. Accordingly, no $B \in L^2(D)$ satisfies $B^2 = A$.

Further Reading In fact, what was shown in [151] is a much more general result from which we have grabbed what was just enough to answer our question. The general result is:

Theorem 14.2.12 *If A is the analytic position operator associated with a domain D, then a necessary and sufficient condition that A have a square root is that $\sqrt{D}$ be disconnected.*

See also [82, 150, 316]. Some related papers are [70, 261].

Answer 14.2.37 We will answer both questions together. Let A be an invertible subnormal operator as in Answer 14.2.36. Then, assume that there is an operator B such that $A = e^B$. Then e^B has always square roots (e.g. $e^{B/2}$) while A does not possess any square root. Accordingly, $A \neq e^B$ for all B.

Answer 14.2.38 Let S be the unilateral shift defined on ℓ^2, then take $A = e^S$ and $B = e^{-S}$. Hence A and B are both subnormal (see the remarks below Answer 14.2.35) and $AB = BA$, in fact, $A^{-1} = B$. Therefore, $AB = BA = I$. Observe in the end that neither A nor B can be quasinormal (see again the last remark below Answer 14.2.35).

Answer 14.2.39 Let S be the usual unilateral shift operator, hence S is subnormal. Then $T := S + I$ (where I designates the identity operator) remains subnormal, however, it is not quasinormal anymore. Set $p(z) = z - 1$. Then

$$p(T) = T - I = S + I - I = S$$

is obviously quasinormal, but T is not quasinormal.

Answer 14.2.40 One main example that comes to mind is the usual shift on ℓ^2 (and isometries in general). Indeed, it is binormal without being normal. But, there are other examples even on $\mathbb{C}^2$, which says that the classes of binormal and normal operators, unlike other classes met above, do not coincide on finite-dimensional spaces. The example to be given is invertible, hence it answers the second question as well. Let

$$T = \begin{pmatrix} 1 & 1 \\ 0 & -1 \end{pmatrix}.$$

Clearly, T is invertible and not normal. But,

$$T^*TT^* = TT^*T^*T \ (= I),$$

i.e. T is binormal.

Answer 14.2.41 ([42]) Let

$$T = \begin{pmatrix} 0 & 0 & 1 \\ 1 & 1 & 0 \\ 1 & -1 & 0 \end{pmatrix}.$$

Then T is invertible. Besides

$$T^*T = \begin{pmatrix} 2 & 0 & 0 \\ 0 & 2 & 0 \\ 0 & 0 & 1 \end{pmatrix} \text{ and } TT^* = \begin{pmatrix} 1 & 0 & 0 \\ 0 & 2 & 0 \\ 0 & 0 & 2 \end{pmatrix}.$$

Since the last two matrices commute, T is binormal. Let us check now that T^2 is not binormal. Readers can easily find that

$$T^2 = \begin{pmatrix} 1 & -1 & 0 \\ 1 & 1 & 1 \\ -1 & -1 & 1 \end{pmatrix}, \ T^2T^{*2} = \begin{pmatrix} 2 & 0 & 0 \\ 0 & 3 & -1 \\ 0 & -1 & 3 \end{pmatrix}$$

and

$$T^{*2}T^2 = \begin{pmatrix} 3 & 1 & 0 \\ 1 & 3 & 0 \\ 0 & 0 & 2 \end{pmatrix}.$$

Since (for instance)

$$T^2T^{*2}T^{*2}T^2 = \begin{pmatrix} * & * & 0 \\ * & * & * \\ * & * & * \end{pmatrix} \neq \begin{pmatrix} * & * & -1 \\ * & * & * \\ * & * & * \end{pmatrix} = T^{*2}T^2T^2T^{*2},$$

it ensures that T^2 cannot be binormal.

Answer 14.2.42 The example was given in [41]. Let

$$A = \begin{pmatrix} 0 & 1 \\ 0 & 0 \end{pmatrix} \text{ and } B = \begin{pmatrix} 1 & 0 \\ 0 & 1 \end{pmatrix}.$$

be both defined on $\mathbb{C}^2$. Then, both A and B are binormal. Nonetheless,

$$T := A + B = \begin{pmatrix} 1 & 1 \\ 0 & 1 \end{pmatrix}$$

is not binormal because

$$T^*TTT^* - TT^*T^*T = \begin{pmatrix} 0 & -2 \\ 2 & 0 \end{pmatrix}.$$

Answer 14.2.43 ([13]) For instance, let:

$$T = \begin{pmatrix} 0 & 1 & 1 \\ 0 & 0 & 1 \\ 0 & 0 & 0 \end{pmatrix}.$$

Clearly $T^3 = 0$ (and $T^2 \neq 0$). However, T is not binormal as

$$T^*T = \begin{pmatrix} 0 & 0 & 0 \\ 0 & 1 & 1 \\ 0 & 1 & 2 \end{pmatrix} \text{ and } TT^* = \begin{pmatrix} 2 & 1 & 0 \\ 1 & 1 & 0 \\ 0 & 0 & 0 \end{pmatrix}$$

and so

$$T^*TTT^* = \begin{pmatrix} * & 0 & * \\ * & * & * \\ * & * & * \end{pmatrix} \neq \begin{pmatrix} * & 1 & * \\ * & * & * \\ * & * & * \end{pmatrix} = TT^*T^*T.$$

Answer 14.2.44 Let S be the unilateral shift on ℓ^2. Then S is obviously binormal, and it is also subnormal. Hence, $S + I$ is subnormal. Now, if $S + I$ were binormal, then according to Theorem 2 in [42], S would be normal, and this is an absurdity. Consequently, $S + I$ is subnormal without being binormal.

Answer 14.2.45 ([42]) On $\mathbb{C}^3$ and for $n \geq 1$ let

$$T_n = \begin{pmatrix} 0 & 0 & \sqrt{2}x_{n+1} \\ x_n & x_n & 0 \\ x_n & -x_n & 0 \end{pmatrix}$$

where $(x_n)_{n \geq 1}$ is a strictly increasing sequence of positive numbers converging to 1. Then, it is seen that

$$T_{n+1}^* T_{n+1} \geq T_n T_n^*, \ \forall n \in \mathbb{N}.$$

Define now on $\mathbb{C}^3 \oplus \mathbb{C}^3 \oplus \mathbb{C}^3 \oplus \cdots$ the operator

$$A = \begin{pmatrix} 0 & 0 & 0 & \cdots & & \cdots \\ T_1 & 0 & 0 & 0 \\ 0 & T_2 & 0 & 0 \\ & \ddots & T_3 & 0 & 0 & \ddots \\ & & & T_4 & \ddots & \ddots \\ \vdots & & & & \ddots & \ddots \end{pmatrix}.$$

As in Answer 14.2.14, we may show that AA^* and A^*A are diagonal operators whereby they necessarily commute, i.e. $A \in (BN)$. Since for all n, $T_{n+1}^* T_{n+1} \geq T_n T_n^*$, we may easily establish the hyponormality of A, that is, $A \in (BN)^+$. There remains to show that $A^2 \notin (BN)$. We find

$$A^2 A^{2*} = \begin{pmatrix} 0 & 0 & 0 & & \cdots & & \cdots \\ 0 & 0 & 0 & & 0 \\ 0 & 0 & T_2 T_1 (T_2 T_1)^* & & 0 & & \ddots \\ & \ddots & 0 & & T_3 T_2 (T_3 T_2)^* & 0 & \ddots \\ & & & & 0 & & \ddots & \ddots \\ \vdots & & & & & & \ddots & \ddots \end{pmatrix}$$

and

$$A^{2*} A^2 = \begin{pmatrix} (T_2 T_1)^* T_2 T_1 & 0 & 0 & \cdots & & \cdots \\ 0 & (T_3 T_2)^* T_3 T_2 & 0 & 0 \\ 0 & 0 & (T_4 T_3)^* T_4 T_3 & 0 & & \ddots \\ & \ddots & & 0 & & \ddots & 0 & \ddots \\ & & & & 0 & & \ddots & \ddots \\ \vdots & & & & & & \ddots & \ddots \end{pmatrix}.$$

In order that $A^2 A^{2*}$ and $A^{2*} A^2$ be commuting, it suffices and it necessitates that (the commutator)

$$[T_{n+1} T_n (T_{n+1} T_n)^*, (T_{n+3} T_{n+2})^* T_{n+3} T_{n+2}] = 0, \ n = 1, 2, \cdots.$$

Since we want $A^2 \notin (BN)$, it is enough to disprove the previous relation for some $n \geq 1$. Choosing say $n = 1$ and $x_1 = 0$, we see that

$$
T_2 T_1 (T_2 T_1)^* = \begin{pmatrix} 0 & * & * \\ * & * & * \\ * & * & * \end{pmatrix} \neq \begin{pmatrix} 2x_3^2(x_5^2 + x_4^2) & * & * \\ * & * & * \\ * & * & * \end{pmatrix} = (T_4 T_3)^* T_4 T_3,
$$

making the commutativity of $A^2 A^{2*}$ and $A^{2*} A^2$ impossible, as wished.

Answer 14.2.46 ([171]) Let

$$
A = \begin{pmatrix} 1 & 1 \\ 0 & 0 \end{pmatrix} \text{ and } W = \begin{pmatrix} \frac{1}{\sqrt{2}} & \frac{1}{\sqrt{2}} \\ 0 & 0 \end{pmatrix}.
$$

Then $A = W|A|$. Now, let

$$
T = \begin{pmatrix} 0 & 0 & 0 & 0 \\ A & 0 & 0 & 0 \\ 0 & I & 0 & 0 \\ 0 & 0 & A & 0 \end{pmatrix}.
$$

Hence $T = V|T|$ where V is a partial isometry (as may readily be checked) given by

$$
V = \begin{pmatrix} 0 & 0 & 0 & 0 \\ W & 0 & 0 & 0 \\ 0 & I & 0 & 0 \\ 0 & 0 & W & 0 \end{pmatrix}.
$$

Also, readers may also check that both T and T^2 are binormal. Therefore, we need only check that $T^3 \neq V^3|T^3|$. We have

$$
T^3 = \begin{pmatrix} 0 & 0 & 0 & 0 \\ 0 & 0 & 0 & 0 \\ 0 & 0 & 0 & 0 \\ A^2 & 0 & 0 & 0 \end{pmatrix}.
$$

Hence

$$
V^3|T^3| = \begin{pmatrix} 0 & 0 & 0 & 0 \\ 0 & 0 & 0 & 0 \\ 0 & 0 & 0 & 0 \\ W^2 & 0 & 0 & 0 \end{pmatrix} \begin{pmatrix} |A^2| & 0 & 0 & 0 \\ 0 & 0 & 0 & 0 \\ 0 & 0 & 0 & 0 \\ 0 & 0 & 0 & 0 \end{pmatrix} = \begin{pmatrix} 0 & 0 & 0 & 0 \\ 0 & 0 & 0 & 0 \\ 0 & 0 & 0 & 0 \\ W^2|A^2| & 0 & 0 & 0 \end{pmatrix}.
$$

So, if $T^3 = V^3|T^3|$, then one would obtain $A^2 = W^2|A^2|$ but this is untrue for

$$A^2 = \begin{pmatrix} 1 & 1 \\ 0 & 0 \end{pmatrix} \neq \begin{pmatrix} \frac{1}{\sqrt{2}} & \frac{1}{\sqrt{2}} \\ 0 & 0 \end{pmatrix} = W^2|A^2|.$$

Accordingly

$$T^3 \neq V^3|T^3|,$$

as suggested.

Answer 14.2.47 The answer is negative for both classes. For the subnormal case, recall that it was shown in (Corollary 2 in [44]) that if T and $T + \lambda$ are in the θ-class, where λ not real, then T is normal.

So, assume for the sake of contradiction that the θ-class contained subnormals. So let T be a subnormal operator which is not normal, hence $T + \lambda$ is subnormal. *Ex hypothesi*, we would then obtain that $T, T + \lambda \in \theta$-class. But this would imply that T is normal, which is manifestly a contradiction.

As for the case of hyponormal operators, recall a surprising result (as described by S. L. Campbell): An operator in the θ-class is subnormal whenever it is hyponormal (Theorem 2 in [45]). Thanks to this result, the counterexample becomes trivial. Indeed, if we assume that every hyponormal operator T is in the θ-class, then it would result that T is subnormal (which is not always true).

Answer 14.2.48 Y. Kato [191] provided many interesting examples of operators in the θ-class, one of them answers our question. It is also a lot simpler (and stronger at the same time) than Campbell-Gellar's.

Let S be the usual unilateral shift operator on ℓ^2 and put $P = I - SS^*$. Define

$$T = \begin{pmatrix} S+i & P \\ 0 & S-i+iP \end{pmatrix}$$

on $\ell^2 \oplus \ell^2$ (where i is the usual complex number). Direct calculations show that T is in the θ-class. This is left to readers. Let us show now that T is not paranormal. Let e_1 be the usual vector in the standard orthonormal basis of ℓ^2 and choose $x = (e_1, ie_1)$. Then

$$Tx = (2ie_1 + e_2, ie_2) \text{ and } T^2x = (-2e_1 + 3ie_2 + e_3, e_2 + ie_3).$$

Accordingly

$$\|Tx\|^2 = 6 > 4\sqrt{2} = \|T^2x\|\|x\|,$$

establishing the non-paranormality of T.

Answer 14.2.49 Let us give a counterexample for both cases. Let S be the usual shift operator on ℓ^2. Then S and S^2 are two commuting quasinormal operators. We claim that $T := S + S^2$ is not in the θ-class. Indeed, it is seen that

$$T^*T = (S^* + S^{*2})(S + S^2) = 2I + S + S^*$$

and

$$T + T^* = S + S^* + S^2 + S^{*2}.$$

Plainly: T is in the θ-class iff $S + S^*$ commutes with $S^2 + S^{*2}$. However, this is impossible as if that were true, then this would yield $SS^{*2} = S^*$ or $SS^* = I$, which is absurd. We leave these simple calculations to be carried out by readers...

Answer 14.2.50 Let S be the usual shift operator on ℓ^2 and put $T = S + I$, where I is the identity operator. Then T is in θ-class while T^2 does not belong to the θ-class. Indeed, readers may check that $S^2 + 2S + I$ is not in the θ-class as carried out in Answer 14.2.49...

Answer 14.2.51 On $\mathbb{C}^2$, consider:

$$A = \begin{pmatrix} 1 & 0 \\ 1 & 1 \end{pmatrix}.$$

Clearly, A is not normal. However, the matrix A is posinormal. There are different ways of seeing that. For example, as in Answer 14.2.52 below. Alternatively, and as in [307]: The matrix

$$B_\lambda = \begin{pmatrix} \lambda & 0 \\ 1 & \lambda \end{pmatrix}$$

(with $\lambda \neq 0$) has as interrupter the matrix

$$P_\lambda = \begin{pmatrix} 1 - |\lambda|^{-2} + |\lambda|^{-4} & -\lambda|\lambda|^{-4} \\ -\bar{\lambda}|\lambda|^{-4} & 1 + |\lambda|^{-2} \end{pmatrix}.$$

In particular, A is posinormal with $P = \begin{pmatrix} 1 & -1 \\ -1 & 2 \end{pmatrix}$ as its interrupter.

(1) The first example is the matrix A defined just above. Another class of such operators is the class of hyponormal operators which are not normal (as is the case of the shift operator).

Let us give another example, a priori weaker as it is defined on an infinite-dimensional space, but it is interesting and has motivated the entire first study of posinormal operators (see [307]). Consider the so-called Cesàro matrix (or operator), that is

$$C = \begin{pmatrix} 1 & 0 & 0 & \cdots \\ \frac{1}{2} & \frac{1}{2} & 0 & \cdots \\ \frac{1}{3} & \frac{1}{3} & \frac{1}{3} & \cdots \\ \vdots & \vdots & \vdots & \vdots \end{pmatrix}$$

defined on ℓ^2. Then C is not normal. By defining

$$D = \begin{pmatrix} \frac{1}{2} & 0 & 0 & \cdots \\ 0 & \frac{2}{3} & 0 & \cdots \\ 0 & 0 & \frac{3}{4} & \cdots \\ \vdots & \vdots & \vdots & \vdots \end{pmatrix}$$

on ℓ^2 as well, a direct computation then yields

$$CC^* = \begin{pmatrix} 1 & \frac{1}{2} & \frac{1}{3} & \cdots \\ \frac{1}{2} & \frac{1}{2} & \frac{1}{3} & \cdots \\ \frac{1}{3} & \frac{1}{3} & \frac{1}{3} & \cdots \\ \vdots & \vdots & \vdots & \vdots \end{pmatrix} = C^*DC$$

making the Cesàro operator posinormal.

Remarks

(a) It is known to many readers that the Cesàro operator is hyponormal. This was first shown in [38]. In fact, it is even subnormal as shown in [204] (another proof could be consulted in [74]).

However, as a direct consequence of posinormality, we have a very simple proof of the hyponormality of the Cesàro operator (borrowed from [307]). It reads: Let $x \in \ell^2$. Then

$$\langle (C^*C - CC^*)x, x \rangle = \langle (C^*C - C^*DC)x, x \rangle = \langle C^*(I - D)Cx, x \rangle,$$

i.e. for all x

$$\langle (I - D)Cx, Cx \rangle \geq 0$$

for $I - D$ is a positive operator.

(b) The Cesàro operator is also co-posinormal, i.e. C^* is posinormal. Details can be consulted in [307].

(2) Let

$$B = \begin{pmatrix} 0 & 0 \\ 1 & 0 \end{pmatrix}.$$

Then B is not posinormal in virtue of Proposition 14.2.10 because ker $B \neq$ ker B^2. Besides, we have already seen that $A = \begin{pmatrix} 1 & 0 \\ 1 & 1 \end{pmatrix}$ is posinormal. Thus, $A - I = B$ is not posinormal.

(3) One simple way is to consider an invertible $A \in B(H)$. As it is known, A is both posinormal and co-posinormal. So, to get the desired counterexample, it suffices to consider any invertible operator which is not normal (and there are plenty of them!).

(4) The matrix A above is just perfect for this answer as well. Indeed, A is posinormal and satisfies $\sigma(A) = \{1\} \subset \mathbb{R}$, and A is plainly not self-adjoint.

(5) Consider again A as above. Then, it may be shown that

$$\|A\| = \sqrt{(3 + \sqrt{5})/2}.$$

Also, $r(A) = 1$ and so A is not normaloid.

Answer 14.2.52 ([94]) On $\mathbb{C}^2$, consider again (the nonnormal):

$$A = \begin{pmatrix} 1 & 0 \\ 1 & 1 \end{pmatrix}.$$

Then, as may be checked, A is (α, β)-normal with $\alpha = \sqrt{(3 - \sqrt{5})/2}$ and $\beta = \sqrt{(3 + \sqrt{5})/2}$.

Chapter 15
Similarity and Unitary Equivalence

15.1 Questions

15.1.1 Two Similar Operators Which Are Not Unitarily Equivalent

First, recall that unitarily equivalent operators have equal spectra. Let us supply two operators $A, B \in B(H)$ having equal spectra yet they are not unitarily equivalent. For example, take

$$A = \begin{pmatrix} 1 & 0 \\ 0 & 2 \end{pmatrix} \text{ and } B = \begin{pmatrix} 1 & 1 \\ 0 & 2 \end{pmatrix}.$$

Clearly, A and B have the same eigenvalues which, in this setting, means that A and B have equal spectra. To see why A and B are not unitarily equivalent, remember that two unitarily equivalent operators are simultaneously (e.g.) self-adjoint. Since A is self-adjoint and B is not, it follows that they cannot be unitarily equivalent.

Since similar operators have the same spectra, the following question becomes interesting.

Question 15.1.1 Give an example of two similar operators which are not unitarily equivalent.

© The Author(s), under exclusive license to Springer Nature Switzerland AG 2022
M. H. Mortad, *Counterexamples in Operator Theory*,
https://doi.org/10.1007/978-3-030-97814-3_15

15.1.2 Two Operators Having Equal Spectra but They Are Not Metrically Equivalent

We say that A, $B \in B(H)$ are metrically equivalent if $\|Ax\| = \|Bx\|$ for all $x \in H$. Equivalently, $A^*A = B^*B$.

> **Question 15.1.2** Find two operators A, $B \in B(H)$ having equal spectra but they are not metrically equivalent.

15.1.3 Two Metrically Equivalent Operators yet They Have Unequal Spectra

> **Question 15.1.3** Find two metrically equivalent operators A, B in $B(H)$ yet they have unequal spectra.

15.1.4 A Non-Self-Adjoint Normal Operator Which Is Unitarily Equivalent to Its Adjoint

> **Question 15.1.4** Give a normal operator A which is unitarily equivalent to its adjoint, but A is not self-adjoint.

15.1.5 Two Commuting Self-Adjoint Invertible Operators Having the Same Spectra yet They Are Not Unitarily Equivalent

> **Question 15.1.5** Find two self-adjoint A and B in $B(H)$ which are also commuting, invertible, having the same spectra and yet they are not unitarily equivalent.

15.1.6 A Matrix Which Is Not Unitarily Equivalent to Its Transpose

In general, a matrix need not be unitarily equivalent to its adjoint. A simple counterexample is to let $A = iI$. What if "adjoint" is replaced by "transpose"?

Question 15.1.6 Find a matrix which is not unitarily equivalent to its transpose.

15.1.7 On Some Similarity Result by J. P. Williams

It was shown in [384] that: If $T, S \in B(H)$ are such that $0 \notin \overline{W(S)}$ (the closure of the numerical range of S) and $S^{-1}TS = T^*$, then $\sigma(T) \subset \mathbb{R}$.

Question 15.1.7 Find two normal operators S, T such that S is also invertible with

$$S^{-1}TS = T^*$$

and yet $\sigma(T) \not\subset \mathbb{R}$ (therefore showing that the assumption $0 \notin \overline{W(S)}$ may not just be dropped even in the event of the normality of both S and T).

15.1.8 Two Self-Adjoint Operators A and B Such That AB Is Not Similar to BA

In general, there are (finite square) matrices A and B, one of them is an orthogonal projection (but surely not both of them as shown in [300]), such that AB is not similar to BA. For example, let

$$A = \begin{pmatrix} 0 & 1 \\ 0 & 0 \end{pmatrix} \text{ and } B = \begin{pmatrix} 0 & 0 \\ 0 & 1 \end{pmatrix}.$$

Then

$$AB = \begin{pmatrix} 0 & 1 \\ 0 & 0 \end{pmatrix} \text{ and } BA = \begin{pmatrix} 0 & 0 \\ 0 & 0 \end{pmatrix},$$

whereby AB cannot be similar to BA (as e.g. their kernels do not share the same dimension).

On the other hand, it is known that if A and B are two (finite) self-adjoint matrices, then AB is always similar to its adjoint BA (see e.g. [300]).

Question 15.1.8 Show that this is not necessarily the case in an infinite-dimensional setting.

15.1.9 Two Self-Adjoint Matrices A and B Such That AB Is Not Unitarily Equivalent to BA

It was shown in [127] that if A and B are two square normal matrices of size $n \times n$, then AB is unitarily equivalent to BA if $n \leq 2$ or rank $A \leq 1$.

Question 15.1.9 Find two 3×3 Hermitian matrices A and B such that AB is not unitarily equivalent to BA.

15.1.10 Two Normal Matrices A and B Such That AB Is Not Similar to BA

In [127], the authors obtained the similarity of AB and BA in the setting of (finite) matrices if A is positive and B is normal. This result is no longer true for operators defined on infinite-dimensional spaces, as the example in Answer 15.1.8 already shows.

Question 15.1.10 ([127]) Show that if $A, B \in B(H)$ are normal, then AB need not always be similar to BA even if $\dim H < \infty$.

Hint: According to a result in [127], the minimal possible counterexample would be 4×4 matrices A and B of rank 3.

15.1.11 Two Self-Adjoint A, B with dim ker AB = dim ker BA yet AB Is Not Similar to BA

In view of the example given in Answer 15.1.8, it is reasonable to ask whether the similarity of AB and BA is achievable when dim ker AB = dim ker BA.

Question 15.1.11 (⊛) Show that this is not always the case.

Further Reading There are many papers dealing with similarity (with its variants) of bounded as well as unbounded operators. We refer readers to [16, 20, 83, 101, 127, 130, 140, 174, 223, 243, 284, 300, 339, 350, 384], and the references therein (and also their citations).

15.1.12 Two Similar Congruent Operators A and B Which Are Not Unitarily Equivalent

Say that $A \in B(H)$ is congruent to $B \in B(H)$ if

$$A = T^* B T$$

for some invertible $T \in B(H)$. In [60], many results on similarity and congruence are established.

Question 15.1.12 Show by an example that two similar and congruent operators A and B are not necessarily unitarily equivalent even when A^2 and B^2 are positive.

15.1.13 Two Quasi-Similar Operators A, B with $\sigma(A) \neq \sigma(B)$

Generalizations of the notions of invertibility and similarity are introduced next:

Definition 15.1.1 We say that $T \in B(H, K)$ is quasi-invertible if T is injective and has dense range.

Say that $A \in B(H)$ and $B \in B(K)$ are quasi-similar if there are quasi-invertible operators $T \in B(H, K)$ and $S \in B(K, H)$ such that

$$TA = BT \text{ and } SB = AS.$$

Remembering that similarity preserves spectra, what about quasi-similarity?

Question 15.1.13 (⊛) Find two quasi-similar operators A and B such that

$$\sigma(A) \neq \sigma(B).$$

Remark S. Clary showed in [63] that quasi-similar *hyponormal* operators do have equal spectra.

15.1.14   Two Quasi-Similar Operators A and B Such That One Is Compact and the Other Is Not

Obviously if an operator A is similar to a compact operator B, then A is compact. What about the analogous question for quasi-similar operators?

Question 15.1.14 Find two quasi-similar operators A and B such that one is compact and the other is not.

15.1.15   Two Similar Subnormal Operators Which Are Not Unitarily Equivalent

Among the important consequences of the Fuglede-Putnam theorem, we have: Any two similar normal operators are unitarily equivalent [294].
    Similarity was then weakened to just quasisimilarity in Lemma 4.1 in [93], and the authors in [284] extended the latter to unbounded operators.

Question 15.1.15 Find two similar hyponormal (resp. subnormal) operators which are not unitarily equivalent.

15.1.16 Two Quasi-Similar Hyponormal Operators Which Are Not Similar

Question 15.1.16 (⊛) Find two quasi-similar hyponormal operators which are not similar.

15.1.17 A Quasinilpotent Operator Not Similar to Its Multiple

In general, a matrix A need not be similar to αA for a certain $\alpha \neq 1$. For example, let $A = \begin{pmatrix} 1 & 0 \\ 0 & 2 \end{pmatrix}$. Then surely A cannot be similar to $2A$ for if they were, then A and $2A$ would have the same trace, and this is not the case. If, however, A is a nilpotent matrix, then A and λA are similar for any $\lambda \neq 0$. What about the case of infinite-dimensional spaces? Well, two cases must be looked at. The case of nilpotence and the case of quasinilpotence (remember that these two classes coincide on a finite-dimensional space). The former is proposed to interested to readers in Question 17.2.10, while the latter is treated next.

Question 15.1.17 (⊛) Let V be the Volterra operator defined on $L^2(0, 1)$. Show that V and λV are similar only if $\lambda = 1$.

Answers

Answer 15.1.1 Let

$$A = \begin{pmatrix} 3 & 1 \\ -2 & 0 \end{pmatrix} \text{ and } B = \begin{pmatrix} 1 & 1 \\ 0 & 2 \end{pmatrix}.$$

To see why A and B are similar, observe that they both have two distinct eigenvalues, namely 1 and 2. Hence they are both diagonalizable, giving rise to two invertible matrices T and S such that

$$T^{-1}AT = \begin{pmatrix} 1 & 0 \\ 0 & 2 \end{pmatrix} \text{ and } S^{-1}BS = \begin{pmatrix} 1 & 0 \\ 0 & 2 \end{pmatrix}.$$

Hence

$$ST^{-1}ATS^{-1} = B,$$

which proves the similarity of A and B because ST^{-1} is invertible.

To see why A is not unitarily equivalent to B, an interesting way to do that is to appeal on [162]: *Let $A = (a_{i,j})$ and $B = (b_{i,j})$ be two square $n \times n$ matrices. If A and B are unitarily equivalent, then*

$$\sum_{i,j=1}^{n} |a_{i,j}|^2 = \sum_{i,j=1}^{n} |b_{i,j}|^2.$$

A digression: This can also be seen by showing that $\|A\| \neq \|B\|$ (for if U is unitary, then $\|U^*AU\| = \|A\|$ as in Exercise 4.3.6 in [256]).

Now, since the sum of the squares of the absolute values of all entries of A (which is 14) is different from the corresponding sum in the case of the matrix B (which is 6), we immediately see that A and B are not unitarily equivalent.

Answer 15.1.2 Take

$$A = \begin{pmatrix} 0 & 1 \\ 0 & 0 \end{pmatrix} \text{ and } B = \begin{pmatrix} 0 & 0 \\ 1 & 0 \end{pmatrix}.$$

Then A and B are not metrically equivalent as plainly $A^*A \neq B^*B$. Observe in the end that the spectra of A and B are both reduced to the singleton $\{0\}$.

Answer 15.1.3 The answer is negative even when A and B are also unitary and self-adjoint. Indeed, let

$$A = \begin{pmatrix} 0 & 1 \\ 1 & 0 \end{pmatrix} \text{ and } B = \begin{pmatrix} 1 & 0 \\ 0 & 1 \end{pmatrix}.$$

Then A and B are self-adjoint and unitary (hence $A^*A = B^*B$). Finally,

$$\sigma(A) = \{-1, 1\} \neq \{1\} = \sigma(B),$$

as wished.

Answer 15.1.4 Let

$$A = \begin{pmatrix} 0 & 1 \\ -1 & 0 \end{pmatrix} \text{ and } U = \begin{pmatrix} 1 & 0 \\ 0 & -1 \end{pmatrix}.$$

Then A and U are both unitary (and U is also self-adjoint). However,

$$UAU^* = A^* \text{ yet } A \neq A^*.$$

Answer 15.1.5 There are counterexamples in a finite-dimensional setting. Indeed, let

$$A = \begin{pmatrix} 1 & 0 & 0 \\ 0 & 1 & 0 \\ 0 & 0 & 3 \end{pmatrix} \text{ and } B = \begin{pmatrix} 1 & 0 & 0 \\ 0 & 3 & 0 \\ 0 & 0 & 3 \end{pmatrix}.$$

Then A and B clearly commute. They are also positive and invertible. In addition, $\sigma(A) = \sigma(B) = \{1, 3\}$. Nonetheless, A and B are not even similar merely because e.g. $\operatorname{tr} A \neq \operatorname{tr} B$.

Answer 15.1.6 Most of the material here is borrowed from [149]. Readers may wish to consult [126] for a related work. See [129] for another example.

Let

$$A = \begin{pmatrix} 0 & 1 & 0 \\ 0 & 0 & 2 \\ 0 & 0 & 0 \end{pmatrix} \text{ and } B = \begin{pmatrix} 0 & 0 & 0 \\ 2 & 0 & 0 \\ 0 & 1 & 0 \end{pmatrix}.$$

Then A and B are unitarily equivalent via the unitary (and self-adjoint) matrix

$$U = \begin{pmatrix} 0 & 0 & 1 \\ 0 & 1 & 0 \\ 1 & 0 & 0 \end{pmatrix},$$

that is, $U^*AU = B$.

Now, we show that A is not unitarily equivalent to B^*. As observed in [149], there are several ways of proving that. For example, if there were some unitary V such that $V^*AV = B^*$, then, upon resolving a system of nine equations in nine unknowns, V would necessarily be of the form

$$V = \begin{pmatrix} 2\alpha & 0 & \beta \\ 0 & \alpha & 0 \\ 0 & 0 & 2\alpha \end{pmatrix}.$$

Clearly V cannot be unitary whichever the values of α and β.

Finally, suppose that $W^*AW = A^*$ for some unitary W (remember that here $A^* = A^t$). Since we already have $U^*AU = B$, it follows that $U^*A^*U = B^*$. Hence

$$U^*W^*AWU = B^*.$$

Since WU is unitary and $(WU)^* = U^*W^*$, we would obtain the unitary equivalence of A and B^*, which, as just observed, is not possible! Thus, A is not unitarily equivalent to its transpose A^*.

Answer 15.1.7 A first example already appeared in [384] in an *infinite-dimensional space*. Here, we give a simple example on $\mathbb{C}^2$ which, apparently, has escaped notice. Consider:

$$S = \begin{pmatrix} 0 & 1 \\ 1 & 0 \end{pmatrix} \text{ and } T = \begin{pmatrix} i & 0 \\ 0 & -i \end{pmatrix}.$$

Then both S and T are even unitary and

$$TS = ST^* = \begin{pmatrix} 0 & i \\ -i & 0 \end{pmatrix}$$

as it is easily checked. Nevertheless,

$$\sigma(T) = \{i, -i\} \not\subset \mathbb{R},$$

as wished.

Answer 15.1.8 ([300] or [127]) First, recall that the null-spaces of two similar operators have the same dimension. What is therefore sought is a pair of self-adjoint operators A and B such that e.g. AB is injective while BA is not. Let A be the diagonal (positive) operator on ℓ^2 whose diagonal is constituted of $1, \frac{1}{2}, \frac{1}{3}, \cdots$. Let $y = (1, \frac{1}{2}, \frac{1}{3}, \cdots)$ and let B be the orthogonal projection onto $\{y\}^\perp$.

Then, it may readily be checked that BA is one-to-one because y is not in the range of A. At the same time,

$$\ker(AB) = \{\alpha y : \alpha \in \mathbb{C}\}.$$

Consequently, AB and BA cannot be similar.

Answer 15.1.9 Take the two self-adjoint matrices (a pair borrowed from [127]):

$$A = \begin{pmatrix} 1 & 0 & 0 \\ 0 & 1 & 0 \\ 0 & 0 & 0 \end{pmatrix}, \ B = i \begin{pmatrix} 0 & -1 & 1 \\ 1 & 0 & -1 \\ -1 & 1 & 0 \end{pmatrix}$$

(A is even an orthogonal projection). Now, readers may easily check that

$$AB = i \begin{pmatrix} 0 & -1 & 1 \\ 1 & 0 & -1 \\ 0 & 0 & 0 \end{pmatrix} \text{ and } BA = i \begin{pmatrix} 0 & -1 & 0 \\ 1 & 0 & 0 \\ -1 & 1 & 0 \end{pmatrix} \ (= (AB)^*).$$

There are several ways of establishing the non-unitary equivalence of AB and BA. For example, by a result in [340] (see [126] for informative historical notes and other interesting results and investigations): Two 3×3 complex matrices S and T

are unitarily equivalent if and only if $\Phi(S) = \Phi(T)$ where $\Phi : M_3(\mathbb{C}) \to \mathbb{C}^7$ is the function defined by:

$$\Phi(X) = (\text{tr } X, \text{tr } X^2, \text{tr } X^3, \text{tr } X^*X, \text{tr } X^*X^2, \text{tr } X^{*2}X^2, \text{tr } X^*X^2X^{*2}X).$$

Therefore, and in order to show AB is not unitarily equivalent to BA, it suffices to check that e.g. tr $X^*X^2X^{*2}X$ has different values for $X = AB$, then for $X = BA$. This is in effect the case for

$$BA(AB)^2(BA)^2AB = \begin{pmatrix} 2 & 0 & -2 \\ -1 & 1 & 0 \\ 1 & -1 & 0 \end{pmatrix}$$

and

$$AB(BA)^2(AB)^2BA = \begin{pmatrix} 4 & 0 & 0 \\ -3 & 1 & 0 \\ 0 & 0 & 0 \end{pmatrix}$$

whereby

$$\text{tr}[BA(AB)^2(BA)^2AB] = 3 \neq 5 = \text{tr}[AB(BA)^2(AB)^2BA],$$

marking the end of the proof.

Remark The unitary equivalence of 2×2 complex matrices can be tested using the map $\Phi : M_2(\mathbb{C}) \to \mathbb{C}^3$ defined by:

$$\Phi(X) = (\text{tr } X, \text{tr } X^2, \text{tr } X^*X)$$

as obtained in [268].

Answer 15.1.10 Let

$$A = \begin{pmatrix} 0 & 0 & 0 & 1 \\ 0 & 1 & 0 & 0 \\ 0 & 0 & 0 & 0 \\ 1 & 0 & 0 & 0 \end{pmatrix} \quad \text{and } B = \begin{pmatrix} 0 & 0 & 0 & 0 \\ 0 & 0 & 0 & 1 \\ 0 & 1 & 0 & 0 \\ 0 & 0 & 1 & 0 \end{pmatrix}.$$

Then A is self-adjoint and B is normal. Moreover,

$$AB = \begin{pmatrix} 0 & 0 & 1 & 0 \\ 0 & 0 & 0 & 1 \\ 0 & 0 & 0 & 0 \\ 0 & 0 & 0 & 0 \end{pmatrix} \quad \text{and } BA = \begin{pmatrix} 0 & 0 & 0 & 0 \\ 1 & 0 & 0 & 0 \\ 0 & 1 & 0 & 0 \\ 0 & 0 & 0 & 0 \end{pmatrix}.$$

These two products obey

$$(AB)^2 = \begin{pmatrix} 0\,0\,0\,0 \\ 0\,0\,0\,0 \\ 0\,0\,0\,0 \\ 0\,0\,0\,0 \end{pmatrix} \neq (BA)^2.$$

Once that is known, assuming that $S^{-1}(BA)S = AB$ for some invertible S or $S(AB) = (BA)S$ would give $S(AB)^2 = (BA)^2 S$ whereby $(BA)^2 S = 0$ and so $(BA)^2 = 0$, a doubtless contradiction! Accordingly, AB and BA cannot be similar.

Answer 15.1.11 ([140]) Let $H = L^2[0, 1]$. Consider the two elements of H defined as

$$g(x) = \begin{cases} -1, \ 0 \leq x \leq 1/2, \\ \ \ 1, \ \ 1/2 < x \leq 1, \end{cases}$$

and

$$f(x) = \begin{cases} \ \ \ \ x, \ \ \ 0 \leq x \leq 1/2, \\ 1 - x, \ 1/2 \leq x \leq 1. \end{cases}$$

Let P be the orthogonal projection onto $\{\lambda f\}$ (the span of f), then set $A = I - P$. Now, for $h \in H$, we define B by

$$(Bh)(x) = \int_{1-x}^{1} h(t)dt.$$

Then, it is straightforward to check the self-adjointness of B.

First, as $Bg = f$, we may easily get that $\ker(AB) = \{\lambda g\}$. Also, a simple reasoning yields $\ker(BA) = \{\lambda f\}$. Thereby,

$$\dim \ker(AB) = \dim \ker(BA) (= 1).$$

If AB were similar to BA, then $(AB)^2$ would again be similar to $(BA)^2$. So, to show that AB is not similar to BA, it suffices therefore to show that $\dim \ker(AB)^2 \neq \dim \ker(BA)^2$. Observe that $\langle f, g \rangle = 0$ which entails that $Ag = g$. Thus,

$$ABAg = Af = 0$$

and so $g \in \ker(BA)^2$ (and so $\dim \ker(BA)^2 \neq 1$).

Now, if $\dim \ker(AB)^2 > 1$, then $\ker(AB)^2$ would contain a vector h (besides g), i.e. $ABABh = 0$. This says that $ABh \in \ker(AB)$ and so $ABh = \lambda g$. Thus, g would end up in the range of AB whereby $g + \lambda f$ would be in the range of B (for some scalar λ) which is constituted of absolutely continuous functions. This is the sought contradiction! Accordingly, AB is not similar to BA.

Answer 15.1.12 ([60]) Let

$$A = \begin{pmatrix} 2 & 2 & 0 & 0 \\ 0 & -2 & 0 & 0 \\ 0 & 0 & 1 & 0 \\ 0 & 0 & 0 & -1 \end{pmatrix} \text{ and } B = \begin{pmatrix} 1 & 1 & 0 & 0 \\ 0 & -1 & 0 & 0 \\ 0 & 0 & 2 & 0 \\ 0 & 0 & 0 & -2 \end{pmatrix}.$$

Then $A^2 \geq 0$ and $B^2 \geq 0$. Also, A and B are similar and congruent to each other. By using $\|A^*A\| = \|A\|^2$ (or else), we can show that

$$\|A\| = \sqrt{2\sqrt{5} + 6} \neq 2 = \|B\|$$

and so A and B cannot be unitarily equivalent.

Answer 15.1.13 The example is borrowed from [161] where more details may be found. Consider the (familiar by now) sequence A_n of $n \times n$ matrices given by

$$A_n = \begin{pmatrix} 0 & 0 & . & . & . & 0 \\ 1 & 0 & 0 & . & & \\ 0 & 1 & 0 & & . & \\ . & . & . & 0 & 0 & . \\ . & & & . & . & . \\ 0 & & & 0 & 1 & 0 \end{pmatrix}.$$

Consider also the sequence B_n (of the same size $n \times n$):

$$B_n = \begin{pmatrix} 0 & 0 & . & . & . & 0 \\ 1/n & 0 & 0 & . & & \\ 0 & 1/n & 0 & & . & \\ . & & . & 0 & 0 & . \\ . & & & . & . & . \\ 0 & & & 0 & 1/n & 0 \end{pmatrix}.$$

Then A_n is (quasi-) similar to B_n. Hence according to Theorem 2.5 in [161], $A := \bigoplus_{n \in \mathbb{N}} A_n$ and $B := \bigoplus_{n \in \mathbb{N}} B_n$ are quasi-similar. Let us find now the spectra of both A and B. It can be shown that $\|A_n^k\| = 1$ for $k < n$. Hence $\|A^k\| = 1$. This gives $r(A) = 1$. Thus,

$$\sigma(A) = \{\lambda \in \mathbb{C} : |\lambda| \leq 1\}.$$

On the other hand, it can be shown that

$$\sigma(B) = \{0\}.$$

Answer 15.1.14 ([161]) Take again the sequences (A_n) and (B_n) which appeared in Answer 15.1.13. Set

$$A := \bigoplus_{n \in \mathbb{N}} A_n \text{ and } B := \bigoplus_{n \in \mathbb{N}} B_n.$$

From Answer 10.2.8 we already know that A is not compact. Since $\|B_n\| \to 0$, it follows that B is compact. Observe in the end that A and B are quasi-similar (see Answer 15.1.13).

Answer 15.1.15 Most of the material here is borrowed from [148] with some adjustments. It is perhaps better to give the operators in their matrix representations. Let

$$R = \begin{pmatrix} 0 & 0 & 0 \cdots & & \cdots \\ 1/\sqrt{2} & 0 & 0 & 0 \\ 0 & 1 & 0 & 0 & \ddots \\ & \ddots & 1 & 0 & 0 & \ddots \\ & & 1 & \ddots & \ddots \\ \vdots & & & \ddots & \ddots \end{pmatrix} \text{ and } S = \begin{pmatrix} 0 & 0 & 0 \cdots & & \cdots \\ 1 & 0 & 0 & 0 \\ 0 & 1 & 0 & 0 & \ddots \\ & \ddots & 1 & 0 & 0 & \ddots \\ & & 1 & \ddots & \ddots \\ \vdots & & & \ddots & \ddots \end{pmatrix},$$

i.e. S is the usual shift and R is a (unilateral) weighted shift, both defined on ℓ^2. Then both S and R are hyponormal (in fact, subnormal).

The similarity of S and R is obtained via the invertible operator

$$T = \begin{pmatrix} 1 & 0 & & & \cdots \\ 0 & \sqrt{2} & 0 & & \cdots \\ 0 & 0 & \sqrt{2} & \ddots & & \ddots \\ & \ddots & \ddots & \sqrt{2} & & \ddots \\ \vdots & & & & \ddots & \ddots \end{pmatrix},$$

that is, $TR = ST$.

In the end, why are not S and R unitarily equivalent? One efficient test here is to observe that S is an isometry while R is not one. Indeed, if there were a unitary $U \in B(\ell^2)$ such that $UR = SU$, then this would imply the untrue statement $R^*R = I$!

Remark In [148], P. R. Halmos said that the example above was due to D. E. Sarason. Another quite similar example (by J. G. Stampfli) also appeared in the very end of [346]. Since both examples are remarkably similar, we omit the second one. However, J. G. Stampfli observed that the two similar subnormal operators there do not need to have similar (minimal) normal extensions. Stampfli's observation applies to the above example as well. Indeed, the normal extension of S cannot be similar to the normal extension of R for if they were, they would be unitarily equivalent given that they are both normal. However, this is impossible for the normal extension of S is unitary while the normal extension of R is not unitary.

Remark The following related result to the above counterexample is of some interest (proofs may be found in Problems 89 and 90 in [148]):
  Let $A, B \in B(\ell^2)$ be two weighted shifts with weights (α_n) and (β_n), respectively.

(1) If $|\alpha_n| = |\beta_n|$ for all n, then A and B are unitarily equivalent.
(2) A necessary and sufficient condition that A and B (with nonzero weights) be similar is that the sequence

$$\left| \frac{\alpha_0 \cdots \alpha_n}{\beta_0 \cdots \beta_n} \right|$$

be bounded away from 0 and from ∞.

Remark What about similar quasinormal operators? Are they necessarily unitarily equivalent? See Question 17.1.7.

Answer 15.1.16 ([63]) Let S be the standard shift on ℓ^2 and let

$$S_n = \begin{pmatrix} 0 & 0 & 0 \cdots & & \cdots \\ 1/n & 0 & 0 & 0 & \\ 0 & 1 & 0 & 0 & \ddots \\ & \ddots & 1 & 0 & 0 & \ddots \\ & & & 1 & \ddots & \ddots \\ \vdots & & & & \ddots & \ddots \end{pmatrix}$$

be also defined on ℓ^2. Next, define on $\ell^2 \oplus \ell^2 \oplus \ell^2 \oplus \cdots$:

$$A = S \oplus S \oplus S \oplus \cdots \text{ and } B = S_1 \oplus S_2 \oplus S_3 \oplus \cdots.$$

  Using basic properties of (infinite) direct sums, it can readily be shown that A and B are both hyponormal. Since each S_n is similar to S (why?), a glance at Theorem 2.5 in [161] yields the quasi-similarity of A and B. But A cannot be similar to B

for A is an isometry whereas B is not bounded below (indeed, similarity preserves boundedness below, as shown in e.g. Exercise 7.3.13 in [256]).

Answer 15.1.17 The proof given here appeared in [109], and we provide a few more details. Notice that this proof is simpler than the one which first appeared in [186].

Recall that the Volterra operator on $L^2(0, 1)$ is quasinilpotent and non-nilpotent, and in particular $\|V^n\| \neq 0$ for all n. We also assume readers have some familiarity with basic results about the Volterra operator (consulting Exercise 9.3.21 in [256] comes in handy).

Now, suppose that V and λV are similar, that is $T^{-1}VT = \lambda V$ for a certain invertible operator $T \in B[L^2(0, 1)]$. Hence $\lambda^n V^n = T^{-1}V^n T$ and $V^n = \lambda^n T V^n T^{-1}$ for all n, thereby

$$|\lambda|^n = \frac{\|T^{-1}V^n T\|}{\|V^n\|} \leq \frac{\|T^{-1}\|\|V^n\|\|T\|}{\|V^n\|} = \|T^{-1}\|\|T\|.$$

On the other hand,

$$|\lambda|^n = \frac{\|V^n\|}{\|T V^n T^{-1}\|} \geq \frac{\|V^n\|}{\|T\|\|V^n\|\|T^{-1}\|} = \frac{1}{\|T\|\|T^{-1}\|}.$$

In other words,

$$\frac{1}{\|T\|\|T^{-1}\|} \leq |\lambda|^n \leq \|T^{-1}\|\|T\|$$

for all n. If $|\lambda| > 1$, upon sending n to ∞ gives a contradiction, and a similar observation yields again a contradiction when $|\lambda| < 1$. Therefore, the only possible choice is $|\lambda| = 1$.

Let us now proceed to show that $\lambda \geq 0$. Since $T^{-1}VT = \lambda V$, clearly

$$T^{-1}(1 - zV)T = (1 - \lambda zV) \text{ or } (1 - \lambda zV)^{-1} = T^{-1}(1 - zV)^{-1}T$$

for any complex number z, thanks to the invertibility of both $1 - \lambda zV$ and $1 - \lambda V$. Also, for $f \in L^2(0, 1)$

$$(1 - \lambda zV)^{-1}f(x) = f(x) + \lambda z \int_0^x e^{\lambda z(x-t)} f(t)dt$$

(see, for instance, Exercise 9.3.21 in [256]). Taking $f(x) = 1$ on $(0, 1)$, and doing some arithmetic imply that $e^{\lambda z x} = T^{-1}(1 - zV)^{-1}T(1)$. By taking $z = -n$, and since $\mathrm{Re}\, V \geq 0$, a similar argument as in Answer 9.2.4 gives $\|(1 + nV)^{-1}\| \leq 1$. Hence

$$\|e^{-\lambda n x}\| \leq \|T^{-1}\|\|(1 + nV)^{-1}\|\|T\| \leq \|T^{-1}\|\|T\|$$

for all $n \geq 0$. Whence necessarily $\mathrm{Re}\,\lambda \geq 0$ (otherwise $\|T^{-1}\|\|T\|$ would be infinite). Because $\lambda^k V = T^{-k}VT^k$ for all $k \geq 2$, it then follows by a similar reasoning that $\mathrm{Re}\,\lambda^k \geq 0$ for all $k \geq 2$. Consequently, $\mathrm{Re}\,\lambda^k \geq 0$ for all $k \geq 1$, and this forces $\lambda \geq 0$. Thus, $\lambda = 1$, as needed.

Chapter 16
The Sylvester Equation

16.1 Basics

Consider the operator equation:

$$AX - XB = C,$$

where $A, B, C \in B(H)$ are given and $X \in B(H)$ is the unknown. This equation is more commonly known as the Sylvester equation [357].

Theorem 16.1.1 ([308]) *Let $A, B \in B(H)$ obey $\sigma(A) \cap \sigma(B) = \varnothing$. Then the equation $AX - XB = C$ has a unique solution X (in $B(H)$) for each $C \in B(H)$.*

An interesting proof of the previous theorem may be found in [219] (or in Exercises 7.3.30 and 7.3.31 in [256]). Readers may also wish to consult the interesting survey devoted to the Sylvester equation [27].

16.2 Questions

16.2.1 The Condition $\sigma(A) \cap \sigma(B) = \varnothing$ Is Not Necessary for the Existence of a Solution to Sylvester's Equation

Question 16.2.1 Show that the condition $\sigma(A) \cap \sigma(B) = \varnothing$ is not necessary for the existence of a solution to Sylvester's equation, i.e. find an $X \in B(H)$ such that $AX - XB = C$, where $A, B, C \in B(H)$ and $\sigma(A) \cap \sigma(B) \neq \varnothing$.

© The Author(s), under exclusive license to Springer Nature Switzerland AG 2022   281
M. H. Mortad, *Counterexamples in Operator Theory*,
https://doi.org/10.1007/978-3-030-97814-3_16

16.2.2   An Equation $AX - XB = C$ Without a Solution X for Some $C \in B(H)$ Where $\sigma(A) \cap \sigma(B) \neq \varnothing$

Question 16.2.2 Give an equation $AX - XB = C$ without a solution X for some $C \in B(H)$ where $\sigma(A) \cap \sigma(B) \neq \varnothing$.

16.2.3   Unitary Equivalence and Sylvester's Equation

W. E. Roth [311] showed in a *finite-dimensional setting* that the similarity of the matrices $\begin{pmatrix} A & C \\ 0 & B \end{pmatrix}$ and $\begin{pmatrix} A & 0 \\ 0 & B \end{pmatrix}$ is equivalent to the existence of a solution X of $AX - XB = C$ (a nice proof appeared in [108]).

It is natural to wonder whether Roth's result remains valid when $\dim H = \infty$.

Question 16.2.3 (Cf. [310]) Let H be a Hilbert space with $\dim H = \infty$. Show that the unitary equivalence of the matrices $\begin{pmatrix} A & C \\ 0 & B \end{pmatrix}$ and $\begin{pmatrix} A & 0 \\ 0 & B \end{pmatrix}$ need not imply the existence of a solution to $AX - XB = C$ in $B(H)$.

16.2.4   Schweinsberg's Theorem and the Sylvester Equation for Quasinormal Operators

Roth's result recalled just above does hold if $\dim H = \infty$ if we further take A and B to be self-adjoint. This is due to M. Rosenblum in [310]. Later, this was generalized by A. Schweinsberg to normal operators (see [328]). It is therefore interesting to know whether this result still holds for the weaker class of quasinormal operators.

Question 16.2.4 Provide an example showing that Rosenblum–Roth's theorem need not hold true for quasinormal operators.

Answers

Answer 16.2.1 Set $A = S^*$ (the shift's adjoint) and $B = 0$. Sylvester's equation then becomes: $S^* X = C$. This equation has always a solution X for any $C \in B(\ell^2)$, given by $X = SC$. Indeed,

$$S^* X = S^* SC = C,$$

and yet

$$\sigma(B) = \{0\} \subset \sigma(S^*) = \{\lambda \in \mathbb{C} : |\lambda| \leq 1\},$$

i.e. the condition $\sigma(B) \cap \sigma(S^*) = \varnothing$ is violated.

Answer 16.2.2 Let S be the shift operator on ℓ^2. Take $A = S$ and $B = 0$ and so obviously $\sigma(B) \subset \sigma(S)$. The equation $SX = C$ does not have a solution X for all $C \in B(\ell^2)$. For instance, if $C = I$ (the identity operator), then if there were an X such that $SX = I$, then it would follow that

$$X = S^* SX = S^*$$

which would then lead to $SS^* = I$!

Answer 16.2.3 [310] Let A be a non-unitary isometry in $B(H)$, i.e. $A^* A = I$ and $C := I - AA^* \neq 0$ (an explicit example would be the shift operator on ℓ^2). The matrix operator

$$\begin{pmatrix} A & C \\ 0 & A^* \end{pmatrix}$$

defined on $B(H \oplus H)$ is unitary as

$$\begin{pmatrix} A & C \\ 0 & A^* \end{pmatrix}^{-1} = \begin{pmatrix} A & C \\ 0 & A^* \end{pmatrix}^* = \begin{pmatrix} A^* & 0 \\ C^* & A \end{pmatrix} = \begin{pmatrix} A^* & 0 \\ C & A \end{pmatrix}.$$

Next, it is seen that

$$\begin{pmatrix} A^* & 0 \\ C & A \end{pmatrix} \begin{pmatrix} A & 0 \\ 0 & 0 \end{pmatrix} \begin{pmatrix} A & C \\ 0 & A^* \end{pmatrix} = \begin{pmatrix} A & C \\ 0 & 0 \end{pmatrix}.$$

This signifies that the matrix operators $\begin{pmatrix} A & 0 \\ 0 & 0 \end{pmatrix}$ and $\begin{pmatrix} A & C \\ 0 & 0 \end{pmatrix}$ are *unitary equivalent* on $B(H \oplus H)$. Consider now the Sylvester equation:

$$AX - XB = C$$

in the case $B = 0$, i.e. $AX = C$. This equation has no solution in $B(H)$. If it had one, then

$$AX = C \Longrightarrow X = A^*C = A^*(I - AA^*) = A^* - A^* = 0$$

and hence $C = 0$, which is the desired contradiction!

Answer 16.2.4 [328] The example is similar to the one in Answer 16.2.3 and so some details are left out. Let S be the shift operator and set $P := I - SS^* \neq 0$. Then

$$\begin{pmatrix} S^* & -I \\ P & S \end{pmatrix} \begin{pmatrix} S & 0 \\ 0 & 0 \end{pmatrix} \begin{pmatrix} S & I \\ 0 & S^* \end{pmatrix} = \begin{pmatrix} S & I \\ 0 & 0 \end{pmatrix}.$$

Obviously, S and 0 are both quasinormal. Nonetheless, for no X does $I = SX - X0$.

Chapter 17
More Questions and Some Open Problems

17.1   More Questions

Question 17.1.1 Give an example showing that a (bounded) linear operator A on a pre-Hilbert space maybe "adjointless."

Question 17.1.2 (See [163]) Inspired by the example in Answer 1.2.16 or else, find a symmetric bilinear mapping that is not open at the origin.

Question 17.1.3 (See Page 105 in [375] for more related results and counterexamples) Find a sequence of invertible (A_n) in $B(H)$ that converges strongly to an invertible $A \in B(H)$ yet (A_n^{-1}) does not converge strongly to A^{-1}.

Question 17.1.4 ([148]) Show that the mapping $A \mapsto A^2$ defined from $B(H)$ into $B(H)$ is not weakly continuous, that is, find a sequence (A_n) in $B(H)$ that converges weakly to $A \in B(H)$ yet (A_n^2) does not converge weakly to A^2.

M. H. Mortad, *Counterexamples in Operator Theory*,
https://doi.org/10.1007/978-3-030-97814-3_17

Question 17.1.5 (Cf. [30]) Let A be a square matrix over the complex field and assume that A has two square roots B and C. Does it follow that B is similar to C?

Question 17.1.6 ([185]) Let $A, B, C \in B(H)$. If $\begin{pmatrix} A & 0 \\ 0 & B \end{pmatrix}$ and $\begin{pmatrix} A & 0 \\ 0 & C \end{pmatrix}$ are unitarily equivalent, is it true that B and C are unitarily equivalent?

Question 17.1.7 *We have already seen that similar subnormal operators do not have to be unitarily equivalent. The corresponding question for quasinormal had resisted solutions for almost 15 years since the paper [192]. It was then solved in the negative in [55] (one of the authors of the previous reference also announced the counterexample in [387]).*

Find two similar quasinormal operators that are not unitarily equivalent.

Question 17.1.8 ([385]) It is known that subnormal operators do not necessarily have square roots. The shift operator on ℓ^2 is a counterexample. Now, what if we assume that a subnormal operator has a certain square root, does this square root have to be subnormal? Give an example showing that the answer is negative.

Question 17.1.9 Let A and B be two commuting subnormal operators. Show that the subnormality of $A + B$ does not guarantee commuting normal extensions of A and B.

Remark The previous question was asked by A. Lubin in [218], and it was answered negatively in [212] after almost four decades.

Question 17.1.10 Can you find a hyponormal operator A such that e^A is not hyponormal?

Question 17.1.11 (Cf. [179] and [195]) Let $A \in B(H)$ be such that

$$|\langle Ax, x \rangle| \leq \langle |A|x, x \rangle, \ \forall x \in H.$$

Does it follow that A is hyponormal (or co-hyponormal or subnormal or quasinormal)?

Question 17.1.12 Find an example of a pair of two operators $A, B \in B(H)$ such that $A^*B = BA^*$ and $AB = BA$, where A is hyponormal but B does not commute with $\varphi(A)$ for some bounded Borel function φ on $\sigma(A)$.

Question 17.1.13 (Cf. e.g. Lemma 3.8 in [53]) (Can you?) Find two quasi-normal operators $A, B \in B(H)$ such that $AB = BA$ but AB is not quasinormal.

Question 17.1.14 On some Hilbert space H, find $A, B \in B(H)$ such that A is positive and injective, AB is positive, $\sigma(B) \subset [0, \infty)$ and yet B is not similar to a bounded self-adjoint operator (see e.g. [332]).

Question 17.1.15 Inspired by the example that appeared in Answer 14.2.45, find an *invertible* $A \in (BN)^+$ for which $A^2 \notin (BN)$.

Question 17.1.16 ([225], cf. [189]) In a finite-dimensional Hilbert space, find two self-adjoint operators A and B such that A is also invertible so that

$$\|[A, B]\| < 1 \text{ and } \|[|A|, B]\| > n$$

where $n \in \mathbb{N}$.

Question 17.1.17 ([189]) Let H be an infinite-dimensional Hilbert space. Show that the map $|\cdot|_{B(H)}$ is not Lipschitz continuous in the operator norm even when restricted to the set of self-adjoint operators.

Question 17.1.18 It is well known that eigenvectors corresponding to distinct eigenvalues for a hyponormal operator T are perpendicular. This is an easy exercise from say [21]. Show that this is not true anymore for the class of weakly normal operators (see [111]).

Question 17.1.19 Find a posinormal operator A such that

$$|A^n| \neq |A|^n$$

for all $n \in \mathbb{N}$.

Question 17.1.20 (Cf. [208]) Give an example of a posinormal operator whose square is not posinormal.

17.2   Some Open Problems

Question 17.2.1 It was conjectured in [111] (cf. [194]) that every weakly normal operator with a real spectrum is self-adjoint. Apparently, and as it was posed, it is still unanswered (no proof and no counterexample) though some positive results appeared in [226] with some extra conditions. What do you think of it?

Question 17.2.2 Recall that if $z \in \mathbb{C}$ and $\operatorname{Re} z^k \geq 0$ for all $k \geq 1$, then this forces $z \geq 0$. What about trying to carry this property over to $B(H)$? In other words, if H is a complex Hilbert space and $T \in B(H)$ is such that $\operatorname{Re} T^k \geq 0$ for all $k \geq 1$, does it follow that $T \geq 0$?

Question 17.2.3 (Fong-Tsui Conjecture) If $T \in B(H)$ is such that $|T| \leq |\operatorname{Re} T|$, is it true that T is self-adjoint? This question appeared in [112]. There, the authors gave partial answers to this conjecture e.g. the fact that it holds when $\dim H < \infty$. Other partial answers appeared in say [252] (see also [222]), but no complete answer has been found yet.

Question 17.2.4 I propose the following conjecture: Can you find two self-adjoint $A, B \in B(H)$ such that $|A| \neq |B|$ and

$$|A| \leq \frac{1}{2}|A + B| \text{ and } |B| \leq \frac{1}{2}|A + B|?$$

Remark Obviously, we should keep away from $\dim H < \infty$ (also A and B must not commute).

Question 17.2.5 (Cf. [257]) Let $A, B \in B(H)$ be both hyponormal (or subnormal or quasinormal). If $AB = BA$, then does it follow that

$$|A + B| \leq |A| + |B|?$$

Question 17.2.6 Consider the complex numbers $a_0, a_1, \cdots, a_n$ and let $\varphi(x) = a_0 + a_1 x + \cdots + a_n x^n$ where $x \in \mathbb{R}$. If $A \in B(H)$ is such that all A^n ($n = 1, \cdots, n$) are hyponormal (or other classes of nonnormal operators), then is it necessary that

$$|\varphi(A)| = |a_0 I + a_1 A + \cdots + a_n A^n| \leq |a_0| I + |a_1||A| + \cdots + |a_n||A^n|?$$

Question 17.2.7 (Cf. Question 13.2.15) Can you find a hyponormal operator T such that

$$|e^T| \neq e^{\operatorname{Re} T} \text{ and } |e^T| \not\leq e^{|T|}?$$

Question 17.2.8 Does Embry's theorem (i.e. Theorem 12.2.4) hold for commuting hyponormal operators? What about other classes of nonnormal operators?

Question 17.2.9 If $A \in B(H)$ is subnormal such that A^n is quasinormal for some integer $n \geq 2$, then A is quasinormal. This was established in [292]. The subnormality assumption was then weakened to hyponormality in [293].
   How about the same question when A is paranormal?

Question 17.2.10 Let $\dim H = \infty$ and let $N \in B(H)$ be nilpotent. Is N always similar to λN where $\lambda \neq 0$?

Part II
Unbounded Linear Operators

Chapter 18
Basic Notions

In this second part of the book, we assume readers are familiar with notions, properties, and notations about the Fourier transform on L^2, Sobolev spaces and inequalities, absolutely continuous functions, and distributions. Readers who are not accustomed to these concepts are therefore invited to consult the very end of this book, where two small chapters are devoted to them. They should also consult [302, 303], and [326] for further reading.

18.1 Basics

Let H be a Hilbert space. It seems appropriate to give the following definition in this part of the book.

Definition 18.1.1 Let A be a linear operator with a domain $D(A)$ (which is a linear subspace of H). We say that A is bounded if

$$\exists \alpha \geq 0, \forall x \in D(A) : \|Ax\| \leq \alpha \|x\|.$$

Otherwise, we say that A is unbounded.

Remark The definition given just above is not, for want of a better word, unanimous. Indeed, readers should be wary that in some textbooks, any operator A with a domain $D(A)$ (where $D(A) \neq H$) is automatically unbounded! The vast majority of operator theorists do not adopt this convention. Naturally, while we understand that *most* unbounded operators that arise in applications are not everywhere defined, considering any linear mapping that is not everywhere defined as unbounded is just unfortunate.

  Let me give a couple of reasons in favor of choosing the definition given above. In my opinion, adopting the other definition would entail some anomalies. For

© The Author(s), under exclusive license to Springer Nature Switzerland AG 2022
M. H. Mortad, *Counterexamples in Operator Theory*,
https://doi.org/10.1007/978-3-030-97814-3_18

instance, the zero operator (or any other bounded operator) when restricted to a linear subspace $D \subset H$ would be called unbounded! On the other hand, everywhere defined unbounded operators (which do exist as in e.g. Question 18.2.2) would be considered bounded! Therefore, two classes of operators would not be in their right place. So, let us just stick to Definition 18.1.1, shall we?

Remark Notice that due to some notational considerations, we may, when needed, use $\mathcal{D}$ (instead of D) to indicate the domain of an operator.

Remark We sometimes write $(A, D(A))$ to indicate an operator A on some domain $D(A)$.

Remark As will be seen, the choice of domains is crucial, and a priori a very small difference in domains can lead to completely opposite conclusions (a butterfly effect here as well?).

Example 18.1.1 Define $M : D(M) \to L^2(\mathbb{R})$ by $Mf(x) = xf(x)$, $x \in \mathbb{R}$, where $D(M) = \{f \in L^2(\mathbb{R}) : xf \in L^2(\mathbb{R})\}$ (recall in passing that M could be defined on other domains). Then M is called a multiplication operator. We show that M is an unbounded operator. Let $I_n = \{x \in \mathbb{R} : |x| \geq n\}$. Then for any $f \in D(M)$, which vanishes outside I_n, we have

$$\|Mf\|_2^2 \geq n^2 \int_{I_n} |f(x)|^2 dx = n^2 \int_{\mathbb{R}} |f(x)|^2 dx = n^2 \|f\|_2^2$$

so that

$$\|Mf\|_2 \geq n\|f\|_2,$$

i.e. M is unbounded.

More generally, if $\varphi : \mathbb{R} \to \mathbb{C}$ is measurable and if $M_\varphi f(x) = \varphi(x)f(x)$ is defined on $D(M_\varphi) = \{f \in L^2(\mathbb{R}) : \varphi f \in L^2(\mathbb{R})\}$, then it may be shown that M_φ is bounded if and only if φ is essentially bounded. For a detailed proof, see e.g. Exercise 10.3.9 in [256].

Example 18.1.2 Let $H^1(\mathbb{R}) = \{f \in L^2(\mathbb{R}) : f' \in L^2(\mathbb{R})\}$ (see the appendix on distributions for more details about such spaces) and define an operator $A : H^1(\mathbb{R}) \to L^2(\mathbb{R})$ by $Af(x) = f'(x)$. Then A is an unbounded operator.

To see that, let $f_n(x) = e^{-n|x|}$ for $n = 1, 2, \cdots$ and $x \in \mathbb{R}$. Then $f_n \in H^1(\mathbb{R})$ for all n. Indeed, let $n \in \mathbb{N}$. Then $f_n \in L^2(\mathbb{R})$ because

$$\|f_n\|_2^2 = \int_{\mathbb{R}} e^{-2n|x|} dx = \frac{1}{2n} + \frac{1}{2n} = \frac{1}{n}$$

and $f'_n \in L^2(\mathbb{R})$ (the distributional derivative) for

$$\|f'_n\|_2^2 = \int_{\mathbb{R}} n^2 e^{-2n|x|} dx = n.$$

Hence

$$\frac{\|f'_n\|_2}{\|f_n\|_2} = n \longrightarrow \infty$$

as n tends to ∞. This is a proof that A is unbounded on $H^1(\mathbb{R})$.

Definition 18.1.2 We say that a linear operator B is an extension of another linear operator A, and we write $A \subset B$, if

$$D(A) \subset D(B) \text{ and } \forall x \in D(A): \ Ax = Bx.$$

Remark If A and B are two operators, then

$$A = B \Longleftrightarrow A \subset B \text{ and } B \subset A.$$

Definition 18.1.3 Let A be a non-necessarily bounded operator with domain $D(A) \subset H$.

(1) The kernel of A (or the null-space of A) is the (linear) subspace

$$\ker A = \{x \in D(A) : Ax = 0\}.$$

(2) The range of A (or image of A) is the (linear) subspace

$$\operatorname{ran} A = A[(D(A)] = \{Ax : x \in D(A)\}.$$

(3) The graph of A, denoted by $G(A)$, is defined by:

$$G(A) = \{(x, Ax) : x \in D(A)\}.$$

Remark In particular, if A is the zero operator restricted to some domain D, then $\ker A = D$.

Next, we define the operators AB and $A + B$ and their natural domains.

Definition 18.1.4 If A and B are two operators with domains $D(A)$ and $D(B)$, respectively, then the product AB is defined by:

$$(AB)x := A(Bx)$$

for x in the domain

$$D(AB) = \{x \in D(B) :\ Bx \in D(A)\} = B^{-1}[D(A)].$$

Similarly, the sum $A + B$ is defined as

$$(A + B)x := Ax + Bx$$

for x in the domain

$$D(A + B) = D(A) \cap D(B).$$

Remark If 0 denotes the zero operator on all of H and A is a linear operator with domain $D(A) \subset H$, then $D(0A) = D(A)$ and $D(A0) = H$. Therefore, $A0 = 0$ whereas $0A \subset 0$ only.

Remark There are many prickly facts about unbounded operators. Readers have the opportunity to see most of them throughout the present book.

Proposition 18.1.1 *Let A, B, and C be three operators with domains $D(A)$, $D(B)$, and $D(C)$, respectively. Then*

$$(A + B) + C = A + (B + C) \text{ and } (AB)C = A(BC)$$

(with equalities holding on e.g. the domains $D(A) \cap D(B) \cap D(C)$ and $D(ABC)$, respectively).

Proposition 18.1.2 *Let A, B, and C be three operators. Then*

$$A \subset B \Longrightarrow CA \subset CB \text{ and } AC \subset BC.$$

The importance of the following definition will be seen when we will define the adjoint of a non-necessarily bounded operator.

Definition 18.1.5 Let A be a linear operator with a domain $D(A) \subset H$. We say that A is densely defined if $\overline{D(A)} = H$.

Remark Obviously, there are non-densely defined operators. For example, let H be a nontrivial Hilbert space. The simplest example perhaps is to define $A = 0$ on $D(A) = \{0\}$. There are many more examples. Another one is to take $H = \ell^2(\mathbb{N})$ say, then consider its standard basis $(e_n)_{n \geq 1}$. Let A be the restriction of the identity operator to $D(A) := \text{span}\{e_2, e_3, \cdots\}$. Then A is not densely defined.

Example 18.1.3 Let $U \in B(H)$ be unitary, and let A be a densely defined operator with domain $D(A)$. Then $D(U^*AU)$ is dense in H. Since $U^* \in B(H)$, we equivalently show that $D(AU)$ is dense. We have

$$D(AU) = \{x \in H : Ux \in D(A)\} = U^{-1}[D(A)].$$

By standard topological arguments (see e.g. Question 16 of the "True or False" Section of Chapter 4 in [255]), we have

$$\overline{D(AU)} = \overline{U^{-1}[D(A)]} = U^{-1}[\overline{D(A)}] = U^{-1}(H) = H.$$

Now, we give a basic definition of commutativity. See Chap. 26 for some more questions on commutativity.

Definition 18.1.6 Let A be a linear operator with domain $D(A)$, and let $B \in B(H)$. We say that B commutes with A if

$$BA \subset AB.$$

In other words, this means that $D(A) \subset D(AB)$ and

$$BAx = ABx, \ \forall x \in D(A).$$

Remark In the previous definition, why not use $BA = AB$? And why not $AB \subset BA$? The first condition is too strong, while the second definition could trivially hold in some cases and without having any kind of pointwise commutativity. For instance, we will see operators A and B such that $D(AB) = \{0\}$, even when $B \in B(H)$, whereby $AB \subset BA$ is always satisfied and regardless of what BA can be.

As in the bounded case, we may also define matrices of unbounded operators, which turns out to be a great tool for dealing with many problems.

Let H and K be two Hilbert spaces and let $A : H \oplus K \to H \oplus K$ (as in the bounded case, we may also use $H \times K$ instead of $H \oplus K$, and we shall intentionally use both notations) be defined by:

$$A = \begin{pmatrix} A_{11} & A_{12} \\ A_{21} & A_{22} \end{pmatrix} \tag{18.1}$$

where $A_{11} \in L(H)$, $A_{12} \in L(K, H)$, $A_{21} \in L(H, K)$, and $A_{22} \in L(K)$ are not necessarily bounded operators. If at least one A_{ij} is unbounded, then A is called a matrix of unbounded operators. If A_{ij} has a domain $D(A_{ij})$ with $i, j = 1, 2$, then

$$D(A) = (D(A_{11}) \cap D(A_{21})) \times (D(A_{12}) \cap D(A_{22}))$$

is the natural domain of A. So if $(x_1, x_2) \in D(A)$, then

$$A \begin{pmatrix} x_1 \\ x_2 \end{pmatrix} = \begin{pmatrix} A_{11}x_1 + A_{12}x_2 \\ A_{21}x_1 + A_{22}x_2 \end{pmatrix}.$$

Remark As in the bounded case, we also allow the abuse of notation $A(x_1, x_2)$ instead of $A\begin{pmatrix} x_1 \\ x_2 \end{pmatrix}$ in many cases.

Not all unbounded operators admit such a decomposition (see e.g. the operator "T^*" in Answer 30.1.9). Here is one characterization.

Proposition 18.1.3 ([361]) *Let $A : D(A) \subset H \to H$ be a linear operator. Then A admits the matrix representation (18.1) iff $PD(A) \subset D(A)$ (or $QD(A) \subset D(A)$), where $P : H \to H$ is defined by $P(x, y) = (x, 0)$ (and $Q : H \to H$ is defined by $Q(x, y) = (0, y)$).*

Let A, B, C, and D be linear operators with domains $\mathcal{D}(A)$, $\mathcal{D}(B)$, $\mathcal{D}(C)$, and $\mathcal{D}(D)$, respectively. Put

$$T = \begin{pmatrix} A & 0 \\ 0 & D \end{pmatrix} \text{ and } S = \begin{pmatrix} 0 & B \\ C & 0 \end{pmatrix}.$$

We call T a diagonal matrix of operators and S an off-diagonal matrix of operators. As we shall observe, these two types of matrices are closely related.

Remark It is fairly easy to see that if at least A or D is unbounded, then so is T. Also, if either B or C is unbounded, S too is unbounded.

Further Reading For more on matrices of unbounded operators, readers may wish to consult the specialized textbook [364]. Another interesting reference is [105]. See also [177, 231, 269, 270, 285], and [388].

18.2 Questions

18.2.1 *An Unbounded Linear Functional That Is Everywhere Defined*

First, we give an example of a bounded linear operator A on a Hilbert space H, which is not everywhere defined. A simple way is to restrict any bounded linear operator $A \in B(H)$ to a linear subspace X of H. Hence, this restriction is not everywhere defined while A stays clearly bounded.

This type of questions could have never been seen (or just seldom) by students in a course on bounded linear operators. Indeed, asking such questions in those courses could have even been regarded as irrelevant. When it comes to unbounded operators, however, this occurs quite often. To better illustrate this point, let A be any operator (even unbounded) with domain $D(A) \subsetneq H$. Setting $B = A - A$, we see that $B = 0$ on $D(A) \cap D(-A) = D(A)$ (and *not on all of H*). Consequently, B is bounded only on $D(A)$ for it is the zero operator on $D(A)$.

Question 18.2.1 Give an example of an unbounded linear functional f : $H \to \mathbb{R}$ (or $\mathbb{C}$), where H is an infinite-dimensional Hilbert space.

18.2.2 An Unbounded Linear Operator That Is Everywhere Defined from H into H

Question 18.2.2 Give an example of an unbounded linear operator $A : H \to H$, where H is an infinite-dimensional Hilbert space.

Further Reading Readers may wish to look at [386].

18.2.3 A Non-densely Defined $T \neq 0$ with $\langle Tx, x \rangle = 0$ for All $x \in D(T)$

Recall that if $T \in B(H)$ where H is a complex Hilbert space and if $\langle Tx, x \rangle = 0$ for all $x \in H$, then necessarily $T = 0$. Is this result still valid for a densely defined operator T? This question has an affirmative answer. That is, if $\langle Tx, x \rangle = 0$ for all $x \in D(T)$, then $Tx = 0$ for each $x \in D(T)$ (remember that this only means that $T \subset 0$). Now, what does happen if the assumption of the density of $D(T)$ is dropped?

Question 18.2.3 Provide a non-densely defined operator T such that $\langle Tx, x \rangle = 0$ for all $x \in D(T)$ and yet $Ty \neq 0$ for some $y \in D(T)$, i.e. $T \not\subset 0$.

18.2.4   A Densely Defined Linear Operator A Satisfying $D(A) = D(A^2) = \cdots = D(A^n) \neq H$

Question 18.2.4 Find a densely defined operator A with domain $D(A) \subset H$ such that

$$D(A) = D(A^2) = \cdots = D(A^n) \neq H$$

where $n \in \mathbb{N}$.

18.2.5   A Densely Defined Operator A Such That $D(A^2) = \{0\}$

As readers have started to observe, the main difficulty when studying unbounded operators is their domains. The first big surprise is the fact that it is quite conceivable that $D(A^2) = \{0\}$, even when A is densely defined. More precisely, if A is densely defined, A^2 could be defined only at the zero vector (hence A^2 is not densely defined). As shall be seen below (e.g. in Questions 20.2.23, 21.2.46, 21.2.47, 21.2.48, and 21.2.49), even under stronger conditions on A, we may still have $D(A^2) = \{0\}$. Thankfully, there are some well-behaved classes for which this is impossible, e.g. unbounded self-adjoint operators or normal ones (yet to be defined).

Question 18.2.5 Give an example of a densely defined operator A such that $D(A^2) = \{0\}$.

18.2.6   A Bounded B (Not Everywhere Defined) and a Densely Defined A Such That $D(BA) \neq D(A)$

It is clear that if $D(B) = H$ (for instance, when $B \in B(H)$), then

$$D(BA) = \{x \in D(A) : Ax \in D(B)\} = D(A).$$

Question 18.2.6 Find a non-everywhere defined bounded operator B such that $D(BA) \neq D(A)$, where A is a densely defined operator, say.

18.2.7 A Densely Defined T with $\operatorname{ran} T \cap D(T) = \{0\}$ yet $D(T^2) \neq \{0\}$

If two linear operators A and B (with domains $D(A)$ and $D(B)$, respectively) satisfy $\operatorname{ran} A \cap D(B) = \{0\}$, then

$$D(BA) = \{x \in D(A) : Ax \in D(B)\} = \{x \in D(A) : Ax = 0\} = \ker A.$$

Question 18.2.7 Find a densely defined operator T such that $\operatorname{ran} T \cap D(T) = \{0\}$ yet $D(T^2) \neq \{0\}$.

18.2.8 Two Densely Defined Operators A and B Such That $D(A + B) = D(A) \cap D(B) = \{0\}$

When dealing with sums of unbounded operators, the main obstacle is the fact that many sums could trivially be defined on $\{0\}$ only! The first example to be given is fairly simple, but readers will see throughout the present book more sophisticated examples.

Question 18.2.8 Give an example of two densely defined operators A and B such that

$$D(A + B) = D(A) \cap D(B) = \{0\}.$$

18.2.9   The Same Symbol T with Two Nontrivial Domains D and D' Such That $D \cap D' = \{0\}$

Question 18.2.9 Find two densely defined operators T with a domain D and T with another domain D' (the same symbol T with different domains) such that $D \cap D' = \{0\}$.

18.2.10   The Sum of a Bounded Operator and an Unbounded Operator Can Be Bounded

Clearly, the sum of two bounded operators (whether they are everywhere defined or not) remains bounded. The sum of two unbounded operators could be bounded (e.g. consider $A - A$, where A is an unbounded operator on some domain). What about the sum of bounded and unbounded operators? If A is unbounded with domain $D(A)$ and $B \in B(H)$, then $A + B$ is never bounded. Indeed, if $A + B$ were bounded, then $A + B - B = A$ would be bounded on $D(A)$, but this is not the case.

Question 18.2.10 Find a bounded operator and an unbounded one whose sum is bounded.

18.2.11   Two Densely Defined T and S Such That $T - S \subset 0$ But $T \not\subset S$

Question 18.2.11 Find two densely defined T and S such that $T - S \subset 0$ but $T \not\subset S$.

18.2.12   Three Densely Defined Operators A, B, and C Satisfying $A(B + C) \not\subset AB + AC$

As regards the distributive laws, one has to be a little more prudent. Indeed, while

$$(A + B)C = AC + BC$$

holds, one only has

$$AB + AC \subset A(B + C).$$

It is easy to see that the equality holds if, for instance, $A \in B(H)$.

Question 18.2.12 (Cf. Question 19.2.31) Find three densely defined operators A, B, and C with domains $D(A)$, $D(B)$ and $D(C)$, respectively, such that

$$A(B + C) \not\subset AB + AC.$$

Answers

Answer 18.2.1 The existence of an everywhere defined unbounded linear operator relies strongly on the axiom of choice (or its variant: Zorn's lemma).

Let $\mathcal{B}$ be a Hamel basis for H, which, as is known, calls on Zorn's lemma (taking an explicit Hilbert space as ℓ^2 or $L^2(\mathbb{R})$ would not make the example more explicit!). We may also suppose that the elements of $\mathcal{B}$ are unit vectors. Now, let x_n be a sequence of distinct elements in $\mathcal{B}$. Finally, define a linear map $f : H \to \mathbb{R}$ by setting $f(x_n) = n$ and $f(x) = 0$ for all other elements x of $\mathcal{B}$. Then f is everywhere defined and it is plainly discontinuous.

Answer 18.2.2 Consider the everywhere discontinuous linear functional $f : H \to \mathbb{R}$ of Answer 18.2.1. Let x_0 be any nonzero vector in H and define $A : H \to H$ by $Ax = f(x)x_0$. Then, A is clearly unbounded and everywhere defined.

Alternatively, and in a slightly different way, we have the following proof (apparently due to Professor Jeffrey Schenker): Let T be a densely defined unbounded operator on a domain D. Let X be the set of all possible extensions T' of T, that is, X contains all operators T' defined on D' such that $D \subset D'$ and T' coincide with T on D. Then, X is not empty and partially ordered by "$\subset$." Moreover, any chain has an upper bound (take unions of domains). By Zorn's lemma, there is a maximal element, noted A. Assume in the end that A is defined on a domain $D' \neq H$. So, pick v in H, which is not in D' and we may define an extension B on $D' + \mathrm{sp}\{v\}$ by mapping v to zero, say. This contradicts the maximality argument. Therefore, the maximal element must be everywhere defined.

Remark Does bijectivity help? The answer is still negative. In Answer 19.2.25, we will see an unbounded operator T that is everywhere defined and such that $T^2 = I$.

Answer 18.2.3 There are many counterexamples. We choose one using matrices of operators. Let H be a Hilbert space with dim $H \geq 2$. Let A be any operator in $B(H)$ and such that $A \neq 0$ (e.g. take $A = I$).

Define T on $D(T) := \{0\} \oplus H$ by:

$$T = \begin{pmatrix} 0 & A \\ 0 & 0 \end{pmatrix}.$$

Observe that T is not densely defined for $\overline{D(T)} \neq H \oplus H$. Now, for any $(x, y) \in D(T) = \{0\} \oplus H$, i.e. for any $(0, y)$ where $y \in H$:

$$\left\langle T \begin{pmatrix} 0 \\ y \end{pmatrix}, \begin{pmatrix} 0 \\ y \end{pmatrix} \right\rangle = \left\langle \begin{pmatrix} Ay \\ 0 \end{pmatrix}, \begin{pmatrix} 0 \\ y \end{pmatrix} \right\rangle = 0$$

and yet $T \not\subset 0$.

Answer 18.2.4 This is easy! Just consider the identity operator I_D restricted to some dense domain D and so $D(I_D) \neq H$. Hence

$$D(I_D^2) = \{x \in D : x \in D\} = D$$

and we may proceed analogously with all other powers of I_D.

Answer 18.2.5 Consider A as being the restriction of the standard Fourier transform to $D(A) = C_0^\infty(\mathbb{R}) \subset L^2(\mathbb{R})$, where $C_0^\infty(\mathbb{R})$ denotes the space of C^∞-functions with a compact support. Then

$$D(A^2) = \{f \in D(A) : Af \in D(A)\} = \{f \in C_0^\infty(\mathbb{R}) : \hat{f} \in C_0^\infty(\mathbb{R})\} = \{0\}$$

as is known from Theorem A.6.

Another example (communicated to me some time ago by Professor Jan Stochel) is to consider the operator A defined by:

$$Af(x) = xf(x)$$

on $D(A)$ constituted of step functions in $L^2(\mathbb{R})$. Then, it is fairly easy to see that $D(A^2) = \{0\}$.

Answer 18.2.6 Let A be the Fourier transform restricted to $D(A) = C_0^\infty(\mathbb{R}) \subset L^2(\mathbb{R})$. Then from Answer 18.2.5,

$$D(A^2) = \{0\} \neq C_0^\infty(\mathbb{R}) = D(A).$$

Answer 18.2.7 The answer is borrowed from [203] though it had a different aim there. This example can further be adapted to produce more interesting examples that might not be known to some readers. Let A be the multiplication operator

$$Af(x) = (1 + x^2)f(x)$$

defined on $D(A) = \{f \in L^2(\mathbb{R}) : (1 + x^2)f \in L^2(\mathbb{R})\}$. Then A is densely defined and unbounded. Let B be the bounded operator defined as

$$Bf = \langle f, g \rangle g$$

where $f \in L^2(\mathbb{R})$ and $g(x) = (1 + x^2)^{-1}$. Then B is of rank 1 (and also self-adjoint). Set $T := BA$ and so T is densely defined because $D(T) = D(A)$. The operator T is defined explicitly by:

$$Tf(x) = BAf(x) = \frac{1}{1 + x^2} \int_{\mathbb{R}} f(x)dx.$$

Notice that there is nothing to be alarmed about in seeing the function f standing alone inside that integral since a priori f is "only" in $L^2(\mathbb{R})$. But, BA must be well defined. To corroborate this fact, we recall that $f \in L^2(\mathbb{R})$, and since $(1 + x^2)f \in L^2(\mathbb{R})$, it follows by the Cauchy–Schwarz inequality that

$$(1 + x^2)^{-1}(1 + x^2)f = f \in L^1(\mathbb{R})$$

making the integral $\int_{\mathbb{R}} f(x)dx$ finite.

Now, $\operatorname{ran} T \cap D(T) = \{0\}$ merely because $g \notin D(T) = D(A)$. In the end, we have

$$D(T^2) = \{f \in D(T) : Tf \in D(T)\} = \{f \in D(A) : BAf \in D(A)\}$$

$$= \left\{ f \in L^2(\mathbb{R}) : (1 + x^2)f, \frac{1}{1 + x^2} \int_{\mathbb{R}} f(x)dx, \int_{\mathbb{R}} f(x)dx \in L^2(\mathbb{R}) \right\}$$

$$\neq \{0\}$$

(witness the function $x \mapsto h(x) = \mathbb{1}_{[0,1]}(x)$).

Answer 18.2.8 Let A be the Fourier transform restricted to the dense subspace $C_0^\infty(\mathbb{R}) \subset L^2(\mathbb{R})$, and let B be the identity operator restricted to $\operatorname{ran} A$. Hence, it is seen that $\operatorname{ran} A$ is dense (directly or as a consequence of Example 20.1.1 below) and so B is densely defined. Finally,

$$D(A + B) = D(A) \cap D(B) = C_0^\infty(\mathbb{R}) \cap \operatorname{ran}(A) = \{0\}$$

because $f \in C_0^\infty(\mathbb{R}) \cap \operatorname{ran}(A)$ forces $f = 0$ (why?).

Answer 18.2.9 Let A and B be two densely defined operators such that $D(A) \cap D(B) = \{0\}$ (as in Answer 18.2.8). Then, let the identity operator restricted to $D(A)$ be denoted by $I_{D(A)}$, and let the identity operator restricted to $D(B)$ be noted $I_{D(B)}$. Then

$$D(I_{D(A)}) \cap D(I_{D(B)}) = D(A) \cap D(B) = \{0\},$$

as wished.

Answer 18.2.10 As an extreme example, let B be defined only on $D(B) := \{0\} \subset H$ (and so B is bounded), and let A be an unbounded operator with domain $D(A) \subset H$. Then $A + B$ is only defined on $\{0\}$, and so it is obviously bounded.

As another example, let $Af(x) = xf(x)$ be defined on $D(A) := C_0^\infty(\mathbb{R})$, and let B be the identity operator defined on the range of the Fourier transform when restricted to $D(A)$. Then, and as in Answer 18.2.8, $D(A + B) = \{0\}$, making $A + B$ trivially bounded. Observe in the end that A is unbounded and that B is bounded.

Remark We ask readers to try to find a densely defined bounded operator B and an unbounded operator A such that $A + B$ is densely defined and bounded. Is this possible?

Answer 18.2.11 Readers will see other counterexamples throughout this book. For example, we have seen in Answer 18.2.8 that there are densely defined operators S and T with $D(S) \cap D(T) = \{0\}$. Then trivially $T - S \subset 0$ and clearly $T \not\subset S$ and $S \not\subset T$ (for otherwise we would have $D(S) \cap D(T) \neq \{0\}$).

Another example is to consider T to be the identity operator restricted to some domain $D \subset H$, and S to be the identity operator restricted to some other domain $D' \subset H$ such that D and D' are not comparable, i.e. $D \not\subset D'$ and $D' \not\subset D$. Then $S - T = 0$ on $D \cap D'$ (we also allow the case $D \cap D' = \{0\}$) and so $S - T \subset 0$. However, $S \not\subset T$ and $T \not\subset S$.

As an explicit example, we may consider $H = L^2(\mathbb{R})$, $D = \{f \in L^2(\mathbb{R}) : e^x f \in L^2(\mathbb{R})\}$, $D' = \{f \in L^2(\mathbb{R}) : e^{-x} f \in L^2(\mathbb{R})\}$. Then $D \not\subset D'$ by considering e.g. $f(x) = e^x . \mathbb{1}_{(-\infty,0]}(x)$, and $D' \not\subset D$ by considering e.g. $g(x) = e^{-x} . \mathbb{1}_{[0,\infty)}(x)$.

Answer 18.2.12 Let $B = -C$, then $B + C = 0$ on $D(B) = D(C)$. Hence

$$D[A(B + C)] = \{x \in D(B + C) : (B + C)x = 0 \in D(A)\} = D(B) = D(C).$$

Also clearly,

$$D(AB) = D(AC).$$

Consequently, to have $A(B + C) \not\subset AB + AC$, it suffices to have $D(B) \not\subset D(AB)$. Now, take $A = B$ (why not?). In other language, the entire "quest" for a counterexample has been reduced to finding a densely defined operator A such that $D(A) \not\subset D(A^2)$. But, hang on, we already have from Answer 18.2.5 a densely defined A such that $D(A^2) = \{0\}$, and so $D(A) \not\subset D(A^2)$, as wished.

Chapter 19
Closedness

19.1 Basics

Having recalled some preliminary notions and results in the previous chapter, we are now in a position to introduce closed and closable operators.

Definition 19.1.1 Let H be a Hilbert space. Let A be a linear operator but non-necessarily bounded. Assume that A has a domain $D(A)$. We say that A is closed if its graph $G(A)$ is closed in $H \times H$. Equivalently,

$$A \text{ is closed} \iff \forall x_n \in D(A) : \begin{cases} x_n \longrightarrow x \\ Ax_n \longrightarrow y \end{cases} \implies \begin{cases} x \in D(A), \\ y = Ax. \end{cases}$$

Remark If $D(A) = H$, by the closed graph theorem we have

$$A \text{ is closed} \iff A \in B(H).$$

Definition 19.1.2 We say that a linear operator A is closable if it possesses a closed extension. Clearly, each closable operator A has a smallest closed extension, which is called its closure, noted $\overline{A}$.

Remark If A is closable, then obviously $A \subset \overline{A}$. It is also clear that A is closed if and only if $A = \overline{A}$.

The next is a characterization of closability using sequences.

Proposition 19.1.1 *Let A be a linear operator with domain $D(A)$. Then A is closable iff for each sequence (x_n) in $D(A)$ such that $x_n \to 0$ and $Ax_n \to y$, then $y = 0$.*

© The Author(s), under exclusive license to Springer Nature Switzerland AG 2022   307
M. H. Mortad, *Counterexamples in Operator Theory*,
https://doi.org/10.1007/978-3-030-97814-3_19

Remark If A is closable, then

$$D(\overline{A}) = \{x \in H : \text{there is } (x_n) \text{ in } D(A) \text{ s.t. } x_n \to x, Ax_n \text{ converges}\}$$

and

$$\overline{A}x = \lim_{n \to \infty} Ax_n, \text{ for } x \in D(\overline{A}).$$

Proposition 19.1.2 *If A is closable, then $G(\overline{A}) = \overline{G(A)}$.*

Definition 19.1.3 Let H be a Hilbert space, and let A be a given linear operator with domain $D(A) \subset H$. The graph norm of A is defined by:

$$\|x\|_{G(A)} = \sqrt{\|x\|^2 + \|Ax\|^2}.$$

The graph inner product on $D(A)$ is defined by:

$$\langle x, y \rangle_{G(A)} = \langle x, y \rangle + \langle Ax, Ay \rangle.$$

Remark The term "graph norm" stands also for other equivalent norms, e.g. for $x \mapsto \|x\| + \|Ax\|$.

There is a practical connection between closed operators and the corresponding graph norms.

Theorem 19.1.3 (See E.g. Theorem 5.1 in [375]) *Let H be a Hilbert space and let A be a linear operator with domain $D(A) \subset H$. Then*

$$A \text{ is closed} \iff (D(A), \|\cdot\|_{G(A)}) \text{ is a Banach space.}$$

The next simple proposition is extremely helpful in many situations.

Proposition 19.1.4 (See E.g. Theorem 5.2 in [375]) *Let A be a bounded linear operator with domain $D(A)$. Then*

$$A \text{ is closed} \iff D(A) \text{ is closed.}$$

Now we say a few words about the invertibility of unbounded operators.

Definition 19.1.4 Let A be an injective operator (not necessarily bounded) from $D(A)$ into H. Then $A^{-1} : \text{ran}(A) \to D(A)$ is called the inverse of A with domain $D(A^{-1}) = \text{ran}(A)$.

Remark We may interpret the previous definition as $AA^{-1} \subset I$ and $A^{-1}A \subset I$.

If the inverse of an unbounded operator is bounded and everywhere defined (e.g. if $A : D(A) \to H$ is closed and bijective), then A is said to be boundedly invertible.

In other words, such is the case if there is a $B \in B(H)$ such that

$$AB = I \text{ and } BA \subset I$$

As is customary, we set $B = A^{-1}$.

Proposition 19.1.5 *If A is boundedly invertible, then it is closed. In general, if A is invertible, then A is closed if and only if A^{-1} is closed.*

The following result is easily seen to hold.

Proposition 19.1.6 *Let A and B be two (boundedly) invertible operators. Then AB is (boundedly) invertible and*

$$(AB)^{-1} = B^{-1}A^{-1}.$$

Next, we give some basic results on the closedness of products and sums of closed operators.

Theorem 19.1.7 *Let A and B be two linear operators on a Hilbert space H. Then*

(1) *AB is closed if A is closed and B is in $B(H)$; or if A is boundedly invertible and B is closed.*
(2) *$A + B$ is closed if $B \in B(H)$ and A is closed.*

Among the useful notions for establishing the closedness (or other concepts) of the sum of two operators is relative boundedness.

Definition 19.1.5 Let B and A be two linear operators with domains $D(B)$ and $D(A)$, respectively (on the same Hilbert space). Assume that $D(A) \subset D(B)$ and

$$\|Bx\| \leq a\|Ax\| + b\|x\|, \ \forall x \in D(A)$$

where "a, b" are positive reals. Then B is said to be relatively bounded with respect to A (or that B is A-bounded). The greatest lower bound of all "a" is called the relative bound of B.

The next result is a criterion for the closedness of the sum of two operators.

Theorem 19.1.8 (Sz-Nagy, [359], See Also [256]) *Let A be a closed operator on $D(A)$. If B is A-bounded with a relative bound "$a < 1$," then $A + B$ is closed on $D(A)$.*

Let us give at the end of this section numerous examples that, besides their own interest, will constitute the cornerstone of the construction of many counterexamples below.

Example 19.1.1 Let $V \in B(H)$ be an isometry, and let A be linear operator with a domain $D(A) \subset H$. Then VA is closed on $D(A)$ if and only if A is closed on

$D(A)$. To see that, observe first that $D(VA) = D(A)$. Since V is an isometry, we have for all $x \in D(A)$: $\|VAx\| = \|Ax\|$. Hence

$$\|VAx\| + \|x\| = \|Ax\| + \|x\|,$$

thereby the graph norms of A and VA coincide. Falling back on Theorem 19.1.3 yields

$$VA \text{ is closed on } D(A) \iff A \text{ is closed on } D(A),$$

as needed.

Example 19.1.2 ([256]) Let $A : D(A) \subset \ell^2 \to \ell^2$ be defined by:

$$Ax = A(x_n) = (nx_n) = (x_1, 2x_2, \cdots, nx_n, \cdots).$$

Then A is closed on $D(A)$. Let (x^k) be a sequence in ℓ^2, which converges to x and such that (Ax^k) converges to some y. That is,

$$\|x^k - x\|_2 \longrightarrow 0 \text{ and } \|Ax^k - y\|_2 \longrightarrow 0.$$

We have

$$|y_n - nx_n| \leq |y_n - nx_n^k| + |nx_n^k - nx_n| = |y_n - nx_n^k| + n|x_n^k - x_n|$$

for all k. Let $\varepsilon > 0$. Since (x^k) converges to x, we have for some $K \in \mathbb{N}$

$$|x_n^k - x_n| \leq \|x^k - x\|_2 < \frac{\varepsilon}{n}$$

when $k \geq K$. Similarly,

$$|y_n - nx_n^k| \leq \|y - Ax^k\|_2 < \varepsilon.$$

Hence, for $k \geq K$, we have

$$|y_n - nx_n| \leq \varepsilon + n \times \frac{\varepsilon}{n} = 2\varepsilon.$$

Consequently, $y_n - nx_n = 0$ for all n, i.e. $Ax = y$ (hence $Ax \in \ell^2$), proving the closedness of A on $D(A)$.

Example 19.1.3 Let (α_n) be a complex sequence. Define a linear operator A by $A(x_n) = (\alpha_n x_n)$ on $D(A) = \{(x_n) \in \ell^2 : (\alpha_n x_n) \in \ell^2\}$. It may then be shown that A is closed as above. See Example 20.1.4 for another proof.

Example 19.1.4 Let (α_n) be a complex sequence and define a linear operator T by:

$$T(x_1, x_2, \cdots) = (0, \alpha_1 x_1, \alpha_2 x_2, \cdots)$$

on $D(T) = \{(x_1, x_2, \cdots) \in \ell^2 : (0, \alpha_1 x_1, \alpha_2 x_2, \cdots) \in \ell^2\}$. Then T is closed on $D(T)$. To see this, we may proceed as before or otherwise write $T = SA$, where S is the shift operator on ℓ^2 and A the closed operator defined in Example 19.1.3. As S is an isometry, Example 19.1.1 tells us that SA is closed on $D(A) = D(T)$.

Example 19.1.5 ([256]) Let $A : D(A) := c_{00} \subset \ell^2 \to \ell^2$ be defined by:

$$Ax = A(x_n) = (nx_n) = (x_1, 2x_2, \cdots, nx_n, \cdots),$$

where c_{00} represents the vector space of all finitely nonzero sequences. Then A is not closed on $D(A)$.

We have to exhibit a sequence (x^k) in c_{00} satisfying $x^k \to x$ and $Ax^k \to y$ and such that e.g. $x \notin c_{00}$ (or $x \in c_{00}$ but $Ax \neq y$).

To show that $x \notin c_{00}$, consider

$$x^k = \left(1, \frac{1}{2^2}, \cdots, \frac{1}{k^2}, 0, 0, \cdots\right)$$

which is clearly in c_{00}. It is then seen that (x^k) converges in ℓ^2 to

$$x = \left(1, \frac{1}{2^2}, \cdots, \frac{1}{k^2}, \frac{1}{(k+1)^2}, \cdots\right) \notin c_{00}.$$

So, there only remains to show that (Ax^k) converges to some $y \in \ell^2$. We may write

$$Ax^k = \left(1, \frac{1}{2}, \cdots, \frac{1}{k}, 0, 0, \cdots\right),$$

and so it is easy to see that (Ax^k) converges to

$$y = \left(1, \frac{1}{2}, \cdots, \frac{1}{k}, \frac{1}{k+1}, \cdots\right),$$

whereby proving the unclosedness of A on c_{00}.

Example 19.1.6 Let $Tf = f'$ be a linear operator defined on the domain

$$D(T) = \{f \in AC[0, 1] : f' \in L^2[0, 1] : f(0) = 0\}.$$

Then T is closed. To show that, we may proceed as in say Example 20.1.5 below. Otherwise, we may call on Proposition B.10. Indeed, the linear functional $F : H^1(0, 1) \to \mathbb{C}$ defined by $F(f) = f(0)$ is continuous w.r.t. the graph norm of

T. Hence $D(T) = \ker F$ is a Hilbert space w.r.t. the graph norm of T, that is, T is closed on $D(T)$ (by Theorem 19.1.3).

We give another proof. Let V be the Volterra operator on $L^2(0, 1)$, that is,

$$Vf(x) = \int_0^x f(t)dt, \, f \in L^2(0, 1).$$

Then $V \in B[L^2(0, 1)]$. Moreover, it can easily be seen that

$$VT \subset I \text{ and } TV = I$$

where I is the identity operator on $L^2(0, 1)$. This signifies that T is boundedly invertible and so T is closed.

Remark Mutatis mutandis, we may show that $Sf = if'$ defined on the domain $D(S) = \{f \in AC[0, 1] : f' \in L^2[0, 1] : f(1) = 0\}$ is also closed.

Now, we give a densely defined non-closable operator.

Example 19.1.7 ([326]) Let H be a Hilbert space, and let D be a dense linear subspace of H. Let $e \neq 0$ be in H. Consider a discontinuous (w.r.t. the strong topology of H) linear functional f on D. Finally, define $T : D \to H$ by $Tx = f(x)e$. Then T is not closable. To see this, let x_n be a sequence in D such that $x_n \to 0$ but $f(x_n) \not\to 0$ (the existence of such a sequence is guaranteed by the discontinuity of f). WLOG, for some $M > 0$ and all n we have $|f(x_n)| \geq M$. Let $y_n = x_n/f(x_n)$. Then

$$\lim_{n\to\infty} y_n = 0 \text{ and } Ty_n = f(y_n)e = e \neq 0,$$

which means that T is not closable.

The following result might be quite handy in some situations (cf. Question 21.2.4).

Example 19.1.8 Let T be a closed operator on D and D'. Then T is also closed on $D \cap D'$. If $D \cap D' = \{0\}$, then T is trivially closed. So, assume that $D \cap D' = \{0\}$, and let $x_n \in D \cap D'$ be such that $x_n \to x$ and $Tx_n \to y$. By the closedness of T on both D and D', we get $x \in D$ and $x \in D'$ and so $x \in D \cap D'$ and $Tx = y$, i.e. T is closed on $D \cap D'$.

Example 19.1.9 Let $\varphi : \mathbb{R} \to \mathbb{C}$ be a measurable function. Define $M_\varphi : D(M_\varphi) \to L^2(\mathbb{R})$ by:

$$M_\varphi f(x) = \varphi(x)f(x)$$

with $x \in \mathbb{R}$, and where $D(M_\varphi) = \{f \in L^2(\mathbb{R}) : \varphi f \in L^2(\mathbb{R})\}$. Then M_φ is densely defined and closed. That M_φ is densely defined is left to readers (see e.g. Exercise 10.3.9. in [256]). There is a proof of the closedness of M_φ, which uses the fact that each L^2-converging sequence has an almost everywhere converging subsequence (see again Exercise 10.3.9. in [256]). The following is an elementary argument: Let $f_n \in D(M_\varphi)$ be such that $f_n \to f$ and $M_\varphi f_n = \varphi f_n \to g$. For $h = D(M_\varphi) = D(M_{\overline{\varphi}})$,

$$\langle \varphi f_n, h \rangle = \langle f_n, \overline{\varphi} h \rangle \longrightarrow \langle g, h \rangle = \langle f, \overline{\varphi} h \rangle = \langle \varphi f, h \rangle.$$

Since $D(M_\varphi)$ is dense, it follows that $g = \varphi f$, that is, $f \in D(M_\varphi)$ and $M_\varphi f = g$, i.e. we have shown the closedness of M_φ.

Example 19.1.10 Let A be the multiplication operator defined on $L^2(\mathbb{R})$ by:

$$Af(x) = xf(x)$$

with domain $C_0^\infty(\mathbb{R}) \subset L^2(\mathbb{R})$. Then A is not closed. Indeed, since $C_0^\infty(\mathbb{R})$ is dense in $L^2(\mathbb{R})$, pick a sequence f_n in $C_0^\infty(\mathbb{R})$ that converges in $L^2(\mathbb{R})$ to $f(x) := 1_{[0,1]}(x)$ say. Observe that $f \notin C_0^\infty(\mathbb{R})$. Now, once it is shown that $xf_n \to xf$ w.r.t. the $L^2(\mathbb{R})$ norm, the non-closedness of A on $C_0^\infty(\mathbb{R})$ follows. Let the reader do the details.

Example 19.1.11 Let A be defined on $D(A) \subset L^2(\mathbb{R})$ (where $D(A)$ is constituted of step functions on $\mathbb{R}$) by $Af(x) = xf(x)$. Then A is not closed on $D(A)$.

Let $L_0^2(\mathbb{R})$ be the space of $L^2(\mathbb{R})$-functions having compact support in $\mathbb{R}$. As it is known (see e.g. Page 24 of [375]), $L_0^2(\mathbb{R})$ is dense in $L^2(\mathbb{R})$ and $D(A)$ is dense in $L_0^2(\mathbb{R})$. Hence A is densely defined. Now, consider some $f \in L_0^2(\mathbb{R})$ *that is not a step function*. Then by density, there is a sequence (f_n) in $D(A)$ such that $f_n \to f$ in L^2. We may also show that $xf_n \to xf$. Since $f \notin D(A)$, it therefore follows that A is not closed on $D(A)$.

The next example deals with some matrices of unbounded operators.

Example 19.1.12 Set

$$T = \begin{pmatrix} A & 0 \\ 0 & D \end{pmatrix} \text{ and } S = \begin{pmatrix} 0 & B \\ C & 0 \end{pmatrix}.$$

It is easy to see that if A and D are closed, then T is closed on $\mathcal{D}(A) \oplus \mathcal{D}(D)$. Indeed, let (x_n, y_n) be in $\mathcal{D}(T) = \mathcal{D}(A) \oplus \mathcal{D}(D)$ such that $(x_n, y_n) \to (x, y)$ such that $T(x_n, y_n) = (Ax_n, Dy_n) \to (z, t)$. By the closedness of both A and D, we may easily obtain the closedness of T. Conversely, it is easy to see if T is closed, then both A and D are closed.

Similarly, if B and C are closed, then S too is closed on $\mathcal{D}(C) \oplus \mathcal{D}(B)$. We may adopt a similar approach as for the diagonal case, or just write:

$$S = \begin{pmatrix} 0 & B \\ C & 0 \end{pmatrix} = \underbrace{\begin{pmatrix} 0 & I \\ I & 0 \end{pmatrix}}_{\mathcal{U}} \underbrace{\begin{pmatrix} C & 0 \\ 0 & B \end{pmatrix}}_{\mathcal{B}}.$$

Then as $\mathcal{B}$ is closed (since B and C are), by also observing that $\mathcal{U}$ is invertible for e.g. $\mathcal{U}^2 = I_{H \oplus H}$ (in fact, $\mathcal{U}$ is a fundamental symmetry), we may say that S is closed on $D(S)$ by Theorem 19.1.7. Finally, and as above, if S is closed, then so are both B and C.

Remark Nonetheless, $\begin{pmatrix} A & B \\ C & D \end{pmatrix}$ need not be closed (it may even fail to be closable) even when all of its entries are closed. Nevertheless, if $(A$ and $D)$ or $(B$ and $C)$ are everywhere defined and bounded and the other pair is closed, then $\begin{pmatrix} A & B \\ C & D \end{pmatrix}$, being a sum of a closed and a bounded (everywhere defined) operator, is itself closed.

Example 19.1.13 (See Also Example 21.1.8) Consider again the operator A that appeared in Example 18.1.2. Then A is closed. By Theorem 19.1.3, A is closed iff $H^1(\mathbb{R})$ is a Banach space with respect to the graph norm of A. Let us therefore show that $H^1(\mathbb{R})$ is in effect complete with respect to the norm

$$\|f\|_{H^1(\mathbb{R})} = (\|f\|_2 + \|f'\|_2)^{\frac{1}{2}}.$$

Let (f_n) be a Cauchy sequence in $H^1(\mathbb{R})$, i.e.

$$\|f_n - f_m\|_{H^1(\mathbb{R})} \longrightarrow 0 \text{ as } n, m \longrightarrow \infty,$$

which, in turn, means that (f_n) and (f'_n) are both Cauchy sequences in $L^2(\mathbb{R})$. Since $L^2(\mathbb{R})$ is complete, there are $f, g \in L^2(\mathbb{R})$ such that

$$\lim_{n \to \infty} \|f_n - f\|_2 = 0 \text{ and } \lim_{n \to \infty} \|f'_n - g\|_2 = 0.$$

It becomes clear that $H^1(\mathbb{R})$ is complete once we show that $g = f'$.

Let $\varphi \in \mathcal{D}(\mathbb{R})$ (remember that in this context, $\mathcal{D}(\mathbb{R}) = C_0^\infty(\mathbb{R})$). Besides, it is independent of n so that by the Cauchy–Schwarz inequality

$$0 \le \left| \int_{\mathbb{R}} \varphi(x)(f(x) - f_n(x))dx \right| \le \|\varphi\|_2 \|f - f_n\|_2 \longrightarrow 0$$

as $n \to \infty$. Therefore,

$$\lim_{n\to\infty} \int_{\mathbb{R}} \varphi(x) f_n(x)dx = \int_{\mathbb{R}} \varphi(x) f(x)dx$$

which precisely means that (f_n) converges to f in $\mathcal{D}'(\mathbb{R})$.

A similar approach leads to the convergence of (f_n') to g in $\mathcal{D}'(\mathbb{R})$. Finally, we infer that (by the continuity of $\frac{d}{dx}$ in $\mathcal{D}'(\mathbb{R})$)

$$f' = \lim_{n\to\infty} f_n' = g$$

by uniqueness of the limit in $\mathcal{D}'(\mathbb{R})$. Since $g \in L^2(\mathbb{R})$, we finally obtain that $f' \in L^2(\mathbb{R})$, and

$$\|f_n - f\|_{H^1(\mathbb{R})} = (\|f_n - f\|_2 + \|f_n' - f'\|_2)^{\frac{1}{2}} \longrightarrow 0$$

when $n \to \infty$. Accordingly, $(H^1(\mathbb{R}), \|\cdot\|_{H^1(\mathbb{R})})$ is a Banach space.

Remark A slightly deeper Fourier Analysis (cf. [215]) tells us that $H^1(\mathbb{R})$ is just an $L^2(\mathbb{R})$ space with a measure that differs from Lebesgue's. Indeed,

$$\widehat{f'(x)} = -it\hat{f}(t).$$

Hence, by the Plancherel theorem

$$\|f\|_{H^1(\mathbb{R})}^2 = \|\hat{f}\|_2^2 + \|\widehat{f'}\|_2^2 = \int_{\mathbb{R}} |\hat{f}(t)|^2(1 + |t|^2)dt.$$

Thus, we may regard $(H^1(\mathbb{R}), dx)$ as $(L^2(\mathbb{R}), (1 + |t|^2)dt)$.

Let us give some more examples about closedness.

Example 19.1.14 Define $Af(x) = if'(x)$ on $C_0^\infty(\mathbb{R})$. Then A is not closed. To see that, let $f \in H^1(\mathbb{R})$ be such that $f \notin C_0^\infty(\mathbb{R})$. So, there is a sequence (f_n) in $C_0^\infty(\mathbb{R})$ such that $f_n \to f$ in $H^1(\mathbb{R})$. Hence $f_n' \to f'$ in $L^2(\mathbb{R})$. This means that $(f_n, if_n') \in G(A)$, which converges to (f, if'). But, $f \notin C_0^\infty(\mathbb{R})$!

What is $\bar{A}$? We know that $A \subset B$, where $Bf(x) = if'(x)$ on $H^1(\mathbb{R})$. Since $C_0^\infty(\mathbb{R})$ is dense in $H^1(\mathbb{R})$ w.r.t. to the graph norm, it results that $\bar{A} = B$.

Example 19.1.15 Define $Af(x) = -f''(x)$ on $C_0^\infty(\mathbb{R})$. Then A is not closed. Just apply a similar argument to that of Example 19.1.14.

Now, on $D(B) := H^2(\mathbb{R}) = \{f \in L^2(\mathbb{R}) : f'' \in L^2(\mathbb{R})\}$ define $Bf(x) = -f''(x)$. Then B is closed. Let us corroborate that using the Fourier transform this time. Let $(f_n, -f_n'') \in D(B) \times L^2(\mathbb{R})$ be converging to $(f, g) \in L^2(\mathbb{R}) \times L^2(\mathbb{R})$. By the Plancherel theorem, we have $\hat{f}_n \to \hat{f}$ and $\widehat{f_n''} = \eta^2 \hat{f}_n \to \hat{g}$ both in $L^2(\mathbb{R})$. By an akin argument as that in Example 19.1.9, we obtain $\eta^2 \hat{f} = \hat{g}$ whereby $f \in H^2(\mathbb{R})$ and $g = -f''$, i.e. B is closed on $H^2(\mathbb{R})$.

Example 19.1.16 ([117]) Consider $B = -id/dx$ and the multiplication operator by x, noted A, on their maximal domains $D(B) = \{f \in L^2(\mathbb{R}) : f' \in L^2(\mathbb{R})\}$ and $D(A) = \{f \in L^2(\mathbb{R}) : xf \in L^2(\mathbb{R})\}$. Then, define now the operator T by:

$$Tf(x) = xf(x) + f'(x)$$

for every $f \in D := D(T) = \{f \in L^2(\mathbb{R}) : xf, f' \in L^2(\mathbb{R})\}$ (clearly, $T = A + iB$). Then T is closed. Indeed, let (f_n) be a sequence in D such that $f_n \to f$ and $Tf_n \to g$. We ought to show that $f \in D$ and $Tf = g$. It then suffices to show the existence of $\lim_{n\to\infty} Bf_n$ and $\lim_{n\to\infty} Af_n$ as this implies that $f \in D(B) \cap D(A) = D$ (for A and B are closed) and $\lim_{n\to\infty} Af_n = Af$ and $\lim_{n\to\infty} Bf_n = Bf$ leading to

$$\lim_{n\to\infty} Tf_n = \lim_{n\to\infty} Af_n + i \lim_{n\to\infty} Bf_n = Af + iBf = Tf.$$

So, let us show the existence of both $\lim_{n\to\infty} Af_n$ and $\lim_{n\to\infty} Bf_n$. Let $h \in D$. Then

$$\|(A + iB)h\|^2 = \|Ah\|^2 + \|Bh\|^2 + \langle Ah, iBh \rangle + \langle iBh, Ah \rangle.$$

Applying an integration by parts yields

$$\langle Ah, iBh \rangle + \langle iBh, Ah \rangle \geq -\|h\|^2$$

and so

$$\|Ah\|^2 + \|Bh\|^2 \leq \|Th\|^2 + \|h\|^2.$$

The convergence of the sequences (Bf_n) and (Af_n) now follows from that of (f_n) and (Tf_n) and by setting $h = f_n - f_m$ (a well known argument). Consequently, T is closed.

19.2   Questions

19.2.1   A Closed Operator Having Any Other Operator as an Extension

Question 19.2.1 Find a closed operator A that has every other operator as an extension.

19.2.2 A Bounded Operator That Is Not Closed, and an Unbounded Operator That Is Closed

Question 19.2.2 Give an example of a bounded operator that is not closed, and another operator that is closed but unbounded.

19.2.3 A Left Invertible Operator That Is Not Closed

Based on the bounded case, we introduce: Let H be a Hilbert space, and let A be an unbounded operator with domain $D(A) \subset H$. We say that A is right invertible if there exists a $B \in B(H)$ such that $AB = I$; and we say that A is left invertible if there is a $C \in B(H)$ such that $CA \subset I$.

Remark Requiring $CA = I$ (where $C \in B(H)$) for the left invertibility is not interesting. Indeed, this gives $D(A) = D(CA) = H$, and so as soon as A is closable, we get $A \in B(H)$!

The importance of the previous notion lies in the following simple result [84].

Proposition 19.2.1 *Let A be an unbounded operator with domain $D(A) \subset H$. If A is left and right invertible simultaneously, then A is boundedly invertible.*

Question 19.2.3 Provide a left invertible operator that is not closed.

19.2.4 A Right Invertible Operator That Is Not Closed

Question 19.2.4 (⊛) Supply a right invertible operator A that is not closed.

19.2.5 A Closable A That Is Injective but $\overline{A}$ Is Not Injective

Let A be a closable operator such that $\overline{A}$ is injective. Then A is obviously injective. Next, we will see that the converse, in general, need not hold.

Question 19.2.5 (⊛) Find a closable A that is injective yet $\overline{A}$ is not injective.

19.2.6 A Closed Densely Defined Unbounded Operator A and a $V \in B(H)$ Such That AV Is Not Densely Defined

If A is densely defined and $B \in B(H)$, then BA is plainly densely defined merely because $D(BA) = D(A)$. Is AB always densely defined?

> **Question 19.2.6** Find a closed densely defined A operator and an everywhere defined and bounded operator V such that AV fails to be densely defined.
> **Hint:** The choice of the "letter V" is not random.

19.2.7 Two Closed Operators A and B Such That BA Is Not Closed

We have already given in Theorem 19.1.7 sufficient conditions for the product of two closed operators to be closed. Another known result, but a little less elementary than Theorem 19.1.7, is as follows.

Theorem 19.2.2 *Let A and B be two densely defined closed operators on a Hilbert space. If* ran A *is closed and* dim ker $A < \infty$, *then AB is closed.*

Remark The above result has been known for quite some time; however, we refer readers to [391] for a new proof of it.

> **Question 19.2.7** Give an example (or more) of two densely defined closed operators A and B such that BA is not closed.

19.2.8 A Compact $B \in B(H)$ and a Closed Operator A Such That BA Is Not Closed

> **Question 19.2.8** Find a compact $B \in B(H)$ and a closed operator A on $D(A) \subset H$ such that BA is not closed.

19.2.9   The Closedness of AB with $B \in B(H)$ Need Not Yield the Closedness of A

It may be shown that if A is a closable densely defined operator such that BA is closed and $B \in B(H)$, then A is closed.

Question 19.2.9 Find some $B \in B(H)$ and a densely defined closable operator A such that AB is closed but A is unclosed.

19.2.10   The Closedness of BA with $B \in B(H)$ Need Not Yield the Closability of A

As observed just above, the closedness of BA, where $B \in B(H)$, does force a closable densely defined operator A to be closed. Do we always have the same conclusion when the closability of A is dropped?

Question 19.2.10 Supply a $B \in B(H)$ and a densely defined non-closable operator A such that BA is closed.

19.2.11   A Left Invertible Operator A and a Closed Operator B Such That AB Is Unclosed

Question 19.2.11 Find a left invertible operator A and a closed operator B such that AB is unclosed.

19.2.12   A Right Invertible Operator A and a Closed Operator B Such That AB Is Unclosed

Question 19.2.12 Find a right invertible operator A and a closed operator B such that AB is unclosed.

19.2.13 Two Closable A and B Such That $\overline{AB} \neq \overline{A}\,\overline{B}$

Question 19.2.13 Find unbounded densely defined closable operators A and B such that

$$\overline{AB} \neq \overline{A}\,\overline{B}.$$

Remark See [318] for related results.

19.2.14 Three Densely Defined Closed Operators A, B, and C Satisfying $\overline{ABC} \neq A\overline{BC}$

Question 19.2.14 Find three closed operators A, B, and C such that

$$\overline{ABC} \neq A\overline{BC}.$$

19.2.15 A Densely Defined Unbounded Closed (Nilpotent) Operator A Such That A^2 Is Bounded and Unclosed

As in the bounded case, we may define nilpotent operators. The definition we are using appeared in [279] (with $n = 2$). It is natural but it has strong assumptions (cf. Question 26.2.22).

Definition 19.2.1 Let A be a non-necessarily bounded operator with a dense domain $D(A)$. We say that A is nilpotent if A^n is well defined and

$$A^n = 0 \text{ on } D(A)$$

for some $n \in \mathbb{N}$ (hence $D(A^n) = D(A)$).

Remark If A is nilpotent with $A^n = 0$, then

$$D(A^n) = D(A^{n-1}) = \cdots D(A^2) = D(A).$$

Question 19.2.15 Find a densely defined, unbounded, and closed operator A for which $A^2 = 0$ on $D(A)$, that is, A is nilpotent (hence A^2 is not closed).

19.2.16   Another Densely Defined Unbounded Closed Operator T Such That T^2 Is Bounded and Unclosed

Question 19.2.16 Give another densely defined, unbounded and closed operator T for which $T^2 = 0$ on $D(T)$, and so T^2 is bounded and unclosed.

19.2.17   A Densely Defined Unbounded and Closed Operator A Such That $D(A) = D(A^2) = \cdots = D(A^n) \neq H$

Question 19.2.17 Find a densely defined unbounded and closed operator A with domain $D(A) \subset H$ such that

$$D(A) = D(A^2) = \cdots = D(A^n) \neq H$$

where $n \in \mathbb{N}$.

19.2.18   A Non-closable Unbounded Nilpotent Operator

Readers have already seen examples of closed nilpotent operators. For instance, the example of Answer 19.2.15 or that of Answer 19.2.16. Is it possible to find a non-closable nilpotent operator?

Question 19.2.18 Find a non-closable unbounded nilpotent operator.

19.2.19   Unbounded Closed (Resp., Unclosable) Idempotent Operators

S. Ôta ([279]) introduced the concept of an unbounded projection or idempotent. Let T be a non-necessarily bounded operator with a dense domain $D(T)$. Say that T is idempotent (or projection) if T^2 is well defined and

$$T^2 = T \text{ on } D(T).$$

Question 19.2.19 Find unbounded closed as well as unbounded non-closable idempotent operators. Provide unclosable everywhere defined idempotent operators as well.

19.2.20   An Unclosed (but Closable) A Such That A^2 Is Closed

We know some conditions from Theorem 19.1.7 making the product of two closed operators closed. What if a product of two densely defined operators is closed, is it true that one of its factors must be closed?

Question 19.2.20 Find an unclosed but closable operator A for which A^2 is closed.

19.2.21   An Unclosed and Closable T Such That T^2 Is Closed (Such That $D(T^2) \neq \{0\}$)

Question 19.2.21 Find an unclosed (and closable) operator T for which T^2 is closed (with $D(T^2) \neq \{0\}$).

19.2.22   A Densely Defined Unclosed Closable A Such That A^2 Is Closed and Densely Defined

In Answers 19.2.20 and 19.2.21, the square of the unclosed operator was not densely defined. Does requiring the square to be densely defined help?

Question 19.2.22 (⊛) Find a densely defined, unclosed, and closable A such that A^2 is closed and densely defined.

19.2.23 An Unbounded Densely Defined Closed Operator A Such That $A^2 \neq 0$ and $A^3 = 0$

Question 19.2.23 (See Question 19.2.30 for the General Case) Find an unbounded densely defined closed operator A with domain $D(A)$ such that $A^2 \neq 0$ and $A^3 = 0$ (where the zero operator in both cases is restricted to $D(A)$).

19.2.24 A Non-closable Operator A Such That $A^2 = 0$ Everywhere on H

If a product AB of two densely defined operators A and B is closed, is it true that at least A or B must be closable? How about when $A = B$?

Question 19.2.24 Let H be a Hilbert space (separable or not). Then there exists a linear operator $A : H \to H$ that is not closable yet $A^2 \in B(H)$ is self-adjoint (downright $A^2 = 0$ everywhere on all of H).

Remark If we want two different unbounded unclosable operators A and B such that $AB = BA = 0$ everywhere on H, that is, $ABx = BAx = 0$, $\forall x \in H$, then we proceed as follows: Let T be a non-closable operator with $D(T) = H$ and such that $T^2 = 0$ everywhere on H. Then define on $H \oplus H$

$$A = \begin{pmatrix} 0 & T \\ 0 & 0 \end{pmatrix} \text{ and } B = \begin{pmatrix} 0 & 0 \\ T & 0 \end{pmatrix}$$

(and so $A \neq B$). Hence

$$AB = \begin{pmatrix} T^2 & 0 \\ 0 & 0 \end{pmatrix} = \begin{pmatrix} 0 & 0 \\ 0 & 0 \end{pmatrix}$$

with $D(AB) = H \oplus H$. Similarly, $BA = \begin{pmatrix} 0 & 0 \\ 0 & 0 \end{pmatrix}$ on $D(BA) = H \oplus H$. Therefore,

$$AB = BA = 0_{H \oplus H}.$$

19.2.25   An Unclosable Operator T Such That $T^2 = I$ Everywhere on H

Assume that a *closable* T obeys $T^2 = I$ everywhere on H. Then $D(T) = H$ and so $T \in B(H)$ (hence, T is invertible as well).

It is therefore natural to ask whether the equation $T^2 = I$ has a non-closable solution.

Question 19.2.25 (See Question 19.2.29 for the General Case) Find a non-closable unbounded operator T such that

$$T^2 = I$$

everywhere on some Hilbert space.

Hint: Use e.g. a matrix of operators containing one of the examples of Question 19.2.24.

19.2.26   An Everywhere Defined Unbounded Operator That Is Neither Injective Nor Surjective

Question 19.2.26 Give an example of an everywhere defined unbounded operator that is neither injective nor surjective.

19.2.27   An Everywhere Defined Unbounded Operator That Is Injective but Non-surjective

Question 19.2.27 Find an everywhere defined unbounded operator that is injective but not surjective.

19.2.28 An Everywhere Defined Unbounded Operator That Is Surjective But Non-injective

Question 19.2.28 Give an example of an everywhere defined unbounded operator that is surjective but non-injective.

19.2.29 A Non-closable Unbounded Operator T Such That $T^n = I$ Everywhere on H and $T^{n-1} \neq I$

Question 19.2.29 Let $n \in \mathbb{N}$ be given. Find a non-closable unbounded operator T such that

$$T^n = I \text{ everywhere on } H \text{ and } T^{n-1} \neq I.$$

19.2.30 An Unclosable Unbounded Operator T Such That $T^n = 0$ Everywhere on H While $T^{n-1} \neq 0$

Question 19.2.30 Let $n \in \mathbb{N}$. Find a non-closable unbounded operator T such that $T^n = 0$ *everywhere* on H but $T^{n-1} \neq 0$.

19.2.31 Three Densely Defined Closed Operators A, B, and C Satisfying $A(B + C) \not\subset AB + AC$

We have already given in Answer 18.2.12 three densely defined operators A, B, and C such that

$$A(B + C) \not\subset AB + AC.$$

Would requiring closedness on them help to establish the full equality?

Question 19.2.31 Find three unbounded densely defined closed operators A, B, and C with domains $D(A)$, $D(B)$, and $D(C)$, respectively, and such that

$$A(B + C) \not\subset AB + AC.$$

19.2.32 A Closed A and a Bounded (Not Everywhere Defined) B Such That A + B Is Unclosed

Question 19.2.32 Find a closed operator A and a bounded operator B such that $A + B$ is not closed.

19.2.33 Two Closed Operators with an Unclosed Sum

Question 19.2.33 (See Answer 21.2.15 for Another Example) Give an example of two densely defined closed operators A and B such that $A + B$ is not closed.

19.2.34 Closable Operators A and B with $\overline{A + B} \neq \overline{A} + \overline{B}$

Question 19.2.34 (See also Question 26.2.18) Find two closable operators A and B such that

$$\overline{A + B} \neq \overline{A} + \overline{B}.$$

Further Reading See [246] and [281] for related results. See also [273] or [375].

19.2.35 Three Densely Defined Closed Operators A, B, and C Satisfying: $A + \overline{B + C} \neq \overline{A + B} + C$

Question 19.2.35 Find three closed operators A, B, and C such that

$$A + \overline{B + C} \neq \overline{A + B} + C.$$

19.2.36 Two Non-closable Operators Whose Sum Is Bounded and Self-Adjoint

Question 19.2.36 Give a pair of two non-closed (or even unclosable) operators A, B such that $A + B$ is closed (even bounded and self-adjoint).

19.2.37 Two Closed Operators Whose Sum Is Not Even Closable

Question 19.2.37 Give a pair of closed operators such that their sum is not even closable.

19.2.38 Two Densely Defined Closed Operators S and T Such That $D(S) \cap D(T) = \{0\}$

Question 19.2.38 (⊛⊛) Give a pair of two densely defined closed operators S and T such that

$$D(S) \cap D(T) = \{0\}.$$

19.2.39 An Unbounded Nilpotent N Such That $I + N$ Is Not Boundedly Invertible

Let $N \in B(H)$ be nilpotent, and let $I \in B(H)$ be the identity operator. Then, it is known that $I \pm N$ are invertible. For example, the inverse of $I - N$ is given by $I + N + N + \cdots + N^p$ if $p + 1$ is the index of nilpotence of N.

What about unbounded nilpotent operators?

Question 19.2.39 Find a closed (or unclosable) nilpotent unbounded operator N such that $I + N$ is not boundedly invertible.

Answers

Answer 19.2.1 This is easy. Let A be a linear operator defined in a Hilbert space H with domain $D(A) = \{0\}$. Then A is trivially closed. Now, let B be *any* operator with domain $D(B) \subset H$. Hence $A \subset B$. Indeed, $D(A) = \{0\} \subset D(B)$ and $Ax = Bx = 0$ for $x = 0$.

Remark Readers might be interested in knowing that unbounded closed densely defined operators can always be diminished in such a way that they remain unbounded, closed, and densely defined on a smaller domain. See [333].

Answer 19.2.2

(1) The identity operator I defined on a (non-closed) subspace of $L^2(\mathbb{R})$ (e.g. on $C_0^\infty(\mathbb{R})$) is bounded but unclosed.
(2) For example, the differential operator $Af = f'$ defined on the domain $D(A) = \{f \in L^2(\mathbb{R}), \ f' \in L^2(\mathbb{R})\}$ is closed but unbounded.

Answer 19.2.3 The simplest example perhaps is to restrict the identity operator on H (noted I_H) to some non-closed domain $D \subset H$ and denote this restriction by I_D. Then I_D is left invertible as

$$I_H I_D = I_D \subset I_H.$$

Since I_D is bounded on an unclosed domain, it follows that I_D is unclosed.

Answer 19.2.4 There is an example of such A that is even everywhere defined in H (is there a more explicit one?). This example was communicated to me by Professor A. M. Davie: Start with B in $B(H)$ such that its range ran(B) is dense but it is not all of H. Let E be a linear subspace of H, which is complementary to ran(B) in the algebraic sense (i.e. ran$(B) + E = H$, without taking closure, while the intersection is $\{0\}$). Then define A on ran(B) by $ABx = x$, and define A on E to be an arbitrary linear mapping of E to H. A then extends by linearity to all of H, and $AB = I$, but A is not bounded (as it is not bounded on ran(B) as if it were, then ran(B) would be closed), so it cannot even be closable.

Answer 19.2.5 The example to be given is due to Professor J. Stochel (personal communication). It is based upon the following lemma due to Jan Stochel et al.

Lemma *Set $A_{e,f}(h + ze) = Ah + zf$ for $h \in D(A)$ and $z \in \mathbb{C}$. If A is closable and $e, f \in H$, then $\overline{A_{e,f}} = \overline{A}_{e,f}$ provided $e \notin D(\overline{A})$.*

Consider an unbounded diagonal operator A with weights $(\lambda_n)_{n=0}^\infty$ restricted to "finite" vectors, i.e. members of the linear span of the orthogonal basis with respect to which A is diagonal. Then take any nonzero complex number z, any $e \notin D(\overline{A})$ (which is possible for $\overline{A}$ is unbounded and closed) and any f in ran$(\overline{A})$, which is not in ran(A) (this is also possible if e.g. $\inf_{n \geq 0} |\lambda_n| > 0$). Then $A_{e,f}$ is injective (because $f \notin$ ran(A)). By the lemma above, $A_{e,f}$ is closable and $\overline{A_{e,f}} = \overline{A}_{e,f}$.

Since $f \in \text{ran}(\overline{A})$, $\overline{A}_{e,f}$ is not injective because if $A(-\frac{1}{z}h) = f$, then $h + ze \neq 0$ and

$$\overline{A}_{e,f}(h + ze) = 0.$$

Answer 19.2.6 (See Answer 28.2.16 for Another Example) Let V be the Volterra operator defined on $L^2[0, 1]$, i.e.

$$(Vf)(x) = \int_0^x f(t)dt,$$

and define $Af = f'$ on $D(A) = \{f \in AC[0, 1] : f' \in L^2[0, 1], f(1) = 0\}$. From the remark below Example 19.1.6, we already know that A is a densely defined closed operator. Let us check that $D(AV)$ is not dense in $L^2[0, 1]$. It is easy to see that

$$D(AV) = \{f \in L^2[0, 1] : Vf \in D(A)\} = \{f \in L^2[0, 1] : \int_0^1 f(t)dt = 0\}$$

which means that $D(AV)$ is reduced to $\{c\}^{\perp}$ (where c represents constant functions). Accordingly, $D(AV)$ cannot be dense in $L^2[0, 1]$.

Remark The operator A given above is not boundedly invertible (why?). So, does it help to have a boundedly invertible A in order that $D(AB)$ be dense if $B \in B(H)$? The answer is again negative as we shall see in Answer 21.2.24 that $D(AB) = \{0\}$ for some $B \in B(H)$ and some densely defined boundedly invertible A.

Answer 19.2.7 Let A be a boundedly invertible operator with an unclosed domain $D(A)$. Let B be its bounded and everywhere defined inverse. Then

$$BA = A^{-1}A = I_{D(A)} \subset I$$

and so BA is not closed.

Another example reads: Let $\mathbf{0}$ be the everywhere defined operator on H. Let A be any closed operator with a dense domain $D(A)$. Then $\mathbf{0}A$ is not closed for

$$\mathbf{0}A = \mathbf{0}_{D(A)} \subset \mathbf{0}$$

where $\mathbf{0}_{D(A)}$ is the $\mathbf{0}$ restricted to $D(A)$.

The last example to be given was communicated to me a while ago by Professor A. M. Davie: On $L^2(S)$, where S is the strip $0 < x < 1$ in the (x, y)-plane, let

$$Af(x, y) = yf(x, y) \text{ and } Bf(x, y) = xf(x, y).$$

If f is such that $xyf(x, y)$ is in L^2 but $yf(x, y)$ is not, then f is in the closure of $D(BA)$ (in the graph norm) but f is not in $D(A)$, and whence not in $D(BA)$.

Answer 19.2.8 We give two examples (in essence, quite similar though).

(1) Let A be defined by $Af(x) = f'(x)$ on the domain

$$D(A) = \{f \in L^2(0, 1) :\ f' \in L^2(0, 1),\ f(0) = 0\}.$$

Then A is closed. Let V be the Volterra operator defined on $L^2(0, 1)$, i.e.

$$(Vf)(x) = \int_0^x f(t)dt,\ \ f \in L^2(0, 1).$$

Then V is compact on $L^2(0, 1)$ and it is easy to see that V is the left inverse of A. Hence $VA \subset I$, that is, $VA = I$ only on the domain $D(A) \subset L^2(0, 1)$. Thus, VA is not closed on $D(A)$.

(2) Let B be defined on ℓ^2 by:

$$B(x_1, x_2, \cdots, x_n, \cdots) = \left(x_1, \frac{1}{2}x_2, \cdots, \frac{1}{n}x_n, \cdots\right).$$

Then B is compact. Now, let A be defined by:

$$A(x_n) = (x_1, 2x_2, \cdots, nx_n, \cdots)$$

on $D(A) = \{(x_n) \in \ell^2 :\ A(x_n) \in \ell^2\}$. Then A is closed (by Example 19.1.3), unbounded, and densely defined as it contains e.g. c_{00}. As before $BA \subset I_{\ell^2}$, that is, BA is bounded on the unclosed subspace $D(A)$ and so BA is not closed.

Remark Let H be a Hilbert space. If A is a closed operator and $B \in B(H)$ is of rank one, then does it follow that BA is closed? The answer is still negative. In fact, such a product can even fail to be closable! See the remark below Answer 21.2.30.

Answer 19.2.9 The key point is to consider $B = 0$ everywhere on a Hilbert space H, then take *any* closable but unclosed densely defined operator A with a domain $D(A) \subset H$. Clearly,

$$D(AB) = \{x \in H : 0 \in D(A)\} = H$$

and $ABx = 0$ for all $x \in H$. Hence $AB \in B(H)$ (and so AB is closed!), yet A was chosen to be unclosed.

Answer 19.2.10 Let A be an unbounded everywhere defined operator (i.e. $D(A) = H$) as in e.g. Answer 18.2.2. Clearly, such an operator cannot be closable. Now, let $B = 0$ everywhere on H and so $BA = 0$ still everywhere on H. This means that $BA \in B(H)$ yet A is not closable.

Answer 19.2.11 Let I_D be the identity operator restricted to a non-closed domain $D \subsetneq H$. Then I_D is left invertible (as in Question 19.2.3). Now, if I_H is the full identity operator on H, then

$$D(I_D I_H) = \{x \in H : x \in D\} = D$$

and so $I_D I_H \subset I_H$. Therefore, $I_D I_H = I_D$ is unclosed even though I_D is left invertible and I_H is everywhere defined and bounded!

Answer 19.2.12 A similar idea to that of Answer 19.2.11 applies. Indeed, let A be a right invertible, which is unclosed (see e.g. Answer 19.2.4). Let $B = I$ be the identity operator on H. Hence AB is unclosed yet A is right invertible.

Answer 19.2.13 The simplest example perhaps is to consider two closed operators A and B whose product AB is unclosed. Then

$$\overline{AB} \neq AB = \overline{A}\,\overline{B}.$$

More precisely,

$$\overline{A}\,\overline{B} = AB \subsetneq \overline{AB}.$$

Let us give a "richer" counterexample (inspired by an example in [153] that was used there for a different purpose). Let

$$A = \frac{d}{dx}, \quad D(A) = \{f \in C^1[0, 1] : f(0) = 0\}$$

and

$$B = -\frac{d}{dx}, \quad D(B) = \{f \in C^1[0, 1] : f(1) = f'(2/5) = 0\}$$

both considered on the Hilbert space $L^2(0, 1)$. Then

$$\overline{B} = -\frac{d}{dx}, \quad D(\overline{B}) = \{f \in H^1(0, 1) : f(1) = 0\}$$

and

$$\overline{A} = \frac{d}{dx}, \quad D(\overline{A}) = \{f \in H^1(0, 1) : f(0) = 0\}.$$

Also

$$AB = -\frac{d^2}{dx^2}, \quad D(AB) = \{f \in C^2[0, 1] : f(1) = f'(0) = f'(2/5) = 0\}.$$

Hence the smallest closed extension of AB is the closed operator

$$\overline{AB} = -\frac{d^2}{dx^2}, \quad D(\overline{AB}) = \{f \in H^2[0, 1] : f(1) = f'(0) = f'(2/5) = 0\}$$

(the closedness of this operator may be seen using Sobolev inequalities and the technique of Example 19.1.8). Now,

$$\overline{A}\,\overline{B} = -\frac{d^2}{dx^2}, \quad D(\overline{A}\,\overline{B}) = \{f \in H^2[0, 1] : f(1) = f'(0) = 0\}.$$

Thus,

$$\overline{AB} \subsetneq \overline{A}\,\overline{B}.$$

Remark See [230] (and [153]) for some conditions that give

$$\overline{BA} = \overline{B}\,\overline{A}$$

whenever A and B are densely defined closable operators.

Answer 19.2.14 Let $A := I$ be the identity operator on H, and let B and C be two closed operators whose product BC is not closed. Then

$$A\overline{BC} = \overline{BC} \neq BC = \overline{B}C = \overline{ABC},$$

as wished.

Answer 19.2.15 ([278]) On ℓ^2, define the *linear* operator A by:

$$Ax = A(x_n) = (x_2, 0, 2x_4, 0, \cdots, \underbrace{nx_{2n}}_{2n-1}, \underbrace{0}_{2n}, \cdots)$$

on the domain $D(A) = \{x = (x_n) \in \ell^2 : (nx_{2n}) \in \ell^2\}$. We may check that $D(A)$ is dense in ℓ^2 and that A is unbounded and closed.

Now, it can readily be verified that $A^2 = 0$ on $D(A^2) = D(A)$. Hence, A^2 is bounded on $D(A)$.

Finally, since we know that $A^2 = 0$ on $D(A)$ and $D(A)$ is not closed, it follows that A^2 is a non-closed operator.

Answer 19.2.16 Let A and B be defined by:

$$Af(x) = (x + |x|)f(x) \text{ and } Bf(x) = (x - |x|)f(x)$$

on their respective domains $D(A) = \{f \in L^2(\mathbb{R}) : (x + |x|)f \in L^2(\mathbb{R})\}$ and $D(B) = \{f \in L^2(\mathbb{R}) : (x - |x|)f \in L^2(\mathbb{R})\}$.

Then both A and B are closed. Besides, $AB = 0$ on $D(AB) = D(B)$ and $BA = 0$ on $D(BA) = D(A)$.

Now, set

$$T = \begin{pmatrix} 0 & A \\ B & 0 \end{pmatrix}$$

which is defined on $D(T) := D(B) \oplus D(A) \subset L^2(\mathbb{R}) \oplus L^2(\mathbb{R})$ (recall that T is closed and unbounded). Hence

$$T^2 = \begin{pmatrix} AB & 0 \\ 0 & BA \end{pmatrix} = \begin{pmatrix} 0 & 0 \\ 0 & 0 \end{pmatrix}$$

on

$$D(T^2) = D(AB) \oplus D(BA) = D(B) \oplus D(A) = D(T),$$

and we are done.

As yet another example, take any closed operator A on a domain $D(A) \subset H$. Consider

$$T = \begin{pmatrix} 0 & A \\ 0 & 0 \end{pmatrix},$$

with $D(T) = H \oplus D(A)$. Now,

$$T^2 = \begin{pmatrix} 0 & 0_{D(A)} \\ 0 & 0_{D(A)} \end{pmatrix} = \begin{pmatrix} 0 & 0_{D(A)} \\ 0 & 0 \end{pmatrix}$$

and so $T^2 = 0$ on $D(T^2) = D(T)$.

Answer 19.2.17 Let A be a densely defined operator such that $A^2 = 0$ on $D(A)$ as in Question 19.2.16 (or Question 19.2.15). Then for any n

$$D(A) = D(A^2) = \cdots = D(A^n) \neq H.$$

In the previous case, observe that only A is closed among all other A^n. So, what if all A^n are closed? Well, we still can obtain a counterexample. Anticipating a little bit, we will see in Answer 19.2.19 a closed operator A such that $A^2 = A$ (hence $D(A^2) = D(A)$) and so A^2 too is closed. Whence $A^3 = A^2 = A$ and so on. Therefore, all A^n are closed and

$$D(A) = D(A^2) = \cdots = D(A^n) \neq H.$$

Answer 19.2.18 Readers will see an alluring example in Answer 19.2.24. Another example might be consulted in [279].

Let us give an example with the means at hand. For example, let A be an unbounded unclosable operator with domain $D(A)$ and set

$$T = \begin{pmatrix} 0 & A \\ 0 & 0 \end{pmatrix}$$

where all zeros being in $B(H)$. Then $D(T) = H \oplus D(A)$ and T is not closable on $D(T)$. It can be checked that $D(T^2) = D(T)$ and that

$$T^2 = \begin{pmatrix} 0 & 0_{D(A)} \\ 0 & 0 \end{pmatrix},$$

i.e. T is nilpotent.

Remark See also the remark below Answer 21.2.1, as well as Answers 26.2.21 and 26.2.22.

Answer 19.2.19 ([263], Cf. the Remark Below Answer 26.2.21) Let A be an unbounded closed operator with domain $D(A) \subset H$, and let I be the identity operator on all of H. Set

$$T = \begin{pmatrix} I & A \\ 0 & 0 \end{pmatrix}$$

and so $D(T) = H \times D(A)$. Then T is densely defined, closed, and unbounded. Since

$$D(T^2) = \{(x, y) \in H \times D(A) : (x + Ay, 0) \in H \times D(A)\} = D(T),$$

we see that

$$T^2 = \begin{pmatrix} I & A \\ 0 & 0 \end{pmatrix} \begin{pmatrix} I & A \\ 0 & 0 \end{pmatrix} = \begin{pmatrix} I & A \\ 0 & 0 \end{pmatrix} = T.$$

In other words, T is idempotent. Once we have seen one example, others come to mind, e.g. A may be replaced by αA say where $\alpha \neq 0$.

For another example, let T be such that $T^2 = T$. If U is unitary, then U^*TU too is a densely defined closed idempotent. The density of $D(U^*TU)$ is already established in Example 18.1.3. Since TU is closed and U^* is invertible, it follows that U^*TU remains closed. Finally, observe that

$$(U^*TU)^2 = U^*TUU^*TU = U^*T^2U = U^*TU.$$

A similar idea applies to the non-closable case. Indeed, define

$$T = \begin{pmatrix} I & A \\ 0 & 0 \end{pmatrix}$$

on $D(T) = H \times D(A)$, where A is *non-closable* this time. Then T too is not closable. Indeed, the non-closability of A gives rise to a sequence (x_n) in $D(A)$ with $x_n \to 0$, $Ax_n \to y$ and $y \neq 0$. Next, $(0, x_n) \in D(T)$, $(0, x_n) \to (0, 0)$ and $T(0, x_n) = (Ax_n, 0) \to (y, 0) \neq (0, 0)$. This proves the non-closability of T. That $T^2 = T$ may be checked as above. Therefore, T is a densely defined unclosable idempotent operator.

Finally, we provide two counterexamples of everywhere defined unclosable idempotent operators. If A is unclosable and $D(A) = H$, then $T = \begin{pmatrix} I & A \\ 0 & 0 \end{pmatrix}$ is an everywhere defined unclosable idempotent operator.

Another example was communicated to me by Professor J. Stochel: Let $f : H \to \mathbb{C}$ be a *discontinuous* linear functional (see Answer 18.2.1). Let e be a vector, which is not in $\ker f$, then set $x_0 = e/f(e)$, whereby $f(x_0) = 1$. Now, define a linear operator A on H by $D(A) = H$ and $Ax = f(x)x_0$ for each $x \in H$. Then for $x \in H$

$$A^2 x = A(f(x)x_0) = f(x)Ax_0 = f(x)f(x_0)x_0 = f(x)x_0 = Ax.$$

Thus, $A^2 = A$ everywhere on the whole of H. Observe in the end that A cannot be closable.

Answer 19.2.20 Take, for example, the usual $L^2(\mathbb{R})$-Fourier transform $\mathcal{F}$ restricted to $C_0^\infty(\mathbb{R})$, noted $\mathcal{F}_0$. Then $\mathcal{F}_0$ is closable but unclosed while $D(\mathcal{F}_0^2) = \{0\}$, that is, $\mathcal{F}_0^2$ is trivially closed on the domain $\{0\}$.

For another example, let $Af(x) = xf(x)$ be the multiplication operator on step functions in $L^2(\mathbb{R})$. Then A is not closed (while being closable) and moreover $D(A^2) = \{0\}$, that is, A^2 is patently closed.

Answer 19.2.21 Consider $T = B \oplus A$ defined on $H \oplus H$, where $B \in B(H)$ and A is a densely defined, closable, and unclosed operator such that $D(A^2) = \{0\}$ (e.g. as in Answer 19.2.20). Then T is not closed (it is closable though), T^2 is closed, and

$$D(T^2) = H \oplus \{0\} \neq \{(0, 0)\},$$

as wished.

Answer 19.2.22 Let T be a densely defined, unbounded, and closed operator T such that $D(T^2) = D(T) \subset H$ (as in Answer 19.2.19 say), then consider the identity operator on H restricted to $D(T)$, noted $I_{D(T)}$. Next, let

$$A = \begin{pmatrix} 0 & T \\ I_{D(T)} & 0 \end{pmatrix}$$

where $D(A) = D(T) \times D(T)$. Clearly, A is closable but unclosed. Let $0_{D(T)}$ and $T_{D(T^2)}$ designate the restrictions of the zero operator and of T to the subspaces $D(T)$ and $D(T^2)$, respectively. We then have

$$A^2 = \begin{pmatrix} 0 & T \\ I_{D(T)} & 0 \end{pmatrix} \begin{pmatrix} 0 & T \\ I_{D(T)} & 0 \end{pmatrix} = \begin{pmatrix} T & 0_{D(T)} \\ 0_{D(T)} & T_{D(T^2)} \end{pmatrix} = \begin{pmatrix} T & 0 \\ 0 & T \end{pmatrix}.$$

Therefore, A^2 is closed on the dense $D(A^2) = D(T) \times D(T)$, as wished.

Answer 19.2.23 Let T be a closed densely defined operator with domain $D(T) \subset H$ such that $T^2 \neq 0$ and $D(T) = D(T^2)$. Perhaps an explicit example is desirable. Let B be a closed operator such that $D(B^2) = D(B) \subset H$ and $B^2 = 0$ on $D(B)$ as in Answer 19.2.16 (or Question 19.2.15), then set $T = \begin{pmatrix} B & 0 \\ 0 & I \end{pmatrix}$, where I is the identity operator on H (hence $D(T) = D(B) \oplus H$). It is seen that T is closed, that $T^2 \neq 0$ and that $D(T^2) = D(T)$.

Now define

$$A = \begin{pmatrix} 0 & T & 0 \\ 0 & 0 & T \\ 0 & 0 & 0 \end{pmatrix}$$

with domain $D(A) := H \oplus D(T) \oplus D(T)$. It may easily be shown that A is closed. Finally,

$$A^2 = \begin{pmatrix} 0 & 0_{D(T)} & T^2 \\ 0 & 0_{D(T)} & 0 \\ 0 & 0 & 0 \end{pmatrix} = \begin{pmatrix} 0 & 0_{D(T)} & T^2 \\ 0 & 0 & 0 \\ 0 & 0 & 0 \end{pmatrix}$$

with $D(A^2) = H \oplus D(T) \oplus D(T)$, and

$$A^3 = \begin{pmatrix} 0 & 0_{D(T)} & 0_{D(T^2)} \\ 0 & 0_{D(T)} & 0_{D(T)} \\ 0 & 0_{D(T)} & 0_{D(T)} \end{pmatrix} = \begin{pmatrix} 0 & 0_{D(T)} & 0_{D(T^2)} \\ 0 & 0 & 0 \\ 0 & 0 & 0 \end{pmatrix} = \begin{pmatrix} 0 & 0_{D(T)} & 0_{D(T)} \\ 0 & 0 & 0 \\ 0 & 0 & 0 \end{pmatrix}$$

(remember that $D(T^2) = D(T)$) with $D(A^3) = H \oplus D(T) \oplus D(T)$, that is, $A^3 = 0$ on $D(A)$, as needed.

Answer 19.2.24 Two examples will be provided. The first example reads: Consider any non-closable operator B with domain $D(B) = H$, then set

$$A = \begin{pmatrix} 0 & B \\ 0 & 0 \end{pmatrix}$$

with $D(A) = H \oplus H$. Clearly, A is unbounded. It is unclosable, as otherwise, $A \in B(H \oplus H)$ by the closed graph theorem. We clearly have

$$A^2 = \begin{pmatrix} 0 & 0 \\ 0 & 0 \end{pmatrix}$$

everywhere on $H \oplus H$.

The second example was communicated to me by Professor J. Stochel: Let $f : H \to \mathbb{C}$ be a *discontinuous* linear functional (see Answer 18.2.1). Let e be a normalized vector in ker f. Now, define a linear operator A on H by $D(A) = H$ and $Ax = f(x)e$ for each $x \in H$. Then for $x \in H$

$$A^2x = A(f(x)e) = f(x)f(e)e = 0.$$

Thus, $A^2 = 0$ everywhere on the whole of H. Accordingly, A^2 is self-adjoint! Therefore, A cannot be closable (see Example 20.1.8 for another proof of the non-closability of A).

Answer 19.2.25 Let A be a non-closable unbounded operator defined on all of H such that $A^2 = 0$ everywhere. Then set

$$T = \begin{pmatrix} A & I \\ I & -A \end{pmatrix},$$

which is defined fully on all of $H \oplus H$. Then T is unclosable and $D(T^2) = H \oplus H$. Since $A - A = 0$ and $A^2 = 0$ both everywhere on all of H, we may write

$$T^2 = \begin{pmatrix} A & I \\ I & -A \end{pmatrix} \begin{pmatrix} A & I \\ I & -A \end{pmatrix} = \begin{pmatrix} A^2 + I & A - A \\ A - A & I + A^2 \end{pmatrix} = \begin{pmatrix} I & 0 \\ 0 & I \end{pmatrix},$$

i.e. $T^2 = I_{H \oplus H}$, as needed.

Remark As mentioned before, $T^2 = I$ tells that T is a bijective (or invertible, not boundedly though) non-closable everywhere defined unbounded operator.

Remark As above, and if for some reason we want two different unbounded non-closable operators A and B such that $AB = BA = I$ everywhere on some Hilbert space, then we proceed as follows: From Answer 19.2.25, we have a non-closable operator T such that $T^2 = I$ everywhere on $H \oplus H$. Setting

$$A = \begin{pmatrix} 0 & T \\ I & 0 \end{pmatrix} \text{ and } B = \begin{pmatrix} 0 & I \\ T & 0 \end{pmatrix},$$

which are everywhere defined on $H \oplus H \oplus H \oplus H$, we see that $A \neq B$ and that

$$AB = \begin{pmatrix} T^2 & 0 \\ 0 & I \end{pmatrix} = \begin{pmatrix} I & 0 \\ 0 & I \end{pmatrix} = \begin{pmatrix} I & 0 \\ 0 & T^2 \end{pmatrix} = BA$$

everywhere.

As for a second example, let T be any unbounded non-closable everywhere defined operator on H and let

$$A = \begin{pmatrix} I & T \\ 0 & I \end{pmatrix}.$$

Then, it is easily seen that A, defined on the entire $H \oplus H$, is bijective and so it is invertible (not boundedly) with an inverse given by $B = \begin{pmatrix} I & -T \\ 0 & I \end{pmatrix}$ as

$$AB = BA = \begin{pmatrix} I & 0 \\ 0 & I \end{pmatrix}.$$

Answer 19.2.26 There are many examples. For instance, let A be any unclosable everywhere defined operator on some Hilbert space H, then define

$$T = \begin{pmatrix} 0 & A \\ 0 & 0 \end{pmatrix}$$

on $D(T) = H \oplus H$. For all $(x, y) \in H \oplus H$, we have

$$T\begin{pmatrix} x \\ y \end{pmatrix} = \begin{pmatrix} Ay \\ 0 \end{pmatrix}.$$

Since $T\begin{pmatrix} x \\ 0 \end{pmatrix} = \begin{pmatrix} 0 \\ 0 \end{pmatrix}$ for some $x \neq 0$, T is not one-to-one. Since

$$\operatorname{ran} T = \operatorname{ran} A \oplus \{0\} \neq H \oplus H,$$

T is not onto either.

Answer 19.2.27 Let T be an everywhere defined unbounded injective operator. For example, consider an unclosable operator T with $D(T) = H$ and such that $T^2 = I$. Now, let $S \in B(H)$ be any injective operator, which is not surjective. Finally, set

$$A := T \oplus S = \begin{pmatrix} T & 0 \\ 0 & S \end{pmatrix},$$

and so $D(A) = H \oplus H$. Then A is unbounded and unclosable. That A is injective is plain. Since ran $S \neq H$, it ensues that

$$\text{ran } A = H \oplus \text{ran } S \neq H \oplus H,$$

that is, A is not surjective.

Answer 19.2.28 A similar idea as in Answer 19.2.27 applies here. Let T be such that $T^2 = I$ (hence T is surjective) and $D(T) = H$, and consider a surjective $S \in B(H)$ that is not injective. Then

$$A = \begin{pmatrix} T & 0 \\ 0 & S \end{pmatrix}$$

has the required properties for A is unbounded, $D(A) = H \oplus H$, ran $A = H \oplus H$ and ker $A \neq \{(0,0)\}$.

Another stronger example is the one from Answer 19.2.4. There, we have an everywhere defined unbounded (non-closable) operator A such that $AB = I$ for some $B \in B(H)$. Clearly, A is surjective. If $A : H \to H$ were injective, A would be bijective and so there would exist a $C : H \to H$ such that $CA = I$. But this would give $C = CAB = B$, and so A would be boundedly invertible, in particular, it would be closed, a contradiction! Accordingly, A is not injective, as wished.

Answer 19.2.29 Let A be a non-closable unbounded operator, which is everywhere defined, i.e. $D(A) = H$, and let $I \in B(H)$ be the identity operator. Inspired by the second example in Answer 2.2.22, consider on $H \oplus H \oplus \cdots \oplus H$ (n times)

$$T = \begin{pmatrix} 0 & I & A & \cdots\cdots & 0 \\ 0 & 0 & I & 0 & \vdots \\ \vdots & & 0 & I & \ddots & \vdots \\ \vdots & & & \ddots & \ddots & 0 \\ 0 & & & & 0 & I \\ I & -A & \cdots\cdots & & 0 & 0 \end{pmatrix}.$$

Observe that $D(T) = H \oplus H \oplus \cdots \oplus H$, which means that T is everywhere defined. Notice also that T is clearly unbounded and not closable. Readers may easily check that

$$T^n = I \oplus I \oplus \cdots \oplus I$$

on $D(T^n) = H \oplus H \oplus \cdots \oplus H$ whereas $T^{n-1} \neq I \oplus I \oplus \cdots \oplus I$. As an illustration, we treat the case $n = 3$. In this case,

$$T = \begin{pmatrix} 0 & I & A \\ 0 & 0 & I \\ I & -A & 0 \end{pmatrix}.$$

Then

$$T^2 = \begin{pmatrix} A & -A^2 & I \\ I & -A & 0 \\ 0 & I & 0 \end{pmatrix} \neq I \oplus I \oplus I \text{ while } T^3 = \begin{pmatrix} I & 0 & 0 \\ 0 & I & 0 \\ 0 & 0 & I \end{pmatrix} = I \oplus I \oplus I,$$

as wished.

Answer 19.2.30 Let A be an unbounded non-closable operator such that $D(A) = H$ such that $A^2 \neq 0$. Now, define

$$T = \begin{pmatrix} 0 & A & A \\ 0 & 0 & A \\ 0 & 0 & 0 \end{pmatrix}$$

and so $D(T) = H \oplus H \oplus H$. Clearly, T is unbounded and not closable. Then

$$T^2 = \begin{pmatrix} 0 & 0 & A^2 \\ 0 & 0 & 0 \\ 0 & 0 & 0 \end{pmatrix}$$

and

$$T^3 = \begin{pmatrix} 0 & 0 & 0 \\ 0 & 0 & 0 \\ 0 & 0 & 0 \end{pmatrix}$$

where all zeros belong to $B(H)$.

To deal with the general case, define on $H \oplus H \oplus H \cdots \oplus H$ (n copies of H) the unbounded non-closable

$$T = \begin{pmatrix} 0 & A & A & \cdots & \cdots & A \\ & 0 & 0 & A & A & \vdots \\ & \vdots & & 0 & A & \ddots & \vdots \\ & \vdots & & & \ddots & \ddots & A \\ & 0 & & & & 0 & A \\ & 0 & 0 & \cdots & \cdots & 0 & 0 \end{pmatrix},$$

where A is unbounded and not closable with $A^{n-1} \neq 0$. Clearly, $D(T) = H \oplus H \oplus \cdots \oplus H$, and as above, it may be checked that

$$T^{n-1} \neq 0 \text{ whereas } T^n = 0$$

everywhere on $D(T^n) = H \oplus H \oplus \cdots \oplus H$.

Answer 19.2.31 Proceeding as in Answer 18.2.12, we end up searching for a closed densely defined A such that $D(A) \not\subset D(A^2)$. So, let

$$Af(x) = e^{x^2} f(x)$$

be defined on $D(A) = \{f \in L^2(\mathbb{R}) : e^{x^2} f \in L^2(\mathbb{R})\}$. Then A is closed and besides $D(A^2) = \{f \in L^2(\mathbb{R}) : e^{2x^2} f \in L^2(\mathbb{R})\}$.

The operator A does satisfy $D(A) \not\subset D(A^2)$. Indeed, if we let $f(x) = e^{-\frac{3}{2}x^2}$, then $f \in D(A)$ whereas $f \notin D(A^2)$.

Remark As in Answer 18.2.12, we could have also considered a densely defined closed A such that $D(A^2) = \{0\}$. However, such examples, even though they exist, they have not been presented yet.

Answer 19.2.32 Let A be the multiplication operator defined by $Af(x) = xf(x)$ on $D(A) = \{f \in L^2(\mathbb{R}) : xf \in L^2(\mathbb{R})\}$ (and so A is closed). Let B be the (bounded!) zero operator defined on $C_0^\infty(\mathbb{R})$. Then

$$(A + B)f(x) = xf(x)$$

for $f \in D(A) \cap C_0^\infty(\mathbb{R}) = C_0^\infty(\mathbb{R})$. Therefore, $A + B$ is plainly unclosed on $C_0^\infty(\mathbb{R})$.

Answer 19.2.33 Let A be a closed operator with a *non-closed* domain $D(A)$, and then take $B = -A$. Then

$$T := A - A = 0 \text{ on } D(A).$$

Since T is bounded on the unclosed $D(A)$, T cannot be closed.

Remark In fact, there are closed operators A and B such that neither AB nor $A+B$ is closed. Just, consider the closed operator A of Answer 19.2.15. Then $A(-A) = -A^2$ is not closed, and neither is $A + (-A) = A - A$.

Answer 19.2.34 (See Also Answer 26.2.18) The counterexample is very simple (a slightly complicated counterexample appeared in [281] even though it was also employed there for *another* aim).

Why not take a closed A with a (non-closed) domain $D(A)$, and then set $B = -A$. So, if 0 denotes the everywhere defined zero operator and $0_{D(A)}$ its restriction to $D(A)$, then

$$\overline{A + B} = 0 \supset 0_{D(A)} = A + B,$$

thereby

$$\overline{A + B} \neq \overline{A} + \overline{B},$$

as wished.

Answer 19.2.35 Let A be a closed operator with a non-closed domain $D(A) \subset H$, then set $B = -A$ and take $C \in B(H)$. Then each of A, B, and C is closed.

Clearly,

$$A + \overline{B + C} = A + B + C = 0_{D(A)} + C = C \text{ on } D(A)$$

and

$$\overline{A + B} + C = 0_H + C = C \text{ on } H,$$

that is, we only have

$$A + \overline{B + C} \subsetneq \overline{A + B} + C.$$

Answer 19.2.36 Let H be a Hilbert space, and let $A : H \to H$ be a non-closable operator as in Question 19.2.24. Set $B = -A$ and so B stays not closable with $D(B) = H$. However,

$$A + B = 0$$

everywhere on H, that is, $A + B$ is bounded and self-adjoint.

Answer 19.2.37 Let $H = L^2[0, 1]$ and define $Af = f'$ on $D(A) = H^1[0, 1]$. Then A is closed. Now, let $Bf = f(0)$ be defined on $H^1[0, 1]$. Then B is not continuous on $H^1[0, 1]$ with respect to the $L^2[0, 1]$-topology. Hence as in Example 19.1.7, $Bf = f(0)g$ (where $g(x) = 1$ for all $x \in [0, 1)$) is not closable on $H^1[0, 1]$. By Proposition B.10 and Theorem 19.1.8, we know that $A + B$ is closed on $H^1[0, 1]$.

Hence $(A + B) - A = B$ on $H^1[0, 1]$, that is, $A + B - A$ is not closable on $H^1[0, 1]$ and yet $A + B$ and $-A$ are both closed.

Remark A very similar argument can be applied in the space $H = L^2(\mathbb{R})$ and by calling on a Sobolev inequality. Indeed, define $Af = f'$ on $D(A) = H^1(\mathbb{R}) = \{f \in L^2(\mathbb{R}) : f' \in L^2(\mathbb{R})\}$ and set $Bf = f(0)g$, where $\|g\|_2 = 1$. Then B is unbounded on $H^1(\mathbb{R})$ as it is so on $C_0^\infty(\mathbb{R})$ (take f_n such that $f_n(0) = n$ and $\|f_n\|_2 = 1$). The rest is obvious.

Answer 19.2.38 (The Example is Borrowed from [369]) Let H be a separable Hilbert space, and let $\{e_n : n \in \mathbb{N}\}$ be an orthonormal basis. Denote by S the closed operator on H defined by $Se_n = ne_n$ for all n (S being defined on its maximal domain, i.e. $D(S) = \{(x_n) \in H : (nx_n) \in H\}$). Then choose a sequence $J_1, J_2, \cdots$ of mutually disjoint infinite subsets of $\mathbb{N}$ and a sequence of vectors $x_1, x_2, \cdots, x_n, \cdots$ such that for all n:

(1) x_n belongs to the closed subspace spanned by the set $\{e_k : k \in J_n\}$.
(2) $x_n \notin D(S)$.
(3) $\|x_n\| = 1$.

(An explicit choice would be to take e.g. $J_n = \{(2k-1)2^{n-1} : k \in \mathbb{N}\}$ and for x_n a scalar multiple of $\sum_{k=1}^\infty k^{-1}e_{(2k-1)2^{n-1}}$). Then, the vectors $\{x_n : n \in \mathbb{N}\}$ are mutually orthogonal for the sets $\{J_n : n \in \mathbb{N}\}$. Since $\|x_n\| = 1$ for all n, there exists an isometry $V \in B(H)$ such that $Ve_n = x_n$ whichever n. Next, define $R = I + \frac{1}{2}V$. Then R will have a bounded everywhere defined inverse R^{-1} (observe that $\|\frac{1}{2}V\| < 1$). Define $T = SR^{-1}$ and so T is clearly closed. Moreover, T is densely defined as

$$D(T) = \{x \in H : R^{-1}x \in D(S)\} = R[D(S)]$$

and so by the continuity of R and the density of $D(S)$ we may write

$$H = R(H) = R[\overline{D(S)}] \subset \overline{R[D(S)]} = \overline{D(T)} \subset H,$$

giving us the density of $D(T)$.

Finally, we show that $D(S) \cap D(T) = \{0\}$ or $D(S) \cap R[D(S)] = \{0\}$. Observe that any element of $D(S)$ is of the form $\sum_{n=1}^\infty \frac{1}{n}\lambda_n e_n$ with $\sum_{n=1}^\infty |\lambda_n|^2 < \infty$. So, let $y \in D(S) \cap R[D(S)]$. Then for some sequences (λ_n) and (μ_n) in ℓ^2, we have

$$y = \sum_{n=1}^\infty \frac{1}{n}\lambda_n e_n = \sum_{n=1}^\infty \frac{1}{n}\mu_n Re_n = \sum_{n=1}^\infty \frac{1}{n}\mu_n(e_n + \frac{1}{2}x_n).$$

By projecting this equation onto the subspace spanned by the vectors $\{e_k : k \in J_n\}$, we find that

$$\frac{1}{2n}\mu_n x_n = \sum_{k \in J_n} \frac{1}{k}(\lambda_k - \mu_k)e_k.$$

Since the right hand side is an element of $D(S)$ and $x_n \notin D(S)$, we are only left with one option, which is $\mu_n = 0$ for all n. This by the bye forces $y = 0$, as wished.

Answer 19.2.39 ([263]) We start with the case of non-closable nilpotent operators. Let N be an unbounded non-closable operator such that $D(N) = H$ and $N^2 = 0$ everywhere on H. Then, $I + N$ cannot be boundedly invertible for it were, it would ensue that $(I + N)^2$ too is boundedly invertible. However,

$$(I + N)^2 = I + 2N + N^2 = I + 2N$$

(all equalities being full) is not even closable, while we all know that boundedly invertible operators must be closed.

Consider now the case of a closed nilpotent operator. The simplest example to think of maybe is

$$N = \begin{pmatrix} 0 & A \\ 0 & 0 \end{pmatrix}$$

defined on $D(N) = H \oplus D(A)$, where A is an unbounded closed operator with domain $D(A)$. If $I_{H \oplus H}$ is the identity operator on $H \oplus H$, then

$$I_{H \oplus H} + N = \begin{pmatrix} I & A \\ 0 & I \end{pmatrix}$$

is not boundedly invertible.

Remark In both cases above, $I + N$ above are invertible (in the sense of Definition 19.1.4).

Chapter 20
Adjoints, Symmetric Operators

20.1 Basics

Definition 20.1.1 Let $A : D(A) \subset H \to K$ be a densely defined linear operator, where H and K are two Hilbert spaces. Then $D(A^*)$ is the set constituted of all $y \in K$ for which there exists a $z \in H$ such that

$$\langle Ax, y \rangle = \langle x, z \rangle, \forall x \in D(A).$$

We set for each $y \in D(A^*)$, $A^*y = z$. The (linear) operator A^* then obtained is called the adjoint of A.

Remarks

(1) Clearly, $D(A^*)$ is a linear subspace of K.
(2) The density of $D(A)$ plays a vital role as regards both the existence and the uniqueness of A^*.
(3) Thanks to the Riesz lemma,

$$y \in D(A^*) \Longleftrightarrow |\langle Ax, y \rangle| \leq \alpha \|x\|$$

for some $\alpha \geq 0$ and all $x \in D(A)$.

Here are some basic properties of the adjoint operator (see e.g. [326] or [375] for proofs).

Theorem 20.1.1 *Let A be a densely defined linear operator. Then*

(1) A^* is closed.
(2) A is closable if and only if $D(A^*)$ is dense. In such case,

$$\overline{A} = A^{**}.$$

M. H. Mortad, *Counterexamples in Operator Theory*,
https://doi.org/10.1007/978-3-030-97814-3_20

(3) If A is closable, then $(\overline{A})^* = A^*$.

(4) Let A be an injective operator with a dense range. Then A^* is invertible and

$$(A^*)^{-1} = (A^{-1})^*.$$

(5) Let A be injective and closable. Then A^{-1} is closable if and only if $\ker(\overline{A}) = \{0\}$. In this case,

$$(\overline{A})^{-1} = \overline{A^{-1}}.$$

(6) We have

$$\ker(A^*) = (\operatorname{ran} A)^{\perp}.$$

If A is closed, then

$$\ker(A) = (\operatorname{ran} A^*)^{\perp}.$$

Proposition 20.1.2 *Let A be a densely defined, and let B be a linear operator such that $A \subset B$. Then B is densely defined and $B^* \subset A^*$.*

Definition 20.1.2 A linear operator A with domain $D(A)$ is called symmetric if

$$\langle Ax, y \rangle = \langle x, Ay \rangle, \forall x, y \in D(A).$$

When A is densely defined, the previous equation is equivalent to $A \subset A^*$.

As in the case of bounded operators, we have the following.

Proposition 20.1.3 *Let A be a densely defined operator with domain $D(A) \subset H$, where H is a (complex) Hilbert space. Then A is symmetric if and only if $\langle Ax, x \rangle \in \mathbb{R}$ for all $x \in D(A)$.*

Definition 20.1.3 Let A be a symmetric operator. We say that A is lower semibounded (resp., upper semibounded) if there is a real number α such that

$$\langle Ax, x \rangle \geq \alpha \|x\|^2 \text{ (resp., } \langle Ax, x \rangle \leq \alpha \|x\|^2)$$

for all $x \in D(A)$.

If A is either lower semibounded or upper semibounded, then A is said to be semibounded.

In particular:

Definition 20.1.4 Let A be a symmetric operator with domain $D(A) \subset H$. We say that A is positive if

$$\langle Ax, x \rangle \geq 0, \forall x \in D(A).$$

The definition of a core is given next (it could have also been given in the previous chapter).

Definition 20.1.5 If A is a closed, a subspace $D \subset D(A)$ is said to be a core for A if $\overline{A|D} = A$.

Now, we gather some rudimentary results concerning the adjoint of products and sums.

Theorem 20.1.4 *Let A and B be two densely defined linear operators defined on a Hilbert space H with domains $D(A)$ and $D(B)$, respectively.*

(1) If $A + B$ is densely defined, then $A^* + B^* \subset (A + B)^*$.
(2) If $B \in B(H)$, then $A^* + B^* = (A + B)^*$.
(3) If BA is densely defined, then $A^* B^* \subset (BA)^*$.
(4) If $B \in B(H)$, then $A^* B^* = (BA)^*$.
(5) If BA is densely defined and A is invertible with $A^{-1} \in B(H)$, then also $A^* B^* = (BA)^*$.

The next two results characterize left and right invertibility via kernels of the operators.

Theorem 20.1.5 ([84]) *A right (resp., left) invertible closed and densely defined operator A such that $\ker(A) \subseteq \ker(A^*)$ (resp., $\ker(A^*) \subseteq \ker(A)$) is boundedly invertible.*
In particular, if A is closed, densely defined, and $\ker(A) = \ker(A^*)$, then

A is left invertible $\iff$ A is right invertible $\iff$ A is boundedly invertible.

Theorem 20.1.6 ([84]) *Assume A is a closed and densely defined operator in H and $B \in B(H)$ is such that BA is boundedly invertible. If either $\ker(A^*) \subseteq \ker(A)$ or $\ker(B) \subseteq \ker(B^*)$, then the operators A and B, and consequently AB, are all boundedly invertible.*

Next, we provide some basic properties about the adjoint operation in the case of matrices of operators.

Let

$$A = \begin{pmatrix} A_{11} & A_{12} \\ A_{21} & A_{22} \end{pmatrix}$$

be defined on

$$D(A) = (D(A_{11}) \cap D(A_{21})) \times (D(A_{12}) \cap D(A_{22})).$$

Even if all A_{ij} $(i, j = 1, 2)$ are densely defined, then A may fail to be densely defined. So, when finding the adjoint of A, we must assume that $D(A)$ is dense and so it possesses a unique adjoint, noted A^*.

Unlike the bounded case, A^* may be quite different from the so-called formal adjoint

$$A_* = \begin{pmatrix} A_{11}^* & A_{21}^* \\ A_{12}^* & A_{22}^* \end{pmatrix}.$$

See Question 30.1.10 for a counterexample. It is, however, worth noticing that we always have

$$A_* \subset A^*.$$

Remark In [388], the authors investigated when $A^* = A_*$, that is, when

$$\begin{pmatrix} A_{11} & A_{12} \\ A_{21} & A_{22} \end{pmatrix}^* = \begin{pmatrix} A_{11}^* & A_{21}^* \\ A_{12}^* & A_{22}^* \end{pmatrix}.$$

Theorem 20.1.7 (See e.g. Proposition 2.6.3 in [364]) *Let A, B, C, and D be densely defined linear operators with domains $D(A)$, $D(B)$, $D(C)$, and $D(D)$, respectively. Set*

$$T = \begin{pmatrix} A & 0 \\ 0 & D \end{pmatrix} \text{ and } S = \begin{pmatrix} 0 & B \\ C & 0 \end{pmatrix}.$$

Then

$$T^* = \begin{pmatrix} A^* & 0 \\ 0 & D^* \end{pmatrix} \text{ and } S^* = \begin{pmatrix} 0 & C^* \\ B^* & 0 \end{pmatrix}.$$

Remark If we know how to find T^*, then we know how to get S^*, and vice versa. Indeed, assume that we know how to calculate T^*. To obtain S^*, write

$$S = \begin{pmatrix} 0 & B \\ C & 0 \end{pmatrix} = \underbrace{\begin{pmatrix} 0 & I \\ I & 0 \end{pmatrix}}_{\mathcal{A}} \underbrace{\begin{pmatrix} C & 0 \\ 0 & B \end{pmatrix}}_{\mathcal{B}}$$

(observe that $D(S) = D(\mathcal{A}B) = D(B)$ for $\mathcal{A}$ is everywhere defined and bounded). Hence

$$S^* = \begin{pmatrix} C^* & 0 \\ 0 & B^* \end{pmatrix} \begin{pmatrix} 0 & I^* \\ I^* & 0 \end{pmatrix} = \underbrace{\begin{pmatrix} C^* & 0 \\ 0 & B^* \end{pmatrix} \begin{pmatrix} 0 & I \\ I & 0 \end{pmatrix}}_{\mathcal{B}^*}$$

by the expression for T^*. Since it can easily be substantiated that $D(\mathcal{B}^*\mathcal{A}) = D(B^*) \oplus D(C^*)$, we infer that

$$S^* = \begin{pmatrix} 0 & C^* \\ B^* & 0 \end{pmatrix}.$$

Corollary 20.1.8 *Let A, B, C, and D be densely defined closable linear operators with domains $\mathcal{D}(A)$, $\mathcal{D}(B)$, $\mathcal{D}(C)$, and $\mathcal{D}(D)$, respectively. If*

$$T = \begin{pmatrix} A & 0 \\ 0 & D \end{pmatrix} \text{ and } S = \begin{pmatrix} 0 & B \\ C & 0 \end{pmatrix},$$

then

$$\overline{T} = \begin{pmatrix} \overline{A} & 0 \\ 0 & \overline{D} \end{pmatrix} \text{ and } \overline{S} = \begin{pmatrix} 0 & \overline{B} \\ \overline{C} & 0 \end{pmatrix}.$$

As usual, we finish by providing several examples.

Example 20.1.1 Let $B \in B(H)$, and let A be a densely defined operator with a domain $D(A)$ and such that $A \subset B$ (hence A is bounded on $D(A)$). Then

(1) $A^* = B^*$: Since $D(A)$ is dense in H, A^* exists and is unique. Hence

$$A \subset B \Longrightarrow B^* \subset A^* \Longrightarrow A^* = B^*$$

because $D(B^*) = H$.

(2) $\overline{A} = B$: Since $A^* = B^*$, it results that

$$\overline{A} = A^{**} = B^{**} = B,$$

as wished.

Remarks

(1) If $B = 0$ is the zero operator on some dense domain $D \subset H$, i.e. $B \subset 0$, then $B^* = 0$, that is, B^* is the zero operator *everywhere* defined in H.

(2) If $B = 0$ is the zero operator on some dense (and unclosed!) domain $D \subset H$, then B is not closed. But, $\overline{B} = 0$, i.e. $\overline{B}$ is the zero operator *everywhere* defined in H.

Example 20.1.2 Let $U \in B(H)$ be unitary, and let A be a densely defined operator with a domain $D(A)$. Then

$$(U^*AU)^* = U^*A^*U.$$

First, recall that $D(U^*AU)$ is dense in H (Example 18.1.3). By calling on Theorem 20.1.4 given that $U, U^{-1} \in B(H)$, it easily follows that

$$(U^*AU)^* = [U^*(AU)]^* = (AU)^*U^{**} = U^*A^*U,$$

as wished.

In the next example, we give another proof why the operator in Example 19.1.7 is unclosable.

Example 20.1.3 Let H be a Hilbert space, and let D be a dense linear subspace of H. Let $e \neq 0$ be in H. Consider $T : D \to H$ defined by $Tx = f(x)e$, where f is a discontinuous linear functional on D (as in Example 19.1.7). Let us find $D(T^*)$ explicitly. We may write

$$\langle Tx, y \rangle = \langle f(x)e, y \rangle = f(x)\langle y, e \rangle.$$

Hence $x \mapsto \langle Tx, y \rangle$ is continuous if and only if $y \perp e$ given that f is discontinuous. Therefore, $D(T^*) = \{e\}^\perp$ and $T^*y = 0$ for $y \in D(T^*)$, whereby T is not closable.

Example 20.1.4 Consider the operator A defined in Example 19.1.3. By finding the adjoint of A (which is also interesting in its own), we may prove the closedness of A as well. Observe first that $D(A)$ is dense in ℓ^2 for it contains c_{00} (the vector space of all finitely nonzero sequences). It is fairly easy to see that $A^*(x_n) = (\overline{\alpha_n}x_n)$ on $D(A^*) = D(A)$. Obviously for all (x_n)

$$A^{**}(x_n) = (\overline{\overline{\alpha_n}}x_n) = (\alpha_n x_n),$$

that is, $\overline{A} = A^{**} = A$.

Example 20.1.5 Let $Tf = if'$ be a linear operator defined on the domain $D(T) = \{f \in AC[0, 1] : f' \in L^2[0, 1], \ f(0) = f(1) = 0\}$.

(1) T is densely defined and closed: First, $D(T)$ contains all polynomials p such that $p(0) = p(1) = 0$. The closure of the latter with respect to the supremum norm is therefore given by $\{f \in C[0, 1] : f(0) = f(1) = 0\}$. By a simple Hölder's inequality, we see that $D(T)$ is in fact dense in $L^2[0, 1]$.

To show that T is closed on $D(T)$, let (f_n) be a sequence in $D(T)$ such that $f_n \to f$ and $Tf_n = if'_n \to g$ (both limits in $L^2[0, 1]$). Next, set $h(x) = -i \int_0^x g(t)dt$ (hence h is absolutely continuous). Hence (for any x in $[0, 1]$)

$$|f_n(x) - h(x)| = \left| \int_0^x [f'_n(t) + ig(t)]dt \right|$$

$$\leq \|f'_n + ig\|_2 \text{ (by the Cauchy-Schwarz inequality)}$$

$$= \|if'_n - g\|_2.$$

By passing to the supremum over $[0, 1]$, it then follows that $\|f_n - h\|_\infty \to 0$ as $n \to \infty$ and hence

$$\|f_n - h\|_2 \leq \|f_n - h\|_\infty.$$

Therefore, $f_n \to h$ in $L^2[0, 1]$. Consequently

$$0 \leq \|f - h\|_2 \leq \|f - f_n\|_2 + \|f_n - h\|_2 \longrightarrow 0.$$

Thus, $f = h$ a.e. Hence we may assume that $f(x) = -i \int_0^x g(t)dt$ for all x. Whence f is absolutely continuous and f_n converges uniformly to $f(x)$ on $[0, 1]$. Moreover, $f(0) = f(1) = 0$ and $f' = -ig \in L^2[0, 1]$. In fine, $f \in D(T)$ and $Tf = g$, i.e. T is closed.

Remark Example 19.1.8 provides an alternative way for showing the closedness of T. Consider

$$D_0(T) = \{f \in AC[0, 1] : f' \in L^2[0, 1], f(0) = 0\}$$

and also

$$D_1(T) = \{f \in AC[0, 1] : f' \in L^2[0, 1], f(1) = 0\}.$$

Clearly

$$D(T) = \{f \in AC[0, 1] : f' \in L^2[0, 1], f(0) = f(1) = 0\} = D_0(T) \cap D_1(T).$$

Since T is closed on $D_0(T)$ and on $D_1(T)$ (cf. Example 19.1.6), it results that T is closed on $D_0(T) \cap D_1(T) = D(T)$.

(2) $[\operatorname{ran}(T)]^\perp = [1]$, where $[1] = \{\alpha \cdot 1\}_{\alpha \in \mathbb{C}}$: To show that $[\operatorname{ran}(T)]^\perp \subset [1]$, we may instead show that $[1]^\perp \subset \operatorname{ran}(T)$. Let $g \in [1]^\perp$ and set $h(x) = \int_0^x g(t)dt$

(with $g \in L^2[0, 1]$). Hence h is absolutely continuous (hence $h \in L^2[0, 1]$). Besides, $h' = g$ a.e. Clearly, $h(0) = 0$ and

$$h(1) = \int_0^1 g(t)dt = \langle 1, g \rangle = 0.$$

Therefore, $h \in D(T)$. Finally,

$$g = T(ih) \in \operatorname{ran}(T),$$

as wished. The other inclusion is left to interested readers.
(3) We have

$$D(T^*) = \{g \in AC[0, 1] : g' \in L^2[0, 1]\}.$$

Let $g \in D(T^*)$, and let $h = T^*g$. Set $\varphi(x) = \int_0^x h(t)dt$. Then for all $f \in D(T)$

$$\int_0^1 if'(x)\overline{g(x)}dx = \langle Tf, g \rangle = \langle f, h \rangle = \int_0^1 f(x)\overline{h(x)}dx = -\int_0^1 f'(x)\overline{\varphi(x)}dx.$$

Hence $\varphi - ig \in [\operatorname{ran}(T)]^\perp = [1]$, that is, $\varphi - ig = c$ (a constant function). Therefore, $g = ic - i\varphi$. Thus, g is absolutely continuous and $g' = -ih \in L^2[0, 1]$ and so $T^*g = h = ig'$. We have shown that $D(T^*) \subset \{g \in AC[0, 1] : g' \in L^2[0, 1]\}$. Conversely, let $g \in AC[0, 1]$ be such that $g' \in L^2[0, 1]$. Let $f \in D(T)$. Since $f(1) = f(0) = 0$, it follows that

$$\langle Tf, g \rangle = \langle f, -ig' \rangle.$$

This means that $g \in D(T^*)$ and that $T^*g = -ig'$.
(4) T is symmetric: Since $D(T) \subset D(T^*)$ and for all $f \in D(T)$, we have $Tf = T^*f$ and so $T \subset T^*$. This obviously means that T is symmetric. Accordingly, T is closed and symmetric (but $D(T) \neq D(T^*)$).

Example 20.1.6 Let $Tf = if'$ be a linear operator defined on the domain $D(T) = \{f \in AC[0, 1] : f' \in L^2[0, 1], \ f(0) = f(1)\}$.

(1) $[\operatorname{ran}(T)]^\perp = [1]$, where $[1] = \{\alpha \cdot 1\}_{\alpha \in \mathbb{C}}$: Just apply a similar method to that of Example 20.1.5.
(2) $D(T) = D(T^*)$: First, it is easy to see that via a simple integration by parts that for all $f, g \in D(T)$

$$\langle Tf, g \rangle = i \int_0^1 f'(x)\overline{g(x)}dx = \int_0^1 f(x)\overline{ig'(x)}dx = \langle f, Tg \rangle.$$

This means that $T \subset T^*$. To prove that $D(T) = D(T^*)$, it suffices to show $D(T^*) \subset D(T)$. So, let $g \in D(T^*)$ and let $h = T^*g$. Set $\varphi(x) = \int_0^x h(t)dt$. Clearly $\varphi(0) = 0$. Also

$$\varphi(1) = \int_0^1 h(t)dt = \langle 1, h \rangle = \langle 1, T^*g \rangle = \langle T1, g \rangle = 0.$$

Now for all $f \in D(T)$

$$i \int_0^1 f'(x)\overline{g(x)}dx = -\int_0^1 f'(x)\overline{\varphi(x)}dx.$$

Hence

$$0 = \int_0^1 [if'(x)\overline{g(x)} + f'(x)\overline{\varphi(x)}]dx = \int_0^1 if'(x)[\overline{g(x) + i\varphi(x)}]dx.$$

Thus, $g + i\varphi$ is perpendicular to $\text{ran}(T)$ and so $g + i\varphi = c$ (a constant function) whereby $g = c - i\varphi$. This tells us that g is absolutely continuous and moreover $g' = -ih \in L^2[0, 1]$. Finally

$$g(0) = g(1) \ (= c).$$

Consequently, $g \in D(T)$ and so $D(T) = D(T^*)$.

Example 20.1.7 Let $Sf = if'$ be a linear operator defined on the domain

$$D(S) = \{f \in AC[0, 1] : f' \in L^2[0, 1]\}.$$

Then S is closed. Indeed, $S = T^*$, where T is the operator introduced in Example 20.1.5.

Remark In the previous three examples, the conclusions remain unchanged if we replace $[0, 1]$ by the interval $[a, b]$.

Example 20.1.8 An alternative way of seeing that the *second* operator A given in Answer 19.2.24 is not closable is to invoke Proposition 4.5 of [352]. There, the authors showed that $D(A^*) = \{e\}^\perp$ (and A^* is the zero operator on $\{e\}^\perp$), that is, A^* is not densely defined and consequently, A is not closable.

Example 20.1.9 ([326]) Let $Tf = if'$ be a linear operator defined on the domain $D(T) = \{f \in AC[0, 1] : f' \in L^2[0, 1], f(0) = f(1) = 0\}$. Let

$$D_0 = \{f \in C^\infty[0, 1], f(0) = f(1) = 0\}.$$

Denote the restriction of T to D_0 by T_0. Then D_0 is a core for T and so $\overline{T_0} = T$.

Indeed, let $f \in D(T)$. As $f' \in L^2(0, 1)$, it follows that there is a sequence (g_n) in $C^\infty[0, 1]$ for which $\|g_n - f'\|_2 \to 0$ as n tends to ∞. Define

$$f_n(x) = \int_0^x g_n(t)dt - x \int_0^1 g_n(t)dt.$$

Clearly, $f_n \in D_0$ (for each n). By the Cauchy-Schwarz inequality, we see that $\|g_n - f'\|_1 \to 0$ as well. Hence

$$\int_0^1 g_n(t)dt \longrightarrow \int_0^1 f'(t)dt = f(1) - f(0) = 0.$$

Therefore,

$$Tf_n = if_n' = ig_n - i \int_0^1 g_n(t)dt \longrightarrow if' = Tf.$$

Now, we may write (for all $x \in [0, 1]$)

$$|f_n(x) - f(x)| = \left| \int_0^x g_n(t)dt - \int_0^x f'(t)dt - x \int_0^1 g_n(t)dt \right|$$

$$\leq \left| \int_0^x (g_n(t) - f'(t))dt \right| + \left| \int_0^1 g_n(t)dt \right|$$

$$\leq \left| \int_0^1 (g_n(t) - f'(t))dt \right| + \left| \int_0^1 g_n(t)dt \right|$$

$$\leq \|g_n - f'\|_1 + \left| \int_0^1 g_n(t)dt \right|.$$

Whence, $\|f_n - f\|_\infty \to 0$ (as n goes to ∞). Therefore, $\|f_n - f\|_2 \to 0$. Consequently, $f_n \to f$ and $T_0 f_n = Tf_n \to Tf$ (w.r.t. $\| \cdot \|_2$), as needed.

Example 20.1.10 Let $n \in \mathbb{N}$ and define each $T_n f(x) = if'(x)$ on $D(T_n) = \{f \in C^n[0, 1] : f(0) = f(1) = 0\}$. Then we have "nested" operators as "$\cdots \subset T_2 \subset T_1$." If we use T and T_0 from Example 20.1.9, then

$$T_0 \subset \cdots \subset T_2 \subset T_1 \subset T.$$

Since the closure of an operator is the smallest closed extension of that operator, it results that

$$T = \overline{T_0} \subset \cdots \subset \overline{T_2} \subset \overline{T_1} \subset \overline{T} = T.$$

Therefore,

$$\overline{T_0} = \cdots = \overline{T_2} = \overline{T_1} = T$$

whereby

$$(T_0)^* = \cdots = (T_2)^* = (T_1)^* = T^*$$

where $D(T^*) = \{f \in AC[0, 1] : f' \in L^2[0, 1]\}$.

Let us now cast a glance at operators defined on $L^2(0, \infty)$.

Example 20.1.11 ([326]) Define $Af(x) = if'(x)$ on the domain $D(A) = \{f \in H^1(0, \infty) : f(0) = 0\}$. Then A is clearly densely defined. Besides, $A^*f(x) = if'(x)$ on $D(A^*) = H^1(0, \infty)$ (so A is clearly symmetric). To corroborate that, let g be in $H^1(0, \infty)$ and let $f \in D(A)$. Hence $f(0) = 0$. Moreover, using an integration by parts

$$\langle Af, g \rangle = \int_0^\infty if'(x)\overline{g(x)}dx = \int_0^\infty f(x)\overline{ig'(x)}dx = \langle f, Ag \rangle$$

which holds for all $f \in D(A)$. Thus, $g \in D(A^*)$ and $A^*g(x) = ig'(x)$.

Conversely, let $g \in D(A^*)$. Let A^b be the restriction of A to functions of $D(A)$ to $[0, b]$, and let A_b be defined by $A_b f(x) = if'(x)$ on $D(A_b) = \{f \in AC[0, b] : f' \in L^2[0, b], \ f(0) = f(b) = 0\}$. From Example 20.1.5, we know that $(A_b)^* f(x) = if'(x)$ on the domain $D[(A_b)^*] = \{f \in AC[0, b] : f' \in L^2[0, b]\}$. Since $A_b \subset A^b$, it follows that $(A^b)^* \subset (A_b)^*$, which implies $(A^b)^*g(x) = ig'(x)$ on $[0, b]$. Also, $(A^b)^*g(x) = A^*g(x)$ for all $x \in [0, b]$. Since this holds for all $b > 0$, it results that $A^*g(x) = ig'(x)$ on the whole of $[0, \infty)$. Finally, because $g, g' \in L^2(0, \infty)$, then $g \in H^1(0, \infty)$.

Remarks

(1) The operator A defined above is also closed. This follows from Proposition B.10 by following the method of Example 19.1.6.
(2) Similarly, we may show that if $Bf(x) = if'(x)$ is defined on $D(B) = \{f \in H^1(-\infty, 0) : f(0) = 0\}$, then $B^* f(x) = if'(x)$ on $D(B^*) = H^1(-\infty, 0)$.

Example 20.1.12 Let $\varphi : \mathbb{R} \to \mathbb{C}$ be a continuous function. Consider $M_\varphi f(x) = \varphi(x)f(x)$ on $D(M_\varphi) = \{f \in L^2(\mathbb{R}) : \varphi f \in L^2(\mathbb{R})\}$. We already know that $D(M_\varphi)$ is dense in $L^2(\mathbb{R})$. Then

$$(M_\varphi)^* = M_{\overline{\varphi}}.$$

First, it is plain that $D(M_\varphi) = D(M_{\overline{\varphi}})$. Now, for all $f, g \in D(M_\varphi)$, we may write

$$\langle M_\varphi f, g \rangle = \int_{\mathbb{R}} \varphi(x) f(x) \overline{g(x)} dx = \int_{\mathbb{R}} f(x) \overline{\overline{\varphi(x)} g(x)} dx = \langle f, M_{\overline{\varphi}} g \rangle$$

whereby $M_{\overline{\varphi}} \subset (M_\varphi)^*$.

Conversely, let $g \in D[(M_\varphi)^*]$ and set $h = (M_\varphi)^* g$. Then for all $f \in D(M_\varphi)$

$$\langle M_\varphi f, g \rangle = \langle f, h \rangle.$$

Thereby

$$\int_{\mathbb{R}} f(x)[\varphi(x) \overline{g(x)} - \overline{h(x)}] dx = 0.$$

By the density of $D(M_\varphi)$, we obtain $\varphi \overline{g} - \overline{h} = 0$. Hence $\overline{\varphi} g = h$ on $\mathbb{R}$. This tells us that g is in $D(M_{\overline{\varphi}})$. Accordingly, $(M_\varphi)^* = M_{\overline{\varphi}}$, as wished.

Remark Another reason why M_φ is closed is

$$(M_\varphi)^{**} = M_{\overline{\overline{\varphi}}} = M_\varphi.$$

Example 20.1.13 Let $\varphi : \mathbb{R} \to \mathbb{R}$ be a continuous function. Consider $M_\varphi f(x) = \varphi(x) f(x)$ on $D(M_\varphi) = \{f \in L^2(\mathbb{R}) : \varphi f \in L^2(\mathbb{R})\}$. Then $C_0^\infty(\mathbb{R})$ is a core for M_φ.

Example 20.1.14 ([326]) Let $Af(x) = -f''(x)$ be defined on $D(A) = \{f \in H^2(\mathbb{R}) : f(0) = f'(0) = 0\}$. Then $A^* f(x) = -f''(x)$ for $f \in D(A^*) = H^2(-\infty, 0) \oplus H^2(0, \infty)$. Observe also that A is densely defined and symmetric.

20.2 Questions

20.2.1 Is There Some Densely Defined Non-closable Operator A Such That $A^*A = I$?

If A is a densely defined closable operator and $A^*A = I$, then clearly $H = D(A^*A) \subset D(A)$ would yield $A \in B(H)$.

Question 20.2.1 Can the equation $A^*A = I$ be satisfied by a densely defined non-closable operator A?

20.2.2 A Densely Defined T Such That $H \neq D(T) + \mathrm{ran}(T^*)$

It is fairly easy to see that if we are given a closed densely defined linear operator $T : D(T) \subset H \to K$, then

$$H = D(T) + \mathrm{ran}(T^*).$$

See e.g. Lemma 1.5 in [350].

Question 20.2.2 Show that this result is no longer true if one drops the closedness of T.

20.2.3 A Densely Defined Symmetric Operator A Such That $\ker(A) \neq \ker(A^*)$

Question 20.2.3 Let A be a densely defined symmetric operator with domain $D(A) \subset H$. Do we have $\ker(A) = \ker(A^*)$?

20.2.4 The Failure of Some Maximality Relations

If $B \in B(H)$ and A is a linear operator with a dense domain $D(A)$ obeying $A \subset B$, then $\overline{A} = B$. In particular, if A is closed, then $A = B$. The proof is simple:

$$A \subset B \Longrightarrow B^* \subset A^* \Longrightarrow A^* = B^*$$

for $D(B^*) = H$. Hence $\overline{A} = A^{**} = B$.

Question 20.2.4 Let A, B be two linear operators on a Hilbert space H. Assume also that $B \in B(H)$. Assume further that A has a domain $D(A)$ and that $A \subset B$.

(1) Do we have $A = B$ if A is densely defined but unclosed?
(2) Do we have $A = B$ if A is closed but non-densely defined?

20.2.5   A Non-closable Symmetric Operator

Obviously, a densely defined symmetric operator is closable. Indeed, if T is such an operator, then $T \subset T^*$ thereby making $D(T^*)$ dense, that is, T is closable. Do non-densely defined symmetric operators enjoy this property?

Question 20.2.5 Find a non-closable operator that is symmetric.

20.2.6   A Densely Defined Operator T Such That $D(T^*) = \{0\}$

If T is densely defined, then it is not automatic that T^* is densely defined. In fact, if T is densely defined, then it may well happen that $D(T^*) = \{0\}$, as will be seen below.

Question 20.2.6 (See Also Question 21.2.27) Find a densely defined operator T such that $D(T^*) = \{0\}$.

20.2.7   A Densely Defined Operator T on a Hilbert Space Such That $T^* = 0$ on $D(T^*) \neq \{0\}$

Question 20.2.7 Find a densely defined operator T on a Hilbert space such that $T^* = 0$ on $D(T^*) \neq \{0\}$.

20.2.8   An Everywhere Defined Linear Operator T Such That $D(T^*) = \{0\}$

In each of the previous two examples above, the operator T was not everywhere defined. Can some everywhere defined operators have a trivial adjoint's domain?

It is worth noticing that if T is an everywhere defined unbounded operator, then $D(T^*)$ is closed. This was shown in e.g. [280].

Question 20.2.8 (❀) Find an operator T on a Hilbert space H such that

$$D(T) = H \text{ and } D(T^*) = \{0\}.$$

20.2.9 A Densely Defined T with $D(TT^*) = D(T^*T) = D(T)$

Question 20.2.9 Find a densely defined operator T on a Hilbert space such that

$$D(TT^*) = D(T^*T) = D(T).$$

20.2.10 Two Densely Defined Operators A and B Such That $(A + B)^* \neq A^* + B^*$

One main obstacle when dealing with densely defined unbounded operators (say A and B) is the fact that in general $(A + B)^* \neq A^* + B^*$. An apparent reason is that $A + B$ could be non-densely defined and so computing $(A + B)^*$ is not possible anymore. What if we assume that $A + B$ is densely defined?

Question 20.2.10 (Cf. Question 20.2.11) Provide densely defined operators A and B such that $A + B$ is densely defined and yet

$$(A + B)^* \neq A^* + B^*.$$

Further Reading See e.g. [156], [246], and [336]. See also [143].

20.2.11 Two Densely Defined A and B (One of Them Is Bounded but Non-everywhere Defined) with $(A + B)^* \neq A^* + B^*$

Question 20.2.11 Provide a densely defined operator A and a *bounded and densely defined* B such that

$$(A + B)^* \neq A^* + B^*.$$

20.2.12   Two Densely Defined Operators A and B Such That $(BA)^* \neq A^*B^*$

Another difficulty when dealing with unbounded operators is that in general $(BA)^* \neq A^*B^*$ as we shall shortly see by giving a counterexample.

Many authors have worked on criteria guaranteeing the equality

$$(BA)^* = A^*B^*$$

to hold (even for the "Banach adjoint," which is not within the scope of this book). For example, we shall see in the next chapter that if A is closed, then $(A^*A)^* = A^*A$. Another result reads [317]: If A and B are densely defined closed linear operators on a Hilbert space, and ran A is closed with finite codimension, then

$$(BA)^* = A^*B^*.$$

Other results may be consulted in [142, 144, 160, 367, 368, 391], and [336].

It is worth noticing in the end that J. Dixmier [91] defined a new product, denoted by "•", for which

$$(AB)^* = B^* \bullet A^*$$

when both A and B are closed. See also [228].

Question 20.2.12 Provide an example of a pair of densely defined operators A and B such that

$$(BA)^* \neq A^*B^*.$$

What about when A is injective?

20.2.13   A Densely Defined Left Invertible Operator T and a Densely Defined Operator S Such That $(ST)^* \neq T^*S^*$

Question 20.2.13 Find a densely defined left invertible operator T and a densely defined operator S such that ST is densely defined and

$$(ST)^* \neq T^*S^*.$$

20.2.14   A Densely Defined Operator A with $(A^2)^* \neq (A^*)^2$

Remember that if A^2 (hence A) is densely defined, then we are only sure of having $(A^*)^2 \subset (A^2)^*$.

Question 20.2.14 Give an example of a linear operator A such that A^2 is densely defined and

$$(A^2)^* \neq (A^*)^2.$$

20.2.15   A Densely Defined Operator T with $(T^2)^* = (T^*)^2$

Question 20.2.15 Give an example of a densely defined operator T such that

$$(T^2)^* = (T^*)^2.$$

20.2.16   Two Bounded (Non-everywhere Defined) or Unbounded Operators A and B Such That $A \subset B$ and $A^* = B^*$

Question 20.2.16 Let A and B be two linear operators such that $A \subset B$. Show that $A^* = B^*$ may take place if A and B are bounded, but they are not both everywhere defined or even if they are unbounded.

20.2.17   A Non-closed Densely Defined Operator T Such That $\operatorname{ran} T \subset D(T)$ and $\operatorname{ran} T^* \subset D(T^*)$

Question 20.2.17 Find a densely defined operator T such that $T \neq T^*$ with $\operatorname{ran} T \subset D(T)$ and $\operatorname{ran} T^* \subset D(T^*)$.

20.2.18 A Closed Densely Defined Operator T Such That ran $T \subset D(T)$ and ran $T^* \subset D(T^*)$

It was shown in [274] (see also [278]) that if a densely defined closed operator T satisfies ran $T \subset D(T^*)$, then T is bounded.

Question 20.2.18 (Cf. Question 20.2.19) Find a densely defined unbounded closed operator T (with $T \neq T^*$) such that ran $T \subset D(T)$ and ran $T^* \subset D(T^*)$.

20.2.19 A Densely Defined Unbounded Closed Operator A Such That ran $A \subset D(A)$ But ran $A^* \not\subset D(A^*)$

Question 20.2.19 Find a densely defined, unbounded, and closed operator A such that ran $A \subset D(A)$ but ran $A^* \not\subset D(A^*)$.

20.2.20 A Densely Defined Closed Operator A with $A + A^*$ Being Densely Defined But Unclosed

Question 20.2.20 Find a densely defined closed operator A such that $A + A^*$ is densely defined but unclosed.

20.2.21 A Closed Operator A Satisfying $AA^* = A^*A + I$

Question 20.2.21 Find a densely defined and closed operator A that satisfies

$$AA^* = A^*A + I.$$

20.2.22 A Closed T with $D(T) = D(T^*)$ but $D(TT^*) \neq D(T^*T)$

Let A and B be two linear operators such that $D(A) = D(B)$. Does it follow that $D(AB) = D(BA)$? The answer is negative in general. For example, if $\mathcal{F}_0$ is the restriction of the $L^2(\mathbb{R})$-Fourier transform to $C_0^\infty(\mathbb{R})$, then set

$$A = \begin{pmatrix} \mathcal{F}_0 & 0 \\ 0 & 0 \end{pmatrix} \text{ and } B = \begin{pmatrix} 0 & 0 \\ \mathcal{F}_0 & 0 \end{pmatrix},$$

where $D(A) = D(B) = C_0^\infty(\mathbb{R}) \oplus L^2(\mathbb{R})$. Then $D(AB) \neq D(BA)$ because $D(AB) = C_0^\infty(\mathbb{R}) \oplus L^2(\mathbb{R})$ and $D(BA) = \{0\} \oplus L^2(\mathbb{R})$.

Is the situation better when both operators are closed, and one of them is the adjoint of the other?

Question 20.2.22 Find a closed and densely defined operator T such that $D(T) = D(T^*)$ but $D(TT^*) \neq D(T^*T)$.

20.2.23 An Unbounded Closed Symmetric and Positive Operator A Such That $D(A^2) = \{0\}$

We have already seen a densely defined operator whose square has a trivial domain. Such operators were not closed. So one might think that it could be impossible for a densely defined closed operator to have a square, whose domain is trivial. In fact, we have a closed symmetric (and positive) operator A with $D(A^2) = \{0\}$. This striking counterexample by P. R. Chernoff came in to simplify a rather complicated construction already obtained by M. Naimark in [271]. For similar impressive counterexamples, see Question 21.2.46 as well as Questions 21.2.47, 21.2.48, and 21.2.49. See also Questions 21.2.8 and 21.2.9.

Notice in the end that a densely defined *boundedly* invertible operator (hence closed) A always satisfies $D(A^2) \neq \{0\}$. To see why, first remember that

$$D(A^2) = \{x \in D(A) : Ax \in D(A)\} = A^{-1}[D(A)],$$

then aiming for a contradiction, suppose that $D(A^2) = \{0\}$. Since $\{0\} = D(A^2) = A^{-1}[D(A)]$, it ensues that

$$\{0\} = A(\{0\}) = AA^{-1}[D(A)] = D(A)$$

(for $AA^{-1} = I$), which is impossible.

Remark The injectivity (alone) of A is just not sufficient as shall be seen in Answer 21.2.46.

Question 20.2.23 (❋❋) Find an example of an unbounded densely defined, closed, symmetric, and positive operator A such that

$$D(A^2) = \{0\}.$$

Answers

Answer 20.2.1 The answer is no. Indeed, let $x \in H$. Then

$$\|Ax\|^2 = \langle Ax, Ax \rangle = \langle A^*Ax, x \rangle = \langle x, x \rangle = \|x\|^2.$$

In particular, A is closable!

Remark Clearly any densely defined A satisfying $AA^* = I$ is necessarily closable.

Answer 20.2.2 Let $T = 0$ on some dense (unclosed) domain $D(T) \subsetneq H$. Then T is not closed (but it is symmetric). Besides $T^* = 0$ everywhere on H and so ran $T^* = \{0\}$. Therefore

$$H \neq D(T) = D(T) + \text{ran}(T^*),$$

as needed.

Answer 20.2.3 No! While we always have ker$(A) \subset$ ker(A^*), the reverse inclusion need not be true. Let $A = 0$ on some dense and non-closed domain D. Then $A^* = 0$ on the whole of H. Therefore, ker$(A^*) = H$ whereas ker$(A) = D(A)$. Therefore, ker$(A^*) \not\subset$ ker(A).

What about the case when A is closed and symmetric? Well, this is still not sufficient. By considering Example 20.1.5, readers may then show that the closed symmetric operator there has a different kernel from that of its adjoint's.

Answer 20.2.4 First, remember that A is bounded on $D(A)$.

(1) The answer is negative! For a counterexample, just consider $A = B_D$ (B restricted to D), where D is a dense (and not closed) subspace of H. Since D is not closed, A, which is bounded on D, cannot be closed. Observe in the end that $A \neq B$ because $D \neq H$!

(2) False! Just consider $A = 0$ (the zero operator) restricted to the trivial domain $D(A) = \{0\}$. Take B to be any bounded operator. Since $A(0) = 0 = B(0)$, we see plainly that $A \subset B$. Finally, it is clear that A is closed on $D(A)$, that $D(A)$ is not dense in H, and that $A \neq B$.

Answer 20.2.5 ([326]) Let T be the non-closable operator of Example 19.1.7, which was defined on $D(T) \subset H$. Define on $H \oplus H$ the operator A by $A(x, 0) = (0, Tx)$ for $(x, 0) \in D(A) = D(T) \oplus \{0\}$. Then A is not closable as T is not. However, A is symmetric as for all $x, y \in D(T)$, we have

$$\langle A(x, 0), (y, 0) \rangle_{H \oplus H} = \langle 0, y \rangle_H + \langle Tx, 0 \rangle_H$$

$$= 0$$

$$= \langle (x, 0), A(y, 0) \rangle_{H \oplus H}.$$

Remark Z. Sebestyén and Zs. Tarcsay (see e.g. [338]) improved a result by the famous M. H. Stone by showing that: A surjective symmetric operator is automatically densely defined.

Answer 20.2.6 There are known examples available in the literature. For instance, see Example 3.4 on Page 105 in [181] or Example 3 on Page 69 in [375] (see also [370]). Most of these examples are not that straightforward.

The simplest example perhaps is the following: Define T on the dense subspace $D(T) := C[0, 1]$ of $L^2[0, 1]$ by $Tf = f(0)$. Then T is densely defined and so T^* is well defined. Moreover, T is not bounded (just take a sequence f_n in $C[0, 1]$ such that $f_n(0) = n$ and $\|f_n\|_2$ is finite, cf. [256]). Since T is defined into $\mathbb{F}$ ($\mathbb{F}$ is either $\mathbb{R}$ or $\mathbb{C}$), its (unique) adjoint is defined on $\mathbb{F}$. It then follows that either $D(T^*) = \{0\}$ or $D(T^*) = \mathbb{F}$. If $D(T^*) = \mathbb{F}$ (so T would be closable), then by the closed graph theorem T^* would become everywhere defined and bounded. Hence $T^{**} = \overline{T}$ too is bounded. But $T \subset \overline{T}$ and so this would mean that T is bounded on $D(T)$, which is impossible. Therefore,

$$D(T^*) = \{0\}$$

Another example borrowed from [382] (Professor D. P. Williams informed me that he saw it in a basic course on unbounded operators in the late 1970s) is to define $T = d/dx$ on

$$D(T) = \{f \in L^2(\mathbb{R}) : f' \text{ exists almost everywhere and } f' \in L^2(\mathbb{R})\}.$$

Let $g \in D(T^*)$ and set $h = T^*g$. Let $f \in D(T)$. Then there exists a sequence of step functions (f_n) such that $f_n \to f$ in $L^2(\mathbb{R})$. Hence

$$\langle Tf, g \rangle = \langle f, h \rangle = \lim_{n \to \infty} \langle f_n, h \rangle = \lim_{n \to \infty} \langle Tf_n, g \rangle = 0.$$

Since $T[D(T)]$ (the range of T) is dense, it follows that $g = 0$. Thus

$$D(T^*) = \{0\}.$$

Answer 20.2.7 This is easy! Let A be a densely defined operator on a Hilbert space H with $D(A^*) = \{0\}$ (as in say Answer 20.2.6). Set

$$T = \begin{pmatrix} A & 0 \\ 0 & 0 \end{pmatrix}$$

on the natural domain $D(T) = D(A) \oplus H$. Then obviously

$$T^* = \begin{pmatrix} A^* & 0 \\ 0 & 0 \end{pmatrix}$$

on

$$D(T^*) = D(A^*) \oplus H = \{0\} \oplus H \neq \{(0, 0)\}.$$

Nevertheless, for all $(x, y) \in D(T^*)$

$$T^*(x, y) = (A^*x, 0) = (0, 0),$$

as wished.

Answer 20.2.8 (An Example Due to S. K. Berberian, See [133]) Let $H = \ell^2$, and let $D(T) \subset \ell^2$ be spanned by the unit vectors $e_k = (0, 0, \cdots, 0, \underbrace{1}_{k}, 0, \cdots)$. Consider the double indexing of (e_k), i.e. $\{e_{kj} : j, k = 1, 2, \cdots\}$. Now, for each k, define

$$T e_{kj} = e_k, \ j = 1, 2, \cdots.$$

Then, extend T linearly to the whole of ℓ^2 (cf. Question 18.2.1) and so $D(T) = \ell^2$. To see why $D(T^*) = \{0\}$, let $y = (y_1, y_2, \ldots) \in D(T^*)$. Then for each k

$$\langle e_{kj}, T^*y \rangle = \langle T e_{kj}, y \rangle = \langle e_k, y \rangle = y_k$$

for $j = 1, 2, \cdots$. By Bessel's inequality, we have

$$\sum_{j=1}^{\infty} |\langle e_{kj}, T^*y \rangle|^2 \leq \|T^*y\|^2$$

and so necessarily

$$0 = \lim_{j \to \infty} \langle e_{kj}, T^* y \rangle = y_k$$

for all k, and so $y = 0$. Consequently $D(T^*) = \{0\}$.

Answer 20.2.9 Obviously any $T \in B(H)$ satisfies the required equalities. Let us give an example of a T that is not everywhere defined. For instance, let $T = I_{D(T)}$ be the identity operator restricted to $D(T) \subset H$. Then $T^* = I$ (the full identity operator over H). Then

$$D(TT^*) = \{x \in H : x \in D(T)\} = D(T)$$

and obviously $D(T^*T) = D(T)$ for $T^* \in B(H)$. Accordingly,

$$D(TT^*) = D(T^*T) = D(T),$$

as wished.

Answer 20.2.10 Let A be the operator defined on the domain $D(A) = \{f \in L^2(\mathbb{R}) : xf \in L^2(\mathbb{R})\}$ by $Af(x) = xf(x)$. Set $B := -A$ and so $D(B) = D(A)$. Since $x \mapsto x$ is real-valued, Example 20.1.12 says that $A^* = A$, hence $B^* = B$ as well. Now,

$$A^* + B^* = A + B = A - A = 0_{D(A)}$$

while

$$(A + B)^* = (A - A)^* = [0_{D(A)}]^* = 0_{L^2(\mathbb{R})}$$

and so

$$(A + B)^* \neq A^* + B^*,$$

as suggested.

Answer 20.2.11 Let A be the operator defined on the domain $D(A) = \{f \in AC[0, 1] : f' \in L^2[0, 1], \ f(0) = f(1)\}$ by $Af(x) = if'(x)$. That $A = A^*$ may be found in say Example 20.1.5. Let $B = 0$ on the domain

$$D(B) = \{f \in AC[0, 1] : f' \in L^2[0, 1], f(0) = f(1) = 0\}.$$

Then, B is clearly bounded and densely defined (and $D(B) \neq L^2[0, 1]$). We have

$$(A + B)f(x) = if'(x)$$

on $D(A + B) = D(A) \cap D(B) = D(B)$. According to Example 20.1.5, we have
$(A + B)^* f(x) = i f'(x)$ with

$$D[(A + B)^*] = \{f \in AC[0, 1] : f' \in L^2[0, 1]\}.$$

Since $A^* = A$ and $B^* = 0$ (on $D(B^*) = L^2[0, 1]$), it results that

$$D(A^*) \cap D(B^*) = \{f \in AC[0, 1] : f' \in L^2[0, 1], \ f(0) = f(1)\},$$

making the equality $(A + B)^* = A^* + B^*$ impossible.

Answer 20.2.12 One obstacle is that BA could just be non-densely defined (even when A and B are). As observed before, we may just consider the $L^2(\mathbb{R})$-Fourier transform restricted to $C_0^\infty(\mathbb{R})$, and noted $\mathcal{F}_0$. Setting $A = B = \mathcal{F}_0$ then implies that

$$D(BA) = D(\mathcal{F}_0^2) = \{0\},$$

and taking the adjoint of $\mathcal{F}_0^2$ therefore would not make sense.

Let us now provide a counterexample in the event of the density of $D(BA)$, hence also answering the second question as A will be one-to-one. The idea here is to exploit the fact that $(BA)^*$ is always closed, while $A^* B^*$ need not be so. More precisely, let B be an unbounded and *boundedly invertible* operator with domain $D(B) \subsetneq H$. Take $A = B^{-1}$, and remember that $A \in B(H)$, that $D(B^*) \neq H$, and that A is injective. Hence

$$D[(BA)^*] = D[(BB^{-1})^*] = D(I^*) = H \neq D[(B^{-1})^* B^*] = D(B^*).$$

Answer 20.2.13 Consider again the restriction of the identity operator noted I_D (which is left invertible), where

$$D = \{f \in AC[0, 1], f' \in L^2(0, 1) : f(0) = f(1) = 0\}.$$

Next, consider the closed operator $Sf(x) = f'(x)$ on the domain $D(S) = \{f \in AC[0, 1], f' \in L^2(0, 1) : f(0) = 0\}$. Now, $SI_D f(x) = f'(x)$ for any f in $D(SI_D) = \{f \in D : f \in D(S)\} = D \cap D(S)$. Hence

$$D(SI_D) = \{f \in AC[0, 1], f' \in L^2(0, 1) : f(0) = f(1) = 0\}.$$

Since, $(I_D)^* = I$, $D(S^*) = \{f \in AC[0, 1], f' \in L^2(0, 1) : f(1) = 0\}$ and $D[(SI_D)^*] = \{f \in AC[0, 1], f' \in L^2(0, 1)\}$, it follows that

$$(SI_D)^* \neq S^* = (I_D)^* S^*,$$

as wished.

Answer 20.2.14 On ℓ^2, define the *densely defined, linear, and closed* operator A by

$$Ax = A(x_n) = (x_2, 0, 2x_4, 0, \cdots, \underbrace{nx_{2n}}_{2n-1}, \underbrace{0}_{2n}, \cdots)$$

on the domain

$$D(A) = \{x = (x_n) \in \ell^2 : (nx_{2n}) \in \ell^2\}$$

(as in Answer 19.2.15). Then

$$A^2 = 0 \text{ on } D(A^2) = D(A).$$

Hence $(A^2)^* = (0_{D(A)})^* = 0$ and the latter is the everywhere defined zero operator, i.e. $D[(A^2)^*] = \ell^2$. Therefore,

$$(A^*)^2 \subset (A^2)^* = 0.$$

If $D[(A^*)^2] = \ell^2$, then $\ell^2 \subset D(A^*)$ or simply $D(A^*) = \ell^2$. Since A^* is closed, the closed graph theorem tells us that A^* would be bounded and everywhere defined. Thus, its adjoint $\overline{A}$ or A would be bounded, and this is absurd. Consequently, $D[(A^*)^2] \neq \ell^2$ and so

$$(A^2)^* \neq (A^*)^2,$$

as wished.

Remark Another example is $A = \begin{pmatrix} 0 & T \\ 0 & 0 \end{pmatrix}$, where T is at least densely defined. We leave the simple details to be provided by interested readers.

Answer 20.2.15 Consider two operators A, B with $B \in B(H)$ and $D(A) \subset H$ such that $A = A^*$ and $B = B^*$. Set $T = \begin{pmatrix} B & A \\ 0 & 0 \end{pmatrix}$ and so

$$T^2 = \begin{pmatrix} B^2 & BA \\ 0 & 0 \end{pmatrix}.$$

Now, $T^* = \begin{pmatrix} B & 0 \\ A & 0 \end{pmatrix}$ on $D(T^*) = D(A) \oplus H$. Hence

$$(T^*)^2 = \begin{pmatrix} B^2 & 0 \\ AB & 0 \end{pmatrix}$$

as

$$D[(T^*)^2] = \{(x, y) \in D(A) \times H : Bx \in D(A), Ax \in H\} = D(AB) \oplus H.$$

Finally,

$$(T^2)^* = \begin{pmatrix} B^2 & 0 \\ AB & 0 \end{pmatrix}$$

where $D[(T^2)^*] = D(AB) \oplus H$, thereupon $(T^2)^* = (T^*)^2$, as wished.

Remark Another example borrowed from [326] reads: Define $Tf(x) = -if'(x)$ on $D(T) = \{f \in H^1(0, 1) : f(0) = f(1) = 0\}$. Then $T^2 f(x) = -f''(x)$ for f in

$$D(T^2) = \{f \in H^2(0, 1) : f(0) = f(1) = f'(0) = f'(1) = 0\}.$$

Now, $T^* f(x) = -if'(x)$ on $D(T^*) = H^1(0, 1)$. Finally, we have $(T^2)^* f(x) = -f''(x)$ for f in

$$D[(T^2)^*] = D[(T^*)^2] = H^2(0, 1).$$

Answer 20.2.16 We already know that if $B \in B(H)$ and A is such that $A \subset B$, then $A^* = B^*$.

In the unbounded case we already have a collection of "nested operators" in Example 20.1.10. Let us give another example. Consider $Af(x) = if'(x)$ on $D(A) = C_0^\infty(0, 1)$, and let $Bf(x) = if'(x)$ on $D(B) = \{f \in AC[0, 1], f' \in L^2(0, 1) : f(0) = f(1) = 0\}$. We know that $D(B^*) = \{f \in AC[0, 1], f' \in L^2(0, 1)\}$. Since $A \subset B$, it results that $B^* \subset A^*$. Hence if $f \in AC[0, 1]$ is such that $f' \in L^2(0, 1)$, then $f \in D(A^*)$. Conversely, let $g \in D(A^*)$. Then for some $h \in L^2(0, 1)$ such that

$$i\langle f', g \rangle = \langle Af, g \rangle = \langle f, h \rangle$$

for all $f \in C_0^\infty(0, 1)$. In a distributional language, this means that the distributional derivative of g is $-ih$. Particularly, g' is an integrable function. Thus, g is in $AC[0, 1]$ and $h = ig'$. Accordingly,

$$D(A^*) = \{f \in AC[0, 1], f' \in L^2(0, 1)\} = D(B^*),$$

as wished.

Answer 20.2.17 This is easy! Let I_D be the identity operator restricted to a non-closed domain $D \subset H$, where H is a Hilbert space. Then

$$\operatorname{ran} I_D = \mathcal{D}(I_D) = D$$

whereas

$$\mathrm{ran}(I_D)^* = \mathcal{D}[(I_D)^*] = H,$$

and that's all!

Answer 20.2.18 Let A be an unbounded closed operator with domain $D(A) \subset H$ and such that $A = A^*$. Define $T = \begin{pmatrix} 0 & A \\ 0 & 0 \end{pmatrix}$. Then $D(T) = H \oplus D(A)$ and so T is densely defined. Moreover, it is closed. Also, plainly $\mathrm{ran}\, T \subset D(T)$.

Now, $T^* = \begin{pmatrix} 0 & 0 \\ A & 0 \end{pmatrix}$ and similarly we easily obtain that

$$\mathrm{ran}(T^*) \subset D(A) \oplus H = D(T^*).$$

Remark In this example, observe that $D(T) \neq D(T^*)$. This is, however, not peculiar to this example. Indeed, according to [278], this is always the case.

Answer 20.2.19 Let T be an unbounded closed operator such that $T = T^*$, and let $B \in B(H)$ be such that $B[D(T)] \not\subset D(T)$. Define now

$$A = \begin{pmatrix} B^* & T \\ 0 & 0 \end{pmatrix}$$

with a domain $D(A) = H \oplus D(T)$. Since A is the sum of an everywhere defined bounded operator and a closed one, A itself is closed. Besides, $A^* = \begin{pmatrix} B & 0 \\ T & 0 \end{pmatrix}$. If $(x, y) \in H \oplus D(T)$, then clearly $B^*x + Ty \in H$. Therefore $\mathrm{ran}\, A \subset D(A)$. Since $B[D(T)] \not\subset D(T)$, readers may easily check that $\mathrm{ran}\, A^* \not\subset D(A^*)$, as needed.

Remark Another counterexample based on a example like the one in Answer 20.2.18 may be found in [279].

Answer 20.2.20 Consider the *closed* operator A defined by:

$$Af(x) = xf'(x) \text{ with domain } D(A) = \{f \in L^2(\mathbb{R}) : xf' \in L^2(\mathbb{R})\}$$

where the derivative is a distributional one. Then

$$A^*f(x) = -xf'(x) - f(x)$$

with $D(A^*) = \{f \in L^2(\mathbb{R}) : xf' \in L^2(\mathbb{R})\}$. Then $D(A) = D(A^*)$ and

$$(A + A^*)f(x) = -f(x)$$

on $D(A) \cap D(A^*) = D(A)$. Thus, $A + A^*$ is densely defined and unclosed.

Answer 20.2.21 ([326]) Let A be defined by:

$$A(x_0, x_1, \cdots) = (x_1, \sqrt{2}x_2, \sqrt{3}x_3, \cdots)$$

on $D(A) = \{(x_n) \in \ell^2 : (\sqrt{n}x_n) \in \ell^2\}$. Then A is closed and besides

$$A^*(x_0, x_1, \cdots) = (0, x_0, \sqrt{2}x_1, \sqrt{3}x_2, \cdots)$$

on $D(A^*) = \{(x_n) \in \ell^2 : (\sqrt{n}x_n) \in \ell^2\}$. It is straightforward to check that $D(AA^*) = D(A^*A)$. Finally, for all $(x_n) \in D(AA^*) = D(A^*A)$

$$AA^*(x_n) = (x_0, 2x_1, 3x_2, \cdots) \text{ and } A^*A(x_n) = (0, x_1, 2x_2, \cdots)$$

whereby

$$AA^*(x_n) = (x_0, x_1, x_2, \cdots) + (0, x_1, 2x_2, \cdots) = (x_n) + A^*A(x_n),$$

that is, $AA^* = I + A^*A$, as needed.

Answer 20.2.22 Let A be a densely defined closed and *unbounded* operator with domain $D(A)$ such that $D(A) = D(A^*) \subset H$. Define T on $H \oplus H$ by:

$$T = \begin{pmatrix} A & I \\ 0 & 0 \end{pmatrix}$$

with domain $D(T) = D(A) \oplus H$. It is plain that T is closed. Also,

$$T^* = \left[\begin{pmatrix} A & 0 \\ 0 & 0 \end{pmatrix} + \begin{pmatrix} 0 & I \\ 0 & 0 \end{pmatrix} \right]^* = \begin{pmatrix} A^* & 0 \\ 0 & 0 \end{pmatrix} + \begin{pmatrix} 0 & 0 \\ I & 0 \end{pmatrix} = \begin{pmatrix} A^* & 0 \\ I & 0 \end{pmatrix}.$$

Since $D(A) = D(A^*)$, it results that $D(T) = D(T^*)$. In addition

$$D(TT^*) = \{(x, y) \in D(A) \times H : (A^*x, x) \in D(A) \times H\} = D(AA^*) \times H$$

and

$$D(T^*T) = \{(x, y) \in D(A) \times H : (Ax + y, 0) \in D(A^*) \times H\}.$$

To see explicitly why $D(TT^*) \neq D(T^*T)$, let α be in H such that $\alpha \notin D(A^*)$. If $x_0 \in D(AA^*) \subset D(A^*) = D(A)$, then $-Ax_0 \in H$. Set $y_0 = -Ax_0 + \alpha$. Then $(x_0, y_0) \in D(AA^*) \times H = D(TT^*)$. Nonetheless, $(x_0, y_0) \notin D(T^*T)$ for

$$Ax_0 + y_0 = Ax_0 - Ax_0 + \alpha = \alpha \notin D(A^*)$$

and this marks the end of the proof.

Answer 20.2.23 ([58]) Let H be a Hilbert space, and let M and N be closed subspaces of H. Consider an isometry V from M onto N. Assume that V is such that $(V - I)M = D$ is a dense subspace of H. Hence $V - I$ is injective and

$$A = i(V + I)(V - I)^{-1}$$

does define a densely defined, closed, and symmetric operator with $D(A) = D$. Then $D(A^2) = \{0\}$ when the ranges of $V + I$ and $V - I$ have a trivial intersection. It suffices therefore to have $M \cap N = \{0\}$. Let us find explicit M, N, and V satisfying these conditions.

Set $H := L^2(S)$, where S is the unit circle. Also, set $M = H^2(S)$ (the Hardy space this time). For V, we take the multiplication operator by a function $\varphi(\theta)$ of modulus one. Finally, let $N = VM = \varphi H^2$. We need to only find an appropriate choice of φ. Recall the following result by Szegö (see e.g. Page 53 in [159]): If f is a nonzero function in H^2, then

$$\int_0^{2\pi} \log |f(\theta)| d\theta > -\infty$$

(and so f cannot vanish on a set of positive measure). Define

$$\varphi(\theta) = \begin{cases} e^{ie^{-\theta-1}}, & 0 < \theta < \pi, \\ -1, & \pi \le \theta \le 2\pi. \end{cases}$$

The noticeable properties of φ are:

(1) $\varphi(\theta) = -1$ on a set of positive measure.
(2) $\varphi(\theta) \ne 1$ but

$$\int_0^{2\pi} \log |f(\theta) - 1| d\theta = -\infty.$$

(3) The imaginary part of φ is positive.

In view of the first property, we have $M \cap N = \{0\}$. Indeed, let $f \in H^2$ be such that also $\varphi f \in H^2$, that is, $\varphi f \in M \cap N$. Hence $(\varphi + 1)f \in H^2$ and it vanishes on a set of positive measure. Thereupon, $(\varphi + 1)f = 0$ and so f is supported on the arc $\pi \le \theta \le 2\pi$. Thus, $f = 0$.

The second property ensures the density of $(V - I)M = (\varphi - 1)H^2$. To see that, let $f \in L^2$ be orthogonal to $(\varphi - 1)H^2$. Hence $(\overline{\varphi} - 1)\overline{g} = f \in H^2$. Whence, by the second property

$$\int_0^{2\pi} \log |f(\theta)| d\theta = -\infty$$

and hence by the Szegö's theorem, we have that $f = 0$ and thereby $g = 0$.

Consequently and with the above choices, $A = i(V + I)(V - I)^{-1}$ gives the desired counterexample. Finally, A is in fact the multiplication operator by $\psi = i(\varphi + 1)(\varphi - 1)^{-1}$ with $D(A) = (\varphi - 1)H^2$. Besides, ψ is positive.

Et voilà, there you have it! A is densely defined, closed, symmetric, and positive yet $D(A^2) = \{0\}$, as wished.

Chapter 21
Self-Adjointness

21.1 Basics

Definition 21.1.1 Let A be a densely defined operator.

(1) Say that A is self-adjoint if $A = A^*$, that is, if A is symmetric and $D(A) = D(A^*)$.
(2) Call A skew-adjoint when $A^* = -A$.
(3) A is called essentially self-adjoint if A is symmetric and its closure $\overline{A}$ is self-adjoint.

Remarks

(1) A self-adjoint operator is necessarily closed. Hence there cannot be any unbounded self-adjoint operator which is everywhere defined.
(2) From the above definition, every self-adjoint operator is symmetric, but a (densely defined) symmetric operator A is not necessarily self-adjoint even when A is closed.

The following theorem gives alternative ways to see when a symmetric operator is self-adjoint.

Theorem 21.1.1 *Let A be a (densely defined) symmetric operator on a Hilbert space H, and let "i" be the usual complex number. Then the following three statements are equivalent:*

(1) *A is self-adjoint.*
(2) *A is closed and $\ker(A^* \pm iI) = \{0\}$.*
(3) $\operatorname{ran}(A \pm iI) = H$.

Corollary 21.1.2 *Let A be a (densely defined) symmetric operator on a Hilbert space H, and let "i" be the usual complex number. Then the following three statements are equivalent:*

© The Author(s), under exclusive license to Springer Nature Switzerland AG 2022
M. H. Mortad, *Counterexamples in Operator Theory*,
https://doi.org/10.1007/978-3-030-97814-3_21

(1) *A is essentially self-adjoint.*
(2) $\ker(A^* \pm iI) = \{0\}$.
(3) $\operatorname{ran}(A \pm iI)$ *are dense.*

Remark There is nothing special about the choice of the complex "i," any complex λ will do! In fact, it was shown in Proposition 3.11 in [326] that if T is a symmetric operator on H such that $T - \lambda I$ is onto and $\operatorname{ran}(T - \bar{\lambda}I)$ is dense in H, then T is self-adjoint.

Let A be an unbounded self-adjoint operator. Is it easy to show that A^2 is self-adjoint? An elementary proof (without using any heavy machinery) is given below. Other proofs, albeit a lot simpler, use more advanced results (see, e.g., Corollary 21.1.5 or Example 29.1.2).

Example 21.1.1 ([11]) Let A be an unbounded self-adjoint operator with domain $D(A) \subsetneq H$. Then A^2 is self-adjoint.

First, observe by Theorem 21.1.1 that both $A + iI$ and $A - iI$ are bijective. Proposition 19.1.5 then implies that $(A \pm iI)^{-1}$ are bounded from H onto $D(A)$.

Now, we show that $D(A^2)$ is dense. To that end, let $y \in H$ and assume that

$$\langle x, y \rangle = 0, \ \forall x \in D(A^2).$$

Hence

$$\langle (A + iI)^{-2}x, y \rangle = 0, \ \forall x \in H$$

and so

$$0 = \langle (A + iI)^{-1}x, [(A + iI)^{-1}]^* y \rangle = \langle (A + iI)^{-1}x, (A - iI)^{-1}y \rangle,$$

still for all $x \in H$. Since $\{(A + iI)^{-1}x : x \in H\} = D(A)$ and $D(A)$ is dense, it follows that

$$(A - iI)^{-1}y = 0,$$

which gives $y = 0$, which makes $D(A^2)$ dense, as needed.

Now, it is clear that

$$A^2 = (A^*)^2 \subset (A^2)^*$$

and so we need only check that $D[(A^2)^*] \subset D(A^2)$. So, take any $y \in D[(A^2)^*]$. Then

$$\exists z \in H : \langle A^2 x, y \rangle = \langle x, z \rangle, \ \forall x \in D(A^2).$$

Hence

$$\langle (A^2 + I)x, y \rangle = \langle x, y + z \rangle, \ \forall x \in D(A^2).$$

If $x \in D(A)$, then $(A + iI)^{-1}x \in D(A^2)$ whereby

$$\langle (A^2 + I)(A + iI)^{-1}x, y \rangle = \langle (A + iI)^{-1}x, y + z \rangle,$$

i.e.

$$\langle (A - iI)x, y \rangle = \langle (A + iI)^{-1}x, y + z \rangle = \langle x, (A - iI)^{-1}(y + z) \rangle.$$

Therefore,

$$\langle Ax, y \rangle = \langle x, (A - iI)^{-1}(y + z) \rangle + i\langle x, y \rangle = \langle x, (A - iI)^{-1}(y + z) - iy \rangle.$$

Thus, $y \in D(A^*) = D(A)$ and

$$Ay = A^*y = (A - iI)^{-1}(y + z) - iy \in D(A).$$

Accordingly, $y \in D(A^2)$, as required.

As regards matrices of unbounded operators, we have the following result which is in fact a consequence of Theorem 20.1.7:

Theorem 21.1.3 (See Proposition 2.6.3 in [364]) *Let A, B, C, and D be densely defined linear operators with domains $\mathcal{D}(A)$, $\mathcal{D}(B)$, $\mathcal{D}(C)$, and $\mathcal{D}(D)$, respectively. Set*

$$T = \begin{pmatrix} A & 0 \\ 0 & D \end{pmatrix} \text{ and } S = \begin{pmatrix} 0 & B \\ C & 0 \end{pmatrix}.$$

Then,

(1) *T is self-adjoint if and only if $A = A^*$ and $D = D^*$.*
(2) *S is self-adjoint if and only if B is closed and $C = B^*$.*

One of the remarkable results of the legendary J. von Neumann is the following:

Theorem 21.1.4 *Let A be a densely defined closed operator. Then A^*A is self-adjoint (and positive) and $D(A^*A)$ is a core for A.*

Corollary 21.1.5 *If A is an unbounded self-adjoint operator, then A^2 too is self-adjoint.*

Remark Obviously, if A is a densely defined closed operator, then AA^* too is self-adjoint and $D(AA^*)$ is a core for A.

Remark More generally, it was shown in Lemma 3.3 in [365] that if A is a densely defined closed operator, then for $k \in \mathbb{N}$, $D[(A^*A)^k]$ is a core for A.

As in the case of bounded self-adjoint (or normal) operators, we have different forms of the spectral theorem for unbounded self-adjoint operators.

Theorem 21.1.6 (Spectral Theorem for Unbounded Self-Adjoint Operators: Integral Form) *Let A be a self-adjoint with domain $D(A) \subset H$. Then there exists a unique spectral measure E on the Borel subsets of $\mathbb{R}$ such that*

$$A = \int_{\sigma(A)} \lambda dE,$$

also written as (for all $x \in D(A)$ and all $y \in H$)

$$\langle Ax, y \rangle = \int_{\sigma(A)} \lambda d\mu_{x,y},$$

where $\mu_{x,y}$, defined by

$$\mu_{x,y}(\Delta) = \langle E(\Delta)x, y \rangle,$$

$x, y \in H$, is a countably additive measure. If f is measurable and finite almost everywhere, then

$$f(A) = \int_{\sigma(A)} f(\lambda) dE.$$

Remark The readers may wish to consult Chapters 4 & 5 in [326] for an exhaustive treatment of the spectral theorem as well as the functional calculus of unbounded self-adjoint operators (among others).

Another form of the spectral theorem is as follows:

Theorem 21.1.7 (Spectral Theorem for Unbounded Self-Adjoint Operators: Multiplication Operator Form, See [302]) *Let H be a separable Hilbert space. Let A be self-adjoint with domain $D(A) \subset H$. Then there exist a σ-finite measure space (X, Σ, μ) and a real-valued function φ in $L^\infty(X, \Sigma, \mu)$ (finite a.e.) such that A is unitarily equivalent to M_φ (the multiplication operator by φ) on $L^2(X, \Sigma, \mu)$.*

As a consequence of the spectral theorem, we have the following theorem:

Theorem 21.1.8 *Let A be self-adjoint with domain $D(A)$. If $A \geq 0$, then there exists a unique self-adjoint $B \geq 0$ such that $B^2 = A$ (B is called the positive square root of A and it is denoted by $\sqrt{A}$).*

As in the bounded case, we have the following proposition:

Proposition 21.1.9 (See, e.g., [337] for a New Proof, and See Also [22]) *If A is an unbounded self-adjoint positive operator and $T \in B(H)$ are such that $TA \subset AT$, then $T\sqrt{A} \subset \sqrt{A}T$, where $\sqrt{A}$ designates the unique self-adjoint positive square root of A.*

Since A^*A is self-adjoint and positive, we may introduce the following notion:

Definition 21.1.2 Let A be a densely defined closed operator. The unique positive square root of A^*A is called the absolute value of A and we write

$$|A| = (A^*A)^{\frac{1}{2}}.$$

Proposition 21.1.10 *Let A be a densely defined closed operator. Then*

$$D(A) = D(|A|) \text{ and } \|Ax\| = \||A|x\|, \ \forall x \in D(A).$$

Let us state now a few results related to relative boundedness (see, e.g., [303] or [326]).

Theorem 21.1.11 (Kato–Rellich) *Let A be a self-adjoint operator on $D(A)$. If B is symmetric and A-bounded with relative bound $a < 1$, then $A + B$ is self-adjoint on $D(A)$.*

The following symmetrized version of the pervious result is sometimes useful.

Theorem 21.1.12 (See, e.g., Theorem 4.5 in [190]) *Let A and B be two symmetric operators such that $D(A) = D(B) = D$ and*

$$\|(A - B)x\| \leq a(\|Ax\| + \|Bx\|) + b\|x\|, \ \forall x \in D,$$

where a and b are positive constants with $a < 1$. Then A is essentially self-adjoint iff B is (in such case, $D(\overline{A}) = D(\overline{B})$). In particular, A is self-adjoint iff B is self-adjoint.

The following weaker version of the Kato–Rellich theorem is also of some interest.

Theorem 21.1.13 (Wüst, See, e.g., [302]) *Let A be a self-adjoint operator on $D(A)$. If B is symmetric and A-bounded with relative bound "$a = 1$," then $A + B$ is essentially self-adjoint on $D(A)$ (or any core of A).*

A generalization of the Kato–Rellich theorem is given next [156].

Theorem 21.1.14 (Hess–Kato) *Let A and B be two densely defined operators with domains $D(A)$ and $D(B)$, respectively. Assume that A is closed. If B is A-bounded*

*and B^* is A^*-bounded with both relative bounds smaller than 1, then $A + B$ is
closed and*

$$(A + B)^* = A^* + B^*.$$

As usual, we finish the section with several examples.

Example 21.1.2 Let $Tf = if'$ be a linear operator defined on the domain $D(T) =$
$\{f \in AC[0, 1] : f' \in L^2[0, 1], \ f(0) = f(1)\}$. Then, from Example 20.1.6, we
already know that T is symmetric and that $D(T) = D(T^*)$. In other words, T is
self-adjoint.

Example 21.1.3 The operator T which appeared in Example 20.1.5 is closed and
symmetric, and however, it cannot be self-adjoint for $D(T) \neq D(T^*)$, i.e., T and
T^* are two different operators.

Example 21.1.4 Let A be an unbounded self-adjoint operator with domain $D(A)$.
Then $T := iA$ is skew-adjoint.

Example 21.1.5 The operator defined by $Af(x) = xf(x)$ is self-adjoint on
$D(A) = \{f \in L^2(\mathbb{R}) : xf \in L^2(\mathbb{R})\}$.

First, recall that A is densely defined. Then, for any $f, g \in D(A)$,

$$\langle Af, g \rangle = \int_{\mathbb{R}} xf(x)\overline{g(x)}dx = \int_{\mathbb{R}} f(x)\overline{xg(x)}dx = \langle f, Ag \rangle.$$

Hence $g \in D(A^*)$ and $Ag = A^*g$, which just means that $A \subset A^*$. Let us show
the reverse inclusion. So, let $g \in D(A^*)$ and set $h_n = A(g\mathbb{1}_{[-n,n]})$, where $n \in \mathbb{N}$.
Clearly, $h_n \in D(A)$, and for all n we have

$$\|h_n\|^2 = \int_{-n}^{n} |xg(x)|^2 dx = \langle Ah_n, g \rangle = \langle h_n, A^*g \rangle \leq \|h_n\| \|A^*g\|$$

as seen by applying the Cauchy–Schwarz inequality. Hence

$$\sup_{n \in \mathbb{N}} \|h_n\| \leq \|A^*g\|.$$

Consequently, $g \in D(A)$, that is, $D(A^*) \subset D(A)$. Since we have already shown
that $A \subset A^*$, it follows that $A = A^*$, i.e., A is self-adjoint.

Meticulous readers have already noticed that Example 21.1.5 follows straightfor-
wardly from Example 20.1.12 (we gave a somehow different proof). Hence, a more
general version of the previous example reads:

Example 21.1.6 Let M_φ be the multiplication operator which appeared in Example 20.1.12. Then M_φ is self-adjoint on $D(M_\varphi)$ if and only if φ is real-valued.

Example 21.1.7 Consider again the operator A of Example 19.1.10. By Example 20.1.13, $C_0^\infty(\mathbb{R})$ is a core for M_x (where M_x is the self-adjoint operator with $D(M_x) = \{f \in L^2(\mathbb{R}) : xf \in L^2(\mathbb{R})\}$). Hence $\overline{A} = M_x$, thereby

$$A^* = (\overline{A})^* = (M_x)^* = M_x.$$

Example 21.1.8 Let $H^1(\mathbb{R}) = \{f \in L^2(\mathbb{R}) : f' \in L^2(\mathbb{R})\}$. Consider the operator $A : H^1(\mathbb{R}) \to L^2(\mathbb{R})$ defined by $Af(x) = if'(x)$. Then A is self-adjoint (hence closed).

One way of seeing that is to use the Fourier transform. Indeed, if $\mathcal{F}$ denotes the $L^2(\mathbb{R})$-Fourier transform, then standard arguments enable us to see that

$$\mathcal{F}^{-1}A\mathcal{F} = M_x \text{ or } A = \mathcal{F}M_x\mathcal{F}^{-1},$$

where M_x denotes the multiplication operator on the maximal domain $D(M_x) = \{f \in L^2(\mathbb{R}) : xf \in L^2(\mathbb{R})\}$. Since M_x is self-adjoint on the previous domain and $\mathcal{F}$ is unitary, it follows that

$$A^* = (\mathcal{F}M_x\mathcal{F}^{-1})^* = (M_x\mathcal{F}^{-1})^*\mathcal{F}^* = \mathcal{F}^{**}M_x\mathcal{F}^{-1} = \mathcal{F}M_x\mathcal{F}^{-1} = A,$$

i.e., A is self-adjoint.

Remark We may also prove the closedness of A by exploiting the closedness of M_x. Indeed, since M_x is closed, $M_x\mathcal{F}^{-1}$ too is closed as $\mathcal{F}^{-1} \in B[L^2(\mathbb{R})]$. Since $\mathcal{F}$ is invertible, $\mathcal{F}M_x\mathcal{F}^{-1}$ (or A) is closed.

Example 21.1.9 As in Example 21.1.8, it may be shown that if A is an unbounded self-adjoint operator and U is unitary, then U^*AU is self-adjoint.

Example 21.1.10 Consider A and B from Example 19.1.15. Then

(1) It may be shown that $\overline{A} = B$. Hence

$$A^* = (\overline{A})^* = B^* = B.$$

This also tells us that A is essentially self-adjoint.
(2) In fact, B is even self-adjoint on $H^2(\mathbb{R})$. This may be seen by using the $L^2(\mathbb{R})$-Fourier transform to obtain (as in Example 21.1.8)

$$\mathcal{F}^{-1}B\mathcal{F} = M_{x^2},$$

where M_{x^2} is self-adjoint.

Example 21.1.11 Define $Af(x) = if'(x)$ on $D(A) = C_0^\infty[0, \infty)$. Then A is densely defined and symmetric. However, A is not essentially self-adjoint. It can be shown that $\overline{A}f(x) = if'(x)$ on $D(\overline{A}) = H^1(0, \infty)$. Hence $\overline{A}$ is clearly not self-adjoint. Alternatively, it suffices to show that $A^* + iI$ is not one-to-one. Observe first that clearly $e^{-x} \in D(A^*)$ and $A^* e^{-x} = -i e^{-x}$. Hence $e^{-x} \in \ker(A^* + iI)$, which is just enough to establish the non-essential self-adjointness of A.

21.2 Questions

21.2.1 Self-Adjoint Operators A with $D(A) \not\subset D(A^2)$

Question 21.2.1 Find an unbounded self-adjoint operator A such that $D(A) \not\subset D(A^2)$.

21.2.2 Symmetric (Closed) Operators Are Not Maximally Self-Adjoint

The following maximality result is easily shown.

Proposition 21.2.1 (Cf. [229]) *Let A and B be two linear operators with domains $D(A)$ and $D(B)$, respectively. If A is densely defined and $D(A^*) \subset D(A)$, then*

$$A \subset B \Longrightarrow A = B$$

whenever $D(B) \subset D(B^*)$.

Corollary 21.2.2 *Self-adjoint operators are maximally symmetric, that is, if $A \subset B$ and A is self-adjoint and B is symmetric, then $A = B$.*

Question 21.2.2 Show that densely defined (closed) symmetric operators are not maximally self-adjoint.

21.2.3 $A^*A \subset AA^* \not\Longrightarrow A^*A = AA^*$

It is known by the maximality of self-adjoint operators that if A is *closed* and densely defined, then

$$A^*A \subset AA^* \Longrightarrow A^*A = AA^*$$

because both AA^* and A^*A are self-adjoint.

Question 21.2.3 Find a densely defined *unclosed* operator A such that $A^*A \subset AA^*$ but $A^*A \neq AA^*$.

21.2.4 Two Self-Adjoint Operators (T, D) and (T, D') Such That T Is Not Self-Adjoint on $D \cap D'$

Question 21.2.4 (Cf. Example 19.1.8) Let (T, D) and (T, D') be self-adjoint operators. Does it follow that T is self-adjoint on $D \cap D'$ (assuming that $D \cap D'$ is dense)?

21.2.5 A Dense Domain $D(A) \subset L^2(\mathbb{R})$ for $Af(x) = xf(x)$ on which A Is Closed and Symmetric but Non-self-adjoint

Consider $Af(x) = xf(x)$ defined formally on $L^2(\mathbb{R})$. We already know that A is self-adjoint on $D(A) = \{f \in L^2(\mathbb{R}) : xf \in L^2(\mathbb{R})\}$. We also know that A is not closed on $C_0^\infty(\mathbb{R})$, while it remains symmetric (which is plain). At the same time, observe that $C_0^\infty(\mathbb{R})$ is too small (w.r.t. $\subset$) compared to $D(A)$. The next question is therefore interesting.

Question 21.2.5 (⊛) Can you find another dense domain $D' \subsetneq D(A)$ on which the symbol A is closed and symmetric without being self-adjoint?

21.2.6 A Dense Domain $D(B) \subset L^2(\mathbb{R})$ for $Bf(x) = if'(x)$ on which B Is Only Closed and Symmetric

Question 21.2.6 We know that $Bf(x) = if'(x)$ is self-adjoint on $H^1(\mathbb{R})$. Give a dense subspace $D \subset L^2(\mathbb{R})$ on which id/dx is only closed and symmetric.

21.2.7 A Closed Densely Defined Non-self-Adjoint A Such That A^2 Is Self-Adjoint

Question 21.2.7 Find an *unbounded*, closed, densely defined, and non-self-adjoint A such that A^2 is self-adjoint.

21.2.8 A Bounded Densely Defined Essentially Self-Adjoint Operator A Such That $D(A^2) = \{0\}$

Question 21.2.8 Provide a (bounded) densely defined essentially self-adjoint operator A such that $D(A^2) = \{0\}$.

21.2.9 An Unbounded Densely Defined Essentially Self-Adjoint Operator A Such That $D(A^2) = \{0\}$

Question 21.2.9 There exists a densely defined and unbounded essentially self-adjoint operator A such that

$$D(A^2) = \{0\}.$$

21.2.10 A Densely Defined Closed Symmetric Operator Which Has Many Self-Adjoint Extensions

A closed symmetric operator may have an infinitude of self-adjoint extensions. For example, in some Hilbert space H, let A be such that $D(A) = \{0\}$. Then A is trivially closed and symmetric. Clearly, each $B_\alpha = \alpha I$ where $\alpha \in \mathbb{R}$ ($I \in B(H)$ being the usual identity operator) constitutes a self-adjoint extension of A. The question is more interesting if the closed symmetric operator is also densely defined. This will be treated next.

Question 21.2.10 Find a densely defined closed symmetric operator which has many self-adjoint extensions.

21.2.11   A Closed and Symmetric Operator Without Self-Adjoint Extensions

Question 21.2.11   Find a closed and symmetric operator without self-adjoint extensions.

21.2.12   An Unbounded Operator A with Domain $D(A)$ Such That $\langle Ax, x \rangle \geq 0$ for All $x \in D(A)$, yet A Is Not Self-Adjoint

We already know that if $A \in B(H)$ is positive, then it is self-adjoint. Can this be carried over to unbounded positive operators?

Question 21.2.12   Find an unbounded (symmetric) densely defined operator A with domain $D(A)$ such that

$$\langle Ax, x \rangle \geq 0$$

for all $x \in D(A)$, yet A is not self-adjoint. Does adding the closedness of A help?

21.2.13   A Densely Defined Operator A Such That Neither A^*A nor AA^* Is Self-Adjoint

Question 21.2.13   Give an example of a non-closed and densely defined operator A such that neither A^*A nor AA^* is self-adjoint.

21.2.14   A Non-closed A Such That A^*A Is Self-Adjoint

We already know that if A is densely defined and closed, then both AA^* and A^*A are self-adjoint. It seems noteworthy to tell readers that quite recently in [335], it was shown that if both AA^* and A^*A are self-adjoint, then A must be closed (another simpler proof may be consulted in [131]). See [87] for another generalization. Some related examples will be given throughout the present manuscript.

Question 21.2.14 Show that it may well occur that A^*A is self-adjoint without assuming that A is closed (obviously such an A has to have a dense domain).

21.2.15 Two Closed Operators A and B Such That $A + B$ Is Unclosed and Unbounded

We have already given an example of two closed operators A and B such that $A+B$ is unclosed (via the usual trick $A = -B$), and so $A + B$ is bounded.

Question 21.2.15 Find two closed operators A and B such that $A + B$ is unclosed and unbounded.

21.2.16 Two Unbounded Self-Adjoint Operators A and B Such That $D(A) \cap D(B) = \{0\}$

Question 21.2.16 (⊛⊛) Find two unbounded self-adjoint operators A and B such that

$$D(A) \cap D(B) = \{0\}.$$

21.2.17 Two Unbounded Self-Adjoint, Positive, and Boundedly Invertible Operators A and B with $D(A) \cap D(B) = \{0\}$

A celebrated result by J. von Neumann [372] states that for a given unbounded self-adjoint operator A in a *separable* Hilbert space H, there is always a unitary $U \in B(H)$ such that

$$D(A) \cap D(U^*AU) = \{0\}.$$

An explicit application (borrowed from [202]) of this theorem is given here. See [202] for other interesting examples.

Remark Another result of similar spirit was shown by K. Schmüdgen in [322]. It reads that if A and B are symmetric operators on invariant dense domains of a separable Hilbert space such that A and B are both unbounded from above (as defined in [322]), then there exist densely defined restrictions A_0 of A and B_0 of B and a unitary U such that $A_0 = U^* B_0 U$.

Question 21.2.17 Find two unbounded self-adjoint, positive, and boundedly invertible operators A and B such that

$$D(A) \cap D(B) = \{0\}.$$

21.2.18 An Unbounded Self-Adjoint Operator A Defined on $D(A) \subset H$ Such That $D(A) \cap D(U^*AU) \neq \{0\}$ for any unitary $U \in B(H)$

It is shown here that the separability assumption in von Neumann's theorem cannot just be dropped.

Question 21.2.18 (⊛ [362]) Find an unbounded self-adjoint operator A defined on $D(A) \subset H$ such that

$$D(A) \cap D(U^*AU) \neq \{0\}$$

for *any unitary* $U \in B(H)$.

21.2.19 $D(A) \cap D(B) = \{0\}$ Is Equivalent to $D(S) \cap D(T) = \{0\}$ where A and B are Positive Boundedly Invertible Self-Adjoint Operators, and S and T Are Closed

Question 21.2.19 Show that the existence of a pair of two unbounded, positive, boundedly invertible self-adjoint operators A and B such that $D(A) \cap D(B) = \{0\}$ is equivalent to the existence of a pair of two densely defined (unbounded) closed operators S and T such that $D(S) \cap D(T) = \{0\}$.

21.2.20   Three Unbounded Self-Adjoint Operators R, S, and T with $D(R) \cap D(S) \neq \{0\}$, $D(R) \cap D(T) \neq \{0\}$, and $D(S) \cap D(T) \neq \{0\}$, yet $D(R) \cap D(S) \cap D(T) = \{0\}$

It is easy to find self-adjoint operators R, S and T such that $D(R) \cap D(S) \cap D(T) = \{0\}$. It suffices to consider two self-adjoint operators S and T say, such that $D(S) \cap D(T) = \{0\}$ and take R to be an arbitrary self-adjoint operator. The next question therefore becomes interesting.

Question 21.2.20 Find three unbounded self-adjoint operators R, S, and T such that $D(R) \cap D(S) \neq \{0\}$, $D(R) \cap D(T) \neq \{0\}$, and $D(S) \cap D(T) \neq \{0\}$, yet

$$D(R) \cap D(S) \cap D(T) = \{0\}.$$

21.2.21   Self-Adjoint Positive Operators C and B Such That $D(C) \cap D(B) = \{0\}$, yet $D(C^\alpha) \cap D(B^\alpha)$ Is Dense for All $\alpha \in (0, 1)$

Question 21.2.21 (⊛) On some Hilbert space H, there are self-adjoint positive operators C and B (with domains $D(C)$ and $D(B)$, respectively) such that $D(C) \cap D(B) = \{0\}$ yet

$$\forall \alpha \in (0, 1) : \overline{D(C^\alpha) \cap D(B^\alpha)} = H.$$

21.2.22   Invertible Unbounded Self-Adjoint Operators A and B Such That $D(A) \cap D(B) = D(A^{-1}) \cap D(B^{-1}) = \{0\}$

Question 21.2.22 (⊛⊛) There are invertible unbounded self-adjoint operators A and B such that

$$D(A) \cap D(B) = D(A^{-1}) \cap D(B^{-1}) = \{0\},$$

where A^{-1} and B^{-1} are not bounded.

21.2.23 Positive Operators *T* and *S* with Dense Ranges Such That $\mathrm{ran}(T^{\frac{1}{2}}) = \mathrm{ran}(S^{\frac{1}{2}})$ and $\mathrm{ran}(T) \cap \mathrm{ran}(S) = \{0\}$

Question 21.2.23 There exist positive operators *T* and *S* with dense ranges such that

$$\mathrm{ran}(T^{\frac{1}{2}}) = \mathrm{ran}(S^{\frac{1}{2}}) \text{ and } \mathrm{ran}(T) \cap \mathrm{ran}(S) = \{0\}.$$

21.2.24 An Unbounded Self-Adjoint Positive Boundedly Invertible *A* and an Everywhere Defined Bounded Self-Adjoint *B* Such That $D(AB) = \{0\}$ and $D(BA) \neq \{0\}$

Question 21.2.24 (⊛) Provide an unbounded self-adjoint positive boundedly invertible operator *A* and an *everywhere defined bounded* self-adjoint *B* such that

$$D(AB) = \{0\},$$

while $D(BA) = D(A)$ is obviously dense.
Hint: Work on $L^2(\mathbb{R})$ using Answer 21.2.17.

21.2.25 An Unbounded Self-Adjoint Positive *C* and a Positive $S \in B(H)$ Such That $D(CS) = \{0\}$, $D(CS^{\alpha}) \neq \{0\}$, and $D(C^{\alpha} S^{\alpha}) \neq \{0\}$ for Each $0 < \alpha < 1$

Question 21.2.25 On some Hilbert space *H*, give an unbounded self-adjoint positive operator *C* and a positive $S \in B(H)$ such that

$$D(CS) = \{0\}, \ D(CS^{\alpha}) \neq \{0\} \text{ and } D(C^{\alpha} S^{\alpha}) \neq \{0\}$$

for each $0 < \alpha < 1$.

21.2.26   An Unbounded Self-Adjoint Positive Operator A on $L^2(\mathbb{R})$ Such That $D(A) \not\subseteq D(\mathcal{F}^* A^\alpha \mathcal{F})$ for Any $0 < \alpha < 1$ Where $\mathcal{F}$ Is the $L^2(\mathbb{R})$-Fourier Transform

Question 21.2.26 (⊛) Find an unbounded self-adjoint and positive operator A on $L^2(\mathbb{R})$ such that

$$D(A) \not\subseteq D(\mathcal{F}^* A^\alpha \mathcal{F})$$

for all $0 < \alpha < 1$, where $\mathcal{F}$ designates the usual $L^2(\mathbb{R})$-Fourier transform.

21.2.27   Another Densely Defined Linear Operator T Such That $D(T^*) = \{0\}$

Question 21.2.27 Based on the example in Answer 21.2.24, find another densely defined operator T such that

$$D(T^*) = \{0\}.$$

21.2.28   A T with $D(T^2) = D(T^*) = D(TT^*) = D(T^*T) = \{0\}$

Question 21.2.28 Find a densely defined operator T such that

$$D(T^2) = D(T^*) = D(TT^*) = D(T^*T) = \{0\}.$$

21.2.29   An Unclosed Operator T Such That TT^* and T^*T Are Closed, but Neither TT^* Nor T^*T Is Self-Adjoint

Question 21.2.29 (⊛) Find an unclosed and densely defined operator T such that TT^* and T^*T are closed, but neither TT^* nor T^*T is self-adjoint.

21.2.30 Yet Another Densely Defined Operator T Such That D(T*) Is Not Dense

Question 21.2.30 Based upon the example which appeared in Answer 18.2.7, find a densely defined operator T such that $D(T^*)$ is not dense.

21.2.31 On the Operator Equation $A^*A = A^2$

We have already mentioned that the equation $A^*A = A^2$ entails the self-adjointness of $A \in B(H)$ (see the discussion just above Question 2.2.24). What about the case of densely defined unbounded operators? In fact, the self-adjointness of A follows from $A^*A = A^2$ if A is *closed* and densely defined (cf. Question 25.1.7). This appeared in [88].

Question 21.2.31 Find a densely defined operator A satisfying $A^*A = A^2$, yet A is not self-adjoint.

21.2.32 A Closed Operator A with $A^*A \subset A^2$ (or $A^2 \subset A^*A$), yet A Is Not Self-Adjoint

Question 21.2.32 Find a closed operator A such that $A^*A \subset A^2$, yet A is not self-adjoint. The same question with the reverse "inclusion" $A^2 \subset A^*A$.

21.2.33 A Rank One Self-Adjoint Operator B and an Unbounded, Self-Adjoint, Positive, and Invertible A Such That B A Is Not Even Closable

Question 21.2.33 Find a rank one self-adjoint operator B and an unbounded, self-adjoint, positive, and boundedly invertible A such that BA is not even closable.

21.2.34 Two Unbounded Self-Adjoint Operators A and B Such That A + B Is Not Self-Adjoint

A priori, many reasons are preventing $A + B$ from being self-adjoint, when A and B are self-adjoint. First, we gather some positive results:

(1) $A + B$ is self-adjoint if, for example, A and B are self-adjoint and $B \in B(H)$.
(2) Also, $A + B$ is self-adjoint if A and B are two unbounded *positive* self-adjoint operators which have commuting spectral measures (see Chap. 26 for more on this notion). For a proof, see Lemma 4.15.1 in [295]. See also Lemma 4.16.1 in the same reference for the case of an infinite sum.
(3) If A and B are unbounded non-necessarily positive, but they "anti-commute strongly" (a notion not defined yet, see Question 26.2.15), then $A + B$ is self-adjoint. This first appeared in [371].
(4) We may also add the Kato–Rellich theorem which provides a criterion for the self-adjointness of the sum.

Remark It is worth noting that if A and B are two unbounded self-adjoint operators having commuting spectral measures, then $A + B$ is densely defined and $\overline{A + B}$ is self-adjoint. See [253] (and [31]) for similar results.

Now, we give some reasons explaining why the sum of two self-adjoint operators may fail to be self-adjoint.

(a) $A + B$ could be densely defined and unclosed, but $\overline{A + B}$ is bounded and self-adjoint.
(b) $A + B$ could be densely defined and unclosed, but $\overline{A + B}$ is unbounded and self-adjoint.
(c) $A + B$ could be closed but non-densely defined.
(d) $A + B$ could be closed, densely defined, and symmetric but non-self-adjoint.

> **Question 21.2.34** Find two unbounded self-adjoint operators A and B such that $A + B$ is not self-adjoint by treating each of the four cases above.

21.2.35 On the Failure of the Essential Self-Adjointness of A + B for Some Closed and Symmetric A and B

It was shown in [254] that if A and B are two unbounded symmetric operators, and if D is a dense linear manifold contained in $D[(A + B)^2]$, then $A + B$ is essentially self-adjoint on each D which is a core for $(A + B)^*\overline{(A + B)}$.

Question 21.2.35 Show that the previous result is not true anymore if one drops the condition on the core D (even if one adds the closedness of both A and B).

21.2.36 A Densely Defined Closed A Such That $A + A^*$ Is Closed, Densely Defined, Symmetric but Non-self-adjoint

As the reader knows or can readily verify, $A + A^*$ is always self-adjoint whenever $A \in B(H)$. How are things in the unbounded case?

Question 21.2.36 Find a densely defined and closed operator A such that $A + A^*$ is closed and densely defined but non-self-adjoint.

21.2.37 A Densely Defined and Closed but Non-symmetric Operator A Such That $A + A^*$ Is Self-Adjoint

Question 21.2.37 Give an example of a densely defined and closed but non-symmetric operator A such that $A + A^*$ is self-adjoint.

21.2.38 Values of λ for which $A + \lambda|A|$ Is (Not) Self-Adjoint or (Not) Closed, Where A Is Closed and Symmetric

Question 21.2.38 (Cf. Questions 21.2.39 and 32.1.3) Let $\lambda \in \mathbb{R}$. Find a self-adjoint A and a closed symmetric non-self-adjoint B such that $D(A) = D(B)$ and

(1) $A + \lambda B$ is self-adjoint (hence closed) for any $-1 < \lambda < 1$.
(2) $A + \lambda B$ is closed but non-self-adjoint for any $|\lambda| > 1$.

21.2.39   An Unbounded Self-Adjoint Operator A Such That $|A| \pm A$ Are Not Even Closed

Question 21.2.39 Give an example of a self-adjoint (unbounded) operator A such that $|A| \pm A$ are not closed.

21.2.40   Two Unbounded Self-Adjoint and Positive Operators A and B Such That $AB + BA$ Is Not Self-Adjoint

Clearly, if $A, B \in B(H)$ are self-adjoint, then $AB+BA$ is self-adjoint. The situation is not as good for two unbounded self-adjoint operators as we will see in the answer to the following question. One hurdle being the fact that $AB + BA$ is likely to be non-densely defined. Even with the density of $D(AB + BA)$, only

$$AB + BA \subset (BA)^* + (AB)^* \subset (AB + BA)^*$$

holds good, i.e., we are only sure that $AB + BA$ is symmetric!

Question 21.2.40 Find two unbounded self-adjoint operators A and B such that $AB + BA$ is not self-adjoint.

21.2.41   A Closed and Densely Defined Operator T Such That $D(T + T^*) = D(TT^*) \cap D(T^*T) = \{0\}$

We have already seen examples of densely defined operators T such that $D(T^*) = \{0\}$. Hence obviously $D(T) \cap D(T^*) = \{0\}$. Nonetheless, such operators were not even closable. Does closedness help anyway?

Question 21.2.41 There exists a densely defined closed operator T such that

$$D(T + T^*) = D(T) \cap D(T^*) = \{0\},$$

and so

$$D(TT^* - T^*T) = D(|T||T^*| - |T^*||T|) = \{0\}.$$

Further Reading The readers may wish to consult the interesting paper [7] in which the authors show that "everything is possible" for the domain intersection $D(T) \cap D(T^*)$ for a closed and densely defined operator T.

21.2.42   Closed S and T with $D(ST) = D(TS) = D(S + T) = \{0\}$

Question 21.2.42 (Cf. Question 32.2.1) Find densely defined closed operators S and T such that

$$D(ST) = D(TS) = D(S + T) = \{0\}.$$

21.2.43   Two Densely Defined Closed Operators A and B Such That $\overline{D(A^*) \cap D(B)} = H$ and $D(A) \cap D(B^*) = \{0\}$

Question 21.2.43 Find in a Hilbert space H two densely defined and closed operators A and B with domains $D(A)$ and $D(B)$, respectively, such that

$$\overline{D(A^*) \cap D(B)} = H \text{ and } D(A) \cap D(B^*) = \{0\}.$$

21.2.44   Two Unbounded Self-Adjoint Positive Invertible Operators A and B Such That $D(A^{-1}B) = D(BA^{-1}) = \{0\}$

Question 21.2.44 There are unbounded self-adjoint positive invertible operators A and B such that

$$D(A^{-1}B) = D(BA^{-1}) = \{0\}$$

(where A^{-1} and B^{-1} are not bounded).

21.2.45   A Densely Defined Unbounded Closed Operator B Such That B^2 and $|B|B$ Are Bounded, Whereas $B|B|$ Is Unbounded and Closed

Question 21.2.45 Give an example of a densely defined unbounded and closed operator B such that B^2 and $|B|B$ are bounded, whereas $B|B|$ is unbounded and closed.

21.2.46   A Closed Operator T with $D(T^2) = D(T^{*2}) = \{0\}$

Question 21.2.46 (❋❋ ([86])) There is a densely defined, unbounded, closed, and one-to-one operator T on some Hilbert space such that

$$D(T^2) = D(T^{*2}) = \{0\}$$

(T^* is also injective).

Remark A priori readers might think that the required example here is weaker than Chernoff's. That was given in Answer 20.2.23. One apparent reason is that the operator there is also symmetric. In fact, the example here is just of a different caliber. Actually, in Answer 20.2.23, it is impossible to have $D(T^{*2}) = \{0\}$. Indeed, when T is symmetric (and densely defined), i.e., $T \subset T^*$, then $T^*T \subset T^{*2}$. By the closedness of T, it results that T^{*2} must be densely defined because T^*T is self-adjoint and in particular densely defined. Thus, $D(T^{*2}) \neq \{0\}$.

21.2.47   A Densely Defined Closed T with $D(T^2) \neq \{0\}$ and $D(T^{*2}) \neq \{0\}$ but $D(T^3) = D(T^{*3}) = \{0\}$

Question 21.2.47 (❋) Find a densely defined closed operator T such that $D(T^2) \neq \{0\}$ and $D(T^{*2}) \neq \{0\}$ but

$$D(T^3) = D(T^{*3}) = \{0\}.$$

21.2.48 A Densely Defined Closed T Such That $\mathcal{D}(T^3) \neq \{0\}$ and $\mathcal{D}(T^{*3}) \neq \{0\}$ yet $\mathcal{D}(T^4) = \mathcal{D}(T^{*4}) = \{0\}$

Question 21.2.48 (⊛) Find a densely defined closed operator T such that $\mathcal{D}(T^3) \neq \{0\}$ and $\mathcal{D}(T^{*3}) \neq \{0\}$ yet

$$\mathcal{D}(T^4) = \mathcal{D}(T^{*4}) = \{0\}.$$

Remark Obviously, $\mathcal{D}(T^3) \neq \{0\}$ will ensure that $\mathcal{D}(T^2) \neq \{0\}$.

21.2.49 A Densely Defined Closed T Such That $\mathcal{D}(T^5) \neq \{0\}$ and $\mathcal{D}(T^{*5}) \neq \{0\}$ While $\mathcal{D}(T^6) = \mathcal{D}(T^{*6}) = \{0\}$

Question 21.2.49 (⊛) There exists a densely defined and closed operator T such that $\mathcal{D}(T^5) \neq \{0\}$ and $\mathcal{D}(T^{*5}) \neq \{0\}$ while

$$\mathcal{D}(T^6) = \mathcal{D}(T^{*6}) = \{0\}.$$

21.2.50 For Each $n \in \mathbb{N}$, There Is a Closed Operator T with $\mathcal{D}(T^{2^n-1}) \neq \{0\}$ and $\mathcal{D}(T^{*2^n-1}) \neq \{0\}$ but $\mathcal{D}\left(T^{2^n}\right) = \mathcal{D}\left(T^{*^{2^n}}\right) = \{0\}$

Question 21.2.50 (⊛) For each $n \in \mathbb{N}$, there is a densely defined and closed operator T (which is an off-diagonal matrix of operators) such that $\mathcal{D}(T^{2^n-1}) \neq \{0\}$ and $\mathcal{D}(T^{*2^n-1}) \neq \{0\}$ whereas

$$\mathcal{D}\left(T^{2^n}\right) = \mathcal{D}\left(T^{*^{2^n}}\right) = \{0\}.$$

21.2.51   Two Unbounded Self-Adjoint Operators *A* and *B* with $D(A) = D(B)$ While $D(A^2) \neq D(B^2)$

It is shown in ([375], Theorem 9.4) that if A and B are two self-adjoint positive operators with domains $D(A)$ and $D(B)$, respectively, then

$$D(A) = D(B) \Longrightarrow D(\sqrt{A}) = D(\sqrt{B}).$$

What about

$$D(A) = D(B) \Longrightarrow D(A^2) = D(B^2)?$$

Question 21.2.51 Find two unbounded self-adjoint operators A and B with domains $D(A)$ and $D(B)$, respectively, such that

$$D(A) = D(B) \text{ while } D(A^2) \neq D(B^2).$$

21.2.52   Self-Adjoint Operators *A* and *B* with $D(A) = D(B)$, but Neither $A^2 - B^2$ nor $AB + BA$ Is Even Densely Defined

As we have just observed, the condition $D(A) = D(B)$ does not entail $D(A^2) = D(B^2)$, even when A and B are self-adjoint. It is worth noting that the condition $D(A) = D(B)$ does not even need to imply that $D(A^2 - B^2)$ (or $D(AB + BA)$) is dense. Before asking the readers to supply a counterexample, we give a simple lemma:

Lemma 21.2.3 ([263]) *Let A and B be two linear operators such that $D(A) = D(B)$. Then*

$$D(AB + BA) = D(A^2 - B^2) \subset D[(A - B)^2] \text{ (or } D[(A + B)^2]).$$

Proof Write

$$A^2 - B^2 + AB - BA \subset (A + B)(A - B).$$

Since $D(A) = D(B)$, it follows that $D(A^2) = D(BA)$ and that $D(B^2) = D(AB)$. Hence

$$D(AB + BA) = D(A^2 - B^2) \subset D[(A + B)(A - B)].$$

But

$$D[(A + B)(A - B)] = D[(A - B)^2]$$

for $D(A + B) = D(A - B)$. The other inclusion which can be shown analogously is left to interested readers.   □

Question 21.2.52 There are self-adjoint positive unbounded operators A and B such that $D(A) = D(B)$, yet neither $A^2 - B^2$ nor $AB + BA$ is densely defined.

21.2.53 On the Impossibility of the Self-Adjointness of AB and BA Simultaneously When B Is Closable (Unclosed) and A Is Self-Adjoint

We start with an example which appeared in [247]:

Example 21.2.1 Consider the following two operators defined by

$$Af(x) = e^{2x} f(x) \text{ and } Bf(x) = (e^{-x} + 1)f(x)$$

on their respective domains

$$D(A) = \{f \in L^2(\mathbb{R}) : e^{2x} f \in L^2(\mathbb{R})\}$$

and

$$D(B) = \{f \in L^2(\mathbb{R}) : e^{-2x} f, e^{-x} f \in L^2(\mathbb{R})\}.$$

Obviously A is self-adjoint, while B is closable without being closed.

The operator BA defined by $BAf(x) = (e^{2x} + e^x)f(x)$ on

$$D(BA) = \{f \in L^2(\mathbb{R}) : e^{2x} f, e^x f \in L^2(\mathbb{R})\}$$

is easily seen to be self-adjoint. The readers may also check that AB is not self-adjoint.

Inspired by this example, we ask the following question:

Question 21.2.53 Is it possible to find two unbounded linear operators A and B, one of them is closable (without being closed) and the other is self-adjoint, such that both BA and AB are self-adjoint?

21.2.54   The Non-self-adjointness of PAP Where P Is an Orthogonal Projection and A Is Self-Adjoint

If $A, P \in B(H)$, A is self-adjoint and P is an orthogonal projection, then PAP is self-adjoint. In fact, only the self-adjointness of P suffices. Is the hypothesis of the orthogonal projection of P (in the case of an unbounded self-adjoint A) sufficient for PAP to be self-adjoint?

Further Reading The readers may wish to consult [143] and [375] for some related results.

Question 21.2.54 Find an unbounded self-adjoint operator A in a Hilbert space H and an orthogonal projection $P \in B(H)$ such that PAP is not self-adjoint.

21.2.55   The Unclosedness of PAP Where P Is an Orthogonal Projection and A Is Self-Adjoint

We have just seen PAP need not be self-adjoint when P is an orthogonal projection and A is an unbounded self-adjoint operator. The example above is rich and it could have other uses. Here, we give an example where PAP is not even closed.

Question 21.2.55 Find an unbounded self-adjoint operator A in a Hilbert space H and an orthogonal projection $P \in B(H)$ such that PAP is not even closed.

21.2.56 Two Self-Adjoint Operators A and B Such That A Is Positive, $B \in B(H)$, and $D\left(A^{1/2}BA^{1/2}\right) = \{0\}$

At the very end of the paper [383], J. P. Williams talked about a simple proof (communicated to him by G. Lumer) of the fact that if $A, B \in B(H)$ are self-adjoint such that $A \geq 0$, then $A^{1/2}BA^{1/2}$ is self-adjoint (respectively, positive) if B is self-adjoint (respectively, positive). The idea of the proof is quite standard for us nowadays. In fact, some of these results were shown again in [158] (which is rather a little unfortunate that the referee and the authors of [158] missed Williams' results). The result was even generalized a while after in [84].

W. Stenger [348] (see also [142]) established some results concerning the self-adjointness of BTB where T is an unbounded self-adjoint operator and where B is some orthogonal projection. Then, he spoke at the end of his paper about Williams being wrong in the above result. His counterexample, however, was for the unbounded case (i.e., Lumer's proof was fine). So, this disagreement should have never taken place.

Having said all that, one can easily find two self-adjoint operators A and B such that $A \geq 0$, $B \in B(H)$, and $A^{1/2}BA^{1/2}$ is not self-adjoint. Even better (or worse? it depends how you perceive it), we have

Question 21.2.56 There are two self-adjoint operators A and B such that $A \geq 0$, $B \in B(H)$, and

$$D\left(A^{1/2}BA^{1/2}\right) = \{0\}.$$

Answers

Answer 21.2.1 In fact, any unbounded self-adjoint operator A has this property. Put another way, it always satisfies $D(A) \not\subset D(A^2)$.

Indeed and more generally, it is shown in, e.g., Lemma 2.1 in [333] that *if H and K are two Hilbert spaces and if $A : D(A) \subset H \to K$ is a densely defined closed operator, then*

$$D(A) = D(A^*A) \Longleftrightarrow A \in B(H, K).$$

So, since $D(A^2) \subset D(A)$ all the time, *any* unbounded self-adjoint operator A necessarily obeys $D(A) \not\subset D(A^2)$.

Remark The previous result tells us that there are not any unbounded self-adjoint nilpotent operators. Indeed, let A be a nilpotent operator with a dense domain $D(A) \subset H$, i.e., $A^n = 0$ for some $n \in \mathbb{N}$. In particular, $D(A) = D(A^2)$. Thus, if A were self-adjoint, we would then obtain $A \in B(H)$!

Answer 21.2.2 Let S be the identity operator restricted to some dense and non-closed domain in a Hilbert space H. Let I be the identity operator defined on the whole of H. Then S is symmetric, I is self-adjoint, and

$$S \subset I \text{ but } S \neq I.$$

What about closed symmetric operators? Well, they would not do either. There are plenty of examples. For example, let $Tf = if'$ be defined on $D(T) = \{f \in AC[0, 1] : f' \in L^2[0, 1], \ f(0) = f(1) = 0\}$. Then T is densely defined, closed, and symmetric. Letting $Af = if'$ be defined on $D(A) = \{f \in AC[0, 1] : f' \in L^2[0, 1], \ f(0) = f(1)\}$ (hence A is self-adjoint), we observe that $T \subset A$ while $T \neq A$.

Answer 21.2.3 Let $A = 0$ on some dense non-closed domain $D(A) \subset H$. Hence A is not closed. Besides, $A^* = 0$ (everywhere defined on H). First, we check the inclusion of domains. Clearly,

$$D(AA^*) = \{x \in H : 0 \in D(A)\} = H,$$

while

$$D(A^*A) = \{x \in D(A) : Ax = 0 \in H\} = D(A).$$

Therefore, $D(A^*A) \subsetneq D(AA^*)$. Since $A^*Ax = AA^*x \ (= 0)$ for all $x \in D(A)$, we see that $A^*A \subset AA^*$ with $A^*A \neq AA^*$.

Answer 21.2.4 (Cf. Question 23.2.4) The answer is negative. Let $Sf(x) = -f'(x)$ be defined on $D(S) = \{f \in L^2(0, 1) : f' \in L^2(0, 1)\}$. Then S is densely defined and closed. Furthermore, $S^* f(x) = f'(x)$ on the domain

$$D(S^*) = \{f \in L^2(0, 1) : f' \in L^2(0, 1), f(0) = f(1) = 0\}$$

so that

$$SS^* f(x) = S^* Sf(x) = -f''(x)$$

with

$$D(SS^*) = \{f \in L^2(0, 1) : f'' \in L^2(0, 1), \ f(0) = f(1) = 0\}$$

and

$$D(S^*S) = \{f \in L^2(0, 1) : f'' \in L^2(0, 1), \ f'(0) = f'(1) = 0\}.$$

Since S is closed and densely defined, both SS^* and S^*S are self-adjoint on their respective domains.

Now, let T be defined by $Tf(x) = -f''(x)$ on

$$D(T) = \{f \in L^2(0, 1) : f'' \in L^2(0, 1), \ f(0) = f(1) = 0 = f'(0) = f'(1)\}.$$

Then T is not self-adjoint (indeed, e.g., $T \subset SS^*$ would imply that $T = SS^*$!). Clearly,

$$D(T) = D(SS^*) \cap D(S^*S).$$

Finally, $-d^2/dx^2$ is self-adjoint on both $D(SS^*)$ and $D(S^*S)$, but it is not so on $D(T)$.

One may also reason as follows: let T be self-adjoint on both D and D' denoted by T_D and $T_{D'}$, respectively. Assume also that D and D' are not comparable and denote T restricted to $D \cap D'$ by $T_{D \cap D'}$. Then, for example, $T_{D \cap D'} \subset T_D$. So if $T_{D \cap D'}$ were self-adjoint, then we would have $T_{D \cap D'} = T_D$ and so $D \cap D' = D$, which is absurd.

Answer 21.2.5 This is not trivial (see [333] for a related work). This example was proposed to me by Professor K. Schmüdgen.

Consider $Af(x) = xf(x)$ on

$$D' = \left\{f \in L^2(\mathbb{R}) : xf \in L^2(\mathbb{R}), \int_{\mathbb{R}} f(x)dx = 0\right\}.$$

The first observation is that the integral $\int_{\mathbb{R}} f(x)dx$ really makes sense. Indeed, since f and xf are both in $L^2(\mathbb{R})$, it follows that f and xf are in $L^2(\mathbb{R})$. Hence, $(1 + |x|)f \in L^2(\mathbb{R})$. Since $1/(1 + |x|)$ is in $L^2(\mathbb{R})$, it follows by invoking the Cauchy–Schwarz inequality that $f \in L^1(\mathbb{R})$.

Second, we show that D' is a dense subspace of $L^2(\mathbb{R})$. That D' is a linear subspace of $L^2(\mathbb{R})$ is plain. To show that D' is dense in $L^2(\mathbb{R})$, we introduce the linear functional B defined from $D(B)$ into $\mathbb{C}$ by

$$Bf(x) = \int_{\mathbb{R}} f(x)dx,$$

where $D(B) = \{f \in L^2(\mathbb{R}) : xf \in L^2(\mathbb{R})\} \ (= D(A))$. Then plainly $D' = \ker B$. Now, the aim is to show that $\ker B$ is dense in $L^2(\mathbb{R})$. This is carried out in two steps.

(1) B is unbounded on $D(B)$: one way of seeing this is to employ sequential continuity. Let $f(x) = \mathbb{1}_{[0,1]}(x)$. Hence $f \in D(B)$. To show that B is not continuous at f, we introduce the sequence $f_n(x) = f(x) - \frac{1}{n}\mathbb{1}_{[1,n+1]}(x)$. Then f_n is in $D(B)$ for each n. Besides, for all $n \in \mathbb{N}$,

$$Bf_n(x) = \int_{\mathbb{R}} f_n(x)dx = \int_{\mathbb{R}} \left[f(x) - \frac{1}{n}\mathbb{1}_{[1,n+1]}(x) \right] dx$$
$$= \int_0^1 dx - \frac{1}{n} \int_1^{n+1} dx = 0.$$

Hence $|Bf_n - Bf| \not\to 0$ because $\int_{\mathbb{R}} f(x)dx = 1$.

(2) By a well-established result on linear functionals (see, e.g., Exercise 2.3.20 in [256]), $\ker B$ ought to be dense in $D(B)$ (w.r.t. the induced topology), i.e., $\overline{\ker B}^{D(B)} = D(B)$, whereas (see, e.g., [255])

$$\overline{\ker B}^{D(B)} = \overline{\ker B}^{L^2(\mathbb{R})} \cap D(B).$$

Ergo, $D(B) \subset \overline{\ker B}^{L^2(\mathbb{R})}$ ($\subset L^2(\mathbb{R})$). Accordingly, passing to the $L^2(\mathbb{R})$-closure yields

$$\overline{\ker B}^{L^2(\mathbb{R})} = \overline{D(B)}^{L^2(\mathbb{R})} = L^2(\mathbb{R}),$$

as needed. Therefore, A is densely defined on D'.

Now, A is clearly symmetric on D' for A is already symmetric on a "larger" subspace.

Finally, we show that A is closed on D' (another proof of the closedness of A may be found in the remark below Answer 21.2.6). In this respect, let f_n be in D' such that $f_n \to f$ and $xf_n \to g$ (both limits in $L^2(\mathbb{R})$). Since A is closed on $D(A)$, we already have $f \in D(A)$ and $g = xf \in L^2(\mathbb{R})$. So we only have to check

$$\int_{\mathbb{R}} f_n(x)dx = 0, \text{ for } n = 1, 2, \cdots \implies \int_{\mathbb{R}} f(x)dx = 0.$$

We have

$$\int_{\mathbb{R}} |f_n(x) - f(x)|dx = \int_{\mathbb{R}} (1 + |x|)\frac{|f_n(x) - f(x)|}{1 + |x|}dx$$
$$\leq \left(\int_{\mathbb{R}} \frac{dx}{(1 + |x|)^2} \right)^{\frac{1}{2}} \left(\int_{\mathbb{R}} (1 + |x|)|f_n(x) - f(x)|^2 dx \right)^{\frac{1}{2}}$$
$$= \sqrt{2}(\|f_n - f\|_2 + \|xf_n - xf\|_2) \longrightarrow 0$$

as $n \to \infty$. But,

$$\left| \int_{\mathbb{R}} f_n(x)dx - \int_{\mathbb{R}} f(x)dx \right| = \left| \int_{\mathbb{R}} [f_n(x) - f(x)]dx \right| \leq \int_{\mathbb{R}} |f_n(x) - f(x)|dx \longrightarrow 0$$

when $n \to \infty$. As $\int_{\mathbb{R}} f_n(x)dx = 0$, it results that $\int_{\mathbb{R}} f(x)dx = 0$. Thus $f \in D'(A)$ and $xf = g$, proving the closedness of A.

Finally, A is not self-adjoint. One way of seeing that reads that if $Tf(x) = xf(x)$ is defined on $D(T) = \{f \in L^2(\mathbb{R}) : xf \in L^2(\mathbb{R})\}$, then T is self-adjoint. So, $A \subset T$ in terms of extensions. If A were self-adjoint, then by the maximality of self-adjoint operators, we would get $A = T$, and this is absurd for $D(T) \not\subset D'$.

Remark The readers may wonder what $D(A^*)$ is anyway? The idea for finding $D(A^*)$ is somewhat similar to that of Example 20.1.12 with a slight modification. Indeed, thanks to the condition $\int_{\mathbb{R}} f(x)dx = 0$, we may show that

$$D(A^*) = \{f \in L^2(\mathbb{R}) : xf + c \in L^2(\mathbb{R}) \text{ for some constant } c\}.$$

Observe that this provides another proof of the non-self-adjointness of A.

Remark There are other related examples (due to Professor Joel Feldman). Indeed, let $Af(x) = (1 + x^2)f(x)$ be defined on

$$D(A) = \left\{ f \in C_0^\infty(\mathbb{R}) : \int_{\mathbb{R}} f(x)dx = \int_{\mathbb{R}} xf(x)dx = 0 \right\}.$$

Then A is symmetric but unclosed. Its closure is $\overline{A}f(x) = (1 + x^2)f(x)$ on

$$D(\overline{A}) = \left\{ f \in L^2(\mathbb{R}) : x^2f \in L^2(\mathbb{R}) : \int_{\mathbb{R}} f(x)dx = \int_{\mathbb{R}} xf(x)dx = 0 \right\}.$$

Notice that A has other closed extensions, for instance, the operator B defined by $Bf(x) = (1 + x^2)f(x)$ on the domain

$$\left\{ f \in L^2(\mathbb{R}) : x^2f \in L^2(\mathbb{R}) : a \int_{\mathbb{R}} f(x)dx + b \int_{\mathbb{R}} xf(x)dx = 0; a, b \in \mathbb{R} \right\}.$$

Also, the operator C defined by $Cf(x) = (1 + x^2)f(x)$ on the maximal domain $D(C) = \{f \in L^2(\mathbb{R}) : x^2f \in L^2(\mathbb{R})\}$ is another closed (in fact, self-adjoint) extension of A.

Answer 21.2.6 Let $Bf(x) = if'(x)$ be defined on the dense domain $D(B) = \{f \in H^1(\mathbb{R}) : f(0) = 0\}$. Notice that the graph norm of B is merely the natural norm of

$H^1(\mathbb{R})$. Besides, the functional $f \mapsto f(0)$ is continuous on $H^1(\mathbb{R})$ as

$$2|f(0)| \leq \|f'\|_2 + \|f\|_2$$

(coming from a Sobolev inequality). Therefore, B is closed. Now, B is densely defined. One way of seeing this is to observe that $D(B)$ contains $D := \{f \in C_0^\infty(\mathbb{R}) : f(0) = 0\}$. A way of establishing the density of D is to observe that the functional $F : f \mapsto f(0)$ is discontinuous on C_0^∞ w.r.t. $\| \cdot \|_2$. Then, it follows that $\ker F = D$ is necessarily dense in $L^2(\mathbb{R})$, thereby $D(B)$ too is dense in $L^2(\mathbb{R})$.

To find $D(B^*)$, first write B as a direct sum of the two operators $B_1 f(x) = if'(x)$ on $D(B_1) = \{f \in H^1(-\infty, 0) : f(0) = 0\}$ and $B_2 f(x) = if'(x)$ on $D(B_2) = \{f \in H^1(0, \infty) : f(0) = 0\}$. Then, from Example 20.1.11, we know that $B_1^* f(x) = if'(x)$ on $D(B_1^*) = H^1(-\infty, 0)$ and $B_2^* f(x) = if'(x)$ on $D(B_2^*) = H^1(0, \infty)$. Accordingly,

$$D(B^*) = H^1(-\infty, 0) \oplus H^1(0, \infty).$$

Therefore, B is symmetric and closed without being self-adjoint.

A similar argument leads to more examples. Indeed, if we define $Cf(x) = if'(x)$ on $D(C) = \{f \in H^1(\mathbb{R}) : f(0) = f(1) = 0\}$. Then C is closed and symmetric. Moreover,

$$D(C^*) = H^1(-\infty, 0) \oplus H^1(0, 1) \oplus H^1(1, \infty).$$

Remark If we consider the operator A of Answer 21.2.5, then we may show that

$$\mathcal{F}^{-1} B \mathcal{F} = A$$

with $\mathcal{F}$ is the usual $L^2(\mathbb{R})$-Fourier transform (where it is plain that $D(\mathcal{F}^{-1} B \mathcal{F}) = D(B\mathcal{F}) = D(A)$). Therefore, it becomes clear that A is closed if and only if B is closed.

Answer 21.2.7 Inspired by Answer 2.2.13, let T be a self-adjoint operator with domain $D(T) \subset H$, and let I be the identity operator on H. Now, let

$$A = \begin{pmatrix} T & I \\ 0 & -T \end{pmatrix}.$$

Then A is closed for it is the sum of a closed operator and an everywhere defined bounded operator. However, A is not self-adjoint because

$$A^* = \left[\begin{pmatrix} T & 0 \\ 0 & -T \end{pmatrix} + \begin{pmatrix} 0 & I \\ 0 & 0 \end{pmatrix} \right]^* = \begin{pmatrix} T & 0 \\ 0 & -T \end{pmatrix} + \begin{pmatrix} 0 & 0 \\ I & 0 \end{pmatrix} = \begin{pmatrix} T & 0 \\ I & -T \end{pmatrix}.$$

As we may indeed check that $D(A^2) = D(T^2) \oplus D(T^2)$, then

$$A^2 = \begin{pmatrix} T & I \\ 0 & -T \end{pmatrix} \begin{pmatrix} T & I \\ 0 & -T \end{pmatrix} = \begin{pmatrix} T^2 & 0 \\ 0 & T^2 \end{pmatrix}$$

which is clearly self-adjoint (and also positive).

Answer 21.2.8 Let

$$L^2_{even}(\mathbb{R}) = \{f \in L^2(\mathbb{R}) : f(x) = f(-x) \text{ almost everywhere in } \mathbb{R}\}.$$

Then $L^2_{even}(\mathbb{R})$ is closed in $L^2(\mathbb{R})$ and so it is in fact a Hilbert space with respect to the induced $L^2(\mathbb{R})$-inner product.

Let $\mathcal{F}$ be the restriction of the L^2-Fourier transform to even-$C_0^\infty(\mathbb{R})$ (that is, even functions in $C_0^\infty(\mathbb{R})$). Set $A = \mathcal{F} + \mathcal{F}^*$ (A is therefore just the Fourier cosine transform) which is defined on even-$C_0^\infty(\mathbb{R})$ as $\mathcal{F}^*$ is defined on the whole of $L^2_{even}(\mathbb{R})$. Then A is densely defined because even-$C_0^\infty(\mathbb{R})$ is dense in $L^2_{even}(\mathbb{R})$. Besides, A is symmetric for

$$A = \mathcal{F} + \mathcal{F}^* \subset \overline{\mathcal{F}} + \mathcal{F}^* \subset (\mathcal{F} + \mathcal{F}^*)^* = A^*.$$

Moreover, $D(A^*) = L^2_{even}(\mathbb{R})$. Hence $\overline{A} = A^*$, that is, A is essentially self-adjoint. Now,

$$A^2 = (\mathcal{F} + \mathcal{F}^*)(\mathcal{F} + \mathcal{F}^*)$$
$$= \mathcal{F}(\mathcal{F} + \mathcal{F}^*) + \mathcal{F}^*(\mathcal{F} + \mathcal{F}^*)$$
$$= \mathcal{F}(\mathcal{F} + \mathcal{F}^*) + \mathcal{F}^*\mathcal{F} + \mathcal{F}^{*2} \text{ (for } \mathcal{F}^* \text{ is everywhere defined).}$$

But,

$$D[\mathcal{F}(\mathcal{F} + \mathcal{F}^*)] = \{f \in \text{even-}C_0^\infty(\mathbb{R}) : (\mathcal{F} + \mathcal{F}^*)f \in \text{even-}C_0^\infty(\mathbb{R})\} = \{0\}$$

by Theorem A.6. Accordingly,

$$D(A^2) = \{0\}.$$

Answer 21.2.9 Let

$$Af(x) = e^{x^2/4} f(x)$$

be defined on $D(A) = \{f \in L^2(\mathbb{R}) : e^{x^2/4} f, e^{x^2/2} \hat{f} \in L^2(\mathbb{R})\}$. Then A is unbounded and densely defined (one needs to check that $D(A)$ is dense in the set of f such that $e^{x^2/4} f$ is in $L^2(\mathbb{R})$, w.r.t. the graph norm). Also, A is symmetric as

$A \subset A^*$ and

$$D(A^*) = \{f \in L^2(\mathbb{R}) : e^{x^2/4} f \in L^2(\mathbb{R})\}.$$

It is well known that A^* is self-adjoint on $D(A^*)$. That is,

$$\overline{A} = (A^*)^* = A^* = \overline{A}^*$$

meaning that A is essentially self-adjoint (hence A cannot be closed).

Now,

$$A^2 f(x) = e^{x^2/2} f(x)$$

on

$$D(A^2) = \{f \in L^2(\mathbb{R}) : e^{x^2/2} f, e^{x^2/2} \hat{f}, e^{x^2/2} \widehat{e^{x^2/4} f} \in L^2(\mathbb{R})\}.$$

By Theorem A.5, $e^{x^2/2} f, e^{x^2/2} \hat{f} \in L^2(\mathbb{R})$ forces $f = 0$ and so

$$D(A^2) = \{0\},$$

as wished.

Answer 21.2.10 Let $T_\alpha f = i f'$ be defined on

$$D(T_\alpha) = \{f \in AC[0, 1], f' \in L^2[0, 1] : f(1) = \alpha f(0)\},$$

where α is a complex number and such that $|\alpha| = 1$. Then each T_α clearly represents an extension of the operator T which is defined by $Tf = i f'$ on the domain

$$D(T) = \{f \in AC[0, 1] : f' \in L^2[0, 1], \ f(0) = f(1) = 0\}.$$

Since T is closed and symmetric (but non-self-adjoint), to answer our question, we need only show that T_α is self-adjoint (for any given α). Let $\alpha \in \mathbb{C}$ be such that $|\alpha| = 1$. First, we show that T_α is symmetric. Let $f, g \in D(T_\alpha)$. Then

$$\langle T_\alpha f, g \rangle = i \int_0^1 f'(x) \overline{g(x)} dx$$

$$= f(1) \overline{g(1)} - f(0) \overline{g(0)} + \int_0^1 f(x) \overline{i g'(x)} dx$$

$$= \underbrace{\alpha \overline{\alpha}}_{=1} f(0) \overline{g(0)} - f(0) \overline{g(0)} + \int_0^1 f(x) \overline{i g'(x)} dx$$

$$= \langle f, T_\alpha g \rangle,$$

and so each T_α is symmetric, i.e., $T_\alpha \subset (T_\alpha)^*$. Now, let g be in $D[(T_\alpha)^*]$, and let λ be in $\mathbb{C}$ such that $e^\lambda = \alpha$. Set $f(x) = e^{\lambda x}$ and so f is clearly in $D(T_\alpha)$ mainly because $f(1) = e^\lambda = \alpha f(0)$.

As observed above, $T \subset T_\alpha$ and so $(T_\alpha)^* \subset T^*$. This tells us that g is in $AC[0, 1]$, $g' \in L^2[0, 1]$, and

$$(T_\alpha)^* g = i g'.$$

Finally, as $f \in D(T_\alpha)$ and g is in $AC[0, 1]$, $g' \in L^2[0, 1]$, we have

$$\langle T_\alpha f, g \rangle - \langle f, T_\alpha g \rangle = f(1)\overline{g(1)} - f(0)\overline{g(0)} = f(0)[\alpha \overline{g(1)} - \overline{g(0)}].$$

Since $f(0) \neq 0$, we get

$$\alpha \overline{g(1)} - \overline{g(0)} = 0 \Longrightarrow \overline{\alpha} g(1) = g(0) \Longrightarrow g(1) = \alpha g(0)$$

(because $|\alpha| = 1$). Thus, $(T_\alpha)^* \subset T_\alpha$ and so each T_α is self-adjoint, as wished.

Answer 21.2.11 Let A be the operator defined by $Af(x) = if'(x)$ on

$$D(A) = \{f \in L^2(0, \infty), f \in AC[0, c] : \forall c > 0, f(0) = 0 \text{ and } f' \in L^2(0, \infty)\}.$$

Then A is densely defined and closed. Besides, $A^* f(x) = if'(x)$ on

$$D(A^*) = \{f \in L^2(0, \infty), f \in AC[0, c] : \forall c > 0 \text{ and } f' \in L^2(0, \infty)\}.$$

Therefore, A is also symmetric. To show that A does not possess any self-adjoint extension, it suffices to show that the deficiency indices n_+ and n_- are not equal. *Recall that*

$$n_+ = \dim \ker(A^* - i) \text{ and } n_- = \dim \ker(A^* + i).$$

Let $f \in \ker(A^* - i)$, i.e., $f'(x) = f(x)$ where $f \in D(A^*)$. Hence $f(x) = Ke^x$ for some constant K. Clearly, such an f cannot be in $D(A^*)$ unless $K = 0$. Thus, $\ker(A^* - i) = \{0\}$ and so $n_+ = 0$.

Similarly, we may show that if $f \in \ker(A^* + i)$, then $f(x) = Ke^{-x}$ for some constant K. Thus, $\ker(A^* + i)$ is spanned by e^{-x} and so $n_- = 1$. Accordingly, A has no self-adjoint extensions.

Further Reading It is worth noticing that a densely defined symmetric operator T on a Hilbert space H has always a self-adjoint extension A on a Hilbert space $G \supset H$ and $D(\overline{T}) = D(A) \cap H$; see Proposition 3.17 of [326] (cf. Question 26.2.24).

Answer 21.2.12 As a first counterexample, consider

$$Af(x) = |x| f(x) \text{ for all } f \in D(A) := C_0^\infty(\mathbb{R}).$$

Obviously, A is symmetric. Now, for all f in $C_0^\infty(\mathbb{R})$, we clearly have

$$\langle Af, f \rangle = \int_{\mathbb{R}} |x| |f(x)|^2 dx \geq 0,$$

yet A is not self-adjoint for it is not closed.

We also asked whether a closed (densely defined) symmetric positive operator must be self-adjoint. The answer is still negative. For instance, the operator B defined by $Bf(x) = (1+x^2)f(x)$ on the domain

$$\left\{ f \in L^2(\mathbb{R}) : x^2 f \in L^2(\mathbb{R}) : a \int_{\mathbb{R}} f(x)dx + b \int_{\mathbb{R}} xf(x)dx = 0, a, b \in \mathbb{R} \right\}$$

(introduced in the second remark below Answer 21.2.5) is a closed densely defined symmetric positive operator but non-self-adjoint. The non-self-adjointness of B follows from the usual observation: $B \subset T$, where T is the self-adjoint operator defined by $Tf(x) = (1+x^2)f(x)$ on $D(T) = \{f \in L^2(\mathbb{R}) : x^2 f \in L^2(\mathbb{R})\}$, and so the maximality of self-adjoint operators would yield the untrue statement $B = T$!

Remark Observe that B is not boundedly invertible. For if it were, B would then be self-adjoint which is not the case.

Answer 21.2.13 There are plenty of examples. For instance, let I_D be the restriction of the identity operator to a dense (unclosed) domain $D \subset H$. Let I be the identity operator on all of H. Then I_D is unclosed and $(I_D)^* I_D = I I_D = I_D$ is not closed and so it cannot be self-adjoint. The same reasoning applies to show that $I_D(I_D)^*$ is not self-adjoint.

Answer 21.2.14 The first example is inspired by one which appeared in [334] (another example will be given below).

Let $T = \frac{d}{dx}$ be defined on $H^1(\mathbb{R}) = \{f \in L^2(\mathbb{R}) : f' \in L^2(\mathbb{R})\}$. Then T is closed (hence T^*T is self-adjoint). Set $A = T|_{H^2(\mathbb{R})}$ where $H^2(\mathbb{R}) = \{f \in L^2(\mathbb{R}) : f'' \in L^2(\mathbb{R})\}$. Then A is not closed and $T^* = A^*$. Because $D(A^*A) = D(T^*T) = H^2(\mathbb{R})$, it follows that $A^*A = T^*T$. Thus, since T^*T is self-adjoint so is A^*A.

As alluded to, we give another example. Let $A = 0$ on some densely defined domain $D(A) \subset H$. Then $A^* = 0$ everywhere on H. Hence $D(AA^*) = \{x \in H : A^*x = 0 \in D(A)\} = H$ and so $AA^*x = 0$ for all $x \in H$, while $A^*Ax = 0$ for all $x \in D(A^*A) = D(A)$.

In other words, this means that $AA^* = 0$, whereas $A^*A \subset 0$. Expressed differently, AA^* is self-adjoint, while A^*A is not even closed.

Answer 21.2.15 Let T be an unbounded unclosed densely defined operator such that T^*T is self-adjoint (as in Answer 21.2.14). Hence $|T|$ is self-adjoint (in particular, it is closed). Next, set $A = |T| + T/2$ and $B = -|T|$. Then A is closed as $T/2$ is $|T|$-bounded with relative bound $1/2 < 1$ (by Theorem 19.1.8). Besides,

$D(A) = D(B) = D(T)$. In fine,

$$A + B = |T| + T/2 - |T| = T/2$$

is unbounded and unclosed on $D(T)$.

Answer 21.2.16 ([56, 57, 137], and [202]) Consider the self-adjoint operator A defined by $A = i\frac{d}{dx}$ on the domain

$$D(A) = \{f \in AC[0, 1] : f' \in L^2[0, 1] \text{ and } f(0) = f(1)\}.$$

Let $(r_n)_{n \geq 0}$ be a dense sequence of rational numbers in $[0, 1]$ and let $0 < \alpha < 1$. Then

$$\int_0^1 \frac{dx}{|x - r_n|^\alpha} = \int_0^{r_n} \frac{dx}{(r_n - x)^\alpha} + \int_{r_n}^1 \frac{dx}{(x - r_n)^\alpha}$$

$$= \frac{1}{1 - \alpha}[r_n^{1-\alpha} + (1 - r_n)^{1-\alpha}]$$

$$= \frac{2}{1 - \alpha}\left[\frac{r_n^{1-\alpha} + (1 - r_n)^{1-\alpha}}{2}\right]$$

$$\leq \frac{2}{1 - \alpha} \times \left(\frac{1}{2}\right)^{1-\alpha} \quad (\text{as } t \mapsto t^{1-\alpha} \text{ is concave for } t > 0)$$

$$= \frac{2^\alpha}{1 - \alpha}.$$

Also,

$$\int_0^1 \frac{dx}{|x - r_n|} = +\infty.$$

Define a function φ by

$$\varphi(x) = \sum_{n=0}^\infty \frac{1}{n!} \times \frac{1}{|x - r_n|^{1/2}}.$$

Observe that φ is positive. Moreover, $\varphi \in L^1[0, 1]$ for

$$\int_0^1 \varphi(x)dx = \sum_{n=0}^\infty \frac{1}{n!}\int_0^1 \frac{dx}{|x - r_n|^{1/2}} \leq \sum_{n=0}^\infty \frac{2\sqrt{2}}{n!} = 2e\sqrt{2}.$$

Therefore, φ is finite for almost every $x \in [0, 1]$. It is also seen that $\varphi \notin L^2_{loc}[0, 1]$ by the density of (r_n) in $[0, 1]$.

Next, define the self-adjoint operator $Bf(x) = \varphi(x)f(x)$ on the domain $D(B) = \{f \in L^2[0, 1] : \varphi f \in L^2[0, 1]\}$. Observe that $Bf \geq ef$ for all $f \in D(B)$. Indeed, as $|x - r_n| \leq 1$, then for each n and so $|x - r_n|^{-1/2} \geq 1$ which implies that

$$\varphi(x) \geq \sum_{n=0}^{\infty} \frac{1}{n!} = e.$$

Finally, we show that

$$D(A) \cap D(B) = \{0\}.$$

Since $D(A) \subset C[0, 1]$, it suffices to show that $C[0, 1] \cap D(B) = \{0\}$. For the sake of contradiction, assume that there is a nonzero function g in the intersection $C[0, 1] \cap D(B)$. Then by the continuity and the "non-zeroness" of g at a point a say, we know that $|g(x)| \geq \varepsilon > 0$ for some $\varepsilon > 0$ and all x in a neighborhood I of a. Hence

$$|g(x)\varphi(x)| \geq \varepsilon|\varphi(x)|$$

on I. Therefore, the function $x \mapsto g(x)\varphi(x)$ would not be in $L^2_{loc}[0, 1]$. Since the latter is not consistent with g being in $D(B)$, it follows that $C[0, 1] \cap D(B) = \{0\}$ or

$$D(A) \cap D(B) = \{0\}.$$

Remark See [137] for a similar example on $L^2(\mathbb{R})$.

Answer 21.2.17 ([202]) Let A be defined by

$$Af(x) = e^{\frac{x^2}{2}} f(x)$$

on $D(A) = \{f \in L^2(\mathbb{R}) : e^{\frac{x^2}{2}} f \in L^2(\mathbb{R})\}$. Then A is self-adjoint, positive, and boundedly invertible. Set $B := \mathcal{F}^* A \mathcal{F}$, where $\mathcal{F}$ denotes the usual L^2-Fourier transform, and so B is boundedly invertible. Moreover,

$$B^* = \left[\mathcal{F}^* A \mathcal{F}\right]^* = \mathcal{F}^* A^* \mathcal{F}^{**} = \mathcal{F}^* A \mathcal{F} = B.$$

That B is positive is also clear. Finally,

$$D(A) \cap D(B) = D(A) \cap D(\mathcal{F}^* A \mathcal{F}) = \{0\}$$

by Theorem A.5.

Answer 21.2.18 ([362]) Let K be a nonseparable Hilbert space and set $H := K \oplus \ell^2(\mathbb{N})$. Then

$$\dim K > \dim \ell^2(\mathbb{N}).$$

Define the diagonal operator T by

$$T e_n = 2^n e_n$$

on

$$D(T) = \left\{ x = (x_n) \in \ell^2(\mathbb{N}) : \sum_{n=1}^{\infty} 4^n |x_n|^2 < \infty \right\}$$

and where (e_n) is the usual orthonormal basis of $\ell^2(\mathbb{N})$. Then T is unbounded and self-adjoint. Next, define the operator $A = I \oplus T$ on H, where I is the identity operator on K. It is plain that A is an unbounded and self-adjoint operator.

Let $U \in B(H)$ be *any* unitary operator. Let $P \in B(H)$ be the projection onto $\ell^2(\mathbb{N})$. By Lemma 2.1 in [362] and $\dim K > \dim \ell^2(\mathbb{N})$, it follows that PU restricted to K, i.e. $PU|_K \to \ell^2(\mathbb{N})$ is not injective. In other language,

$$\exists x, y \in K, x \neq 0 : \ U(x, 0) = (y, 0).$$

Accordingly,

$$D(A) \cap D(U^* A U) \neq \{0\}.$$

Answer 21.2.19 We have already seen unbounded, positive, boundedly invertible self-adjoint operators A and B with $D(A) \cap D(B) = \{0\}$.

Let us show the other implication. Assume as in Question 19.2.38 that we are given a pair of densely defined closed operators S and T such that $D(S) \cap D(T) = \{0\}$. Once that is known, consider now $|S|$ and $|T|$ (or their squares $S^* S$ and $T^* T$) which are self-adjoint, positive, and they satisfy

$$D(|S|) \cap D(|T|) = D(S) \cap D(T) = \{0\}.$$

If we want boundedly invertible operators as well, set $A = S^* S + I$ and $B = T^* T + I$. Then A and B remain self-adjoint and positive. Besides, A and B are boundedly invertible (as in say Theorem 13.13 in [314]). Finally,

$$D(A) \cap D(B) = D(S^* S) \cap D(T^* T) \subset D(S) \cap D(T) = \{0\},$$

i.e., $D(A) \cap D(B) = \{0\}$, as required.

Remark It is worth noticing that K. Schmüdgen [321] (cf. [369]) obtained that every unbounded self-adjoint T has two closed and symmetric restrictions A and B such that

$$D(A) \cap D(B) = \{0\} \text{ and } D(A^2) = D(B^2) = \{0\}.$$

Following [8], this result will be referred to as the van Daele–Schmüdgen theorem in the present manuscript.

The fascinating result by K. Schmüdgen (later generalized by Brasche–Neidhardt in [34], see also [8]) also dealt with higher powers.

Answer 21.2.20 Let A and B be two self-adjoint operators such that $D(A) \cap D(B) = \{0\}$. Since $D(A) = D(|A|)$, it follows that also $D(|A|) \cap D(B) = \{0\}$. Now set

$$R = \begin{pmatrix} A & 0 & 0 \\ 0 & B & 0 \\ 0 & 0 & A \end{pmatrix}, \ S = \begin{pmatrix} A & 0 & 0 \\ 0 & |A| & 0 \\ 0 & 0 & B \end{pmatrix} \text{ and } T = \begin{pmatrix} B & 0 & 0 \\ 0 & |A| & 0 \\ 0 & 0 & A \end{pmatrix},$$

where $D(R) = D(A) \oplus D(B) \oplus D(A)$, $D(S) = D(A) \oplus D(A) \oplus D(B)$, and $D(T) = D(B) \oplus D(A) \oplus D(A)$. Hence

$$D(R) \cap D(S) = D(A) \oplus \{0\} \oplus \{0\}, \ D(R) \cap D(T) = \{0\} \oplus \{0\} \oplus D(A)$$

and

$$D(S) \cap D(T) = \{0\} \oplus D(A) \oplus \{0\}.$$

In other words, $D(R) \cap D(S) \neq \{0\}$, $D(R) \cap D(T) \neq \{0\}$, and $D(S) \cap D(T) \neq \{0\}$, yet

$$D(R) \cap D(S) \cap D(T) = \{(0, 0, 0)\},$$

as needed.

Answer 21.2.21 ([202]) Let $\alpha \in (0, 1)$, and let A and B be as in Answer 21.2.16, i.e., A and B are self-adjoint and $D(A) \cap D(B) = \{0\}$, where B is also positive.

First, we show that

$$D(A) \cap D(B^\alpha) = D(A).$$

Define

$$g_\alpha(x) = \sum_{n=1}^{\infty} \frac{1}{(n!)^\alpha} \times \frac{1}{|x - r_n|^{\alpha/2}},$$

where (r_n) is a dense rational sequence in $[0, 1]$. Put

$$\beta = \sum_{n=1}^{\infty} \frac{1}{(n!)^{\alpha}} < \infty.$$

We then have

$$[g_{\alpha}(x)]^2 = \beta^2 \left(\sum_{n=1}^{\infty} \frac{1}{\beta(n!)^{\alpha}} \times \frac{1}{|x - r_n|^{\alpha/2}} \right)^2$$

$$\leq \beta^2 \sum_{n=1}^{\infty} \frac{1}{\beta(n!)^{\alpha}} \times \frac{1}{|x - r_n|^{\alpha}}$$

$$= \beta \sum_{n=1}^{\infty} \frac{1}{(n!)^{\alpha}} \times \frac{1}{|x - r_n|^{\alpha}},$$

where we have used Jensen's inequality (see e.g. [379]). Hence, as in Answer 21.2.16, we obtain

$$\int_0^1 [g_{\alpha}(x)]^2 dx = \beta \sum_{n=1}^{\infty} \frac{1}{(n!)^{\alpha}} \int_0^1 \frac{dx}{|x - r_n|^{\alpha}} \leq \frac{2^{\alpha} \beta^2}{1 - \alpha} < \infty.$$

We now show that $D(A) \subset D(B^{\alpha})$ for each $\alpha \in (0, 1)$. Clearly,

$$[f(x)]^{\alpha} = \left(\sum_{n=1}^{\infty} \frac{1}{n!} \times \frac{1}{|x - r_n|^{1/2}} \right)^{\alpha} \leq \sum_{n=1}^{\infty} \frac{1}{(n!)^{\alpha}} \times \frac{1}{|x - r_n|^{\alpha/2}} = g_{\alpha}(x)$$

by the subadditivity property $(a+b)^{\alpha} \leq a^{\alpha} + b^{\alpha}$ (where $a, b \geq 0$). By remembering that $B = M_f$ (the multiplication operator by f), it follows that

$$D(Mg_{\alpha}) \subset D(M_{f^{\alpha}}) = D[(M_f)^{\alpha}].$$

Accordingly, for all $\alpha \in (0, 1)$,

$$D(A) \subset D(B^{\alpha}) \text{ and so } D(A) \cap D(B^{\alpha}) = D(A).$$

Setting $C = |A| + I$ whereby $D(C) = D(|A|) = D(A)$, it is seen that $D(C) \cap D(B) = \{0\}$. Finally, for all $\alpha \in (0, 1)$,

$$D(C) = D(C) \cap D(B^{\alpha}) \subset D(C^{\alpha}) \cap D(B^{\alpha}).$$

In other words, $D(C^{\alpha}) \cap D(B^{\alpha})$ is dense whichever $\alpha \in (0, 1)$, as required.

Remark The same method gives more counterexamples. Indeed, another pair is B (the same B) and $C := A^2 + I$. Indeed, we also have in such case

$$D(C) \subset D(A) \subset D[(M_f)^\alpha].$$

Answer 21.2.22 ([202], Section 5 with an Idea due to Paul R. Chernoff) We only give the main ideas of this construction.

Let $A = e^{-T}$, where $T = id/dx$ and A is defined on its maximal domain, say. Then A is a nonsingular (unbounded) positive self-adjoint operator. Now, set $B = VAV$, where V is the multiplication operator by

$$v(x) = \begin{cases} -1, & x < 0, \\ 1, & x \geq 0. \end{cases}$$

Then V is a fundamental symmetry. Moreover, A and B obey

$$D(A) \cap D(B) = D(A^{-1}) \cap D(B^{-1}) = \{0\}.$$

As observed, this is not obvious and it is carried out in several steps (we refer the readers to [202] for more details).

Answer 21.2.23 ([107], [202]) Let A and B be two self-adjoint positive operators such that $A, B \geq I$ and $D(A) \cap D(B) = \{0\}$ (an explicit pair is available say in Answer 21.2.17). Then A^{-1} and B^{-1} are positive (bounded) with disjoint dense ranges. Now, set

$$T = A^{-1} + B^{-1} \text{ and } S = 2A^{-1} + B^{-1}.$$

It is easy to see that T and S are injective. Hence T^{-1} and S^{-1} are densely defined and positive.

To show that $\operatorname{ran}(T) \cap \operatorname{ran}(S) = \{0\}$, let $t \in \operatorname{ran}(T) \cap \operatorname{ran}(S)$ and so $t = Tx = Sy$. Therefore

$$A^{-1}(x - 2y) = B^{-1}(y - x) \in D(A) \cap D(B) = \{0\}$$

and so $x = y = 0$. Thus, $t = 0$ as wished.

To conclude, since $0 \leq T \leq S \leq 2T$, it follows by Theorem 1 in [92] that

$$\operatorname{ran}(T^{\frac{1}{2}}) = \operatorname{ran}(S^{\frac{1}{2}}).$$

Answer 21.2.24 In fact, we have a slightly better counterexample than what is suggested. Consider again the operators T and A:

$$Tf(x) = e^{\frac{x^2}{2}} f(x)$$

defined on $D(T) = \{f \in L^2(\mathbb{R}) : e^{\frac{x^2}{2}} f \in L^2(\mathbb{R})\}$ and $A := \mathcal{F}^* T \mathcal{F}$.

Clearly T is boundedly invertible (hence so is A) and

$$Bf(x) := T^{-1}f(x) = e^{\frac{-x^2}{2}} f(x)$$

is defined from $L^2(\mathbb{R})$ onto $D(T)$.

We already know that $D(AB)$ is trivial if $D(A) \cap \text{ran}(B)$ is so and if B is also one-to-one (which is the case). But,

$$D(A) \cap \text{ran}(B) = D(A) \cap D(T) = \{0\}$$

as this is already available to us from Answer 21.2.17. Accordingly,

$$D(AB) = \{0\}.$$

Obviously, $D(BA) = D(A)$ and that finishes the answer.

Answer 21.2.25 Consider the operators A and B of Answer 21.2.16, and hence $D(A) \cap D(B) = \{0\}$. From Answer 21.2.21, the same pair also obeys

$$D(A) \cap D(B^\alpha) = D(A),$$

where $\alpha \in (0, 1)$. By looking closely at the operator B, we see that it is also boundedly invertible, denote its (bounded) inverse by S. Then

$$D(AS) = \{0\}$$

because $D(A) \cap \text{ran}(S) = D(A) \cap D(B) = \{0\}$. Next, we claim that $D(AS^\alpha) = \{f \in L^2(\mathbb{R}) : S^\alpha f \in D(A)\} \neq \{0\}$. To see this, first observe that

$$D(A) \cap \text{ran}(S^\alpha) = D(A) \cap D(B^\alpha) = D(A).$$

Now, let $g \in D(A)$ with $g \neq 0$. Since $D(A) \subset \text{ran}(S^\alpha)$, it results that $g = S^\alpha f \in D(A)$ for some (nonzero) $f \in L^2(\mathbb{R})$. That is, $f \in D(AS^\alpha)$ and so $D(AS^\alpha) \neq \{0\}$.

To get a pair satisfying all claims together, we use a pair such that $D(C^\alpha) \cap D(B^\alpha)$ is dense as in Answer 21.2.21. In the latter, we set $C = |A| + I$ (hence $D(C) = D(A)$). Therefore,

$$D(CS) = D(AS) = \{0\} \text{ and } D(AS^\alpha) = D(CS^\alpha) \neq \{0\}.$$

Finally,

$$D(C^\alpha S^\alpha) \neq \{0\}$$

follows from the density of $D(C^\alpha) \cap D(B^\alpha)$ and similar arguments as above.

Answer 21.2.26 ([202]) Consider *iterum*

$$Af(x) = e^{\frac{x^2}{2}} f(x)$$

on $D(A) = \{f \in L^2(\mathbb{R}) : e^{\frac{x^2}{2}} f \in L^2(\mathbb{R})\}$. Then A is self-adjoint, positive (and also boundedly invertible). Let $0 < \alpha < 1$ and choose $a > 1$ with $\alpha \geq a^{-1}$. Then $e^{-ax^2/2}$ is in $D(A)$ for

$$e^{x^2/2} e^{-ax^2/2} = e^{-(a-1)x^2/2}$$

is plainly in $L^2(\mathbb{R})$. But (employing the Fourier transform of a Gaussian, see, e.g., [215]),

$$\left(\mathcal{F} e^{-ax^2/2}\right)(y) e^{\frac{\alpha}{2}y^2} = \frac{1}{\sqrt{a}} e^{\frac{(\alpha - a^{-1})}{2} y^2}$$

is not in $L^2(\mathbb{R})$ thereby

$$e^{\frac{-ax^2}{2}} \notin D(\mathcal{F}^* A^\alpha \mathcal{F}).$$

Answer 21.2.27 Consider again the operators A and B introduced in Answer 21.2.24, that is,

$$Af(x) = e^{\frac{x^2}{2}} f(x)$$

on $D(A) = \{f \in L^2(\mathbb{R}) : e^{\frac{x^2}{2}} f \in L^2(\mathbb{R})\}$ and $B := \mathcal{F}^* A \mathcal{F}$. We then found that $D(BA^{-1}) = \{0\}$.

Finally, set $T := A^{-1}B$. Then T is clearly densely defined because $D(T) = D(B)$ as also $A^{-1} \in B[L^2(\mathbb{R})]$. Thus,

$$D(T^*) = D[(A^{-1}B)^*] = D(BA^{-1}) = \{0\},$$

as needed.

Answer 21.2.28 ([259]) Let A and B be as in Answer 21.2.27, and set $T := A^{-1}B$. Then T is densely defined and $D(T^*) = \{0\}$. Hence plainly

$$D(TT^*) = \{0\}.$$

Now,

$$D(T^*T) = \{f \in D(T) : Tf \in D(T^*)\} = \{f \in D(A^{-1}B) : A^{-1}Bf = 0\}.$$

Since $A^{-1}B$ is one-to-one, it follows that $D(T^*T) = \{0\}$.

Finally,

$$D(T^2) = D(A^{-1}BA^{-1}B) = D[(BA^{-1})B] = \{f \in D(B) : Bf = 0\}$$

and so $D(T^2) = \{0\}$ by the injectivity of B.

Answer 21.2.29 We give three types of counterexamples.

(1) A trivial counterexample is to consider an unclosed and densely defined operator T such that $D(T^*T) = D(TT^*) = \{0\}$ as in Question 21.2.28. Then TT^* and T^*T are both trivially closed, and clearly none of TT^* and T^*T is self-adjoint for none of them is densely defined.

(2) What if we suppose that $D(T^*T) \neq \{0\}$ and $D(TT^*) \neq \{0\}$? The answer is still no! Let A be a densely defined operator such that $D(A^*A) = D(AA^*) = \{0\}$, and then set $T = \begin{pmatrix} 0 & A \\ 0 & 0 \end{pmatrix}$ with the dense domain $D(T) = H \oplus D(A)$. Therefore,

$$D(T^*T) = H \oplus \{0\} \neq \{(0,0)\} \text{ and } D(TT^*) = \{0\} \oplus H \neq \{(0,0)\}.$$

The non-self-adjointness of both TT^* and T^*T is clear.

(3) Another natural question is: what if we further assume that TT^* and T^*T are both densely defined? The answer is still negative (and this was kindly pointed out to me by Professor Zs. Tarcsay): let A be a densely defined closed operator such that it is positive but non-self-adjoint in a Hilbert space H. Assume further that $\operatorname{ran}A$ is dense in H. Define an inner product space on $\operatorname{ran}A$ by

$$\langle Ax, Ay \rangle_A = \langle Ax, y \rangle, x, y \in D(A).$$

Denote the completion of this pre-Hilbert space by H_A. Define the canonical embedding operator $T : H_A \supseteq \operatorname{ran}A \to H$ by

$$T(Ax) := Ax, \ x \in D(A).$$

Then T is closable because $D(A) \subset D(T^*)$. Indeed, for all $y \in D(A)$, then

$$|\langle T(Ax), y \rangle|^2 = |\langle Ax, y \rangle|^2$$
$$\leq \langle Ay, y \rangle \langle Ax, x \rangle$$
$$= \langle Ay, y \rangle \langle Ax, Ax \rangle_A$$

(with $x \in D(A)$) and where we have used the generalized Cauchy–Schwarz inequality. In addition, for any $x, y \in D(A)$, we have that

$$\langle T(Ax), y \rangle = \langle Ax, y \rangle = \langle Ax, Ay \rangle_A$$

from which we get

$$T^*y = Ay \in H_A, \ y \in D(A).$$

Hence $D(A) \subset D(TT^*)$ and also

$$TT^*y = T(Ay) = Ay, \ y \in D(A).$$

This means that $A \subset TT^*$. Now, let $y \in D(TT^*)$. Then for some $x \in D(A)$, $T^*y = Ax = T^*x$. Hence

$$x - y \in \ker T^* \subseteq (\operatorname{ran} T)^{\perp} = (\operatorname{ran} A)^{\perp} = \{0\}.$$

Therefore, $y = x \in D(A)$, establishing $A = TT^*$.

To obtain the other case, let $S = T^*$ be restricted to $D(A)$. Then S is densely defined and closable and obeys $T \subset T^{**} \subset S^*$. Accordingly,

$$A = TT^* = S^*S.$$

Remark The construction of H_A and T is borrowed from [331].

Answer 21.2.30 We have already seen such examples. Howbeit, and for the sake of variety of examples, we provide a different example which is also needed for another purpose (see the remark below this answer). Consider again the operators A and B introduced in Answer 18.2.7. We already know that BA is densely defined. Besides A and B are self-adjoint and so $(BA)^* = AB$ as B is everywhere defined and bounded. So if we set $T = BA$, then T is densely defined and

$$D(T^*) = D(AB) = \{f \in L^2(\mathbb{R}) : \langle f, g \rangle g \in L^2(\mathbb{R}), \langle f, g \rangle \in L^2(\mathbb{R})\}$$

and so

$$D(T^*) = \{f \in L^2(\mathbb{R}) : \langle f, g \rangle = 0\} = \{g\}^{\perp}.$$

Therefore, $D(T^*)$ cannot be dense.

Remark This tells us that BA need not be closable even when $B \in L^2(\mathbb{R})$ is of rank one and self-adjoint and A too is self-adjoint. Notice that a weaker counterexample on a Banach space already appeared in [275].

Answer 21.2.31 As mentioned, we must sidestep closedness. The most trivial example is to consider a densely defined unclosed operator A such that

$$D(A^2) = D(A^*A) = \{0\},$$

as in say Question 21.2.28. Then $A^*A = A^2$ is trivially satisfied but A is not self-adjoint since A is not even closed.

Answer 21.2.32 To deal with the first case, consider any closed, densely defined, and symmetric operator A^* *which is not self-adjoint*. Then $A^* \subset A$ and so $A^*A \subset A^2$.

As for the second case, consider any closed, densely defined, and symmetric operator A *which is not self-adjoint*. A similar observation yields $A^2 \subset A^*A$.

Still in the second case, we may even consider a closed, symmetric and positive operator A such that $D(A^2) = \{0\}$ (Question 20.2.23). Then trivially $A^2 \subset A^*A$ (for $D(A^2) \subset D(A^*A)$) and A is not self-adjoint.

Answer 21.2.33 Consider again the operators A and B introduced in Answer 18.2.7. Then the required conditions are all fulfilled.

To see why BA is not closable on $D(A)$, we could argue by using sequences etc. Alternatively, if BA were closable, then $D[(BA)^*]$ would be dense. However, from Answer 21.2.30, we know that $D[(BA)^*] = D(AB)$ is not dense in $L^2(\mathbb{R})$. Therefore, BA is not closable, as wished.

Answer 21.2.34

(1) Let A be an unbounded self-adjoint operator. If $B = -A$, then $A + B$ is densely defined and unclosed on $D(A) \subset H$, so it cannot be self-adjoint. Nonetheless,

$$\overline{A + B} = 0 \in B(H).$$

(2) Consider any unbounded self-adjoint operator T with domain $D(T)$. Then if we set $A = \begin{pmatrix} T & 0 \\ 0 & T \end{pmatrix}$ and $B = \begin{pmatrix} 0 & T \\ T & 0 \end{pmatrix}$, then

$$A + B = \begin{pmatrix} T & T \\ T & T \end{pmatrix}$$

is an unclosed unbounded operator on $D(T) \oplus D(T)$ (see Answer 30.1.6 below for further details). Therefore, $A + B$ is unbounded and non-self-adjoint. In the end, observe that $A + B$ is essentially self-adjoint by Wüst's theorem, that is, $\overline{A + B}$ is self-adjoint.

(3) Use again the counterexample of Question 21.2.17. Then A and B are unbounded and self-adjoint (even positive) operators. In addition,

$$D(A + B) = D(A) \cap D(B) = \{0\}.$$

Thus, $A + B$ is trivially closed, but it cannot be self-adjoint because it is not even densely defined.

(4) Let T be a closed, densely defined, symmetric but non-self-adjoint operator. Then $|T|$ is self-adjoint. Now, set

$$A = |T| + \frac{T}{3} \text{ and } B = -|T|.$$

Then $D(A) = D(B)$. Obviously, B is self-adjoint. To see why A is self-adjoint, observe that the symmetric $T/3$ is $|T|$-bounded with relative bound $1/3$, and so by the Kato–Rellich theorem A is self-adjoint. But

$$A + B = |T| + \frac{T}{3} - |T| = \frac{T}{3}$$

is densely defined, closed, and solely symmetric, that is, it is not self-adjoint on $D(T)$.

Answer 21.2.35 ([254]) Let A be a closed and symmetric operator such that $D(A^2) = \{0\}$. Hence A is not self-adjoint. Setting $B = A$, we see that

$$D[(A + B)^2] = D(4A^2) = D(A^2) = \{0\}.$$

Hence $D[(A + B)^2]$ cannot contain any dense set D. Finally,

$$\overline{A + B} = \overline{A + A} = 2A$$

is not self-adjoint.

Answer 21.2.36 Let S and T be two unbounded self-adjoint operators such that $S + T$ is closed but non-self-adjoint (see Answer 21.2.34, Case 21.2.34). Now, set $A = \begin{pmatrix} 0 & T \\ S & 0 \end{pmatrix}$. Then A is obviously closed and densely defined. However,

$$A + A^* = \begin{pmatrix} 0 & S + T \\ S + T & 0 \end{pmatrix}$$

is densely defined, closed (and symmetric), and non-self-adjoint.

Answer 21.2.37 Let T be an unbounded self-adjoint operator with domain $D(T)$. Setting $A = \begin{pmatrix} 0 & T \\ 0 & 0 \end{pmatrix}$, we see that A is closed, but it is not symmetric and yet

$$A + A^* = \begin{pmatrix} 0 & T \\ T & 0 \end{pmatrix}$$

is clearly self-adjoint.

Answer 21.2.38 In fact, we can find plenty of them. Let B be any closed, symmetric but non-self-adjoint operator. Set $A = |B|$ and so A is self-adjoint (and also positive). Recall that $D(A) = D(|B|) = D(B)$. The answer relies on results about relative boundedness.

(1) Clearly, for all $x \in D(A)$,

$$\|Bx\| = \||B|x\| = \|Ax\|$$

and so

$$\|\lambda Bx\| = |\lambda|\||B|x\| = |\lambda|\|Ax\|.$$

Hence

$$\|\lambda Bx\| \leq |\lambda|\|Ax\| + a\|x\|$$

for all $x \in D(A)$ (and all positive a). Assume now that $|\lambda| < 1$, and so λB is A-bounded with relative bound smaller than 1. Since B is symmetric and A is self-adjoint, the Kato–Rellich theorem allows us to establish the self-adjointness of $A + \lambda B$, as required.

(2) Let $|\lambda| > 1$, and then write

$$A + \lambda B = \lambda\left(\frac{1}{\lambda}A + B\right).$$

Since $\lambda \neq 0$, the closedness or self-adjointness of $A + \lambda B$ is equivalent to that of $\lambda^{-1}A + B$. We may write for all $x \in D(B)$

$$\left\|\frac{1}{\lambda}Ax\right\| = \left\|\frac{1}{\lambda}|B|x\right\| = \frac{1}{|\lambda|}\|Bx\| \leq \frac{1}{|\lambda|}\|Bx\| + a\|x\|$$

(for all positive a). This means that $\lambda^{-1}A$ is B-bounded with relative bound $|\lambda|^{-1} < 1$. Since B is closed, Theorem 19.1.8 tells us that $A + \lambda B$ is closed on $D(B)$.

To treat the non-self-adjoint case, let $|\lambda| > 1$. As above, the self-adjointness of $|B| + \lambda B$ is equivalent to that of $|B|/\lambda + B$. Now, we write for all $x \in D(B)$

$$\left\|\left(\frac{1}{\lambda}|B| + B - B\right)x\right\| = \left\|\frac{1}{\lambda}|B|x\right\| = \frac{1}{|\lambda|}\|Bx\|$$

$$\leq \frac{1}{|\lambda|}\|Bx\| + \frac{1}{|\lambda|}\left\|\left(\frac{1}{\lambda}|B| + B\right)x\right\|$$

$$\leq \frac{1}{|\lambda|}\left(\|Bx\| + \left\|\left(\frac{1}{\lambda}|B| + B\right)x\right\|\right) + a\|x\|$$

for all $a \geq 0$. Since $1/|\lambda| < 1$, Theorem 21.1.12 tells us that $|B|/\lambda + B$ is self-adjoint if and only if B is self-adjoint. Since B is not self-adjoint, neither is $|B|/\lambda + B$.

Answer 21.2.39 Let T be an unbounded, self-adjoint, and positive operator with domain $D(T)$.

(1) Set $A = -T$ and so $|A| = T$. Hence

$$|A| + A = T - T = 0_{D(T)}$$

is clearly unclosed.

(2) Set $A = T$ and so $|A| = T$. Hence

$$|A| - A = T - T = 0_{D(T)}$$

is not closed either.

Answer 21.2.40 There are, a priori, three main obstacles preventing $AB + BA$ from being self-adjoint:

(1) A quite fine counterexample is to take two unbounded self-adjoint and positive (even invertible) operators A and B such that $D(A) \cap D(B) = \{0\}$. Then $AB + BA$ is not self-adjoint merely because it is not even densely defined because

$$D(AB + BA) \subset D(B) \cap D(A) = \{0\}.$$

(2) Let A be any unbounded, self-adjoint, positive and boundedly invertible operator with domain $D(A) \subset H$, then set $B := A^{-1}$. Hence $B \in B(H)$ is positive. Therefore

$$AB + BA = AA^{-1} + A^{-1}A = I + A^{-1}A \subset 2I,$$

that is, $AB + BA$, which is defined and bounded on $D(A)$, fails to be closed and so it fails to be self-adjoint.

(3) We have saved the best for last. We use an idea which appeared in the last case in Answer 21.2.34 as well as the powerful tool of matrices of operators. Thanks to all that, we will reduce the problem of the sum of two products into the sum of two operators.

   Let T be a densely defined closed symmetric operator which is *not self-adjoint*, with domain $D(T) \subset H$. Then $|T| + T/2$ is self-adjoint by the Kato–Rellich theorem. Recall also that $-|T|$ is self-adjoint. Let I be the identity operator on H. Now, define

$$B = \begin{pmatrix} 0 & I \\ I & 0 \end{pmatrix} \text{ and } A = \begin{pmatrix} |T| + T/2 & 0 \\ 0 & -|T| \end{pmatrix}$$

on $H \oplus H$ and $D(A) = D(T) \oplus D(T)$, respectively. Clearly, B is a fundamental symmetry and A is self-adjoint. Besides,

$$AB = \begin{pmatrix} 0 & |T| + T/2 \\ -|T| & 0 \end{pmatrix} \text{ and } BA = \begin{pmatrix} 0 & -|T| \\ |T| + T/2 & 0 \end{pmatrix}.$$

Observe in passing that $D(AB) = D(BA)$ and that both AB and BA are closed, in fact $(AB)^* = BA$ and $(BA)^* = AB$. In the end,

$$AB + BA = \frac{1}{2} \begin{pmatrix} 0 & T \\ T & 0 \end{pmatrix}$$

is densely defined, closed, symmetric, and not self-adjoint on $D(T) \oplus D(T)$.

Answer 21.2.41 Let A and B be two unbounded self-adjoint operators on a Hilbert space H with $D(A+B) = D(A) \cap D(B) = \{0\}$. Then define on $H \times H$ the operator T by

$$T = \begin{pmatrix} 0 & A \\ B & 0 \end{pmatrix} \text{ and so } T^* = \begin{pmatrix} 0 & B \\ A & 0 \end{pmatrix}.$$

It is plain that T is closed. Moreover,

$$T + T^* = \begin{pmatrix} 0 & A+B \\ A+B & 0 \end{pmatrix}.$$

Therefore,

$$D(T) \cap D(T^*) = D(T + T^*) = D(A + B) \times D(A + B) = \{(0, 0)\}.$$

Thus,

$$D(TT^* - T^*T) = D(TT^*) \cap D(T^*T) \subset D(T) \cap D(T^*) = \{(0, 0)\}.$$

Since $D(|T|) = D(T)$ and $D(|T^*|) = D(T^*)$ for T and T^* are closed, it follows as above that

$$D(|T||T^*| - |T^*||T|) = \{(0, 0)\},$$

as needed.

Answer 21.2.42 Consider two densely defined, closed, and symmetric restrictions A and B of some self-adjoint operator such that

$$D(A) \cap D(B) = \{0\} \text{ and } D(A^2) = D(B^2) = \{0\}$$

as in van Daele–Schmüdgen theorem. Then set

$$T = \begin{pmatrix} 0 & A \\ B & 0 \end{pmatrix} \text{ and } S = \begin{pmatrix} 0 & B \\ A & 0 \end{pmatrix}.$$

Both S and T are closed (moreover, $T^* \subset S$) and they do satisfy

$$D(ST) = D(TS) = D(S + T) = \{0\},$$

as required.

Answer 21.2.43 Let T be a closed and densely defined operator such that $D(T) \cap D(T^*) = \{0\}$ (as in Question 21.2.41). Set $A = |T|$ (hence A is self-adjoint) and $B = T$. Then

$$\overline{D(A^*) \cap D(B)} = \overline{D(|T|) \cap D(T)} = \overline{D(T)} = H$$

and

$$D(A) \cap D(B^*) = D(T) \cap D(T^*) = \{0\},$$

as wished.

Answer 21.2.44 Let A and B be two unbounded self-adjoint operators such that

$$D(A) \cap D(B) = D(A^{-1}) \cap D(B^{-1}) = \{0\},$$

where A^{-1} and B^{-1} are not bounded (an explicit pair on $L^2(\mathbb{R})$ already appeared in Answer 21.2.22). Then

$$D(A^{-1}B) = \{x \in D(B) : Bx \in D(A^{-1})\} = \{x \in D(B) : Bx = 0\}$$

and so

$$D(A^{-1}B) = \{0\}$$

for B is one-to-one. Similarly, we may show that

$$D(BA^{-1}) = \{0\},$$

as suggested.

Answer 21.2.45 Let A be an unbounded self-adjoint and positive operator with domain $D(A)$. Let

$$B = \begin{pmatrix} 0 & A \\ 0 & 0 \end{pmatrix}$$

be defined on $H \oplus D(A)$. Then B is closed and $B^2 = 0$ on $H \oplus D(A)$. Also,

$$|B| = \begin{pmatrix} 0 & 0 \\ 0 & A \end{pmatrix} \text{ and so } |B|B = \begin{pmatrix} 0 & 0_{D(A)} \\ 0 & 0 \end{pmatrix},$$

that is, $|B|B$ is bounded on $H \oplus D(A)$. However,

$$B|B| = \begin{pmatrix} 0 & A^2 \\ 0 & 0 \end{pmatrix}$$

is clearly unbounded and closed on $H \oplus D(A^2)$.

Answer 21.2.46 ([86]) Let A and B be two unbounded self-adjoint operators such that $D(A^{-1}B) = D(BA^{-1}) = \{0\}$ where A^{-1} and B^{-1} are not bounded. Then define

$$T = \begin{pmatrix} 0 & A^{-1} \\ B & 0 \end{pmatrix}$$

on $D(T) := D(B) \oplus D(A^{-1}) \subset L^2(\mathbb{R}) \oplus L^2(\mathbb{R})$. Since A^{-1} and B are closed, we know that T is closed on $D(T)$. Incidentally, T is one-to-one. Now,

$$T^2 = \begin{pmatrix} 0 & A^{-1} \\ B & 0 \end{pmatrix} \begin{pmatrix} 0 & A^{-1} \\ B & 0 \end{pmatrix} = \begin{pmatrix} A^{-1}B & 0 \\ 0 & BA^{-1} \end{pmatrix}.$$

Hence

$$D(T^2) = D(A^{-1}B) \oplus D(BA^{-1}) = \{0\} \oplus \{0\} = \{(0,0)\},$$

as needed.

Finally, since

$$T^* = \begin{pmatrix} 0 & B \\ A^{-1} & 0 \end{pmatrix},$$

because A^{-1} and B are self-adjoint (also observe that T^* too is injective), we get

$$T^{*2} = \begin{pmatrix} BA^{-1} & 0 \\ 0 & A^{-1}B \end{pmatrix}$$

on

$$D(T^{*2}) = D(BA^{-1}) \oplus D(A^{-1}B) = \{(0,0)\}$$

and that marks the end of the proof.

Answer 21.2.47 ([259]) Let A and B be self-adjoint operators in $L^2(\mathbb{R})$ such that $D(AB) = \{0_{L^2(\mathbb{R})}\}$ but $D(BA) \neq \{0_{L^2(\mathbb{R})}\}$ as in Question 21.2.24. Recall that A is one-to-one. Now, set

$$T = \begin{pmatrix} 0 & A \\ B & 0 \end{pmatrix}.$$

Hence

$$T^2 = \begin{pmatrix} 0 & A \\ B & 0 \end{pmatrix} \begin{pmatrix} 0 & A \\ B & 0 \end{pmatrix} = \begin{pmatrix} AB & 0 \\ 0 & BA \end{pmatrix}$$

and so $D(T^2) = \{0_{L^2(\mathbb{R})}\} \oplus D(BA) \neq \{(0_{L^2(\mathbb{R})}, 0_{L^2(\mathbb{R})})\}$. Finally,

$$T^3 = \begin{pmatrix} AB & 0 \\ 0 & BA \end{pmatrix} \begin{pmatrix} 0 & A \\ B & 0 \end{pmatrix} = \begin{pmatrix} 0 & ABA \\ BAB & 0 \end{pmatrix}.$$

Obviously, $D(BAB) = \{0_{L^2(\mathbb{R})}\}$. Since

$$D(ABA) = \{x \in D(A) : Ax \in D(AB) = \{0_{L^2(\mathbb{R})}\}\} = \ker A$$

and A is injective, it results that we too have $D(ABA) = \{0_{L^2(\mathbb{R})}\}$. Accordingly, $D(T^3) = \{(0_{L^2(\mathbb{R})}, 0_{L^2(\mathbb{R})})\}$. Finally, as

$$T^* = \begin{pmatrix} 0 & B \\ A & 0 \end{pmatrix},$$

then we may similarly show that $D(T^{*2}) \neq \{0\}$ and $D(T^{*3}) = \{0\}$, marking the end of the proof.

Answer 21.2.48 ([259]) Let A and B be two unbounded self-adjoint operators such that

$$\mathcal{D}(A^{-1}B) = \mathcal{D}(BA^{-1}) = \{0\}$$

where A^{-1} and B^{-1} are not bounded. Now, define

$$S = \begin{pmatrix} 0 & A^{-1} \\ B & 0 \end{pmatrix}$$

on $\mathcal{D}(S) := \mathcal{D}(B) \oplus \mathcal{D}(A^{-1}) \subset L^2(\mathbb{R}) \oplus L^2(\mathbb{R})$. Then S is densely defined and closed. In addition, we already know from Answer 21.2.46 that $\mathcal{D}(S^2) = \mathcal{D}(S^{*2}) = \{0\}$. Notice now that we may write

$$S = \underbrace{\begin{pmatrix} A^{-1} & 0 \\ 0 & B \end{pmatrix}}_{C} \underbrace{\begin{pmatrix} 0 & I \\ I & 0 \end{pmatrix}}_{D} \text{ and } S^* = DC$$

because C and D are self-adjoint and D is a fundamental symmetry. Now, define T on $L^2(\mathbb{R}) \oplus L^2(\mathbb{R}) \oplus L^2(\mathbb{R}) \oplus L^2(\mathbb{R})$ by

$$T = \begin{pmatrix} 0 & C \\ D & 0 \end{pmatrix}$$

where $\mathbf{0}$ is the zero matrix of operators on $L^2(\mathbb{R}) \oplus L^2(\mathbb{R})$. Then,

$$T^2 = \begin{pmatrix} CD & \mathbf{0} \\ \mathbf{0} & DC \end{pmatrix}, \ T^3 = \begin{pmatrix} \mathbf{0} & CDC \\ DCD & \mathbf{0} \end{pmatrix}$$

and

$$T^4 = \begin{pmatrix} CDCD & \mathbf{0} \\ \mathbf{0} & DCDC \end{pmatrix} = \begin{pmatrix} S^2 & \mathbf{0} \\ \mathbf{0} & S^{*2} \end{pmatrix}.$$

Also, since $T^* = \begin{pmatrix} \mathbf{0} & D \\ C & \mathbf{0} \end{pmatrix}$, we also have

$$T^{*2} = \begin{pmatrix} DC & \mathbf{0} \\ \mathbf{0} & CD \end{pmatrix}, \ T^{*3} = \begin{pmatrix} \mathbf{0} & DCD \\ CDC & \mathbf{0} \end{pmatrix}$$

and

$$T^{*4} = \begin{pmatrix} DCDC & \mathbf{0} \\ \mathbf{0} & CDCD \end{pmatrix} = \begin{pmatrix} S^{*2} & \mathbf{0} \\ \mathbf{0} & S^2 \end{pmatrix}.$$

Finally, observe that

$$\mathcal{D}(T^2) = \mathcal{D}(S) \oplus \mathcal{D}(S^*) = \mathcal{D}(B) \oplus \mathcal{D}(A^{-1}) \oplus \mathcal{D}(A^{-1}) \oplus \mathcal{D}(B) \neq \{0_{[L^2(\mathbb{R})]^4}\}$$

and that

$$\mathcal{D}(T^3) = \mathcal{D}(B) \oplus \mathcal{D}(A^{-1}) \oplus \{0\} \oplus \{0\} \neq \{0_{[L^2(\mathbb{R})]^4}\},$$

but

$$\mathcal{D}(T^4) = \mathcal{D}(S^2) \oplus \mathcal{D}(S^{*2}) = \{0_{[L^2(\mathbb{R})]^4}\}.$$

The corresponding relations about the adjoints' domains may be checked similarly. The proof is therefore complete.

Answer 21.2.49 ([259]) We shall avoid already known details as similar cases have already been handled. First, choose A and B as in Answer 21.2.24, that is, $\mathcal{D}(AB) = \{0\}$ and $\mathcal{D}(BA) = \mathcal{D}(A)$, where A and B are self-adjoint operators and B is bounded and everywhere defined. Then, let

$$C = \begin{pmatrix} B & 0 \\ 0 & A \end{pmatrix} \text{ and } D = \begin{pmatrix} 0 & I \\ I & 0 \end{pmatrix}.$$

Then C and D are self-adjoint. Setting $S = CD = \begin{pmatrix} 0 & B \\ A & 0 \end{pmatrix}$, we see that $S^* = \begin{pmatrix} 0 & A \\ B & 0 \end{pmatrix}$. Finally, define T on $[L^2(\mathbb{R})]^4$ by

$$T = \begin{pmatrix} \mathbf{0} & C \\ D & \mathbf{0} \end{pmatrix} \text{ and so } T^* = \begin{pmatrix} \mathbf{0} & D \\ C & \mathbf{0} \end{pmatrix}$$

with $\mathbf{0}$ being the zero matrix of operators on $L^2(\mathbb{R}) \oplus L^2(\mathbb{R})$. Hence

$$T^6 = \begin{pmatrix} S^3 & \mathbf{0} \\ \mathbf{0} & S^{*3} \end{pmatrix} \text{ and } T^{*6} = \begin{pmatrix} S^{*3} & \mathbf{0} \\ \mathbf{0} & S^3 \end{pmatrix}.$$

Accordingly,

$$\mathcal{D}(T^6) = \mathcal{D}(S^3) \oplus \mathcal{D}(S^{*3}) = \{(0,0)\} = \mathcal{D}(T^{*6})$$

due to the assumptions on A and B. However,

$$T^5 = \begin{pmatrix} \mathbf{0} & CDCDC \\ DCDCD & \mathbf{0} \end{pmatrix}, \quad T^{*5} = \begin{pmatrix} \mathbf{0} & DCDCD \\ CDCDC & \mathbf{0} \end{pmatrix}$$

and, as can simply be checked,

$$\mathcal{D}(CDCDC) = \{0\} \text{ but } \mathcal{D}(DCDCD) \neq \{0\}.$$

Consequently, $\mathcal{D}(T^5) \neq \{0\}$. In the end, a similar idea yields

$$\mathcal{D}(T^{*5}) \neq \{0\},$$

as wished.

Answer 21.2.50 ([259]) We use a proof by induction. The statement is true for $n = 2$ as was seen before. Assume now that there is a closed $T = \begin{pmatrix} 0 & A \\ B & 0 \end{pmatrix}$ such that $\mathcal{D}(T^{2^n-1}) \neq \{0\}$ and $\mathcal{D}(T^{*2^n-1}) \neq \{0\}$ with $\mathcal{D}(T^{2^n}) = \mathcal{D}(T^{*2^n}) = \{0\}$. Now, write

$$T = \begin{pmatrix} 0 & A \\ B & 0 \end{pmatrix} = \underbrace{\begin{pmatrix} A & 0 \\ 0 & B \end{pmatrix}}_{C} \underbrace{\begin{pmatrix} 0 & I \\ I & 0 \end{pmatrix}}_{\tilde{B}}.$$

Next, set $S = \begin{pmatrix} 0 & C \\ \tilde{B} & 0 \end{pmatrix}$ and so

$$S^{2^{n+1}} = \begin{pmatrix} (C\tilde{B})^{2^n} & 0 \\ 0 & (\tilde{B}C)^{2^n} \end{pmatrix} = \begin{pmatrix} T^{2^n} & 0 \\ 0 & T^{*2^n} \end{pmatrix}$$

and so $\mathcal{D}(S^{2^{n+1}}) = \{0\}$. On the other hand,

$$S^{2^{n+1}-1} = \begin{pmatrix} 0 & (C\tilde{B})^{2^n-1}C \\ \tilde{B}(C\tilde{B})^{2^n-1} & 0 \end{pmatrix}.$$

Since $\tilde{B}$ is everywhere defined and bounded, it results that

$$\mathcal{D}(\tilde{B}(C\tilde{B})^{2^n-1}) = \mathcal{D}((C\tilde{B})^{2^n-1}) = \mathcal{D}(T^{2^n-1}) \neq \{0\},$$

which leads to $\mathcal{D}(S^{2^{n+1}-1}) \neq \{0\}$, as wished.

The case of adjoints may be treated in a similar manner, therefore leaving it to interested readers.

Answer 21.2.51 ([263]) First, notice that there could be a simpler counterexample, but here we construct one using Question 20.2.22. Let T be a closed and densely defined operator such that $D(T) = D(T^*)$ but $D(TT^*) \neq D(T^*T)$. Then TT^* and T^*T are both self-adjoint and positive, as are $|T|$ and $|T^*|$. Moreover,

$$D(|T|) = D(T) = D(T^*) = D(|T^*|).$$

However,

$$D(|T|^2) = D(T^*T) \neq D(TT^*) = D(|T^*|^2),$$

as needed.

Answer 21.2.52 ([263]) First, observe that $D(A) = D(B)$ yields $D(AB + BA) = D(A^2 - B^2)$. So, it suffices to exhibit A and B with the claimed properties such that, e.g., $D(A^2 - B^2)$ is not dense.

Consider a closed, densely defined, symmetric, and positive operator T such that $D(T^2) = \{0\}$, then set $A = T/2 + |T|$ and $B = |T|$. That A and B are positive is plain. Also, $D(A) = D(B)$ and B is self-adjoint. As for the self-adjointness of A, one needs to call on the Kato–Rellich theorem.

Now, by the lemma preceding the question, if $A^2 - B^2$ were densely defined, so would be $D[(A - B)^2]$. However,

$$D[(A - B)^2] = D(T^2) = \{0\},$$

and so $A^2 - B^2$ is not densely defined.

Answer 21.2.53 Among the remarkable results shown in [87], *a closable operator T, such that T^2 is self-adjoint, is closed.*

Let B be a solely closable operator with domain $D(B)$, and let A be a self-adjoint operator with domain $D(A)$. Then define

$$T = \begin{pmatrix} 0 & B \\ A & 0 \end{pmatrix}$$

on $D(T) = D(A) \oplus D(B)$. Clearly, T is closable but unclosed. Moreover,

$$T^2 = \begin{pmatrix} BA & 0 \\ 0 & AB \end{pmatrix}.$$

So, if BA and AB were both self-adjoint, it would ensue that T^2 is self-adjoint which, given the closability of T, would yield the closedness of T. But, this is absurd, and so one of AB and BA at least must be non-self-adjoint.

Answer 21.2.54 The idea of the counterexample is borrowed from [375]. Let B be an unbounded self-adjoint operator with domain $D(B) \subset H$, and let C be *a symmetric operator which is B-bounded with relative bound smaller than one.* Define on $H \oplus H$ the operators

$$T = \begin{pmatrix} 0 & B \\ B & 0 \end{pmatrix}, \quad S = \begin{pmatrix} C & 0 \\ 0 & C \end{pmatrix} \quad \text{and} \quad P = \begin{pmatrix} I & 0 \\ 0 & 0 \end{pmatrix}$$

where I is the usual identity operator on H. Then T is self-adjoint on $D(T) = D(B) \oplus D(B)$, S is symmetric on $D(S) = D(C) \oplus D(C)$, and P is an orthogonal projection on $H \oplus H$. Set

$$A = T + S = \begin{pmatrix} C & B \\ B & C \end{pmatrix}.$$

Then A is self-adjoint. To see that, let $(x, y) \in D(B) \oplus D(B)$. We have

$$\begin{pmatrix} C & 0 \\ 0 & C \end{pmatrix} \begin{pmatrix} x \\ y \end{pmatrix} = \begin{pmatrix} Cx \\ Cy \end{pmatrix} \quad \text{and} \quad \begin{pmatrix} 0 & B \\ B & 0 \end{pmatrix} \begin{pmatrix} x \\ y \end{pmatrix} = \begin{pmatrix} By \\ Bx \end{pmatrix}.$$

Hence for all $(x, y) \in D(B) \oplus D(B)$, we have

$$\|S(x, y)\| = \|Cx\| + \|Cy\|$$
$$\leq a\|Bx\| + a\|By\|$$
$$= a\|T(x, y)\|$$
$$\leq a\|T(x, y)\| + b\|(x, y)\|$$

for all $b \geq 0$. Therefore, S is T-bounded with relative bound $a < 1$.

Since T is self-adjoint and S is symmetric, the Kato–Rellich theorem establishes the self-adjointness of $A = S + T$ on $D(T)$.

Now, as $P \in B(H \oplus H)$, then

$$D(PAP) = D(AP) = D(C) \cap D(B) \oplus H = D(B) \oplus H$$

and so

$$PAP = \begin{pmatrix} I & 0 \\ 0 & 0 \end{pmatrix} \begin{pmatrix} C & B \\ B & C \end{pmatrix} \begin{pmatrix} I & 0 \\ 0 & 0 \end{pmatrix} = \begin{pmatrix} C & B \\ 0_{D(B)} & 0_{D(B)} \end{pmatrix} \begin{pmatrix} I & 0 \\ 0 & 0 \end{pmatrix}$$

$$= \begin{pmatrix} C & 0 \\ 0_{D(B)} & 0 \end{pmatrix}.$$

Hence

$$PAP \subset \tilde{C} := \begin{pmatrix} C & 0 \\ 0 & 0 \end{pmatrix}.$$

If PAP were self-adjoint, then given that $\tilde{C}$ is symmetric (because C is), the usual maximality argument gives $PAP = \tilde{C}$, and so C would be self-adjoint. This is the sought contradiction and so PAP cannot be self-adjoint.

Remark The readers may also show that if C is not essentially self-adjoint, then neither is PAP. Also, if C has no self-adjoint extension, then PAP does not have a self-adjoint extension either.

Answer 21.2.55 Let B be a densely defined closed operator with domain $D(B) \subset H$, and then define on $H \oplus H$

$$A = \begin{pmatrix} 0 & B \\ B^* & 0 \end{pmatrix} \text{ and } P = \begin{pmatrix} I & 0 \\ 0 & 0 \end{pmatrix}.$$

Obviously, A is self-adjoint on $D(B^*) \oplus D(B)$ and P is an orthogonal projection on $H \oplus H$. The readers may then check that

$$PAP = \begin{pmatrix} 0_{D(B^*)} & 0 \\ 0 & 0 \end{pmatrix},$$

and so PAP is unclosed on $D(PAP) = D(B^*) \oplus H$.

Answer 21.2.56 We give an example based on others previously constructed. Let A and B be two unbounded self-adjoint operators with $D(A^{1/2}B) = \{0\}$. This easily follows from Answer 21.2.24 by renaming A there by $A^{1/2}$ here. More precisely, $A^{1/2} = \mathcal{F}^* T \mathcal{F}$, where $Tf(x) = e^{\frac{x^2}{2}} f(x)$ with $D(T) = \{f \in L^2(\mathbb{R}) : e^{\frac{x^2}{2}} f \in L^2(\mathbb{R})\}$, and $Bf(x) = e^{\frac{-x^2}{2}} f(x)$ is defined on $L^2(\mathbb{R})$.

Since $D(A^{1/2}B) = \{0\}$ and since $A^{1/2}$ is one-to-one, it results that

$$D(A^{1/2}BA^{1/2}) = \{f \in D(A^{1/2}) : A^{1/2}f = 0\} = \{0\}$$

(in particular, $A^{1/2}BA^{1/2}$ cannot be self-adjoint).

Chapter 22
(Arbitrary) Square Roots

22.1 Basics

Definition 22.1.1 Let A and B be linear operators. Say that B is a square root of A if $B^2 = A$.

Examples 22.1.1

(1) The identity operator cannot have unbounded closable square roots (cf. Questions 22.2.2 and 22.2.3).
(2) We have seen unclosable square roots of the identity operator.
(3) We have also seen closed and unclosable operators whose square roots are themselves (idempotent operators).

22.2 Questions

22.2.1 Compact Operators Having Unbounded Square Roots

We have seen unbounded operators A such that A^2x is only defined for $x = 0$. Hence A^2 is trivially compact. Can we have finer examples?

Question 22.2.1 There are many nonzero compact operators that possess unbounded square roots.

22.2.2 A Bounded Invertible Operator Without Any Closed Square Root

Question 22.2.2 Find a bounded and invertible operator without any closed square root.

22.2.3 A Non-closable Operator Without Any Closable Square Root

Question 22.2.3 Find a non-closable operator without any closable square root whatsoever.

22.2.4 An Operator S with a Square Root T but T* Is Not a Square Root of S*

It is known that $B \in B(H)$ is a square root of some $A \in B(H)$ if and only if B^* is a square root of A^*. This is not always the case for unbounded operators.

Question 22.2.4 Find an operator S with a square root T but T^* is not a square root of S^*.

22.2.5 An Operator S with a Square Root T but $\overline{T}$ Is Not a Square Root of $\overline{S}$

Question 22.2.5 Find an operator S with a square root T but $\overline{T}$ is not a square root of $\overline{S}$.

22.2.6 A Densely Defined Closed Operator T Such That T^2 Is Densely Defined and Non-closable

Question 22.2.6 Give a densely defined closed operator T such that T^2 is densely defined and unclosable.

Remark Put in other words, the foregoing example says that densely defined non-closable operators may possess densely defined closed square roots.

22.2.7 Square Roots of a Self-Adjoint Operator Need Not Have Equal Domains

We have already observed that if A and B are two self-adjoint positive operators with domains $D(A)$ and $D(B)$, respectively, then

$$D(A) = D(B) \implies D(\sqrt{A}) = D(\sqrt{B}).$$

It is therefore natural to wonder whether this property remains valid for arbitrary square roots, that is, if A and B are square roots of some self-adjoint S, i.e. $A^2 = B^2 = S$, is it true that $D(A) = D(B)$?

Question 22.2.7 Find unbounded operators S, A, and B, where S is self-adjoint, such that $A^2 = B^2 = S$ yet $D(A) \neq D(B)$.

Answers

Answer 22.2.1 For example, let T be an unbounded square root of $0 \in B(H)$ (as in Answer 19.2.24). Then T is a square root of the compact operator $0 \in B(H)$.

Let us give nonzero compact operators with unbounded square roots. Let $B \in B(H)$ be a square root of some compact operator C. It is then seen that $T \oplus B$ is always an unbounded square root of the nonzero compact operator $0 \oplus C$.

Answer 22.2.2 Let A be as in Answer 14.2.36, hence A is bounded and invertible. Assume that there is a closed operator B such that $B^2 = A$. Then

$$D(B^2) = D(A) = H \subset D(B)$$

whereby $D(B) = H$. The closed graph theorem then tells us that B is bounded, but this is not consistent with Answer 14.2.36. Thus, A does not possess any closed square root.

Answer 22.2.3 ([263]) Let A be a non-closable (unbounded) operator such that $A^2 = 0$ everywhere on H as in Answer 19.2.24. Assume now that B is a closable square root of A, that is, $B^2 = A$. Hence $B^4 = A^2 = 0$ everywhere on H. Therefore,

$$H = D(B^4) \subset D(B).$$

This signifies that B would be everywhere defined on H, and by remembering that B is closable, it would result that B is everywhere defined and bounded. Hence A too would be bounded and this is the wanted contradiction. Accordingly, the non-closable A does not possess any closable square root.

Answer 22.2.4 ([263]) Consider a non-closable operator A with $D(A) = H$ and $D(A^*) = \{0\}$ (e.g. Berberian's example given in Answer 20.2.8). Set

$$T = \begin{pmatrix} 0 & A \\ 0 & 0 \end{pmatrix}$$

where $D(T) = H \oplus H$. Then T is unbounded and $T^2 = 0$ on $H \oplus H$, i.e. T is a square root of $0 \in B(H \oplus H)$. Hence

$$T^* = \begin{pmatrix} 0 & 0 \\ A^* & 0 \end{pmatrix}$$

on $D(T^*) = \{0\} \oplus H$. Clearly

$$D[(T^*)^2] = \{(0, y) \in D(T^*) : T^*(0, y) \in D(T^*)\} = D(T^*).$$

Therefore, T^* cannot be a square root of $0^* = 0 \in B(H \oplus H)$ for

$$D[(T^*)^2] \neq H \oplus H.$$

Another example is to let A to be an unbounded closed operator with domain $D(A) \subset H$ and to let I to be the identity operator in H. Then set

$$T = \begin{pmatrix} I & A \\ 0 & -I \end{pmatrix}$$

which is closed. Hence

$$T^2 = \begin{pmatrix} I & A \\ 0 & -I \end{pmatrix} \begin{pmatrix} I & A \\ 0 & -I \end{pmatrix} = \begin{pmatrix} I & 0 \\ 0 & I_{D(A)} \end{pmatrix} := S$$

where $I_{D(A)}$ is the identity operator restricted to $D(A)$. Since

$$T^* = \begin{pmatrix} I & 0 \\ A^* & -I \end{pmatrix},$$

it ensures that

$$T^{*2} = \begin{pmatrix} I_{D(A^*)} & 0 \\ 0 & I \end{pmatrix}$$

meaning that T^* is not a square root of $S^* = \begin{pmatrix} I & 0 \\ 0 & I \end{pmatrix}$.

Answer 22.2.5 Simply consider the second example of Answer 22.2.4. Then, by considering $(T^*)^*$ or else, it is seen that

$$\overline{T} = \begin{pmatrix} I & \overline{A} \\ 0 & -I \end{pmatrix}.$$

Therefore,

$$\overline{T}^2 = \begin{pmatrix} I & 0 \\ 0 & I_{D(\overline{A})} \end{pmatrix}$$

and so $\overline{T}$ cannot be a square root of $\overline{S} = \begin{pmatrix} I & 0 \\ 0 & I \end{pmatrix}$.

Answer 22.2.6 We need to find a densely defined closed operator T such that T^2 is also densely defined but $(T^2)^*$ is not densely defined. Let A and B two self-adjoint operators (B is also everywhere defined and bounded) defined on $L^2(\mathbb{R})$ and such that

$$D(AB) = \{0\} \text{ and } D(BA) = D(A)$$

(cf. Answer 21.2.24).
    Now, define

$$T = \begin{pmatrix} B & A \\ 0 & 0 \end{pmatrix}$$

on $D(T) := L^2(\mathbb{R}) \oplus D(A)$. Clearly T is closed on $D(T)$. Moreover,

$$T^2 = \underbrace{\begin{pmatrix} B^2 & BA \\ 0 & 0 \end{pmatrix}}_{\text{noted } S}$$

because

$$D(T^2) = \{(f, g) \in L^2(\mathbb{R}) \times D(A) : (Bf + Ag, 0) \in (L^2(\mathbb{R}), D(A))\}$$
$$= L^2(\mathbb{R}) \oplus D(A)$$
$$= D(S).$$

Obviously T^2 is densely defined. Since $B \in B[L^2(\mathbb{R})]$ and A and B are self-adjoint, it follows that

$$(T^2)^* = \left[\begin{pmatrix} B^2 & 0 \\ 0 & 0 \end{pmatrix} + \begin{pmatrix} 0 & BA \\ 0 & 0 \end{pmatrix} \right]^* = \begin{pmatrix} B^2 & 0 \\ 0 & 0 \end{pmatrix} + \begin{pmatrix} 0 & 0 \\ AB & 0 \end{pmatrix},$$

that is,

$$(T^2)^* = \begin{pmatrix} B^2 & 0 \\ AB & 0 \end{pmatrix}.$$

Thus,

$$D[(T^2)^*] = \{0\} \oplus L^2(\mathbb{R})$$

which is not dense, as required.

Answer 22.2.7 ([263]) We give three square roots of the same self-adjoint (even positive) operator, whose domains are pairwise different. Let T be any unbounded self-adjoint operator with domain $D(T) \subsetneq H$ and set $S = \begin{pmatrix} T & 0 \\ 0 & T \end{pmatrix}$ where $D(S) = D(T) \oplus D(T)$. Then both $A := \begin{pmatrix} 0 & T \\ I & 0 \end{pmatrix}$ and $B := \begin{pmatrix} 0 & I \\ T & 0 \end{pmatrix}$ constitute square roots of S yet

$$D(A) = H \oplus D(T) \neq D(T) \oplus H = D(B).$$

By taking T to be further positive, it is seen that $\begin{pmatrix} \sqrt{T} & 0 \\ 0 & \sqrt{T} \end{pmatrix}$, where $\sqrt{T}$ is the unique positive square root of T, is yet another square root of S whose domain differs from both $D(A)$ and $D(B)$.

Remark Would the conclusion be different if we assumed that the square roots are self-adjoint? The answer is yes. Indeed, let A and B be two self-adjoint square roots of some (necessarily self-adjoint and positive) S, i.e. $A^2 = B^2 = S$. Then $D(A^2) = D(B^2)$ and so

$$D(A) = D(|A|) = D(\sqrt{A^2}) = D(\sqrt{B^2}) = D(|B|) = D(B).$$

Chapter 23
Normality

23.1 Basics

Definition 23.1.1 A densely defined operator A is said to be normal if

$$\|Ax\| = \|A^*x\|, \ \forall x \in D(A) = D(A^*).$$

Here is an equivalent definition.

Proposition 23.1.1 *A densely defined A is normal if and only if A is closed and $AA^* = A^*A$.*

The next result is trivial.

Proposition 23.1.2 *A normal symmetric operator is automatically self-adjoint.*

As a direct consequence of Proposition 21.2.1 (just above Question 21.2.2), we have the following.

Corollary 23.1.3 *Normal operators are maximally normal, that is, if $A \subset B$ and both A and B are normal, then $A = B$.*

Remark It is worth noticing that a spectral theorem for unbounded normal operators does exist (recall that normal operators constitute the largest known class of operators enjoying a spectral theorem). See e.g. [66, 314], or [326].

Let us give a few examples.

Example 23.1.1 Consider the operator A defined in Example 19.1.3. Then A is normal.

Example 23.1.2 Let M_φ be the multiplication operator, which appeared in Example 20.1.12. Then M_φ is always normal on $D(M_\varphi)$ (i.e. for any φ).

© The Author(s), under exclusive license to Springer Nature Switzerland AG 2022
M. H. Mortad, *Counterexamples in Operator Theory*,
https://doi.org/10.1007/978-3-030-97814-3_23

Example 23.1.3 Let A be an unbounded normal operator, and let $U \in B(H)$ be unitary. Then $T := U^*AU$ is normal. Indeed, the closedness of T is clear to readers by now. Since U is unitary, it follows by the normality of A that

$$T^*T = U^*A^*UU^*AU = U^*A^*AU = U^*AA^*U = U^*AUU^*A^*U = TT^*,$$

whereby T is normal, as wished.

23.2 Questions

23.2.1 A Normal T* Such That T Is Not Normal

It is clear that if T is a normal (unbounded) operator, then so is its adjoint T^*. Indeed,

$$T^*(T^*)^* = T^*T^{**} = T^*\overline{T} = T^*T = TT^* = \overline{T}T = (T^*)^*T^*$$

by the normality (and the closedness) of T. The converse holds if one further assumes that T is closed. More precisely, if T is closed such that T^* is normal, then T too is normal.

Question 23.2.1 Find a closable densely defined nonnormal operator T such that T^* is normal.

23.2.2 A T Such That TT* = T*T yet T Is Not Normal

The aim is to see why a densely defined (unclosed) operator T obeying $TT^* = T^*T$ cannot be considered normal. In fact, it is shown that the equality $TT^* = T^*T$, even when occurring on a dense common domain, is not sufficient to force T to be closed. See [87] for related results.

Question 23.2.2 Find a densely defined T such that $TT^* = T^*T$ and yet T is not normal.

23.2.3 An Unbounded Densely Defined Closed Nonnormal Operator T Such That $D(T) = D(T^*)$ and $D(TT^*) = D(T^*T)$

Recall that an unbounded normal operator T does satisfy the two domain conditions $D(T) = D(T^*)$ and $D(TT^*) = D(T^*T)$. What about the converse?

Question 23.2.3 Find an unbounded closed and densely defined operator T such that $D(T) = D(T^*)$ and $D(TT^*) = D(T^*T)$, yet T is not normal.

Remark Technically, if we did not require T to be unbounded, then any $T \in B(H)$ (which is not normal) would do.

23.2.4 A Nonnormal Densely Defined Closed S that Satisfies $T \subset S^*S$ and $T \subset SS^*$

The following result was shown in [254].

Theorem 23.2.1 *Let S and T be two densely defined unbounded operators on a Hilbert space H with respective domains $D(S)$ and $D(T)$. Assume that*

$$\begin{cases} T \subset S^*S, \\ T \subset SS^*. \end{cases}$$

Let $D \subset D(T)$ be dense and a core, for example, for S^*S. If S is closed, then S is normal.

Remark After I proved the foregoing theorem, I came across a result due to Stochel–Szafraniec (Theorem 5 in [356]), which I was not aware of. There, the authors had already shown a similar result to mine. One of their conditions is to take S^*S, restricted to D, to be essentially self-adjoint. But they then obtained in their proof that D is a core for S^*S. In fact, these two conditions are equivalent.

A slight generalization of the previous result is as follows.

Theorem 23.2.2 ([254]) *Let $R, S,$ and T be three densely defined unbounded operators on a Hilbert space with respective domains $D(R), D(S),$ and $D(T)$. Assume that*

$$\begin{cases} T \subset R, \\ T \subset S. \end{cases}$$

Assume further that R and S are self-adjoint. Let $D \subset D(T)$ be dense and a core, for instance, for S. Then $R = S$.

Is the condition on the core that important?

Question 23.2.4 Find a densely defined operator T and a closed and densely defined operator S such that

$$\begin{cases} T \subset S^*S, \\ T \subset SS^*, \end{cases}$$

yet S is not normal.

23.2.5   Two Unbounded Self-Adjoint Operators A and B, B Is Positive, Such That $\overline{AB}$ Is Normal Without Being Self-Adjoint

Let us first give a theorem in which we gather results about the self-adjointness of the normal product of two self-adjoint operators.

Theorem 23.2.3 *Let A and B be two self-adjoint operators. Set $N = AB$ and $M = BA$.*

(i) *If $A, B \in B(H)$, one of them is positive and N (resp., M) is normal, then N (resp., M) is self-adjoint. That is, $AB = BA$ ([2, 236], and [305]).*

(ii) *If only $B \in B(H)$, $B \geq 0$, and N (resp., M) is normal, then N (resp., M) is self-adjoint. Also $BA \subset AB$ (resp., $BA = AB$), [236].*

(iii) *If $B \in B(H)$, $A \geq 0$, and N (resp., M) is normal, then N (resp., M) is self-adjoint. Also $BA \subset AB$ (resp., $BA = AB$), [145].*

(v) *If both A and B are unbounded and N is normal, then it is self-adjoint whenever $B \geq 0$, [236].*

By looking closely at the cases of the theorem above, we see that there are cases that are not covered. For example, if both A and B are unbounded and $\overline{N}$ is normal, then does N need to be essentially self-adjoint when $B \geq 0$?

Question 23.2.5 (❀❀) Find an unbounded self-adjoint operator A and a positive unbounded self-adjoint operator B such that $\overline{AB}$ is normal, but it is not self-adjoint, i.e. AB is not essentially self-adjoint.

Further Reading See also [183]. We may add to the above list the following result (which might be postponed by readers who are not already familiar with unbounded hyponormal operators and the spectrum of unbounded operators, both yet to be defined).

Theorem 23.2.4 ([84]) *Let A and B be two self-adjoint operators such that B is bounded and positive. If BA is hyponormal, then both BA and AB are self-adjoint (and $AB = BA$) whenever $\sigma(BA) \neq \mathbb{C}$.*

The following related result was also shown in [84] ("half of it" may be seen as a consequence of the foregoing theorem).

Proposition 23.2.5 *Let $A, B \in B(H)$ be two self-adjoint operators such that B is positive. If BA (or AB) is hyponormal, then both BA and AB are self-adjoint, that is, $AB = BA$.*

Remark Notice in the end that the previous proposition improves a similar result obtained in [175]. There, the author showed the same result by assuming, in addition to the above hypotheses, that B had a dense range.

23.2.6   Two Unbounded Self-Adjoint Operators A and B Such That A Is Positive, AB is Normal But Non-self-adjoint

Another case not covered by Theorem 23.2.3 just above is answered next in the negative:

> **Question 23.2.6 ([240])** Find two unbounded self-adjoint operators A and B such that A is positive, AB is normal but non-self-adjoint.

Answers

Answer 23.2.1 The counterexample is simple once we have noticed that the main issue is closedness. So let T be the restriction of the identity operator to some dense (non-closed) domain in some Hilbert space H. Then, T is unclosed and so it cannot be normal. But, $T^* = I$ is the full identity operator on H and so it is normal.

Answer 23.2.2 By Question 21.2.28, we know that there is a densely defined operator T such that

$$D(TT^*) = D(T^*T) = \{0\}.$$

Hence, trivially $TT^* = T^*T$ (coinciding only on $\{0\}$). However, T cannot be normal as it is unclosed.

Let us give another example where $TT^* = T^*T$ coincide on some dense domain. Let $T = I_{D(T)}$ be the identity operator restricted to $D(T) \subset H$. Then $T^* = I$, and so

$$D(TT^*) = \{x \in H : x \in D(T)\} = D(T)$$

and obviously $D(T^*T) = D(T)$. Hence

$$TT^*f = T^*Tf$$

for all $f \in D(T^*T) = D(TT^*) = D(T)$. Such T cannot be normal as:

(1) T is unclosed...
(2) Or as $D(T) \neq D(T^*)$...
(3) Or $T \subset I$ and if T were normal, it would ensure that $T = I$ by the maximality of normal operators...
(4) Or...

Answer 23.2.3 An example already appeared in Answer 20.2.21. Let us give another example. Let A be a densely defined closed operator with a domain $D(A)$ such that $D(A) = D(A^*) \subset H$ (cf. Answer 26.2.19 below). Define T on $H \oplus H$ by

$$T = \begin{pmatrix} 0 & A/2 \\ -A^* & 0 \end{pmatrix}$$

with a domain $D(T) = D(A^*) \oplus D(A) = D(A) \oplus D(A)$. Then T is closed and

$$T^* = \begin{pmatrix} 0 & -A \\ A^*/2 & 0 \end{pmatrix}.$$

Hence

$$T^*T = \begin{pmatrix} AA^* & 0 \\ 0 & \frac{A^*A}{4} \end{pmatrix} \text{ and } TT^* = \begin{pmatrix} \frac{AA^*}{4} & 0 \\ 0 & A^*A \end{pmatrix}.$$

Finally, it is plain that $D(TT^*) = D(T^*T)$ and that T is not normal.

Remark Observe also that

$$T^*T - TT^* = \frac{3}{4} \begin{pmatrix} AA^* & 0 \\ 0 & -A^*A \end{pmatrix}$$

is self-adjoint on $D(AA^*) \oplus D(A^*A)$.

Answer 23.2.4 ([254]) Let S be defined by $Sf(x) = f'(x)$ on $D(S) = \{f \in L^2(0, 1) : f' \in L^2(0, 1)\}$. Then S is densely defined and closed, but it is not normal. Indeed, $S^* f(x) = -f'(x)$ on

$$D(S^*) = \{f \in L^2(0, 1), f' \in L^2(0, 1) : f(0) = f(1) = 0\}$$

and so S is not normal for $D(S) \neq D(S^*)$. Moreover,

$$SS^* f(x) = S^* Sf(x) = -f''(x)$$

with

$$D(SS^*) = \{f \in L^2(0, 1), f'' \in L^2(0, 1) : f(0) = f(1) = 0\}$$

and

$$D(S^* S) = \{f \in L^2(0, 1), f'' \in L^2(0, 1) : f'(0) = f'(1) = 0\}.$$

Now, let T be defined by $Tf(x) = -f''(x)$ on

$$D(T) = \{f \in L^2(0, 1), f'' \in L^2(0, 1) : f(0) = f(1) = 0 = f'(0) = f'(1)\}.$$

Then

$$\begin{cases} T \subset S^* S, \\ T \subset SS^*, \end{cases}$$

T and S are densely defined, S is closed, and, as we have just seen, S is not normal.

Answer 23.2.5 ([236]) Let us consider the operators A and B defined as:

$$A = -i\frac{d}{dx} : D(A) \rightarrow L^2(\mathbb{R}), B = |x| : D(B) \rightarrow L^2(\mathbb{R})$$

where $D(B) = \{f \in L^2(\mathbb{R}) : xf \in L^2(\mathbb{R})\}$ and $D(A) = H^1(\mathbb{R})$. Then $N := AB$ is defined by:

$$Nf(x) = -i(|x|f(x))' = -i|x|f'(x) - i \ \text{sign}(x)f(x)$$

for f in $D(N) = \{f \in L^2(\mathbb{R}) : |x|f, -i(|x|f)' \in L^2(\mathbb{R})\}$.
  Then N has the required properties. Let us carry this out in several steps.

(1) **The operator N is densely defined and not closed:** The operator N is densely defined since it contains $C_0^\infty(\mathbb{R})$. It is not closed for it does not equal its closure, which is to be found below.

(2) **The operator N' defined by:**

$$N'f(x) = -i|x|f'(x) - i \ \mathrm{sign}(x)f(x)$$

on $D(N') = \{f \in L^2(\mathbb{R}) : |x|f' \in L^2(\mathbb{R})\}$ **is a closed extension of** N:

Clearly, N' is an extension of N. We need to check that N' is closed on this domain with respect to the graph norm of N'. Take $(f_n, N'f_n) \in G(N')$ such that $(f_n, N'f_n) \to (f, g)$. Since $f_n \to f$ in L^2, then in the distributional sense we have $f'_n \to f'$. On $\mathbb{R}\setminus\{0\}$, we have $|x|f'_n \to |x|f'$ in the distributional sense again. By uniqueness of the limit, one derives $N'f = |x|f'$ for almost every x, hence we have a full equality in $L^2(\mathbb{R})$. This tells us that N' is closed on $D(N')$.

(3) **The closure $\overline{N}$ of N is N':** This follows from the inclusions:

$$C_0^\infty(\mathbb{R}\setminus\{0\}) \subset D(N) \subset D(\overline{N}).$$

Hence $D(N)$ is dense in $D(\overline{N})$ with respect to the graph norm of $\overline{N}$.

(4) $N^*g = -i|x|g'$ **with** $D(N^*) = \{f \in L^2(\mathbb{R}) : |x|f' \in L^2(\mathbb{R})\}$. Let $f \in C_0^\infty(\mathbb{R}\setminus\{0\})$, and let $g \in L^2(\mathbb{R})$. We have

$$\langle Nf, g\rangle = \int_{\mathbb{R}} (|x|f(x))'\overline{ig(x)}dx = ((|x|f)', \overline{ig})$$

since $(|x|f)' \in C_0^\infty(\mathbb{R}\setminus\{0\})$. By the definitions of the distributional derivative and the product of distributions, since $|x|$ is C^∞ on $\mathbb{R}\setminus\{0\}$ one has

$$((|x|f)', \overline{ig}) = -(|x|f, -i\overline{g}') = (f, i|x|\overline{g}').$$

We also have $\langle f, h\rangle = (f, \overline{h})$, where $h \in L^2(\mathbb{R})$. Consequently $h = -i|x|g'$ as a distribution. But h is in $L^2(\mathbb{R})$, and so $|x|g' \in L^2(\mathbb{R})$. Therefore, $N^*g = -i|x|g'$ on

$$D(N^*) = \{g \in L^2(\mathbb{R}) : |x|g' \in L^2(\mathbb{R})\}.$$

(5) $\overline{N}$ **is normal but it is not self-adjoint:** First, recall that $D(N^*) = D(\overline{N}^*)$. Clearly $\overline{N}$ is not self-adjoint (it is not even symmetric for $\overline{N} - \overline{N}^* \subseteq \pm i$). However, it is normal as

$$\overline{N}\,\overline{N}^* f(x) = \overline{N}(-i|x|f'(x)) = -i(-i|x||x|f'(x))' = -x^2 f''(x) - 2xf'(x)$$

and

$$\overline{N}^*\,\overline{N} f(x) = \overline{N}^*[-i(|x|f(x))'] = -x^2 f''(x) - 2xf'(x)$$

Readers may check in the end that

$$D(\overline{N}\,\overline{N}^*) = \{f \in L^2(\mathbb{R}) : xf', x^2 f'' \in L^2(\mathbb{R})\}$$

and that $D(\overline{N}^* \overline{N}) = D(\overline{N}\,\overline{N}^*)$.

Remark It is worth noticing in passing that the operator $\overline{N}$ is unitarily equivalent to $M = M_+ \oplus M_-$, where M_+ is defined on $L^2(\mathbb{R})$ by $M_+ f(s) = (s - \frac{1}{2}i) f(s)$, and M_- is defined on $L^2(\mathbb{R})$ by $M_- f(s) = (s + \frac{1}{2}i) f(s)$ on their maximal domains. The required unitary transformation is given by:

$$Uf = U_+ f_+ \oplus U_- f_-$$

where f_+ is the restriction of f to $\mathbb{R}^+$ and f_- is the restriction of f to $\mathbb{R}^-$. The operator U_+ is defined by $U_+ = \mathcal{F}^{-1} V$, where $\mathcal{F}^{-1}$ is the inverse L^2-Fourier transform and $V : L^2(\mathbb{R}^+) \to L^2(\mathbb{R})$ is the unitary operator defined by:

$$(Vf)(t) = e^{\frac{t}{2}} f(e^t),$$

whereas U_- is defined by $U_- = \mathcal{F}^{-1} W$, where $W : L^2(\mathbb{R}^-) \to L^2(\mathbb{R})$ defined by:

$$(Wf)(t) = e^{-\frac{t}{2}} f(e^{-t}).$$

This is not obvious, and details may be consulted in [237] or [240].

Answer 23.2.6 ([240]) Let $B = -i \frac{d}{dx}$. Then B is self-adjoint on $H^1(\mathbb{R})$. Take the self-adjoint positive $Af(x) = (1 + |x|) f(x)$ with the following domain:

$$D(A) = \{f \in L^2(\mathbb{R}) : (1 + |x|) f \in L^2(\mathbb{R})\}.$$

First, we check that $M := AB$ is normal, i.e. M is closed and $MM^* = M^*M$. That M is closed follows from the (bounded) invertibility of A and the closedness of B.

For f in $D(M) = \{f \in L^2(\mathbb{R}) : (1 + |x|) f' \in L^2(\mathbb{R})\}$, one has

$$Mf(x) = -i(1 + |x|) f'(x).$$

To find its adjoint, we proceed as in Answer 23.2.5. We then find

$$M^* f(x) = \mp i f(x) - i(1 + |x|) f'(x).$$

Clearly

$$MM^* f(x) = M^* M f(x) = -(1 + |x|)(\mp 2 f'(x) - (1 + |x|) f''(x)).$$

The previous equations combined with

$$D(MM^*) = D(M^*M) = \{f \in L^2(\mathbb{R}) : (1+|x|)f' \in L^2(\mathbb{R}), (1+|x|)f'' \in L^2(\mathbb{R})\}$$

say that AB is a normal operator. Observe in the end that M is not self-adjoint.

Chapter 24
Absolute Value. Polar Decomposition

24.1 Basics

As in the case of bounded operators, closed operators too enjoy a polar decomposition.

Theorem 24.1.1 *Let A be a densely defined closed operator. Then, there exists a partial isometry U with initial space $(\ker A)^{\perp}$ and final space $\overline{\operatorname{ran} A}$ such that*

$$A = U|A|.$$

Just as in the case of their bounded counterparts, nicer classes of unbounded operators still have a more informative polar decomposition.

Theorem 24.1.2 *Let A be a normal operator with a domain $D(A)$. Then*

$$A = U|A| = |A|U$$

where $U \in B(H)$ is unitary.
In particular, if A is self-adjoint, then U may be taken to be a fundamental symmetry.

24.2 Questions

24.2.1 Closed Densely Defined Operators S and T Such That $S \subset T$ but $|S| \not\subset |T|$

Question 24.2.1 Find closed densely defined operators S and T such that $S \subset T$ but $|S| \not\subset |T|$.

© The Author(s), under exclusive license to Springer Nature Switzerland AG 2022   451
M. H. Mortad, *Counterexamples in Operator Theory*,
https://doi.org/10.1007/978-3-030-97814-3_24

24.2.2   A Closed Operator T and an Unclosed Operator A Such That $|T^*| = |T| = |A|$

When A is closed, then we know how to compute $|A|$. When A is unclosed, it is, in general, impossible to define $|A|$.

Question 24.2.2 There is a closed and densely defined operator T and a non-closed operator A such that

$$|T^*| = |T| = |A|.$$

24.2.3   A Non-closed A with $D(A) \neq D(|A|)$

We already know that if A is densely defined and closed, then $D(A) = D(|A|)$.

Question 24.2.3 Find a densely defined non-closed A with $D(A) \neq D(|A|)$.

24.2.4   A Non-closed Densely Defined Operator A Without Any Polar Decomposition

Question 24.2.4 Exhibit a non-closed and densely defined operator A that does not have a polar decomposition (in terms of partial isometries).

24.2.5   A Non-closed Densely Defined Operator A and a Unitary $U \in B(H)$ Such That $A \subset U|A|$

Question 24.2.5 Find a non-closed and densely defined operator A and a unitary $U \in B(H)$ such that

$$A \subset U|A|.$$

24.2.6 A Non-closed Densely Defined Operator A and a Unitary U ∈ B(H) Such That A ⊂ U|A| and UA ⊂ |A|

> **Question 24.2.6** Find a non-closed and densely defined operator A and a unitary $U \in B(H)$ such that
> $$A \subset U|A| \text{ and } UA \subset |A|.$$

24.2.7 A Closed and Symmetric A Such That $|A^n| \neq |A|^n$

There are classes of unbounded operators A for which $|A^n| = |A|^n$ for all $n \in \mathbb{N}$ (see e.g. [172] or [365]). In particular, the last equations hold for unbounded normal operators. What about the class of closed symmetric operators?

> **Question 24.2.7** Find a closed and symmetric A such that
> $$|A^n| \neq |A|^n$$
> for some $n \in \mathbb{N}$.

24.2.8 A Self-Adjoint B ∈ B(H) and a Self-Adjoint Positive A Such That BA Is Closed Without Being Self-Adjoint

> **Question 24.2.8** Find a bounded self-adjoint operator B and a self-adjoint and positive A such that BA is closed without being self-adjoint.

Answers

Answer 24.2.1 Let T be a densely defined closed symmetric operator, i.e. $T \subset T^*$. If $|T| \subset |T^*|$ were true, then $|T| = |T^*|$ by the self-adjointness of both $|T|$ and $|T^*|$. Hence T would be normal. But a normal symmetric operator is automatically self-adjoint. So, for a counterexample to the question, it suffices to consider any densely defined closed symmetric operator that is not self-adjoint.

Answer 24.2.2 Let A and T be as in the example of Answer 21.2.14. Then T is normal as it is unitarily equivalent via the L^2-Fourier transform to the multiplication operator (on the appropriate domain) by a complex-valued function. Hence $|T^*| = |T|$. Since $A^*A = T^*T$, it results that

$$|T^*| = |T| = |A|.$$

Answer 24.2.3 Obviously, we are not concerned with examples where it is impossible to compute $|A|$. Once that is said, the example that comes to mind is again that of Answer 21.2.14. Recall that $T = \frac{d}{dx}$ is closed when defined on $H^1(\mathbb{R}) = \{f \in L^2(\mathbb{R}) : f' \in L^2(\mathbb{R})\}$. So, if A is the restriction of T to $H^2(\mathbb{R}) = \{f \in L^2(\mathbb{R}) : f'' \in L^2(\mathbb{R})\}$, then A is not closed. By Question 24.2.2, we know that $|T| = |A|$. Therefore (as T is closed)

$$H^1(\mathbb{R}) = D(T) = D(|T|) = D(|A|) \neq D(A) = H^2(\mathbb{R}),$$

as wished.

Answer 24.2.4 As already observed, the primary handicap is that we cannot always find $|A|$ in the event of the unclosedness of A. But, in order that the answer be exhaustive, we need to treat the situations where it is possible to compute $|A|$. So, let A and T be such that $|T| = |A|$ and $D(A) \neq D(T)$, where T is closed and A is not so. Now, if the non-closed A admitted a polar decomposition, then we would have

$$A = U|A|$$

for some partial isometry $U \in B[L^2(\mathbb{R})]$. But then

$$D(A) = D(U|A|) = D(|A|) = D(|T|) = D(T),$$

and this impossible. Accordingly, no such polar decomposition exists.

Answer 24.2.5 The question as riddling as it seems to be uses somewhat similar arguments as before. Indeed, consider yet again two operators A and T such that $|T| = |A|$ and $D(A) \subsetneq D(T)$.

Since T is normal, we know from the (unitary) polar decomposition that

$$T = U|T| = |T|U$$

for some *unitary* $U \in B[L^2(\mathbb{R})]$. Thus,

$$A \subset T = U|T| = U|A|,$$

as wished.

Answer 24.2.6 Replace the normal T that appeared in Answer 21.2.14 by the self-adjoint

$$Tf(x) = if'(x)$$

on $H^1(\mathbb{R}) = \{f \in L^2(\mathbb{R}) : f' \in L^2(\mathbb{R})\}$. Set $A = T|_{H^2(\mathbb{R})}$, where $H^2(\mathbb{R}) = \{f \in L^2(\mathbb{R}) : f'' \in L^2(\mathbb{R})\}$. Then A is not closed and $T = A^*$. Hence $A^*A = T^2$ is self-adjoint. Since T is self-adjoint,

$$T = U|T| = |T|U$$

for some *fundamental symmetry* $U \in B[L^2(\mathbb{R})]$. Now, as in Answer 24.2.5, $A \subset U|A|$. Left multiplying by the unitary and self-adjoint U, we get the other "inclusion" $UA \subset |A|$.

Answer 24.2.7 Let A be a closed, symmetric, and positive operator as in Question 20.2.23. Then $D(A^2) = \{0\}$ and so "$|A^2|$" does not make sense whereas $|A|^2$ is self-adjoint.

Remark It would be interesting to find (if this is possible) a counterexample in the case of the density of $D(A^2)$. Interested readers should give this a go!

Answer 24.2.8 Let H be a Hilbert space, and let T be an unbounded closed non-self-adjoint operator with a domain $D(T) \subset H$. Then $T = U|T|$ for some partial isometry $U \in B(H)$. Also, $\begin{pmatrix} 0 & U^* \\ U & 0 \end{pmatrix}$ is self-adjoint and bounded and $\begin{pmatrix} |T| & 0 \\ 0 & 0 \end{pmatrix}$ is clearly self-adjoint, unbounded, and positive. Nevertheless,

$$\begin{pmatrix} 0 & U^* \\ U & 0 \end{pmatrix} \begin{pmatrix} |T| & 0 \\ 0 & 0 \end{pmatrix} = \begin{pmatrix} 0 & 0 \\ U|T| & 0 \end{pmatrix} = \begin{pmatrix} 0 & 0 \\ T & 0 \end{pmatrix}$$

is closed without being self-adjoint.

Chapter 25
Unbounded Nonnormal Operators

25.1 Questions

25.1.1 An $S \in B(H)$ Such That $0 \notin \overline{W(S)}$, and an Unbounded Closed Hyponormal T Such That $ST \subset T^*S$ but $T \neq T^*$

Recall the following definition: An unbounded densely defined operator A is called hyponormal if:

(1) $D(A) \subset D(A^*)$,
(2) $\|A^*x\| \leq \|Ax\|$ for all $x \in D(A)$.

The following result was shown in [83]:

Theorem 25.1.1 *Let S be a bounded operator on a $\mathbb{C}$-Hilbert space H such that $0 \notin \overline{W(S)}$. Let T be an unbounded and closed hyponormal operator with a dense domain $D(T) \subset H$. If $ST^* \subset TS$, then T is self-adjoint.*

The condition $ST^* \subset TS$ in the foregoing result is not purely conventional, i.e. we may not obtain the desired result by merely assuming instead that $ST \subset T^*S$, even with a slightly stronger condition (i.e. symmetricity in lieu of hyponormality).

Question 25.1.1 Find an operator $S \in B(H)$ with $0 \notin \overline{W(S)}$, and an unbounded closed symmetric T such that $ST \subset T^*S$ and yet $T \neq T^*$.

Remark What about assuming further that T is boundedly invertible in the previous question? This hypothesis does yield the self-adjointness of T (if we keep all of the other assumptions). This was also shown [83].

© The Author(s), under exclusive license to Springer Nature Switzerland AG 2022
M. H. Mortad, *Counterexamples in Operator Theory*,
https://doi.org/10.1007/978-3-030-97814-3_25

25.1.2   A Closed A with $A^*AA \subset AA^*A$ Does Not Yield the Quasinormality of A

Recall that a densely defined closed operator A is said to be quasinormal if $A^*AA = AA^*A$. The previous equality may be weakened to $AA^*A \subset A^*AA$ (see [172], cf. [365]). What about the reverse "inclusion"?

Question 25.1.2 (Cf. [172]) Find a closed and densely defined operator A such that

$$A^*AA \subset AA^*A$$

but A is not quasinormal.

25.1.3   An Unbounded Paranormal Operator T Such That $D(T^*) = \{0\}$

First, recall that a linear operator $A : D(A) \subset H \to H$ is said to be paranormal if

$$\|Ax\|^2 \le \left\| A^2x \right\| \|x\|$$

for all $x \in D(A^2)$. This is clearly equivalent to $\|Ax\|^2 \le \|A^2x\|$ for all *unit* vectors $x \in D(A^2)$.

It is plain that hyponormal operators are closable. Indeed, if A is hyponormal, then $D(A) \subset D(A^*)$ gives the density of $D(A^*)$. This also says that e.g. quasinormal are closable. In particular, if A is symmetric (densely defined) or quasinormal say, then

$$D(A^*) \ne \{0\}.$$

What about paranormal operators? A. Daniluk was the first to find a counterexample in [81] of a densely defined paranormal operator T that is not closable. His example was not too complicated, but it did require quite a few results before getting to the final counterexample (readers may consult [81] for further details). It is worth noticing that paranormal operators appearing there have an invariant domain, and this is the essence of its difficulty.

Next, we provide a somehow different counterexample in the sense that in our case the domain of the adjoint is trivial!

Question 25.1.3 ([264]) Find a densely defined paranormal operator T such that

$$D(T^*) = \{0\}.$$

25.1.4   A Closable Paranormal Operator Whose Closure Is Not Paranormal

Again, A. Daniluk found in [81] a densely defined closable paranormal operator T for which $\overline{T}$ was not paranormal. Next, we give a different counterexample.

Question 25.1.4 (⊛) ([264]) Find a closable paranormal operator whose closure fails to remain paranormal.

25.1.5   A Densely Defined Closed Operator T Such That Both T and T^* Are Paranormal and $\ker T = \ker T^* = \{0\}$ but T Is Not Normal

As mentioned just before Question 14.2.10, T. Ando showed in [5] that $T \in B(H)$ is normal if and only if both T and T^* are paranormal and such that $\ker T = \ker T^*$. This was then improved by T. Yamazaki and M. Yanagida in [393] who proved the same result but without the requirement $\ker T = \ker T^*$. Can this result be extended to unbounded densely defined closed operators?

Question 25.1.5 Find a densely defined closed operator T such that both T and T^* are paranormal and one-to-one (so $\ker T = \ker T^*$) but T is not normal.

25.1.6   Q-Normal Operators

A densely defined closed operator A satisfying

$$AA^* = qA^*A,$$

where $q \in (1, \infty)$, is said to be q-normal.

It is not hard to see (see e.g. [282]) that A is q-normal if and only if

$$\|A^*x\| = \sqrt{q}\|Ax\|, \ \forall x \in D(A) = D(A^*).$$

Notice in the end that from Answer 21.2.41, there is a densely defined closed operator A such that $D(AA^*) \cap D(A^*A) = \{0\}$. Hence

$$AA^* = qA^*A$$

holds *for all* $q \in \mathbb{C}$! It is therefore more interesting to find an example satisfying the previous equation on some common dense domain. This will be seen just below.

Question 25.1.6 Find a densely defined and closed operator A that satisfies

$$AA^* = qA^*A$$

where $q \in (1, \infty)$.

Further Reading For more properties, applications, and further investigations, readers may be interested in consulting: [50, 61, 200, 282, 283, 286, 329], and [397].

25.1.7   On the Operator Equations $A^*A = A^n$, $n \geq 3$

In [88], the following result was established.

Theorem 25.1.2 *Let $n \in \mathbb{N}, n \geq 2$. Let H be a complex Hilbert space, and let $A \in B(H)$. Then A is a solution of the equality*

$$A^n = A^*A$$

if and only if:

- $A = A^*$ *(if $n = 2$).*

- *There is a family $P_1, \ldots, P_n \in B(H)$ of orthogonal projections such that $P_j P_k = 0, (j \neq k)$ such that*

$$A = \sum_{k=1}^{n} e^{\frac{2k\pi i}{n}} P_k$$

(if $n \geq 3$).

Question 25.1.7 (⊛) (Cf. Question 21.2.31) Can you find a densely defined closed *unbounded* operator A such that $A^*A = A^n$ with $n \geq 3$?

Answers

Answer 25.1.1 Consider any unbounded densely defined closed symmetric operator T that is *not self-adjoint*. Let $S = I$, i.e. the identity operator on the entire Hilbert space. Then $0 \notin \overline{W(S)}$. Finally, it is plain that

$$T = ST \subset T^* = T^*S.$$

Answer 25.1.2 ([172]) Take a closed densely defined operator A such that $D(A^2) = \{0\}$ (we have seen quite a few by now). Hence

$$D(A^*AA) = \{x \in D(A^2) : A^2x \in D(A^*)\} = \{0\}.$$

Since A is closed, it possesses a polar decomposition $A = U|A|$, where $U \in B(H)$ is a partial isometry. Among the properties of this decomposition, we recall (see e.g. [326]):

$$A = |A^*|U \text{ and } A^* = U^*|A^*| = |A|U^*.$$

Therefore,

$$\begin{aligned} AA^*A &= U|A|A^*U|A| \\ &= U|A|U^*|A^*|U|A| \\ &= UA^*|A^*|U|A| \\ &= UA^*A|A| \\ &= U|A|^2|A| \\ &= U|A|^3. \end{aligned}$$

Since $U|A|^3$ is densely defined, it follows that AA^*A is densely defined. Hence $A^*AA \subset AA^*A$. Now, if A were quasinormal, then we would have $A^*AA = AA^*A = U|A|^3$. But this is clearly not coherent with $D(A^*AA) = \{0\}$.

To recap, $A^*AA \subset AA^*A$ but $A^*AA \neq AA^*A$, as wished.

Answer 25.1.3 Let T be a densely defined operator T such that

$$D(T^2) = D(T^*) = \{0\}$$

as in Answer 21.2.28. Then for $x \in D(T^2) = \{0\}$

$$\|Tx\|^2 = \|T^2x\|\|x\| = 0,$$

i.e. T is trivially paranormal.

Answer 25.1.4 Let $\mathcal{F}$ be the usual $L^2(\mathbb{R})$-Fourier transform, and let A be its restriction to the dense subspace $C_0^\infty(\mathbb{R})$, then define

$$T = \begin{pmatrix} 0 & A \\ \frac{1}{2}A & 0 \end{pmatrix}$$

where $D(T) = C_0^\infty(\mathbb{R}) \oplus C_0^\infty(\mathbb{R})$. Then T is densely defined. It is closable for e.g.

$$\begin{pmatrix} 0 & A \\ \frac{1}{2}A & 0 \end{pmatrix} \subset \begin{pmatrix} 0 & \mathcal{F} \\ \frac{1}{2}\mathcal{F} & 0 \end{pmatrix}$$

and the latter is clearly closed (in fact everywhere defined and bounded!).

Since $D(A^2) = \{0\}$, it ensues that $D(T^2) = \{(0,0)\}$. So T is trivially paranormal. Therefore, there only remains to show that $\overline{T}$ is not paranormal. Clearly,

$$\overline{T} = \begin{pmatrix} 0 & \mathcal{F} \\ \frac{1}{2}\mathcal{F} & 0 \end{pmatrix}.$$

Hence

$$\overline{T}^2 = \begin{pmatrix} 0 & \mathcal{F} \\ \frac{1}{2}\mathcal{F} & 0 \end{pmatrix}\begin{pmatrix} 0 & \mathcal{F} \\ \frac{1}{2}\mathcal{F} & 0 \end{pmatrix} = \begin{pmatrix} \frac{1}{2}\mathcal{F}^2 & 0 \\ 0 & \frac{1}{2}\mathcal{F}^2 \end{pmatrix}$$

(defined now on all of $L^2(\mathbb{R}) \oplus L^2(\mathbb{R})$). If $\overline{T}$ were paranormal, it would ensue that

$$\forall(f,g) \in L^2(\mathbb{R}) \oplus L^2(\mathbb{R}) : \|\overline{T}(f,g)\|^2 \leq \|\overline{T}^2(f,g)\|\|(f,g)\|.$$

In particular, this would be true for some $(0, g)$ with $\|g\|_2 \neq 0$, where $\| \cdot \|_2$ denotes the usual $L^2(\mathbb{R})$-norm. That is, we would have

$$\|\mathcal{F}g\|_2^2 \leq \frac{1}{2}\|\mathcal{F}^2g\|_2\|g\|_2.$$

By the Plancherel theorem $\|\mathcal{F}g\|_2 = \|g\|_2$, and by the general theory, $\mathcal{F}^2g(x) = g(-x)$. Thus the previous inequality would become

$$\|g\|_2^2 \leq \frac{1}{2}\|g\|_2\|g\|_2 \text{ or merely } 1 \leq \frac{1}{2}$$

which is absurd. Accordingly, $\overline{T}$ is not paranormal, as wished.

Answer 25.1.5 ([264]) From Question 21.2.46, we have a densely defined unbounded closed operator T for which:

$$D(T^2) = D(T^{*2}) = \{0\}.$$

Hence, both T and T^* are trivially paranormal (as in say Answer 25.1.3).

We also know from Answer 21.2.46 that both T and T^* are one-to-one and so

$$\ker T = \ker T^* (= \{0\}).$$

However, T cannot be normal for it were, T^2 would too be normal, in particular it would be densely defined, which is impossible here.

Answer 25.1.6 The example to be given is borrowed from Exercise 22 on Pages 56–57 in [326] (see also [282]). Let $q \in (1, \infty)$ and define

$$A(\cdots, x_{-1}, x_0, \boldsymbol{x_1}, \cdots) = (\cdots, qx_{-2}, \sqrt{q}x_{-1}, \boldsymbol{x_0}, \cdots)$$

on $D(A) = \{(x_n) \in \ell^2(\mathbb{Z}) : (q^{-n/2}x_n) \in \ell^2(\mathbb{Z})\}$. Then it may be shown that

$$A^*(\cdots, x_{-1}, \boldsymbol{x_0}, x_1, \cdots) = (\cdots, \sqrt{q}x_0, \boldsymbol{x_1}, q^{-1/2}x_2, \cdots)$$

on

$$D(A^*) = \{(x_n) \in \ell^2(\mathbb{Z}) : (q^{(1-n)/2}x_n) \in \ell^2(\mathbb{Z})\} = D(A).$$

In the end, let $(x_n) \in D(A)$. Then

$$\begin{aligned}
\|A^*(x_n)\|^2 &= \cdots + q^2|x_{-1}|^2 + q|x_0|^2 + |x_1|^2 + q^{-1}|x_2|^2 + \cdots \\
&= q(\cdots + q|x_{-1}|^2 + |x_0|^2 + q^{-1}|x_1|^2 + \cdots) \\
&= q\|A(x_n)\|^2,
\end{aligned}$$

as needed.

Answer 25.1.7 ([88]) The answer is negative. Let A be a closed and densely defined operator, which obeys $A^*A = A^n$, where $n \geq 3$. Then

$$A^*A = A^n \implies AA^*A = A^{n+1} \implies AA^*A = A^nA \implies AA^*A = A^*AA,$$

showing the quasinormality of A. It then follows that $D(A) \subset D(A^*)$ (as e.g. A is hyponormal). Hence

$$D\left(A^2\right) \subseteq D\left(A^*A\right) = D\left(A^n\right)$$

or merely

$$D\left(A^2\right) = D\left(A^n\right).$$

Also

$$D\left(A^3\right) \subseteq D\left(A^*AA\right) = D\left(A^{n+1}\right)$$

so that

$$D\left(A^2\right) = D\left(A^{n+1}\right).$$

Now, since A is closed, it follows that A^2 is closed as it is already quasinormal (see e.g. Proposition 5.2 in [351]). Also, the quasinormality of A yields that of A^2 (by Corollary 3.8 in [172], say) and so A^2 is hyponormal. Therefore,

$$D\left(A^2\right) \subseteq D[\left(A^2\right)^*]$$

and

$$D\left(A^2\right) = D\left(A^4\right).$$

In the end, according to Corollary 2.2 in [360], it follows that A^2 is everywhere defined and bounded on H. Hence $D(A) = H$ and so the closed graph theorem now intervenes to make $A \in B(H)$, as coveted.

Chapter 26
Commutativity

26.1 Basics

We have already introduced the concept of commutativity of a bounded operator with an unbounded one. In the case of two densely defined operators A and B, things are more delicate. Indeed, a major obstacle is that $AB - BA$ may only be defined on $\{0\}$ as we have already observed on many occasions. The remark below the integral form of the spectral theorem for (bounded) self-adjoint operators is the definition we are adopting in the unbounded case. Notice that we have already touched on this definition implicitly in some cases.

Definition 26.1.1 Two (unbounded) self-adjoint (or normal) operators A and B are said to strongly commute if their corresponding spectral measures commute.

Theorem 26.1.1 ([302]) *Let A and B be two self-adjoint operators. Then the following statements are equivalent:*

(1) A and B strongly commute.
(2) If $\operatorname{Im} \lambda$ and $\operatorname{Im} \mu$ are nonzero, then

$$R(\lambda, A)R(\mu, B) = R(\mu, B)R(\lambda, A).$$

(3) For all $x, y \in \mathbb{R}$, $e^{iyB}e^{ixA} = e^{ixA}e^{iyB}$.

Proposition 26.1.2 *Let A and B be two self-adjoint (or normal) operators.*

1. If $A, B \in B(H)$, then $AB = BA$ iff A and B strongly commute.
2. If $B \in B(H)$, then $BA \subset AB$ iff A and B strongly commute.

It is important for the readers to know that an expression of the type $AB = BA$ (i.e., the pointwise commutativity) does not mean the strong commutativity of A and B even when these operators are (unbounded) self-adjoint; see Question 26.2.8 (cf. Question 26.2.7).

© The Author(s), under exclusive license to Springer Nature Switzerland AG 2022
M. H. Mortad, *Counterexamples in Operator Theory*,
https://doi.org/10.1007/978-3-030-97814-3_26

The following result due to K. Schmüdgen (see, e.g., Lemma 5.31 in [326]) is of great interest for Answer 26.2.8 (as well as for another counterexample below).

Lemma 26.1.3 *Let X and Y be bounded self-adjoint operators on a Hilbert space H such that $\ker(X) = \ker(Y) = \{0\}$, and let Q be the projection onto the closure of $[X, Y]H$ (where $[X, Y]$ designates the commutator of X and Y). Set $D = XY(I - Q)H$. Then $A := X^{-1}$ and $B := Y^{-1}$ are self-adjoint operators obeying:*

1. *$D \subseteq D(AB) \cap D(BA)$ and $AB = BAx$ for all $x \in D$.*
2. *If $QH \cap XH = QH \cap YH = \{0\}$, then D is a core for A and B.*

Remark Notice that if A and B are two unbounded normal operators which commute strongly, then A and B necessarily commute *pointwise* on some common core (see, e.g., Corollary 5.28 in [326]).

26.2 Questions

26.2.1 $\overline{BA} \subset \overline{AB} \not\Rightarrow BA \subset AB, B \in B(H)$

Let $B \in B(H)$, and let A be a closable operator such that $BA \subset AB$. Then it may be shown that $\overline{BA} \subset \overline{AB}$ (see, e.g., Lemma 2.8 of [363]). What about the converse?

Question 26.2.1 Find a closable operator A and a $B \in B(H)$ such that $\overline{BA} \subset \overline{AB}$ but $BA \not\subset AB$.

26.2.2 A Self-Adjoint $B \in B(H)$ and an Unclosed A with $BA \subset AB$ but $f(B)A \not\subset Af(B)$ for Some Continuous Function f

It was shown in say ([236], Proposition 1) that if $BA \subset AB$ where $B \in B(H)$ is self-adjoint and A is closed, then $f(B)A \subset Af(B)$ for any continuous function f on $\sigma(B)$. In fact, this result was stated with the assumption "A being unbounded and self-adjoint," but by looking closely at its proof, we see that only the closedness of A was needed.

Question 26.2.2 (⊛ **Cf. Question 26.2.1**) Find a bounded self-adjoint B in some Hilbert space H and an unclosed densely defined A such that $BA \subset AB$ but $f(B)A \not\subset Af(B)$ for some continuous function f on $\sigma(B)$.

26.2.3 A Self-Adjoint $S \in B(H)$ and a Non-closable T with $ST \subset TS$ but $f(S)T \not\subset Tf(S)$ for Some Continuous Function f

In the previous counterexample, the operator A was unclosed but it was closable. It might therefore not be too interesting to find a counterexample in the case of non-closability, but the idea of the construction of the counterexample has a certain pedagogic interest.

Question 26.2.3 Find a bounded self-adjoint S in some Hilbert space H and a non-closable densely defined T such that $ST \subset TS$ but $f(S)T \not\subset Tf(S)$ for some continuous function f on $\sigma(S)$.

26.2.4 An Unbounded Self-Adjoint Operator Commuting with the L^2-Fourier Transform

Question 26.2.4 Give an example of an unbounded self-adjoint operator which commutes with the Fourier transform (on $L^2(\mathbb{R})$).

26.2.5 A Closed Symmetric A and a Unitary U Such That $A \subsetneq U^*AU$

Let $U \in B(H)$ be unitary, and let A be an unbounded self-adjoint operator such that $UA \subset AU$. Then $UA = AU$. Indeed, as $UA \subset AU$ and A is self-adjoint, we obtain, upon passing to adjoints, that $U^*A \subset AU^*$. Right and left multiplying by U then yield $AU \subset UA$ and so $AU = UA$, as wished.

Can we weaken the self-adjointness of A to closedness+symmetricity?

Question 26.2.5 (⊛) Find a closed and symmetric A, as well as a unitary $U \in B(H)$ such that

$$A \subsetneq U^*AU,$$

that is, $UA \subset AU$ but $UA \neq AU$.

26.2.6   A Self-Adjoint A and B Where B Is a Bounded Multiplication Operator and A Is a Differential Operator which Commute Strongly

Let A be defined by $Af(x) = -if'(x)$ on the domain $D(A) := H^1(\mathbb{R}) = \{f \in L^2(\mathbb{R}) : f' \in L^2(\mathbb{R})\}$. Then A is self-adjoint. Now, define B on $L^2(\mathbb{R})$ by

$$Bf(x) = \varphi(x) f(x),$$

with φ being essentially bounded over $\mathbb{R}$. Under some appropriate conditions on φ, we may show that $BA \subset AB$ forces $B = cI$. Is it still possible to find a self-adjoint operator which commutes with some bounded (nontrivial) multiplication operator?

Question 26.2.6 Give self-adjoint operators A and B, where B is a bounded multiplication operator and A is a differential operator which commute strongly.

26.2.7   Is the Product of Two "Commuting" Unbounded Self-Adjoint (Respectively, Normal) Operators Always Self-Adjoint (Respectively, Normal)?

Let A and B be two self-adjoint where at least one of them is unbounded. Is AB self-adjoint? What about BA? To obtain the self-adjointness we need at least some form of commutativity given that just in the bounded case, we already know that AB is self-adjoint iff $AB = BA$, whenever $A, B \in B(H)$ are self-adjoint.

In the case where say $B \in B(H)$, the main issue then is the closedness of BA (while the closedness of AB is clear). On the other hand, the expression $BA \subset AB$ a priori only gives that BA is symmetric because $AB = (BA)^*$.

So, to establish the self-adjointness of AB, it seems impossible to obtain that without using some form of the spectral theorem either directly or by other auxiliary means (which also require the spectral theorem for their proofs anyway). Let us summarize some of these results in one theorem:

Theorem 26.2.1 *Let A and B be two strongly commuting self-adjoint (respectively, normal) operators. Then $\overline{AB}$ and $\overline{BA}$ are both self-adjoint (respectively, normal).*

In particular, if we further assume that $B \in B(H)$ (and so strong commutativity becomes $BA \subset AB$), then AB and $\overline{BA}$ are both self-adjoint (respectively, normal).

Remark The preceding result also holds for positive and self-adjoint B and A. In other words, if A and B are two strongly commuting self-adjoint positive operators, then $\overline{AB}$ and $\overline{BA}$ are both positive. In particular, if we also assume that $B \in B(H)$, then AB and $\overline{BA}$ are both positive.

Due to the maximality of self-adjoint or normal operators, we have the following corollary:

Corollary 26.2.2 *Let A and B be two self-adjoint (respectively, normal) operators such that $B \in B(H)$ and $BA \subset AB$. Then*

$$AB = \overline{BA}.$$

In the case of self-adjointness, $BA \subset AB$ iff AB is self-adjoint.

See, for example, [31] and [183] for proofs of the above results and for further results. The proof mainly uses the fact that two commuting spectral measures generate a joint spectral measure. See [28] (or [326]) for more details.

Question 26.2.7 Find a self-adjoint (respectively, normal) $B \in B(H)$ and a self-adjoint A such that $BA \subset AB$ yet BA is not self-adjoint (respectively, not normal).

Next, find two unbounded injective self-adjoint (respectively, normal) operators A and B such that $AB = BA$ on some subspace yet AB is not self-adjoint (respectively, not normal).

26.2.8 Two Unbounded Self-Adjoint Operators A and B which Commute Pointwise on Some Common Core but A and B Do Not Commute Strongly: Nelson-Like Counterexample

One of the most striking counterexamples in (unbounded) operator theory is the so-called Nelson's example. It says that an expression of the type $AB = BA$ does not necessarily mean that A and B strongly commute. More precisely, E. Nelson showed in [272] that there exists a pair of two essentially self-adjoint operators A and B on some common domain D such that

(1) $A : D \to D, B : D \to D$
(2) $ABx = BAx$ for all $x \in D$
(3) But $e^{it\overline{A}}$ and $e^{is\overline{B}}$ do not commute, i.e., A and B do not strongly commute

Remark Apparently, the first textbook to include Nelson's example is [302]. The same example is developed in detail in [325], pp. 257–258. The example to be given is perhaps the simplest one in the literature. It is due to K. Schmüdgen (in [323], see also [326]).

Note in the end that there are results giving conditions implying the strong commutativity of $\overline{A}$ and $\overline{B}$, for instance [119]. Another result is given below:

Theorem 26.2.3 ([273]) *Let A and B be two semi-bounded operators in a Hilbert space H, and let D be a dense linear manifold contained in the domains of AB, BA, A^2, and B^2 such that $ABx = BAx$ for all $x \in D$. If the restriction of $(A + B)^2$ to D is essentially self-adjoint, then A and B are essentially self-adjoint, and $\overline{A}$ and $\overline{B}$ strongly commute.*

Question 26.2.8 (❋❋) Find two unbounded self-adjoint operators A and B which commute pointwise on some common core, yet A and B do not commute strongly.

26.2.9 A Densely Defined Closed T Such That $T^2 = T^{*2}$ yet T^2 Is Not Self-Adjoint

If $T \in B(H)$, then T^2 is self-adjoint if and only if $T^2 = T^{*2}$. Is this always the case when T is closed and densely defined? A priori, we are only sure that T^2 is symmetric if $T^2 = T^{*2}$ and $D(T^2)$ is dense. Indeed,

$$T^2 = T^{*2} \subset (T^2)^*.$$

Question 26.2.9 Find a densely defined closed operator T such that $T^2 = T^{*2}$ yet T^2 is not self-adjoint.

26.2.10 Is There a Normal T Such That $(T^2)^* \neq T^{*2}$? What About When T^2 Is Normal?

We have already seen a densely defined operator T such that $(T^2)^* \neq T^{*2}$ and another one such that $(T^2)^* = T^{*2}$; see Questions 20.2.14 and 20.2.15.

Question 26.2.10 Can you find a normal operator T such that $(T^2)^* \neq T^{*2}$? What about when only T^2 is normal? What about when T^2 is normal and T is quasinormal?

Remark Notice that this question could have been asked in the chapter on normality, but it resembles in some sense the one of Question 26.2.9.

26.2.11 $BA \subset T \nRightarrow BA = T$ *Even if* $B \in B(H)$, *A and T Are All Self-Adjoint*

It was shown in [90] that if T, A. and B are (unbounded) self-adjoint operators, then

$$T \subset AB \Longrightarrow T = AB$$

(and so A and B strongly commute). We refer to this result as the Devinatz–Nussbaum–von Neumann theorem. What about the "inclusion" $BA \subset T$ when all operators are self-adjoint? The authors in [90] obtained the self-adjointness of BA by further assuming that $B^{-1} \in B(H)$.

Question 26.2.11 Let $B \in B(H)$ be self-adjoint, and let A be self-adjoint with a domain $D(A)$ such that $BA \subset T$, where T is also self-adjoint. Does it follow that $BA = T$?

26.2.12 $T \subset AB \nRightarrow T = AB$ *Where T, A, and B Are Normal*

Can we expect the Devinatz–Nussbaum–von Neumann theorem to hold good for normal operators? The answer is not as we will see in the answer to the question below.

Let us say a few words about some similar situations. Devinatz–Nussbaum showed the following result:

Theorem 26.2.4 ([89]) *If A, B, and N are unbounded normal operators obeying* $N = AB = BA$, *then A and B strongly commute.*

The following related results were obtained in [253]:

Corollary 26.2.5 *Let A and B be two unbounded normal operators. If B is boundedly invertible and BA = AB, then A and B strongly commute whenever AB is densely defined.*

Proposition 26.2.6 *Let A and B be two unbounded self-adjoint operators such that B is boundedly invertible. If AB ⊂ BA, then A and B strongly commute whenever AB is densely defined.*

Question 26.2.12 Find three normal operators A, B, and T such that $T \subset AB$ yet A and B do not necessarily (strongly) commute.

26.2.13   Two Unbounded Self-Adjoint Operators A and B which Commute Strongly Where B Is a Multiplication Operator and A Is a Differential Operator

There are various ways to show that two differential operators on some domains, on which they are normal, commute strongly. Here we apply Theorem 26.2.4 just above to the operators $A = \frac{\partial}{\partial x}$ and $B = \frac{\partial}{\partial y}$ defined on $D(A) = \{f \in L^2(\mathbb{R}^2) : Af \in L^2(\mathbb{R}^2)\}$ and $D(B) = \{f \in L^2(\mathbb{R}^2) : Bf \in L^2(\mathbb{R}^2)\}$, respectively. Then, both A and B are normal and their product $T := AB$ is even *self-adjoint* on

$$D(T) = \left\{ f \in L^2(\mathbb{R}^2) : \frac{\partial^2}{\partial x \partial y} f \in L^2(\mathbb{R}^2) \right\},$$

as it is unitarily equivalent via the $L^2(\mathbb{R}^2)$-Fourier transform to the multiplication operator by the real-valued function $(\lambda, \eta) \mapsto \lambda \eta$. Therefore, A commutes strongly with B.

Question 26.2.13 Give two unbounded self-adjoint operators A and B which commute strongly, where B is a multiplication operator and A is a differential operator.

26.2.14   On the Failure of the Ôta–Schmüdgen Criterion of Strong Commutativity for Normal Operators

Let T be a densely defined closable operator in a Hilbert space H, and define the operator Θ_T (formally) on $H \oplus H$ by

$$\Theta_T = \begin{pmatrix} 0 & T^* \\ T & 0 \end{pmatrix}.$$

Then, it was demonstrated in [285] that *if S and T are self-adjoint operators, then S strongly commutes with T if and only if Θ_S strongly commutes with Θ_T.*

Question 26.2.14 Find a self-adjoint S and a skew-adjoint (hence normal) T such that S strongly commutes with T, yet Θ_S does not strongly commute with Θ_T.

26.2.15   Anti-Commutativity and Exponentials

First we give a definition:

Definition 26.2.1 If A and B are linear operators such that $B \in B(H)$, then we say that B anticommutes with A if $BA \subset -AB$. In the case both A and B are unbounded and self-adjoint, then we say that A and B are (strongly) anti-commuting if

$$e^{itA} B \subset B e^{-itA}, \ \forall t \in \mathbb{R}.$$

See [289] and [371] for more on this topic.

As in the case of (strong) commutativity, one is tempted to think that two unbounded self-adjoint operators A and B strongly anticommute if

$$e^{isA} e^{itB} = e^{itB} e^{-isA}, \ \forall s, t \in \mathbb{R}?$$

Question 26.2.15 Show that this is not necessarily the case.

26.2.16 A Densely Defined Unbounded Operator Without a Cartesian Decomposition

Recall that *any* $T \in B(H)$ has a Cartesian decomposition, i.e., $T = A + iB$ where $A, B \in B(H)$ are self-adjoint.

Now, we give a definition of the Cartesian decomposition of a densely defined operator. Notice that the definition of the Cartesian decomposition could be different elsewhere.

Definition 26.2.2 ([281]) Let T be a densely defined operator with domain $D(T) \subset H$. If there exist densely defined symmetric operators A and B with domains $D(A)$ and $D(B)$, respectively, and such that

$$T = A + iB \text{ with } D(A) = D(B),$$

then T is said to have a Cartesian decomposition.

Remark (Cf. Question 26.2.16) When $D(T) \subset D(T^*)$ say, then $T = A + iB$ where

$$A = \operatorname{Re} T = \frac{T + T^*}{2} \text{ and } B = \operatorname{Im} T = \frac{T - T^*}{2i}.$$

Call A and B the real and the imaginary parts of T, respectively.

The next result concerns the Cartesian decomposition of (unbounded) normal operators.

Theorem 26.2.7 (See, e.g., 5.30 in [326]) *If A and B are strongly commuting self-adjoint operators on a Hilbert space H, then $N = A + iB$ is a normal operator and $N^* = A - iB$.*

Each normal operator is of this form, i.e., if N is normal, then

$$A := \frac{\overline{N + N^*}}{2} \text{ and } B := \frac{\overline{N - N^*}}{2i}$$

are strongly commuting self-adjoint operators such that $N = A + iB$.

A natural question to ask is whether every densely defined operator has a Cartesian decomposition.

Question 26.2.16 Find a densely defined unbounded operator without a Cartesian decomposition.

26.2.17   Are There Unbounded Self-Adjoint Operators A and B with $A + iB \subset 0$?

Question 26.2.17 Are there unbounded self-adjoint operators A and B with $A + iB \subset 0$?

26.2.18   Two Unbounded Self-Adjoint Operators A and B Such That $\overline{A + iB} \neq \overline{A} + i\overline{B}$

Question 26.2.18 Find two unbounded self-adjoint operators A and B such that $T = A + iB$ is not closed. Hence

$$\overline{A + iB} \neq \overline{A} + i\overline{B}.$$

26.2.19   A Closed Operator T Such That $D(T) = D(T^*)$ but $(T + T^*)/2$ and $(T - T^*)/2i$ Are Not Essentially Self-Adjoint

If T is normal, then both $(T + T^*)/2$ and $(T - T^*)/2i$ are essentially self-adjoint. Does the closedness of T suffice to obtain the essential self-adjointness of the real and the imaginary parts? The answer is negative. Indeed, let T be a densely defined closed operator such that

$$D(T + T^*) = D(T) \cap D(T^*) = \{0\},$$

as in Question 21.2.41. Hence, neither $(T + T^*)/2$ nor $(T - T^*)/2i$ can be essentially self-adjoint for they are not even densely defined. Observe that in this example we have $D(T) \neq D(T^*)$. Would assuming that $D(T) = D(T^*)$ help now?

Question 26.2.19 ([375]) Find a closed operator T such that $D(T) = D(T^*)$ but $(T + T^*)/2$ and $(T - T^*)/2i$ are not essentially self-adjoint.

26.2.20 An Unbounded Normal T Such That Both $(T + T^*)/2$ and $(T - T^*)/2i$ Are Unclosed

Let T be an unbounded normal operator. Then $\operatorname{Re} T = (T + T^*)/2$ or $\operatorname{Im} T = (T - T^*)/2i$ need not be closed. There are easy counterexamples (for instance, take $T^* = -T$ in the former case and $T^* = T$ in the latter). Observe that the first example does give the closedness of the imaginary part, whereas the second example gives the closedness of the real part. The next question therefore becomes interesting.

Question 26.2.20 Find a normal T such that *neither* $(T + T^*)/2$ *nor* $(T - T^*)/2i$ is closed.

26.2.21 Closed Nilpotent Operators Having a Cartesian Decomposition Are Always Everywhere Defined and Bounded

In the remark below Answer 21.2.1, we observed that there are not any unbounded nilpotent self-adjoint operators according to Definition 19.2.1, and using the following lemma:

Lemma 26.2.8 ([333]) *If H and K are two Hilbert spaces and if $A : D(A) \subset H \to K$ is a densely defined closed operator, then*

$$D(A) = D(A^*A) \iff A \in B(H, K).$$

If A is normal, then $D(A^2) = D(A^*A)$. So if A is nilpotent and normal (so $D(A) = D(A^2)$), then necessarily $A \in B(H)$!

Are there unbounded closed symmetric operators? The answer is negative even for a larger class of operators.

Question 26.2.21 Show that there are not any closed densely defined nilpotent operators having a Cartesian decomposition.

26.2.22 An Unbounded "Nilpotent" Closed Operator Having Positive Real and Imaginary Parts

The following result was obtained in [114]:

Theorem 26.2.9 *Let $T = A + iB$, where A and B are self-adjoint (one of them is also positive), $D(A) = D(B)$, and $D(AB) = D(BA)$. If $T^2 \subset 0$, then $T \in B(H)$ is normal, and so $T = 0$ everywhere on H.*

Notice that according to Definition 19.2.1, a nilpotent T necessarily satisfies $D(T^2) = D(T)$. So $T^2 \subset 0$ does not necessarily mean a nilpotent T, but the question still has some interest.

Question 26.2.22 Find densely defined closed symmetric positive operators A and B such that $D(A) = D(B)$, $D(AB) = D(BA)$, and $T = A + iB$ for which $T^2 \subset 0$ yet $T \not\subset 0$.

26.2.23 Two Commuting Unbounded Normal Operators Whose Sum Fails to Remain Normal

Let A and B be normal operators. We have already pointed out that when $A, B \in B(H)$, then $A + B$ is normal whenever $AB = BA$. What if only $B \in B(H)$ is the sum $A + B$ normal? The answer is yes. If $BA \subset AB$, then $A + B$ is normal (see, e.g., [251]).

What about when both A and B are unbounded, normal, and strongly commuting? In this case, $\overline{A + B}$ is normal (see, e.g., [253]).

Further Reading More results on the normality of the sum of normal operators may be consulted in [31, 251, 253], and [254].

Question 26.2.23 Find two (strongly) commuting unbounded normal operators A and B such that $A + B$ is not normal.

26.2.24 A Formally Normal Operator Without Any Normal Extension

Recall that a densely defined T is said to be formally normal if

$$\|Tx\| = \|T^*x\|, \forall x \in D(T) \subset D(T^*).$$

Question 26.2.24 (⊛⊛) Find a formally normal operator without any normal extension.

Further Reading The existence of a formally normal operator without a normal extension was first obtained in [64]; see also [281, 353–355], and [358].

Answers

Answer 26.2.1 Let $H = L^2(\mathbb{R})$, and let $B = \mathcal{F}$ be the usual L^2-Fourier transform. Consider also $A = I_{C_0^\infty}$, that is, the identity operator restricted to $C_0^\infty(\mathbb{R}) \subset L^2(\mathbb{R})$. Then $\overline{A} = I$, i.e., the full identity operator on $L^2(\mathbb{R})$.

Now, clearly $B\overline{A} \subset \overline{A}B$ (in fact, $B\overline{A} = \overline{A}B$). However, $BA \not\subset AB$ for if $BA \subset AB$ were true, we would have

$$D(BA) = D(A) = C_0^\infty(\mathbb{R}) \subset D(AB) = \{f \in L^2(\mathbb{R}) : \mathcal{F}f \in C_0^\infty(\mathbb{R})\},$$

which is evidently untrue. Indeed, let f be any *nonzero* function in $C_0^\infty(\mathbb{R})$. Then $f \in L^2(\mathbb{R})$ but $\mathcal{F}f \notin C_0^\infty(\mathbb{R})$ (why?), i.e., $f \notin D(AB)$, as needed.

Answer 26.2.2 (This Spectacular Counterexample Was Kindly Communicated to Me by Professor R. B. Israel) Let H be $L^2(\mathbb{R})$, and let $D(A) \subset L^2(\mathbb{R})$ be constituted of rational functions without poles on $\mathbb{R}$ and vanishing at $\pm\infty$, then define $Af(x) = xf(x)$ on $D(A)$. Next, define the self-adjoint bounded operator B on $L^2(\mathbb{R})$:

$$Bf(x) = \frac{1}{1+x^2}f(x).$$

That $D(A)$ is dense in $L^2(\mathbb{R})$ may be found in Exercise 6.3.2 in [3]. It could also be shown that A is not closed on $D(A)$ (as in say Example 19.1.10).

Now, B commutes with A because

$$BAf(x) = ABf(x) = \frac{x}{1+x^2}f(x)$$

for any f in $D(A)$ (since also $D(A) \subset D(AB)$).

Nonetheless, for say $f(x) = e^x$ defined on $\mathbb{R}$, $f(B)A \not\subset Af(B)$. That is, e^B does not commute with A. To see that, recall first that

$$e^B f(x) = e^{1/(1+x^2)}f(x)$$

for $f \in L^2(\mathbb{R})$. If $f \in D(A)$ with $f \neq 0$, then $e^B f \notin D(A)$ for $e^B f$ is never a rational function. Thus,

$$D(e^B A) = D(A) \not\subset D(Ae^B),$$

as wished.

Answer 26.2.3 Choose A and B as in Answer 26.2.2, and so $BA \subset AB$ but $e^B A \not\subset Ae^B$ (where $B \in B[L^2(\mathbb{R})]$ is self-adjoint and A is unclosed on $D(A)$). Let C be a densely defined non-closable operator, for instance, one such that $D(C^*) = \{0\}$. Now, define

$$T = \begin{pmatrix} A & 0 \\ 0 & C \end{pmatrix} \text{ and } S = \begin{pmatrix} B & 0 \\ 0 & 0 \end{pmatrix}$$

with $D(T) = D(A) \times D(C)$ and $D(S) = L^2(\mathbb{R}) \times L^2(\mathbb{R})$. Obviously, T is densely defined and S is (bounded) self-adjoint. Besides,

$$D(T^*) = D(A^*) \times D(C^*) = D(A^*) \times \{0\},$$

whereby T is not closable. It is also easy to see that $D(T) \subset D(TS)$, and since $ST(f, g) = TSf(f, g)$ for all (f, g) in $D(T)$, it follows that $ST \subset TS$.

However, $e^S T \not\subset Te^S$. To see why, remember that $e^S = \begin{pmatrix} e^B & 0 \\ 0 & I \end{pmatrix}$. Taking $(f, 0) \in D(A) \times D(C) = D(T)$, with $f \neq 0$, it is seen that

$$e^B(f, 0) = (e^B f, 0) \notin D(A) \times D(C),$$

by Answer 26.2.2.

Answer 26.2.4 Let $\mathcal{F}$ be the Fourier transform on $L^2(\mathbb{R})$. Then $\mathcal{F}$ is unitary and so

$$\mathcal{F} = \int_0^{2\pi} e^{i\lambda} dE_\lambda$$

in terms of some unique spectral measure E_λ. Now, set

$$A = \int_{\mathbb{R}} \lambda dE_\lambda,$$

which is unbounded and self-adjoint. By construction, it strongly commutes with $\mathcal{F}$, that is,

$$\mathcal{F}A \subset A\mathcal{F}.$$

Another way of constructing a commuting operator with $\mathcal{F}$ is (an idea communicated to me a while ago by Professor A. M. Davie): the Fourier transform $\mathcal{F}$ satisfies $\mathcal{F}^4 = I$, and it has four eigenvalues $1, -1, i$, and $-i$. In fact,

$$\sigma_p(\mathcal{F}) = \sigma(\mathcal{F}) = \{1, -1, i, -i\}$$

(see, e.g., [256]).

Then, the entire space $L^2(\mathbb{R})$ is the direct sum of the four corresponding eigenspaces. An operator A on $L^2(\mathbb{R})$ commutes with $\mathcal{F}$ if and only if each of the eigenspaces is invariant under A. So one constructs such an A just by specifying an operator on each eigenspace.

Answer 26.2.5 Both counterexamples here are due to B. Fuglede in [118]:

(1) Consider $H = L^2([0, 1] \times \mathbb{R})$ and set $A_p = -i\frac{\partial}{\partial x}$ on H with $f(0, y) = f(1, y)$ for almost every $y < p$ and $f(0, y) = f(1, y) = 0$ for almost every $y > p$. The domain of A_p may now be chosen in a way that A_p becomes closed and symmetric. Next, define the unitary transformation U by

$$Uf(x, y) = f(x, y + 1).$$

In the end,

$$U^* A_p U = A_{p+1} \supsetneq A_p.$$

(2) Consider two closed symmetric operators T and T' such that $T \subset T'$. Consider an operator A_p on $\ell^2(\mathbb{Z})$ whose graph consists of pairs $(x, y) \in \ell^2(\mathbb{Z}) \times \ell^2(\mathbb{Z})$ for which (x_n, y_n) is in the graph of T' for every $n < p$ and in the graph of T for every $n \geq p$. Then A_p is closed and symmetric for every p.

Now, let $U(x_n) = (x_{n+1})$ be defined on $\ell^2(\mathbb{Z})$ (the bilateral shift). Then U is unitary and

$$U^* A_p U = A_{p+1} \supsetneq A_p.$$

Answer 26.2.6 It is simpler to work on $L^2(\mathbb{R}^2)$. Define the self-adjoint operators A and B as follows:

$$Af(x, y) = -i\frac{\partial}{\partial x} f(x, y) = -if_x'(x, y) \text{ and } Bf(x, y) = e^{-|y|} f(x, y)$$

on $D(A) = \{f \in L^2(\mathbb{R}^2) : f_x' \in L^2(\mathbb{R}^2)\}$ and $D(B) = L^2(\mathbb{R}^2)$, respectively. Then

$$D(AB) = \{f \in L^2(\mathbb{R}^2) : e^{-|y|} f_x' \in L^2(\mathbb{R}^2)\}.$$

The latter also implies that $D(BA) = D(A) \subset D(AB)$. In the end, for any $f \in D(A)$, we have

$$BAf(x, y) = ABf(x, y) = -ie^{-|y|} f'_x(x, y),$$

meaning that $BA \subset AB$, which amounts to strong commutativity.

Answer 26.2.7 As already observed, the main issue is the closedness of BA. As a simple counterexample, let $B = 0$ everywhere on H and let A be any unbounded self-adjoint operator. Then clearly $BA \subset AB$, and BA is not self-adjoint.

Let us treat the second case. From Answer 21.2.44, we have a pair of two self-adjoint injective operators A and B

$$D(AB) = D(BA) = \{0\}$$

(where A^{-1} there has been renamed here as A). However, neither AB nor BA can be self-adjoint for none of them is densely defined.

Answer 26.2.8 ([326]) Let S be the usual shift on ℓ^2, i.e.,

$$S(x_0, x_1, \cdots, x_n, \cdots) = (0, x_0, x_1, \cdots)$$

for $x = (x_n) \in \ell^2$. Then $X := S + S^*$ and $Y := -i(S - S^*)$ are clearly bounded and self-adjoint on ℓ^2. Moreover, $[X, Y]H = \mathbb{C}e_0$, where $e_0 = (1, 0, \cdots)$. Then it can be checked that

$$\ker(X) = \ker(Y) = \{0\} \text{ and } QH \cap XH = QH \cap YH = \{0\}.$$

Now, set $A := X^{-1}$ and $B := Y^{-1}$, which are well defined in view of the injectivity of X and Y. Then, A and B are (unbounded) self-adjoint operators. By Lemma 26.1.3, A and B commute on the common core $D := XY(I - Q)H$. Since the resolvents $R(0, A) = X$ and $R(0, B) = Y$ plainly do not commute, it results that A and B do not commute strongly, and this marks the end of the proof.

Answer 26.2.9 ([87]) In Answer 21.2.46, we found a closed densely defined operator T such that

$$D(T^2) = D(T^{*2}) = \{0\}$$

(and so T^2 cannot be self-adjoint).

What about assuming that $D(T^2)$ is dense? This is still not enough. Indeed, consider two self-adjoint operators A and B such that $AB = BA$ on some common core, but A and B do not commute strongly.

To get the appropriate example, we choose Schmüdgen's example (see Answer 26.2.8). Then set

$$T = \begin{pmatrix} 0 & A \\ B & 0 \end{pmatrix}$$

with $D(T) = D(B) \oplus D(A)$. Clearly, T is closed and $T^* = \begin{pmatrix} 0 & B \\ A & 0 \end{pmatrix}$ with $D(T^*) = D(A) \oplus D(B)$. Hence

$$T^2 = \begin{pmatrix} AB & 0 \\ 0 & BA \end{pmatrix} = \begin{pmatrix} BA & 0 \\ 0 & AB \end{pmatrix} = T^{*2}.$$

To conclude, T^2 cannot be self-adjoint for if it were, then this would lead to the self-adjointness of AB, which in turn would yield strong commutativity of A and B, which is impossible.

Answer 26.2.10 If T is normal, then $(T^2)^* = T^{*2}$. We already know that $T^{*2} \subset (T^2)^*$ for any T, whenever T^2 is densely defined. So, we only have to show that $D[(T^2)^*] \subset D(T^{*2})$ when T is normal. Since T^2 is normal, it follows that $D[(T^2)^*] = D(T^2)$. By the normality of T, we have

$$D(T^2) = D(T^*T) = D(TT^*) = D(T^{*2}),$$

which finishes the proof in this case.

Now, if T^2 is normal without much information about T, then $(T^2)^* = T^{*2}$ need not hold in general. It suffices to consider some particular non-closable T such that $T^2 = 0$ everywhere. For example, consider a non-closable everywhere defined operator A such that e.g. $D(A^*) = \{0\}$ as in Answer 20.2.8. By taking $T = \begin{pmatrix} 0 & A \\ 0 & 0 \end{pmatrix}$, it is seen that $T^2 = 0 \in B(H \oplus H)$ is downright self-adjoint. Whence, $(T^2)^* = 0$ on the whole of $H \oplus H$, while $D(T^{*2}) = \{0\} \oplus H$. In other words, $(T^2)^* \neq T^{*2}$.

Nevertheless, things are better if T satisfies $D(T^*T) \subset D(TT^*)$[1] and if e.g. T^2 is normal (or just $D[(T^2)^*] = D(T^2)$).

Since $D(T^*T) \subset D(TT^*)$, upon passing to the positive square root, we obtain $D(T) \subset D(T^*)$ (why?). So, there only remains to check that $D[(T^2)^*] \subset D(T^{*2})$. We have

$$D[(T^2)^*] = D(T^2) \subset D(T^*T) \subset D(TT^*) \subset D(T^{*2}),$$

as wished.

---

[1] This is the case when T is, e.g., quasinormal, as in say [365]; see also Question 32.1.5.

Remark See [87] for the general case of T^n, $n \in \mathbb{N}$.

Answer 26.2.11 The answer is no. The main issue is closedness. Indeed, since T is closed, and BA is not necessarily closed, there is no reason why BA should be equal to T. As an explicit counterexample, take B to be the bounded inverse of some boundedly invertible self-adjoint operator A and $T = I_H$ (the identity operator on the whole of H). Then

$$BA = A^{-1}A = I_{D(A)} \subset I_H$$

and plainly $I_{D(A)} \neq I_H$.

Answer 26.2.12 The answer is negative. Indeed, just in the naive case of unitary operators, we have that a product of *any* two unitary operators is always unitary, even when the two factors of the product do not commute.

Answer 26.2.13 Define A and B as follows:

$$Af(x, y) = -i\frac{\partial}{\partial x}f(x, y) = -if'_x(x, y)$$

on $D(A) = \{f \in L^2(\mathbb{R}^2) : f'_x \in L^2(\mathbb{R}^2)\}$, and

$$Bf(x, y) = (y^2 + 1)f(x, y)$$

on $D(B) = \{f \in L^2(\mathbb{R}^2) : (1+y^2)f \in L^2(\mathbb{R}^2)\}$. Then A and B are both unbounded and self-adjoint on their respective domains. Next we have

$$D(AB) = \{f \in L^2(\mathbb{R}^2) : (1 + y^2)f, (1 + y^2)f'_x \in L^2(\mathbb{R}^2)\}$$

and

$$D(BA) = \{f \in L^2(\mathbb{R}^2) : (1 + y^2)f'_x \in L^2(\mathbb{R}^2)\}.$$

Hence $D(AB) \subset D(BA)$, and for all $f \in D(AB)$, we have

$$ABf(x, y) = BAf(x, y) = (1 + y^2)f'_x(x, y).$$

Therefore, $AB \subset BA$. Since $B^{-1} \in B[L^2(\mathbb{R}^2)]$, it follows by say Proposition 26.2.6 (just above Question 26.2.12) that both A and B strongly commute.

Answer 26.2.14 Let S be an unbounded self-adjoint operator with a domain $D(S)$. Set $T = iS$ where $i^2 = -1$. Then T is plainly skew-adjoint. Additionally, S strongly commutes with T.

Now, the self-adjoint

$$\Theta_S = \begin{pmatrix} 0 & S \\ S & 0 \end{pmatrix} \text{ and } \Theta_T = \begin{pmatrix} 0 & -iS \\ iS & 0 \end{pmatrix}$$

cannot strongly commute. In fact, they do not even pointwise commute for the simple reason that

$$\Theta_S \Theta_T = \begin{pmatrix} iS^2 & 0 \\ 0 & -iS^2 \end{pmatrix} \neq \begin{pmatrix} -iS^2 & 0 \\ 0 & iS^2 \end{pmatrix} = \Theta_T \Theta_S.$$

Remark Observe in passing that $D(\Theta_S) = D(\Theta_T)$ and that $\Theta_S \Theta_T$ is skew-adjoint, yet both Θ_S and Θ_T are self-adjoint.

Answer 26.2.15 ([289]) The relation

$$e^{isA} e^{itB} = e^{itB} e^{-isA}, \ \forall s, t \in \mathbb{R}$$

forces $A = 0$. Indeed, the previous equation implies

$$e^{-itB} e^{isA} e^{itB} f = e^{-isA} f, \ \forall f \in H.$$

By differentiating against t and then setting $t = 0$, we obtain

$$e^{isA} Bf = Be^{isA} f.$$

Hence A and B (strongly) commute (as in say Lemma 1.1 in [371]). By the same lemma, we may also obtain

$$e^{isA} e^{itB} = e^{itB} e^{isA}, \ \forall s, t \in \mathbb{R}.$$

Combining the previous equation with the very first equation yields $e^{isA} = I$ for all $s \in \mathbb{R}$. Thus, $A = 0$.

Answer 26.2.16 The main idea is to recall that a densely defined operator T with domain $D(T)$ has a Cartesian decomposition (as in Definition 26.2.2) if and only if $D(T) \subset D(T^*)$. So, to get a counterexample, we need to avoid the condition $D(T) \subset D(T^*)$.

Without delay, let us give a counterexample. Let T be a densely defined operator such that $D(T^*) = \{0\}$. Now, assume that there were two densely defined symmetric operators A and B (with $D(A) = D(B)$) such that $T = A + iB$. Then, it would result that

$$A - iB \subset A^* - iB^* \subset T^*$$

and so $D(A) \cap D(B) \subset D(T^*) = \{0\}$, that is $D(A) \cap D(B) = \{0\}$. Since $D(A) = D(B)$, we would obtain $D(A) = D(B) = \{0\}$. Thus, we would obtain $D(T) = \{0\}$ which is a contradiction. Consequently, a densely defined T with, e.g., $D(T^*) = \{0\}$ does not have any Cartesian decomposition.

Answer 26.2.17 The answer is yes in general and never in some other circumstances. For example, let A and B be two unbounded self-adjoint operators such that $D(A) \cap D(B) = \{0\}$. With this condition, it is plain that $A + iB \subset 0$.

There are other cases when this is impossible, for example, when $D(A) = D(B)$. Indeed, $A + iB \subset 0$ yields $A = -iB$, and given the fact that both A and B are self-adjoint (in fact, only closedness+symmetricity suffice here), we obtain $A = B = 0$ everywhere on H.

Another instance is when A commutes strongly with B (A and B still being self-adjoint), and hence $A + iB$ would be normal and so by maximality, we get $A + iB = 0$ everywhere on H. Hence $D(A) \cap D(B) = H$ and so $D(A) = D(B) = H$ which, by the closedness of A and B, yields $A, B \in B(H)$, and so $A = B = 0$.

Answer 26.2.18 Let S be an unbounded self-adjoint operator with a domain $D(S) \subsetneq H$, where H is a complex Hilbert space. Set

$$A = \begin{pmatrix} 0 & S \\ S & 0 \end{pmatrix} \text{ and } B = \begin{pmatrix} 0 & iS \\ -iS & 0 \end{pmatrix}$$

where $i^2 = -1$. Then both A and B have $D(S) \times D(S)$ as their common domain. Moreover, they are both clearly self-adjoint. If we denote the zero operator on $D(S)$ by $0_{D(S)}$, then

$$T = A + iB = \begin{pmatrix} 0 & 0_{D(S)} \\ 2S & 0 \end{pmatrix};$$

the latter being unclosed on $D(S) \times D(S) \subsetneq H \times H$. Consequently,

$$\overline{A + iB} \neq A + iB = \overline{A} + i\overline{B},$$

as wished.

Remark A different example appeared in [281]. It is also stronger for it had a different aim.

Answer 26.2.19 Let S be a closed, symmetric but non-self-adjoint operator. Then $|S|$ is self-adjoint. Set $T = |S| + iS/2$. We need to check that T is closed and that $D(T) = D(T^*)$. Since S is closed, we know that $D(S) = D(|S|)$, and so $D(T) = D(S)$.

To show that T is closed, the readers may check (as already done before) that $iS/2$ is $|S|$-bounded with relative bound $1/2$, and then Theorem 19.1.8 gives the closedness of $|S| + iS/2$.

Now, we see that for any $x \in D(|S|) = D(S) \subset D(S^*)$,

$$\left\| -i\frac{1}{2}S^*x \right\| \leq \left\| \frac{1}{2}Sx \right\| \leq \frac{1}{2}\||S|x\| + a\|x\|$$

for all $a \geq 0$. This means that $iS/2$ is $|S|$-bounded with relative bound $1/2$ and that $-iS^*/2$ is $|S|$-bounded with relative bound $1/2$. Since $|S|$ is self-adjoint, Theorem 21.1.14 allows us to write

$$T^* = (|S| + iS/2)^* = |S| - iS^*/2$$

leading to

$$D(T^*) = D(|S|) \cap D(S^*) = D(S) = D(T).$$

Finally, we see that $(T - T^*)/2i$ is not essentially self-adjoint because $(T - T^*)/2i = S/2$.

To get a counterexample for the real part, consider $T = S/2 + i|S|$ and we ask the readers to reason similarly.

Answer 26.2.20 Let A be an unbounded self-adjoint operator with a domain $D(A)$, and let B be an unbounded skew-adjoint operator (for example, let $B = iA$) and so B is normal. Then both $A - A^*$ and $B + B^*$ are unclosed.

Set

$$T = \begin{pmatrix} A & 0 \\ 0 & B \end{pmatrix}.$$

Clearly T is normal on $D(A) \oplus D(B)$. Moreover,

$$\text{Re}\, T = \frac{1}{2}\begin{pmatrix} A + A^* & 0 \\ 0 & B + B^* \end{pmatrix} \quad \text{and} \quad \text{Im}\, T = \frac{1}{2i}\begin{pmatrix} A - A^* & 0 \\ 0 & B - B^* \end{pmatrix}.$$

By a glance at Example 19.1.12, we see that $\text{Re}\, T$ is unclosed for $B + B^*$ is unclosed and that $\text{Im}\, T$ is unclosed because $A - A^*$ is unclosed.

Answer 26.2.21 Let A be a densely defined closed operator with domain $D(A) \subset H$ such that $A^n = 0$ on $D(A)$ for some n. Suppose also that $D(A) \subset D(A^*)$ and so A does have a Cartesian decomposition. It is seen that

$$D(A) = D(A^2) \subset D(A^*A) \subset D(A)$$

whereby $D(A^*A) = D(A)$. Since A is closed, Lemma 26.2.8 (just above Question 26.2.21) yields $A \in B(H)$.

Remark By the arguments above, there cannot be unbounded densely defined closed idempotent operators T such that $D(T) \subset D(T^*)$.

Answer 26.2.22 To obtain such an example, recall that we already have a densely defined closed symmetric positive operator A such that $D(A^2) = \{0\}$. Let $B = A$ and set $T = A + iA = (1 + i)A$. Then $D(T^2) = D(A^2) = \{0\}$ and so $T^2 \subset 0$ trivially. Observe in the end that $T \not\subset 0$, i.e., T does not vanish on $D(T)$.

Answer 26.2.23 The only issue is closedness. Let A be an unbounded normal with a domain $D(A)$. Set $B = -A$, and then B too is normal. Moreover, A plainly strongly commutes with B. However, $A + B$, which is the zero operator on $D(A)$, is not closed, and so it cannot be normal.

Remark The counterexample in the case of the non-strong commutativity of A and B is not important given that it is already known not to hold when $A, B \in B(H)$ are normal (where $AB \neq BA$). Despite all that, we give an example which might be interesting in this book and elsewhere.

Let A be the multiplication operator by x and let $B = d/dx$, both defined on their maximal domains $D(A)$ and $D(B)$ in $L^2(\mathbb{R})$, respectively (see Example 19.1.16). Then A is self-adjoint and B is normal. Define now the operator $T := A + B = A + i(-iB)$ by

$$Tf(x) = xf(x) + f'(x)$$

for every $f \in D(T) = \{f \in L^2(\mathbb{R}) : xf, f' \in L^2(\mathbb{R})\}$. We already know from Example 19.1.16 that T is closed. Such a T cannot, however, be normal. Indeed, if T were normal, then $D(T) = D(T^*)$. Hence

$$\frac{T + T^*}{2} \subset A \text{ and } \frac{T - T^*}{2i} \subset -iB.$$

By the maximality of self-adjoint operators, we obtain

$$A = \frac{\overline{T + T^*}}{2} \text{ and } -iB = \frac{\overline{T - T^*}}{2i}.$$

By Theorem 26.2.7 (just above Question 26.2.16), A and $-iB$ would be strongly commuting, but this is not the case for they do not even commute pointwise. Therefore $A + B$ is densely defined and closed but non-normal.

Answer 26.2.24 ([324] or [326]) Let A and B be as in Answer 26.2.8 and consider again $D = XY(I - Q)H$. Set $Tx = Ax + iBx$ for x in $D(T) := D$. We claim that T has the required properties.

Let $x \in D$. If $y \in D$, then

$$\langle Ty, x \rangle = \langle Ay + iBy, x \rangle = \langle y, Ax - iBx \rangle$$

gives $x \in D(T^*)$. Using Lemma 26.1.3, we may write

$$\|Ax \pm iBx\|^2 = \langle Ax \pm iBx, Ax \pm iBx \rangle$$
$$= \|Ax\|^2 + \|Bx\|^2 \mp i\langle Ax, Bx \rangle \pm i\langle Bx, Ax \rangle$$
$$= \|Ax\|^2 + \|Bx\|^2 \mp i\langle BAx, x \rangle \pm i\langle ABx, x \rangle$$
$$= \|Ax\|^2 + \|Bx\|^2,$$

thereby T is formally normal, i.e., $\|Tx\| = \|T^*x\|$ for all $x \in D(T) \subset D(T^*)$.

Assume now that T has a normal extension N on a bigger Hilbert space which we denote by K. Firstly, we show that $T^*|_D \subset N^*$. So, let $y \in D \subset D(T) \subset D(N) = D(N^*)$. Also, let P be the projection of K onto H. As

$$\langle x, T^*y \rangle = \langle Tx, y \rangle = \langle Nx, y \rangle = \langle x, PN^*y \rangle$$

for all $x \in D(T)$, then $T^*y = PN^*y$. But

$$\|Ty\| = \|T^*y\| = \|PN^*y\| \leq \|N^*y\| = \|Ny\| = \|Ty\|$$

for T and N are formally normal. Hence $\|PN^*y\| = \|N^*y\|$. This entails $N^*y \in H$ and so $T^*y = N^*y$. This means that $T^*|_D \subseteq N^*$.

For the last step of the proof, let $\tilde{A}$ and $\tilde{B}$ be the closures of the operators $(N + N^*)/2$ and $(N - N^*)/2i$, respectively. Since $T \subset N$ and $T^*|_D \subset N^*$, we obtain $A|_D \subset \tilde{A}$. Since D is a core for A by Lemma 26.1.3, $A|_D$ is essentially self-adjoint. By Proposition 1.17 in [326], H is a reducing subspace for $\tilde{A}$ and $A = \overline{A|_D}$ and we can write

$$\tilde{A} = A \oplus A_1 \text{ on } G = H \oplus H^\perp$$

where A_1 is some operator on $H^\perp$. In a similar manner, we may write $\tilde{B} = B \oplus B_1$. Therefore, $\tilde{A}$ and $\tilde{B}$ are strongly commuting self-adjoint operators. This entails that A and B strongly commute. This, however, contradicts the result in Answer 26.2.8.

Chapter 27
The Fuglede–Putnam Theorems and Intertwining Relations

27.1 Basics

We have already spoken about the Fuglede–Putnam theorem in the bounded case. In this chapter, we explore the validity of different versions of it in the unbounded case as well as some intertwining relations. We also include some brief historical notes.

As is known, the Fuglede–Putnam theorem is the second salient result in operator theory, at least as far as normal operators are concerned. The most common unbounded version reads the following:

Theorem 27.1.1 *If $T \in B(H)$ and A and B are normal and unbounded, then*

$$TA \subset BT \Longleftrightarrow TA^* \subset B^*T.$$

Remark Let $B \in B(H)$, and let A be a densely defined operator such that $\overline{A}$ is normal. If $BA \subset AB$, then $BA^* \subset A^*B$. This easy observation appeared in [31]. What about the converse? Well, the example of Answer 26.2.1, where $\overline{A} = I$ is normal and so $A^* = I$, is a counterexample to the reverse implication here too.

Remark The Fuglede–Putnam theorem has many applications. The most tremendous one is the fact that it improves the statement of the spectral theorem of normal operators. To cite only a little amount of its applications, we refer the readers to [2, 54, 144, 145, 183, 242, 245, 251, 284, 294, 305], and [394].

It is worth noticing that the problem leading to this theorem was first mooted by J. von Neumann [373] who had already established it in a finite-dimensional setting. B. Fuglede was the first one to answer this problem affirmatively in [117] and in the case $A = B$. It is important to tell the readers that P. R. Halmos obtained in [147] almost simultaneously as B. Fuglede a quite different proof of the theorem above. More precisely, at the end of August 1949, B. Fuglede communicated his proof to P. R. Halmos at the Boulder meeting of the American Mathematical Society. Halmos' proof dealt with the all bounded version, and however, P. R. Halmos indicated that

© The Author(s), under exclusive license to Springer Nature Switzerland AG 2022
M. H. Mortad, *Counterexamples in Operator Theory*,
https://doi.org/10.1007/978-3-030-97814-3_27

only minor modifications were needed to adapt his proof to the more general case of unbounded operators.

Then C. R. Putnam [294] proved the above version. S. K. Berberian [19] amazingly noted that the two versions were equivalent.

There are different proofs of the Fuglede–Putnam theorem. Perhaps the most elegant proof is the one due to M. Rosenblum [309]. An equally elegant proof (in the bounded case), unfortunately, unknown even to many specialists, is due to C. R. Putnam himself in [298]. This paper also contains another extraordinarily short proof of a generalization of the Fuglede–Putnam theorem due to G. Weiss in [376]. For other proofs, see, e.g., [299] and [304].

In the end, the readers may wish to consult [265], which is a quite interesting monograph about the Fuglede–Putnam theorem.

By analogy to the bounded case, we define the following:

Definition 27.1.1 Say that an operator T intertwines two operators A and B when $TA \subset BT$.

27.2   Questions

27.2.1   A Boundedly Invertible Positive Self-Adjoint Unbounded Operator A and an Unbounded Normal Operator N Such That $AN^* = NA$ but $AN \not\subset N^*A$

As far as I can tell, the first version of the Fuglede–Putnam theorem including only unbounded operators is as follows:

Theorem 27.2.1 ([236]) *If A is a closed symmetric operator and if N is an unbounded normal operator, then*

$$AN \subset N^*A \implies AN^* \subset NA$$

whenever $D(N) \subset D(A)$.

It is therefore natural to wonder whether $AN^* \subset NA$ implies $AN \subset N^*A$ when N is normal and unbounded and A is closed. The answer is negative even under stronger conditions.

Question 27.2.1 (⊛) Find a boundedly invertible and positive self-adjoint unbounded operator A and an unbounded normal operator N such that

$$AN^* = NA \text{ but } AN \not\subset N^*A.$$

In other terms, NA is self-adjoint, while N^*A is not (cf. Question 32.2.2).

27.2.2   A Closed T and a Normal M Such That $TM \subset MT$ but $TM^* \not\subset M^*T$ and $M^*T \not\subset TM^*$

Question 27.2.2 (⊛) Find a closed T and an unbounded normal M such that $TM \subset MT$ but $TM^* \not\subset M^*T$ and $M^*T \not\subset TM^*$.

27.2.3   A Normal $B \in B(H)$ and a Densely Defined Closed A Such That $BA \subset AB$ Yet $B^*A \not\subset AB^*$

Let $B \in B(H)$ be normal, and let A be closed and densely defined (and unbounded). If $BA \subset AB$, does it follow that $B^*A \subset AB^*$, or equivalently $BA^* \subset A^*B$? B. Fuglede himself asked this question in [117]. This is a natural and interesting question for it is true when $A \in B(H)$. B. Fuglede found a strong counterexample in [118] to be given below. However, Fuglede's counterexample is not well known as it was, apparently, missed by some writers and referees alike. See, e.g., [172] and [180] for more details.

Question 27.2.3 Find a normal $B \in B(H)$ and a densely defined closed A with domain $D(A) \subset H$ such that $BA \subset AB$ yet $B^*A \not\subset AB^*$.

27.2.4   A Self-Adjoint T and a Unitary B with $BT \subset TB^*$ but $B^*T \not\subset TB$

If $B, T \in B(H)$ and B is normal, then $BT = TB^*$ entails $B^*T = TB$ which is the (all) bounded version of the Fuglede–Putnam theorem. What about replacing " $=''$ by " $\subset''$? The answer is negative even when B is unitary. Consider again a unitary $U \in B(H)$ and a closed (and symmetric) A such that $UA \subset AU$ and $U^*A \not\subset AU^*$ (as in Question 26.2.5), and then let

$$B = \begin{pmatrix} U & 0 \\ 0 & U^* \end{pmatrix} \text{ and } T = \begin{pmatrix} 0 & A \\ 0 & 0 \end{pmatrix}.$$

Obviously B is even unitary on $H \oplus H$ and T is closed on $H \oplus D(A)$. Moreover, the products BT and TB^* are well defined. More precisely,

$$BT = \begin{pmatrix} 0 & UA \\ 0 & 0 \end{pmatrix} \text{ and } TB^* = \begin{pmatrix} 0 & AU \\ 0 & 0 \end{pmatrix}$$

on the respective domains $D(BT) = D(T) = H \oplus D(A) = H \oplus D(UA)$ and $D(TB^*) = H \oplus D(AU)$. As $UA \subset AU$, it follows that $BT \subset TB^*$.

Since

$$B^*T = \begin{pmatrix} 0 & U^*A \\ 0 & 0 \end{pmatrix} \text{ and } TB = \begin{pmatrix} 0 & AU^* \\ 0 & 0 \end{pmatrix},$$

and since $U^*A \not\subset AU^*$, we plainly obtain $B^*T \not\subset TB$, and we are done with this example.

Now, if T is self-adjoint and if $B \in B(H)$, then $BT \subset TB$ does imply that $B^*T \subset TB^*$ (just take adjoints). One may therefore wonder whether $BT \subset TB^*$ implies $B^*T \subset TB$ in the case that T is self-adjoint and B is normal.

Question 27.2.4 (✳) Show that this is untrue.

27.2.5   A Closed Operator Which Does Not Commute with Any (Nontrivial) Everywhere Defined Bounded Operator

Question 27.2.5 (✳✳) In some Hilbert space H, find a closed operator T which does not commute with any bounded operator $S \in B(H)$, except for scalar operators (i.e., those of the type αI for some scalar α).

27.2.6   A Self-Adjoint and a Closed Operators Which Are Not Intertwined by Any Bounded Operator Apart from the Zero Operator

Question 27.2.6 There are a self-adjoint operator A and a densely defined closed operator B which are not intertwined by any (everywhere defined) bounded operator except the zero operator. Also, the same pair A and B in the opposite order cannot be intertwined either by any bounded operator except the zero operator.

27.2.7 A Self-Adjoint Operator A and a Closed Symmetric Restriction of A Not Intertwined by Any Bounded Operator Except the Zero Operator

Question 27.2.7 There are a self-adjoint operator A and a densely defined closed symmetric operator B (with $B \subset A$) which are not intertwined by any (everywhere defined) bounded operator except the zero operator.

27.2.8 Two Densely Defined Closed Operators A and B Not Intertwined by Any Densely Defined Closed (Nonzero) Operator

Question 27.2.8 There are two densely defined closed operators A and B which are not intertwined by any densely defined closed operator apart from the zero operator.

27.2.9 On the Failure of a Generalization to Unbounded Operators of Some Similarity Result by M.R. Embry

Inspired by the bounded case, define the numerical range of a non-necessarily bounded operator T by

$$W(T) = \{\langle Tx, x \rangle : x \in D(T), \|x\| = 1\}.$$

Embry's theorem (Theorem 12.2.4) was generalized to the case of two unbounded operators A and B in [244]. An interesting corollary is given below:

Corollary 27.2.2 *Let $T \in B(H)$ be such that $0 \notin W(T)$. If A is an unbounded normal operator with $TA \subset A^*T$, then A is self-adjoint.*

Is the previous corollary still valid if T is unbounded (self-adjoint if necessary)? The answer is negative even in the case of a full equality.

Question 27.2.9 ([244], cf. [101]) Find three unbounded operators T, A, and B such that T is self-adjoint, positive, and boundedly invertible (hence $0 \notin W(T)$); A and B are strongly commuting and normal (hence $AB = BA$), $TB = AT$ but $A \neq B$.

27.2.10   Are There Two Normal Operators A and $B \in B(H)$ Such That $BA \subset 2AB$ with AB Is Normal?

Question 27.2.10 (Cf. Question 32.2.7) Are there two normal operators A and $B \in B(H)$ such that $BA \subset 2AB \neq 0$ with AB being normal?

27.2.11   On a Result About Commutativity by C. R. Putnam

C. R. Putnam showed in [294] (cf. [76] for a version over Banach spaces), using the Fuglede theorem, that if A is a normal operator and $B, C \in B(H)$ are such that $BA + C \subset AB$ and $CA \subset AC$, then $C = 0$, whereby $BA \subset AB$.

Question 27.2.11 ([294]) Is the conclusion of the previous result true if A is not required to be normal?

Answers

Answer 27.2.1 ([250]) Define the following operators A and N by

$$Af(x) = (1 + |x|)f(x) \text{ and } Nf(x) = -i(1 + |x|)f'(x)$$

(with $i^2 = -1$), respectively, on the domains

$$D(A) = \{f \in L^2(\mathbb{R}) : (1 + |x|)f \in L^2(\mathbb{R})\}$$

and

$$D(N) = \{f \in L^2(\mathbb{R}) : (1 + |x|)f' \in L^2(\mathbb{R})\}.$$

Clearly, A is self-adjoint on $D(A)$.

As for N, it can be shown that it is normal on $D(N)$ (see Answer 23.2.6). We can find that

$$N^* f(x) = -i \operatorname{sign}(x)f(x) - i(1 + |x|)f'(x)$$

with

$$D(N^*) = \{f \in L^2(\mathbb{R}) : (1 + |x|)f' \in L^2(\mathbb{R})\}$$

and where "sign" is the usual sign function.

Doing some arithmetic allows us to find that

$$AN^* f(x) = NAf(x) = -i(1 + |x|)\text{sign}(x)f(x) - i(1 + |x|)^2 f'(x)$$

for any f in the *equal* domains

$$D(AN^*) = D(NA) = \{f \in L^2(\mathbb{R}) : (1 + |x|)f \in L^2(\mathbb{R}), \ (1 + |x|)^2 f' \in L^2(\mathbb{R})\}$$

and so

$$NA = AN^* = (NA)^*$$

(since $A^{-1} \in B[L^2(\mathbb{R})]$). Nonetheless, $AN \not\subset N^*A$ for $ANf(x) = -i(1 + |x|)^2 f'(x)$, whereas

$$N^*Af(x) = -2i\,\text{sign}(x)(1 + |x|)f(x) - i(1 + |x|)^2 f'(x),$$

that is,

$$ANf(x) \neq N^*Af(x).$$

Accordingly, $AN \not\subset N^*A$.

Answer 27.2.2 ([262]) Consider

$$M = \begin{pmatrix} N^* & 0 \\ 0 & N \end{pmatrix} \text{ and } T = \begin{pmatrix} 0 & 0 \\ A & 0 \end{pmatrix}$$

where N is normal with domain $D(N)$ and A is closed with domain $D(A)$ and such that $AN^* = NA$ but $AN \not\subset N^*A$ and $N^*A \not\subset AN$ as defined in Answer 27.2.1. Clearly, M is normal and T is closed. Observe that $D(M) = D(N^*) \oplus D(N)$ and $D(T) = D(A) \oplus L^2(\mathbb{R})$. Now,

$$TM = \begin{pmatrix} 0 & 0 \\ A & 0 \end{pmatrix} \begin{pmatrix} N^* & 0 \\ 0 & N \end{pmatrix} = \begin{pmatrix} 0_{D(N^*)} & 0_{D(N)} \\ AN^* & 0 \end{pmatrix} = \begin{pmatrix} 0 & 0_{D(N)} \\ AN^* & 0 \end{pmatrix}.$$

Likewise

$$MT = \begin{pmatrix} N^* & 0 \\ 0 & N \end{pmatrix} \begin{pmatrix} 0 & 0 \\ A & 0 \end{pmatrix} = \begin{pmatrix} 0 & 0 \\ NA & 0 \end{pmatrix}.$$

Since $D(TM) = D(AN^*) \oplus D(N) \subset D(NA) \oplus L^2(\mathbb{R}) = D(MT)$, it ensues that $TM \subset MT$. Now, it is seen that

$$TM^* = \begin{pmatrix} 0 & 0 \\ A & 0 \end{pmatrix} \begin{pmatrix} N & 0 \\ 0 & N^* \end{pmatrix} = \begin{pmatrix} 0 & 0_{D(N^*)} \\ AN & 0 \end{pmatrix}$$

and

$$M^*T = \begin{pmatrix} N & 0 \\ 0 & N^* \end{pmatrix} \begin{pmatrix} 0 & 0 \\ A & 0 \end{pmatrix} = \begin{pmatrix} 0 & 0 \\ N^*A & 0 \end{pmatrix}.$$

Since $ANf \neq N^*Af$ for any $f \neq 0$, we infer that $TM^* \not\subset M^*T$ and $M^*T \not\subset TM^*$.

Answer 27.2.3 In fact, the answer is also negative even when B is unitary. We already have a counterexample from Question 26.2.5. There, we have seen a unitary $U \in B(H)$ (where $H = L^2([0, 1] \times \mathbb{R})$) and a closed and *symmetric* A such that $UA \subset AU$ and $A \neq U^*AU$. The latter just means that $U^*A \not\subset AU^*$, as needed.

Answer 27.2.4 ([18]) The example to be given is a slight modification of the arguments used in the preamble of this question. Consider a unitary $U \in B(H)$ and a closed A such that $UA \subset AU$ and $U^*A \not\subset AU^*$. Consider

$$B = \begin{pmatrix} U & 0 \\ 0 & U^* \end{pmatrix} \text{ and } T = \begin{pmatrix} 0 & A \\ A^* & 0 \end{pmatrix}.$$

Then B is unitary and T is self-adjoint on $D(A^*) \oplus D(A)$. Besides,

$$BT = \begin{pmatrix} 0 & UA \\ U^*A^* & 0 \end{pmatrix} \text{ and } TB^* = \begin{pmatrix} 0 & AU \\ A^*U^* & 0 \end{pmatrix}.$$

Since $UA \subset AU$, it ensues that $U^*A^* \subset A^*U^*$. Therefore, $BT \subset TB^*$. Since $U^*A \not\subset AU^*$ is equivalent to $UA^* \not\subset A^*U$, the readers may easily check that

$$B^*T \not\subset TB$$

because $D(B^*T) = D(T) \not\subset D(TB)$, as wished.

Remark It is worth noticing that if B is a normal operator and if A is a closed densely defined operator with $\sigma(A) \neq \mathbb{C}$ (yet to be defined), then

$$BA \subset AB \implies g(B)A \subset Ag(B)$$

for any bounded complex Borel function g on $\sigma(B)$. In particular, we have $B^*A \subset AB^*$. This appeared in [172].

Answer 27.2.5 ([117]) Let A be the multiplication operator by x, and let $B = -id/dx$, both defined on their maximal domains $D(A)$ and $D(B)$ in $L^2(\mathbb{R})$, respectively (see Example 19.1.16, and recall that A and B are self-adjoint). Define now the operator $T := A + iB$ by

$$Tf(x) = xf(x) + f'(x)$$

for every $f \in D := D(T) = \{f \in L^2(\mathbb{R}) : xf, f' \in L^2(\mathbb{R})\}$. We already know that T is closed (Example 19.1.16).

If $\lambda \in \mathbb{C}$, then the equation $Tf = \lambda f$ may be expressed as

$$f'(x) + (x - \lambda)f(x) = 0.$$

It is known that the only solutions of the previous equation are given by $f(x) = Kf_\lambda(x)$, where K is a constant and

$$f_\lambda(x) = e^{-\frac{(x-\lambda)^2}{2}}$$

(observe that $f_\lambda \in D$ for any λ). Hence λ is a simple eigenvalue for T with cf_λ being the associated eigenvector (where $c \in \mathbb{C}$).

Assume now for the sake of contradiction that T commutes with an everywhere defined bounded operator S, that is, $ST \subset TS$. This means that $STf = TSf$ for all $f \in D(T) \subset D(TS)$. In particular,

$$TSf_\lambda = \lambda Sf_\lambda, \ \forall \lambda \in \mathbb{C}.$$

This actually signifies that λ is an eigenvalue for T with Sf_λ being an eigenvector (in case $Sf_\lambda \neq 0$). Therefore, we may write $Sf_\lambda = c_\lambda f_\lambda$, where c_λ is some complex number depending on λ only.

Let us show that c_λ is in fact a differentiable function of λ. Since c_λ is bounded (as S is), we deduce by Liouville's theorem that c_λ is a constant function, that is, c_λ becomes just a constant c. First, it may be shown (in H) that

$$\frac{f_\lambda - f_\mu}{\lambda - \mu} \longrightarrow g_\mu, \ \text{when } \lambda \to \mu.$$

As a consequence, if h is any fixed element of $L^2(\mathbb{R})$, then the complex function $\lambda \mapsto \langle f_\lambda, h \rangle$ is differentiable at any λ. The objective now is to show that c_λ is differentiable at *any* given point μ. Since $Sf_\lambda = c_\lambda f_\lambda$, it follows that

$$c_\lambda \langle f_\lambda, f_\mu \rangle = \langle c_\lambda f_\lambda, f_\mu \rangle = \langle Sf_\lambda, f_\mu \rangle = \langle f_\lambda, S^* f_\mu \rangle.$$

Here both $\langle f_\lambda, f_\mu \rangle$ and $\langle f_\lambda, S^* f_\mu \rangle$ are differentiable. Therefore,

$$c_\lambda = \frac{\langle f_\lambda, S^* f_\mu \rangle}{\langle f_\lambda, f_\mu \rangle}$$

is differentiable at any point λ provided that $\langle f_\lambda, f_\mu \rangle \neq 0$. This is particularly the case at $\lambda = \mu$. Finally, as $Sf_\lambda = cf_\lambda$ for all λ, we infer that

$$Sf = cf, \ \forall f \in L^2(\mathbb{R})$$

given that S is bounded and that the set of all finite linear combinations of the elements f_λ is everywhere dense in $L^2(\mathbb{R})$.

Answer 27.2.6 ([18]) Let $H = L^2(\mathbb{R}) \oplus L^2(\mathbb{R})$, let A be any unbounded self-adjoint operator with domain $D(A) \subset H$, and let B be a closed operator such that $D(B^2) = \{0\} = D(B^{*2})$ (as in Question 21.2.46). Let $T \in B(H)$. Clearly,

$$TA \subset BT \implies TA^2 \subset BTA \subset B^2 T.$$

Hence

$$D(TA^2) = D(A^2) \subset D(B^2 T) = \{x \in H : Tx \in D(B^2) = \{0\}\} = \ker T.$$

Since A^2 is densely defined, it follows that

$$H = \overline{D(A^2)} \subset \overline{\ker T} = \ker T \subset H,$$

whereby $\ker T = H$, that is, $T = 0$, as required.

Now, we deal with the second part of the question. Plainly,

$$SB \subset AS \implies S^* A \subset B^* S^*.$$

As before, we obtain

$$S^* A^2 \subset B^{*2} S^*.$$

Similar arguments as above then yield $S^* = 0$ or merely $S = 0$, as needed.

Answer 27.2.7 As in the remark below Answer 21.2.19, simply take any unbounded self-adjoint operator A, then consider either of its two closed symmetric restrictions, and denote it by B, where also $D(B^2) = \{0\}$. Finally, consider $T \in B(H)$ such that $TA \subset BT$. Then the readers should comfortably obtain $T = 0$ as carried out above.

Answer 27.2.8 ([18]) Let $H = L^2(\mathbb{R}) \oplus L^2(\mathbb{R})$, and let A be a densely defined closed operator with domain $D(A) \subset L^2(\mathbb{R}) \oplus L^2(\mathbb{R})$ such that $A^2 = 0$ on $D(A^2) =$

$D(A)$ (see, e.g., Question 19.2.16). Let B be a closed operator satisfying $D(B^2) = \{0\}$. Let T be a closed operator. We have

$$TA \subset BT \implies TA^2 \subset B^2T.$$

But

$$D(TA^2) = \{x \in D(A^2) : 0 \in D(T)\} = D(A) \text{ and } D(B^2T) = \ker T.$$

Hence

$$D(A) \subset \ker T \subset L^2(\mathbb{R}) \oplus L^2(\mathbb{R}).$$

Passing to the closure (w.r.t. $L^2(\mathbb{R}) \oplus L^2(\mathbb{R})$) implies that

$$\ker T = L^2(\mathbb{R}) \oplus L^2(\mathbb{R})$$

because $\ker T$ is closed as T is closed. Therefore, $Tx = 0$ for all $x \in D(T)$, i.e., $T \subset 0$. Accordingly, as T is bounded on $D(T)$ and also closed, then $D(T)$ becomes closed and so $D(T) = H$, that is, $T = 0$ everywhere.

Answer 27.2.9 ([244]) Consider again the following operators T and A defined by

$$Tf(x) = (1 + |x|)f(x) \text{ and } Af(x) = -i(1 + |x|)f'(x)$$

on their respective domains

$$D(T) = \{f \in L^2(\mathbb{R}) : (1 + |x|)f \in L^2(\mathbb{R})\}$$

and

$$D(A) = \{f \in L^2(\mathbb{R}) : (1 + |x|)f' \in L^2(\mathbb{R})\}.$$

Then from Answer 27.2.1, we already know that $TA^* = AT$. Since A is normal so is A^*, and besides $AA^* = A^*A$. Obviously, $0 \notin W(T)$ for T is self-adjoint positive and boundedly invertible on $D(T)$. In the end, observe that $A \neq A^*$.

Answer 27.2.10 ([54]) In fact, we show that AB is normal if and only if $|\lambda| = 1$, whenever A and B are normal, $B \in B(H)$ and $BA \subset \lambda AB \neq 0$ where $\lambda \in \mathbb{C}$. First, since A is closed and B is bounded, AB is automatically closed.

Since A is normal, so is λA. Hence the Fuglede–Putnam theorem gives

$$BA \subset \lambda AB \implies BA^* \subset \bar{\lambda}A^*B \text{ or } \lambda B^*A \subset AB^*.$$

Using the above "inclusions," we have on the one hand

$$(AB)^* AB \supset B^* A^* AB$$

$$= B^* A A^* B \text{ (since } A \text{ is normal)}$$

$$\supset \frac{1}{\bar{\lambda}} B^* A B A^*$$

$$\supset \frac{1}{\bar{\lambda}\lambda} B^* B A A^*$$

$$= \frac{1}{|\lambda|^2} B^* B A A^*.$$

Since A and AB are closed, and B is bounded, all of $(AB)^* AB$, $A^* A$, and $B^* B$ are self-adjoint so that "adjointing" the previous inclusion yields

$$(AB)^* AB \subset \frac{1}{|\lambda|^2} A A^* B^* B.$$

As $|\lambda|$ is real, the conditions of the Devinatz–Nussbaum–von Neumann theorem (just above Question 26.2.11) are all met, whereby

$$(AB)^* AB = \frac{1}{|\lambda|^2} A A^* B^* B.$$

On the other hand, we may write

$$AB(AB)^* \supset ABB^* A^*$$

$$= AB^* B A^* \text{ (because } B \text{ is normal)}$$

$$\supset \lambda B^* A B A^*$$

$$= B^* (\lambda AB) A^*$$

$$\supset B^* B A A^*.$$

As above, we obtain

$$AA^* B^* B \supset AB(AB)^*,$$

and by the Devinatz–Nussbaum–von Neumann theorem once more, we end up with

$$AB(AB)^* = AA^* B^* B.$$

Accordingly, we clearly see that AB is normal iff $|\lambda| = 1$, completing the proof.

Answer 27.2.11 ([294]) Without the normality of A, Putnam's result fails to hold even when all operators are in the form of 2×2 matrices. For instance, let

$$A = \begin{pmatrix} 1 & 1 \\ 0 & 1 \end{pmatrix}, \quad B = \begin{pmatrix} 2 & 0 \\ 0 & 1 \end{pmatrix} \text{ and } C = \begin{pmatrix} 0 & 1 \\ 0 & 0 \end{pmatrix}.$$

Then $BA + C = AB$ and $CA = AC$. Howbeit, $C \neq 0$, i.e., $BA \neq AB$, as wished.

Chapter 28
Commutators

28.1 Basics

It is not easy to give a meaningful and universal definition of the "commutator" of two unbounded operators. For example and as already observed, even when A and B are unbounded self-adjoint operators, $AB - BA$ could be defined only on $\{0\}$. Besides, having $AB - BA = 0$ does not mean that A and B strongly commute, as we have already observed counterexamples previously.

For the purpose of this chapter, however, we content ourselves with the following relatively basic definition:

Definition 28.1.1 Let A and B be two non-necessarily bounded operators. The operator $AB - BA$ is called the commutator of A and B. It may be denoted occasionally by $[A, B]$.

28.2 Questions

28.2.1 Two Operators A, B, One of Them Is Unbounded, with $BA - AB \subset I$

Let I be the usual identity operator. Linear operators A and B obeying the Heisenberg canonical commutation relation

$$AB - BA = I$$

play a crucial role in the formulation of quantum mechanics.

It is well known to readers that no (finite) square matrices A and B need to satisfy $AB - BA = I$. This is seen via a very simple trace argument. So, perhaps there are

© The Author(s), under exclusive license to Springer Nature Switzerland AG 2022
M. H. Mortad, *Counterexamples in Operator Theory*,
https://doi.org/10.1007/978-3-030-97814-3_28

bounded operators A, $B \in B(H)$ (where dim $H = \infty$) such that $AB - BA = I$. The answer is still negative. Indeed, assuming that $AB - BA = I$ would imply that at least one of the two operators A and B must be unbounded. The widespread proof is due to H. Wielandt (see Theorem 13.6 in [314]) though this result was first obtained by A. Wintner. See again [314] for more details and further comments. Another much simpler proof reads (see, e.g., Remark 3.2.9 in [184]): if $AB - BA = I$, i.e., $AB = BA + I$, then we have

$$\sigma(AB) = \sigma(BA + I) = \sigma(BA) + \{1\}$$

which is definitely not consistent with $\sigma(AB) \cup \{0\} = \sigma(BA) \cup \{0\}$.

Notice in the end that there are bounded but *non-everywhere defined* operators A and B such that $AB - BA \subset I$. For instance, take two densely defined bounded operators A and B with $D(A) \cap D(B) = \{0\}$ (as in Answer 18.2.8). Then

$$D(AB) \cap D(BA) \subset D(B) \cap D(A) = \{0\},$$

and so $BA - AB \subset I$ is trivially satisfied. By the same token, we may even choose two self-adjoint operators (both unbounded this time) A and B such that $D(A) \cap D(B) = \{0\}$.

Question 28.2.1 Find two operators A and B, one of them is unbounded, such that $BA - AB$ is densely defined and

$$BA - AB \subset I.$$

28.2.2   On Some Theorem of F. E. Browder About Commutators of Unbounded Operators

F. E. Browder showed in [36] that if B is in $B(H)$ and A is an unbounded positive self-adjoint operator such that for some $n \in \mathbb{N}$,

$$BA^n - A^n B \subset B_1 A^{n-1},$$

$$B_1 A^n - A^n B_1 \subset B_2 A^{n-1},$$

where B_1, $B_2 \in B(H)$, then $AB - BA$ is bounded.

See [36] for other similar results.

Question 28.2.2 Show that the condition of the self-adjointness of A in the aforementioned result may not just be dropped even when B is taken to be self-adjoint and unitary.

28.2.3 Are There Two Closable Unbounded Operators A and B Such That $AB - BA$ is Everywhere Defined?

In practice, when A and B are unbounded, the "commutator" $AB - BA$ is rarely everywhere defined. Let us give a simple instance of that. Let $A = B$ be defined as in Answer 19.2.24. Then $D(A) = D(B) = H$ and so

$$AB - BA = A^2 - A^2 = 0$$

everywhere on H. However, as we already know, A (hence B) is not closable.

Question 28.2.3 Can you find two closable unbounded operators A and B such that $AB - BA$ is everywhere defined?

28.2.4 Two Self-Adjoint Operators A and B Such That $AB + BA$ Is Bounded, While $AB - BA$ Is Unbounded

It is extremely simple to find self-adjoint operators A and B such that $AB - BA$ is bounded while $AB + BA$ is unbounded. Merely take an unbounded self-adjoint operator A, and then set $B = A$. Hence

$$AB - BA = A^2 - A^2 = 0_{D(A^2)}$$

is bounded, while $AB + BA = 2A^2$ is obviously unbounded.

Question 28.2.4 Provide a pair of self-adjoint operators A and B such that $AB + BA$ is bounded whereas $AB - BA$ is unbounded.

28.2.5   Two Self-Adjoint Positive Operators (One of Them Is Unbounded) A and B Such That AB − BA Is Unbounded

Question 28.2.5 Find two self-adjoint positive operators (one of them is unbounded) A and B such that $AB - BA$ is unbounded.

28.2.6   On the Positivity of Some Commutator

If A and B are bounded and self-adjoint, then from the proof of Putnam's theorem (see the remark below Theorem 4.2 on Page 61 in [80]), the inequality $[B, iA] \geq \alpha I$ is impossible for any $\alpha > 0$. Then in [287], this result was extended to unbounded self-adjoint operators under the extra condition $D(B) \subset D(A)$.

Question 28.2.6 Show that if the domains $D(A)$ and $D(B)$ are not comparable, then the preceding generalization need not remain true.

28.2.7   Two Unbounded and Self-Adjoint Operators A and B Such That D(A) = D(B) and A² − B² Is Bounded but AB − BA Is Unbounded

Question 28.2.7 There are two unbounded and self-adjoint operators A and B such that $D(A) = D(B)$ and $A^2 - B^2$ is bounded but $AB - BA$ is unbounded.

28.2.8   Two Densely Defined Unbounded and Closed Operators C and B Such That CB − BC Is Bounded and Unclosed, While |C|B − B|C| Is Unbounded and Closed

Question 28.2.8 Give an example of two densely defined unbounded and closed operators C and B such that $CB - BC$ is bounded and unclosed, while $|C|B - B|C|$ is unbounded and closed.

28.2.9 Two Unbounded Injective Self-Adjoint Operators A and B Such That $AB - BA$ Is Bounded, While $|A|B - B|A|$ Is Unbounded

On $L^2(\mathbb{R})$, consider the two operators

$$Af(x) = if'(x) \text{ and } Bf(x) = \varphi(x)f(x)$$

(where φ is an almost everywhere differentiable real-valued function) on their maximal domains. Then both A and B are unbounded self-adjoint operators. Moreover, $|A|B - B|A|$ is bounded whenever $AB - BA$ is bounded (i.e., when φ' is in $L^\infty(\mathbb{R})$). As noted by A. McIntosh in [225], $|A|B - B|A|$ is defined by

$$(|A|B - B|A|)f(x) = \frac{1}{\pi}\text{p.v.} \int (x-y)^{-2}(\varphi(x) - \varphi(y))f(y)dy,$$

which is a one-dimensional L^2 case of a more general result by A.-P. Calderón [40]. The great T. Kato then asked whether this result could be established in an abstract setting, i.e., whether $|A|B - B|A|$ is bounded whenever $AB - BA$ is bounded. The first (and seemingly the only) counterexample is due to A. McIntosh in [225]. The counterexample we are about to give here is new and simpler in some sense than McIntosh's. Moreover, in our case, both $AB - BA$ and $|A|B - B|A|$ are even closed. Still more interesting, in our case both A and B are injective (this last observation was overlooked in [258]).

Question 28.2.9 (⊛⊛) Give an example of two unbounded injective self-adjoint operators A and B such that $AB - BA$ is bounded (and closed), while $|A|B - B|A|$ is unbounded (and also closed).

28.2.10 Two Unbounded (Injective) Self-Adjoint Operators A and B Such That $AB - BA$ Is Bounded and Commutes with B (and A) Yet $|A|B - B|A|$ Is Unbounded

A. McIntosh asked in [225] whether $|A|B - B|A|$ is bounded if $AB - BA$ is bounded and commutes with B, say. His question remained unanswered in his paper (and elsewhere). It is worth stressing that his counterexample is not applicable to answer this last question in the negative.

However, our counterexample above is fortunately also a counterexample to McIntosh's question as we shall see, but a couple of things must first be made clear to readers.

First, it is not clear what A. McIntosh meant by the commutativity of $AB - BA$ and B? Indeed, $AB - BA$ is bounded, but it cannot be everywhere defined as A and B are unbounded self-adjoint operators (see Answer 28.2.3). So, we cannot apply Definition 18.1.6 here. He probably meant some kind of pointwise commutativity. Second, A. McIntosh did not require that $AB - BA$ be densely defined, and so a counterexample, in this case, is technically not excluded.

Question 28.2.10 (Cf. Question 29.2.11) Find two unbounded injective self-adjoint operators A and B such that $AB - BA$ is bounded and commutes with B (and A) yet $|A|B - B|A|$ is unbounded.

28.2.11 Two Unbounded Self-Adjoint, Positive, and Boundedly Invertible Operators A and B Such That $AB - BA$ Is Bounded While $\sqrt{A}\sqrt{B} - \sqrt{B}\sqrt{A}$ Is Unbounded

Question 28.2.11 (⊛ (Cf. Question 32.2.5)) Find a pair of unbounded self-adjoint, positive, and boundedly invertible operators A and B such that $AB - BA$ is bounded while $\sqrt{A}\sqrt{B} - \sqrt{B}\sqrt{A}$ is unbounded.

28.2.12 A Densely Defined Closed Operator T Such That $|T^*||T| - |T||T^*|$ Is Unbounded

Question 28.2.12 Find a densely defined and closed T such that $|T^*||T| - |T||T^*|$ is unbounded.

28.2.13 A Densely Defined and Closed Operator T Such That $TT^* - T^*T$ Is Unbounded and Self-Adjoint but $|T||T^*| - |T^*||T|$ Is Bounded and Unclosed

Question 28.2.13 There is a densely defined and closed operator T such that $TT^* - T^*T$ is unbounded but $|T||T^*| - |T^*||T|$ is bounded (even zero on some domain).

28.2.14 A Densely Defined and Closed Operator T Such That $TT^* - T^*T$ Is Bounded Whereas $|T||T^*| - |T^*||T|$ Is Unbounded

Question 28.2.14 There is a densely defined and closed operator T such that $TT^* - T^*T$ is bounded whereas $|T||T^*| - |T^*||T|$ is unbounded.

28.2.15 Self-Adjoint Operators A and B Such That $\frac{1}{2}|\langle \overline{[A, B]}f, f\rangle| \not\leq \|Af\|\|Bf\|$

Question 28.2.15 Find self-adjoint operators A and B with domains $D(A)$ and $D(B)$, respectively, such that

$$\frac{1}{2}|\langle Cf, f\rangle| \leq \|Af\|\|Bf\|, \ \forall f \in D(A) \cap D(B) \cap D(C)$$

does not hold where $C = \overline{[A, B]}$.

28.2.16 Unbounded Skew-Adjoint Operators Cannot Be Universally Commutable

Following [137], we say that a skew-adjoint operator A is universally commutable in the classical sense if for all skew-adjoint operators B, $AB - BA$ is essentially skew-adjoint, i.e., $\overline{AB - BA}$ is skew-adjoint.

Question 28.2.16 Show that unbounded skew-adjoint operators cannot be universally commutable.

Answers

Answer 28.2.1 Let $Af(x) = xf(x)$ and $Bf(x) = f'(x)$ be defined on $D(A) = L^2(0, 1)$ and $D(B) = \{f \in L^2(0, 1) : f' \in L^2(0, 1)\}$, respectively. Then $A \in B[L^2(0, 1)]$. Hence for $f \in D(AB) \cap D(BA)$, we have

$$BAf(x) - ABf(x) = f(x) + xf'(x) - xf'(x) = f(x)$$

and so $BA - AB \subset I$, as required.

Answer 28.2.2 We will find a fundamental symmetry $B \in B(H)$ (on some appropriate Hilbert space H) and a densely defined, symmetric, positive, and closed operator A with a domain $D(A) \subset H$ satisfying both conditions of Browder's theorem and yet $AB - BA$ is densely defined, unbounded, and closed.

Let T be a densely defined closed symmetric unbounded positive operator with $D(T) \subset L^2(S)$, where S is the unit circle, and such that $D(T^2) = \{0\}$ (as in Answer 20.2.23). Let I be the identity operator on $L^2(S)$. Let

$$B = \begin{pmatrix} 0 & I \\ I & 0 \end{pmatrix} \text{ and } A = \begin{pmatrix} T & 0 \\ 0 & 2T \end{pmatrix}$$

be both defined on $L^2(S) \oplus L^2(S)$ and $D(A) = D(T) \oplus D(T)$, respectively. Observe that B is a fundamental symmetry and that A is densely defined, closed, symmetric, and positive. Moreover,

$$D(A^2) = D(T^2) \oplus D(T^2) = \{(0, 0)\}.$$

Hence $D(A^n) = \{(0, 0)\}$ for all $n \geq 3$. So

$$D(BA^n - A^n B) = D(B_1 A^n - A^n B_1) = \{(0, 0)\}$$

for any $B_1 \in B(L^2(S) \oplus L^2(S))$, i.e., the two "inclusions"

$$BA^n - A^n B \subset B_1 A^{n-1} \text{ and } B_1 A^n - A^n B_1 \subset B_2 A^{n-1}$$

are trivially satisfied for any $B_1, B_2 \in B(L^2(S) \oplus L^2(S))$. Nonetheless,

$$AB - BA = \begin{pmatrix} 0 & T \\ 2T & 0 \end{pmatrix} - \begin{pmatrix} 0 & 2T \\ T & 0 \end{pmatrix} = \begin{pmatrix} 0 & -T \\ T & 0 \end{pmatrix}$$

is densely defined, *unbounded*, and closed on $D(T) \oplus D(T)$.

Answer 28.2.3 In fact, $AB - BA$ is never everywhere defined as long as A and B are closable and unbounded. In that regard, assume $D(AB - BA) = H$ for some Hilbert space H. Then

$$H = D(AB - BA) \subset D(A) \cap D(B)$$

and so $D(A) \cap D(B) = H$. The only way out is $D(A) = D(B) = H$. Since A and B are both closable, it follows that $A, B \in B(H)$!

Remark It remains unknown to me (if this is any interesting) whether one can find two unbounded unclosable operators A and B such that $AB - BA = I$ *everywhere* on H.

Answer 28.2.4 Let T be a self-adjoint operator with a domain $D(T)$. Set

$$A = \begin{pmatrix} 0 & T \\ T & 0 \end{pmatrix} \text{ and } B = \begin{pmatrix} -T & 0 \\ 0 & T \end{pmatrix}.$$

Then A and B are self-adjoint. Moreover,

$$AB = \begin{pmatrix} 0 & T^2 \\ -T^2 & 0 \end{pmatrix} = -\begin{pmatrix} 0 & -T^2 \\ T^2 & 0 \end{pmatrix} = -BA.$$

This makes $AB + BA$ bounded (the zero operator on $D(T^2) \oplus D(T^2)$).
Nevertheless,

$$AB - BA = \begin{pmatrix} 0 & 2T^2 \\ -2T^2 & 0 \end{pmatrix}$$

is plainly unbounded.

Answer 28.2.5 Let A be defined by $Af(x) = -f''(x)$ on the domain $D(A) = \{f \in L^2(0, 1), f'' \in L^2(0, 1) : f(0) = f(1) = 0\}$.

Then A is unbounded, self-adjoint, and positive. Define on $L^2(0, 1)$ the bounded and positive operator $Bf(x) = xf(x)$. We then have for all $f \in D(BA) \cap D(AB)$

$$ABf(x) - BAf(x) = -2f'(x).$$

Accordingly, $AB - BA$ is plainly unbounded.

Answer 28.2.6 On $L^2(\mathbb{R})$, define the operators $Af(x) = xf(x)$ and $Bf(x) = -if'(x)$ on $D(A) = \{f \in L^2(\mathbb{R}) : xf \in L^2(\mathbb{R})\}$ and $D(B) = \{f \in L^2(\mathbb{R}) : f' \in L^2(\mathbb{R})\}$, respectively. Then

$$[B, iA] = 1$$

say on some dense domain.

Answer 28.2.7 Let T be an unbounded and self-adjoint operator, and then take $A = \begin{pmatrix} 0 & T \\ T & 0 \end{pmatrix}$ and $B = \begin{pmatrix} T & 0 \\ 0 & -T \end{pmatrix}$ both defined on $D(T) \oplus D(T)$. Hence $A^2 = B^2 = \begin{pmatrix} T^2 & 0 \\ 0 & T^2 \end{pmatrix}$ and so $A^2 - B^2$ is bounded (on $D(A^2)$), while

$$AB - BA = \begin{pmatrix} 0 & -2T^2 \\ 2T^2 & 0 \end{pmatrix}$$

is obviously unbounded.

Answer 28.2.8 Let B be as in Answer 21.2.45 and set $C = B$. Then $CB - BC = 0_{D(B^2)}$ is clearly bounded and unclosed. By a glance at Answer 21.2.45 once again, we easily see that

$$|B|B - B|B| = \begin{pmatrix} 0 & -A^2 \\ 0 & 0 \end{pmatrix}$$

which is closed (and unbounded) on $D(|B|B - B|B|) = H \oplus D(A^2)$.

Answer 28.2.9 ([258]) The example to be given will be defined in the end on $L^2(\mathbb{R}) \oplus L^2(\mathbb{R}) \oplus L^2(\mathbb{R}) \oplus L^2(\mathbb{R})$.

Let R, S, and T be three self-adjoint operators (yet to be chosen) on a certain Hilbert space H, with domains $\mathcal{D}(R)$, $\mathcal{D}(S)$, and $\mathcal{D}(T)$, respectively. Assume also that S is positive. Now, define on $H \oplus H$ the operators

$$A = \begin{pmatrix} 0 & S \\ S & 0 \end{pmatrix} \text{ and } B = \begin{pmatrix} T & 0 \\ 0 & R \end{pmatrix}$$

where $\mathcal{D}(A) = \mathcal{D}(S) \oplus \mathcal{D}(S)$ and $\mathcal{D}(B) = \mathcal{D}(T) \oplus \mathcal{D}(R)$. Hence

$$AB - BA = \begin{pmatrix} 0 & SR - TS \\ ST - RS & 0 \end{pmatrix}.$$

Since $|A| = \begin{pmatrix} S & 0 \\ 0 & S \end{pmatrix}$, it results that

$$|A|B - B|A| = \begin{pmatrix} ST - TS & 0 \\ 0 & SR - RS \end{pmatrix}.$$

To obtain the appropriate operators, let C and D be such that

$$\mathcal{D}(CD) = \mathcal{D}(DC) = \{0_{L^2(\mathbb{R})}\}$$

(as those in, e.g., Answer 21.2.44). Next, define

$$S = \begin{pmatrix} C & 0 \\ 0 & 2C \end{pmatrix} \text{ and } T = \begin{pmatrix} 0 & D \\ D & 0 \end{pmatrix}$$

(and so S is self-adjoint and positive as needed). Observe also that S and T are injective for C and D are so (see again Answer 21.2.44). Now, clearly

$$ST = \begin{pmatrix} 0 & CD \\ 2CD & 0 \end{pmatrix} \text{ and } TS = \begin{pmatrix} 0 & 2DC \\ DC & 0 \end{pmatrix}.$$

Hence

$$\mathcal{D}(ST) = \mathcal{D}(TS) = \{0_{[L^2(\mathbb{R})]^2}\}.$$

This says that $AB - BA$ is bounded on $L^2(\mathbb{R}) \oplus L^2(\mathbb{R}) \oplus L^2(\mathbb{R}) \oplus L^2(\mathbb{R})$. In fact, $AB - BA$ is trivially bounded as it is only defined on $\{0\}$ and $AB - BA$ is therefore closed.

In order that $|A|B - B|A|$ be unbounded, it suffices to exhibit a self-adjoint R such that $SR - RS$ is unbounded. For this purpose, consider $R = \begin{pmatrix} 0 & I \\ I & 0 \end{pmatrix}$ where I is the usual identity operator (hence R is a fundamental symmetry). Therefore,

$$SR = \begin{pmatrix} 0 & C \\ 2C & 0 \end{pmatrix}, \ RS = \begin{pmatrix} 0 & 2C \\ C & 0 \end{pmatrix}$$

and

$$SR - RS = \begin{pmatrix} 0 & -C \\ C & 0 \end{pmatrix}$$

which is patently unbounded, as looked forward to.

In conclusion, observe that A and B are in effect one-to-one.

Answer 28.2.10 Let A and B be two unbounded injective self-adjoint operators such that $D(AB - BA) = \{0\}$ and $|A|B - B|A|$ is unbounded (as in Answer 28.2.9). As already observed, we need only check "some commutativity" of $AB - BA$ with say B. Set $C = AB - BA$ (and so $D(C) = \{0\}$). Then $D(BC) = \{0\}$ and

$$D(CB) = \{f \in D(B) : Bf = 0\} = \{0\}$$

since B is injective. Therefore, $BC = CB$ trivially.

Answer 28.2.11 ([258]) Consider A and $B := \mathcal{F}^* A \mathcal{F}$ introduced in Answer 21.2.17 and so $D(A) \cap D(B) = \{0\}$, making $AB - BA$ trivially bounded.

The unique positive self-adjoint square root of A is given by

$$\sqrt{A} f(x) = e^{\frac{x^2}{4}} f(x)$$

on $D(\sqrt{A}) = \{f \in L^2(\mathbb{R}) : e^{\frac{x^2}{4}} f \in L^2(\mathbb{R})\}$. Hence $\sqrt{B} := \mathcal{F}^* \sqrt{A} \mathcal{F}$.

There "only" remains to check that $\sqrt{B}\sqrt{A} - \sqrt{A}\sqrt{B}$ is unbounded. A way of seeing that is to consider the action of this operator on some Gaussian functions. More precisely, let $f_0(x) = e^{-ax^2}$, where $a > 0$ is yet to be determined. Then $f_0 \in L^2(\mathbb{R})$ (for all $a > 0$) as

$$\|f_0\|_2^2 = \sqrt{\frac{\pi}{2a}}.$$

Now, it may be checked that

$$\mathcal{F}^* \sqrt{A} \mathcal{F} \sqrt{A} f_0(x) = \frac{2}{\sqrt{5 - 4a}} e^{-\frac{4a-1}{5-4a}x^2},$$

a calculation valid for $1/4 < a < 5/4$. Similarly, we find that

$$\sqrt{A} \mathcal{F}^* \sqrt{A} \mathcal{F} f_0(x) = \frac{1}{\sqrt{1 - a}} e^{-\frac{1-5a}{4(a-1)}x^2}$$

valid for the values of a in $(1/5, 1)$. Therefore, for $1/4 < a < 1$, we have

$$(\sqrt{B}\sqrt{A} - \sqrt{A}\sqrt{B}) f_0(x) = \frac{2}{\sqrt{5 - 4a}} e^{-\frac{4a-1}{5-4a}x^2} - \frac{1}{\sqrt{1 - a}} e^{-\frac{1-5a}{4(a-1)}x^2}.$$

Thus,

$$
\|(\sqrt{B}\sqrt{A} - \sqrt{A}\sqrt{B})f_0\|_2^2 = \frac{4\sqrt{\pi}}{5-4a}\frac{1}{\sqrt{\frac{2(4a-1)}{5-4a}}} + \frac{\sqrt{\pi}}{(1-a)\sqrt{\frac{1-5a}{2(a-1)}}}
$$

$$
- \frac{4\sqrt{\pi}}{\sqrt{5-4a}\sqrt{1-a}} \times \frac{1}{\sqrt{\frac{1-5a}{4(a-1)} + \frac{4a-1}{5-4a}}},
$$

where $1/4 < a < 1$. Upon sending $a \to 1/4$ say, we readily see that the quantity $\|(\sqrt{B}\sqrt{A} - \sqrt{A}\sqrt{B})f_0\|_2/\|f_0\|_2$ goes to infinity, making $\sqrt{B}\sqrt{A} - \sqrt{A}\sqrt{B}$ unbounded, as wished.

Answer 28.2.12 Let A and B be two self-adjoint and positive operators such that $B \in B(H)$ as in Answer 28.2.5. Set

$$
T = \begin{pmatrix} 0 & A \\ B & 0 \end{pmatrix} \text{ and so } T^* = \begin{pmatrix} 0 & B \\ A & 0 \end{pmatrix}.
$$

Besides, T is closed on $H \oplus D(A)$ and also

$$
|T| = \begin{pmatrix} B & 0 \\ 0 & A \end{pmatrix} \text{ and } |T^*| = \begin{pmatrix} A & 0 \\ 0 & B \end{pmatrix}.
$$

Therefore,

$$
|T^*||T| - |T||T^*| = \begin{pmatrix} AB - BA & 0 \\ 0 & BA - AB \end{pmatrix}
$$

Since $AB - BA$ is unbounded, it ensues that $|T^*||T| - |T||T^*|$ too is unbounded and this finishes the answer.

Answer 28.2.13 Let A be an unbounded, self-adjoint, and positive operator with a domain $D(A) \subset H$. Define T on $D(T) = D(A) \oplus H$ by

$$
T = \begin{pmatrix} 0 & 0 \\ A & 0 \end{pmatrix} \text{ and so } T^* = \begin{pmatrix} 0 & A \\ 0 & 0 \end{pmatrix}.
$$

Hence

$$
TT^* = \begin{pmatrix} 0 & 0 \\ 0 & A^2 \end{pmatrix} \text{ and } T^*T = \begin{pmatrix} A^2 & 0 \\ 0 & 0 \end{pmatrix}.
$$

Also,

$$|T||T^*| = \begin{pmatrix} 0 & 0_{D(A)} \\ 0 & 0 \end{pmatrix} \text{ and } |T^*||T| = \begin{pmatrix} 0_{D(A)} & 0 \\ 0 & 0 \end{pmatrix}.$$

Thus,

$$TT^* - T^*T = \begin{pmatrix} -A^2 & 0 \\ 0 & A^2 \end{pmatrix}$$

is clearly unbounded (and self-adjoint), whereas $|T||T^*| - |T^*||T|$ is bounded.

Answer 28.2.14 Let A and B be two self-adjoint and positive operators with domains $D(A)$ and $D(B)$, respectively. Set $T = \begin{pmatrix} 0 & \sqrt{A} \\ \sqrt{B} & 0 \end{pmatrix}$. Hence

$$TT^* - T^*T = \begin{pmatrix} A - B & 0 \\ 0 & B - A \end{pmatrix}$$

and

$$|T||T^*| - |T^*||T| = \begin{pmatrix} \sqrt{B}\sqrt{A} - \sqrt{A}\sqrt{B} & 0 \\ 0 & \sqrt{A}\sqrt{B} - \sqrt{B}\sqrt{A} \end{pmatrix}.$$

To get the desired counterexample, we need to choose A and B such that $D(A) \cap D(B) = \{0\}$ and so $TT^* - T^*T$ becomes trivially bounded. At the same time, the same pair must make $\sqrt{A}\sqrt{B} - \sqrt{B}\sqrt{A}$ unbounded (and thus $|T||T^*| - |T^*||T|$ too is unbounded). Such a pair, however, has already appeared in Answer 28.2.11 and we are done.

Answer 28.2.15 ([110]) Let $H = L^2[0, 1]$. Define $Af(x) = if'(x)$ on $D(A) = \{f \in AC[0, 1] : f' \in L^2[0, 1], \ f(0) = f(1)\}$, and let $Bf(x) = xf(x)$ for $f \in L^2[0, 1]$. Then A and B are self-adjoint (B is even bounded). Moreover, $[A, B] \subset iI$ where the domain of $[A, B]$ is constituted of all absolutely continuous functions f such that $f' \in L^2[0, 1]$ and $f(0) = f(1) = 0$. Now, clearly $C = \overline{[A, B]} = iI$ *everywhere* on $L^2[0, 1]$.

Finally, let $f(x) = 1$ on $[0, 1]$ (and so $f \in D(A) \cap D(B) \cap D(C)$). Hence

$$|\langle Cf, f \rangle| = 1 \text{ and } Af = 0$$

and so

$$\frac{1}{2}|\langle Cf, f \rangle| \leq \|Af\|\|Bf\|, \ \forall f \in D(A) \cap D(B) \cap D(C)$$

is definitely violated.

Answer 28.2.16 ([137]) Let A be any unbounded skew-adjoint operator with a domain $D(A) \subsetneq H$. Choose $a \in H$ such that $a \notin D(A)$ (hence $a \neq 0$) and consider the orthogonal projection onto the span of a, noted P. Now, set $B = iP$. Then B is a bounded skew-adjoint operator. Also, $D(AB) = \{a\}^{\perp}$ and so $\overline{D(AB)} \neq H$. Finally, because

$$D(AB - BA) \subset D(AB),$$

$AB - BA$ cannot be densely defined.

Chapter 29
Spectrum

29.1 Basics

The spectrum of unbounded (closed) operators is defined as its analogue of bounded operators. It seems noteworthy, be that as it may, to tell readers that there exist different definitions of the spectrum in the case of unclosed operators.

Definition 29.1.1 Let A be an operator on a complex Hilbert space H. The resolvent set of A, denoted by $\rho(A)$, is defined by:

$$\rho(A) = \{\lambda \in \mathbb{C} : \lambda I - A \text{ is bijective and } (\lambda I - A)^{-1} \in B(H)\}.$$

The complement of $\rho(A)$, denoted by $\sigma(A)$,

$$\sigma(A) = \mathbb{C} \setminus \rho(A)$$

is called the spectrum of A.

Also, $(\lambda I - A)^{-1}$ is called the resolvent of A in λ, and it is denoted by $R(\lambda, A)$ (i.e. as in the bounded case).

Remark It is clear that $\lambda \in \rho(A)$ iff there exists a $B \in B(H)$ such that

$$(\lambda I - A)B = I \text{ and } B(\lambda I - A) \subset I.$$

Proposition 29.1.1 *Let A be a closed operator. Then $\sigma(A)$ is a closed subset of $\mathbb{C}$.*

The next result is important.

Proposition 29.1.2 *Let A be a closed operator, and let $\lambda \in \mathbb{C}$. If $\lambda I - A$ is bijective, then $(\lambda I - A)^{-1} \in B(H)$.*

The coming simple result is sometimes useful.

M. H. Mortad, *Counterexamples in Operator Theory*,
https://doi.org/10.1007/978-3-030-97814-3_29

Proposition 29.1.3 *Let A be a linear operator with $\rho(A) \neq \varnothing$. Then A is closed.*

Corollary 29.1.4 *Let A be a linear operator. If A is not closed, then $\sigma(A) = \mathbb{C}$.*

As we are already aware, there are closed (and densely defined) operators A such that A^2 is not closed. The next result tells us when powers of closed operators remain closed.

Proposition 29.1.5 (See e.g. [221]) *If A is closed and has a nonempty resolvent set, then A^n is closed for all $n \in \mathbb{N}$.*

Corollary 29.1.6 *If A is closed such that A^n is unclosed for some $n \in \mathbb{N}$, then $\sigma(A) = \mathbb{C}$.*

Remark Recently, it was obtained in [87] among other results that if A is a closable non-closed operator in a Hilbert space such that A^n is closed for some $n \in \mathbb{N}$, then

$$\sigma(A^n) = \mathbb{C}.$$

Here are some straightforward properties of the spectrum.

Proposition 29.1.7 *If A is closed, $\lambda \in \sigma(A)$, and $\alpha \in \mathbb{C}$, then*

$$\lambda + \alpha \in \sigma(A + \alpha I) \text{ and } \alpha\lambda \in \sigma(\alpha T).$$

We will also make use of the following result.

Proposition 29.1.8 (See [221], See also [209] and Example 29.1.1) *Let $n \in \mathbb{N}$ and let A be a closed operator, then*

$$\sigma(A^n) = \{\lambda^n : \lambda \in \sigma(A)\}.$$

Remark Proposition 29.1.8 also holds for polynomials (as in e.g. Theorem 9.6 on Page 326 of [361]).

As in the bounded case, we have the following.

Proposition 29.1.9 *If A is a densely defined closed operator and $\lambda \in \mathbb{C}$, then $\lambda \in \sigma(A)$ if and only if $\overline{\lambda} \in \sigma(A^*)$.*

Proposition 29.1.10 (See [31] for a New Proof of This Well-Known Result) *Let A be a self-adjoint operator. Then $\sigma(A) \subset \mathbb{R}$.*

In fact, we have the following.

Proposition 29.1.11 *Let A be a densely defined closed symmetric operator. Then A is self-adjoint if and only if $\sigma(A) \subset \mathbb{R}$.*

We finish by giving a few examples.

Example 29.1.1 Let A be a closed operator with a domain $D(A) \subset H$. Then

$$\sigma(A^2) = \{\lambda^2 : \lambda \in \sigma(A)\}.$$

Let $\lambda \in \mathbb{C}$ be such that $\lambda^2 \notin \sigma(A^2)$. Then $A^2 - \lambda^2 I$ has a bounded inverse. Hence for some $B \in B(H)$, we have

$$B(A^2 - \lambda^2 I) \subset (A^2 - \lambda^2 I)B = I.$$

But $A^2 - \lambda^2 I = (A - \lambda I)(A + \lambda I)$ and so $(A - \lambda I)(A + \lambda I)B = I$. Hence $A - \lambda I$ is surjective. To see why $A - \lambda I$ is injective, let $x \in D(A)$ be such that $Ax = \lambda x$. Hence $Ax \in D(A)$, i.e. $x \in D(A^2)$. Therefore, $A^2 x = \lambda^2 x$, i.e. $(A^2 - \lambda^2 I)x = 0$ and so $x = 0$ for $A^2 - \lambda^2 I$ is injective. Accordingly, $A - \lambda I$ is bijective and so $(A - \lambda I)^{-1} \in B(H)$ or $\lambda \notin \sigma(A)$.

Conversely, let $\lambda \in \mathbb{C}$ be such that $\lambda^2 \in \sigma(A^2)$. We need to show that either $\lambda \in \sigma(A)$ or $-\lambda \in \sigma(A)$. Assuming that this is not the case implies that both $A - \lambda I$ and $A + \lambda I$ have bounded inverses. Hence $A^2 - \lambda^2 I = (A - \lambda I)(A + \lambda I)$ too has a bounded inverse. Thus, $\lambda^2 \notin \sigma(A^2)$, which is the sought contradiction.

We already know by different means that the square of a self-adjoint operator remains self-adjoint. Here is another approach:

Example 29.1.2 Let A be an unbounded self-adjoint operator. Then A^2 is self-adjoint.

We already know that $D(A^2)$ is dense. We also know that A^2 is symmetric. Now, let $\lambda \in \sigma(A^2)$. Since A is self-adjoint, $\sigma(A) \subset \mathbb{R}$ and so by Example 29.1.1, $\sigma(A^2) \subset \mathbb{R}$. This says that A^2 is closed, and so Proposition 29.1.11 gives the self-adjointness of A^2, as needed.

Example 29.1.3 Every nonempty (unbounded) closed subset K of $\mathbb{C}$ may be regarded as the spectrum of some unbounded normal operator on ℓ^2:

Indeed, by the separability of $\mathbb{C}$, K too is separable and so

$$\overline{\{\alpha_n : n \in \mathbb{N}\}} = K.$$

Next, define the unbounded normal operator $A(x_n) = (\alpha_n x_n)$ on the domain $D(A) = \{(x_n) \in \ell^2 : (\alpha_n x_n) \in \ell^2\}$. By following a similar idea to that used in the bounded case (see e.g. Exercise 7.3.5 in [256]), we may show that

$$\sigma(A) = \overline{\{\alpha_n : n \in \mathbb{N}\}} = K,$$

as wished.

Example 29.1.4 ([326]) Let $\varphi : \mathbb{R} \to \mathbb{C}$ be a continuous function and consider the multiplication operator $M_\varphi f(x) = \varphi(x) f(x)$ on $D(M_\varphi) = \{ f \in L^2(\mathbb{R}) : \varphi f \in L^2(\mathbb{R}) \}$. Then

$$\sigma(M_\varphi) = \overline{\operatorname{ran} \varphi} = \overline{\varphi(\mathbb{R})}$$

(in particular, $\sigma(M_x) = \mathbb{R}$). To see that, let $\lambda \in \operatorname{ran} \varphi = \varphi(\mathbb{R})$ and so $\lambda = \varphi(x_0)$ for some $x_0 \in \mathbb{R}$. If $\varepsilon > 0$, the continuity of φ tells us that for some (open) neighborhood J of x_0 we have $|\varphi(x) - \varphi(x_0)| < \varepsilon$ for all $x \in J$. If $\mathbb{1}_J$ designates the indicator function of J, then

$$\| (M_\varphi - \lambda I) \mathbb{1}_J \| \le \varepsilon \| \mathbb{1}_J \|.$$

So, if λ were in $\rho(M_\varphi)$, then we would have

$$\| \mathbb{1}_J \| = \| R(\lambda, M_\varphi)(M_\varphi - \lambda I) \mathbb{1}_J \| \le \varepsilon \| R(\lambda, M_\varphi) \| \| \mathbb{1}_J \|$$

which is impossible when $\varepsilon \| R(\lambda, M_\varphi) \| < 1$. Therefore, $\lambda \in \sigma(M_\varphi)$ and so we have shown that $\varphi(\mathbb{R}) \subset \sigma(M_\varphi)$. Since M_φ is closed, we know that $\sigma(M_\varphi)$ is closed whereby

$$\overline{\varphi(\mathbb{R})} \subset \sigma(M_\varphi).$$

Conversely, let $\lambda \notin \overline{\varphi(\mathbb{R})}$. Then there exists a certain $\varepsilon > 0$ such that $|\varphi(x) - \lambda| \ge \varepsilon$ for all $x \in \mathbb{R}$. Set $\psi(x) = 1/(\varphi(x) - \lambda)$ and so ψ is bounded on $\mathbb{R}$ giving rise to an everywhere defined and bounded (multiplication) operator M_ψ on $L^2(\mathbb{R})$. In other words,

$$M_\psi f(x) = \psi(x) f(x) = \frac{1}{\varphi(x) - \lambda} f(x), \, f \in L^2(\mathbb{R}).$$

Whence $M_\varphi M_\psi = M_\psi M_\varphi = I$. Consequently, $\lambda \notin \sigma(M_\varphi)$, as needed.

Example 29.1.5 (Cf. [326]) Let $T_\alpha f = if'$ be defined on the domain $D(T_\alpha) = \{ f \in AC[0, 1] : f' \in L^2[0, 1], \, f(1) = \alpha f(0) \}$, where $\alpha \in \mathbb{C}$ is nonzero. Then

$$\sigma(T_\alpha) = \sigma_p(T_\alpha) = \{ -i \ln \alpha + 2k\pi : k \in \mathbb{Z} \}.$$

In particular, for the self-adjoint T_1, we have

$$\sigma(T_1) = \sigma_p(T_1) = \{ 2k\pi : k \in \mathbb{Z} \}.$$

29.2 Questions

29.2.1 A Densely Defined Closed Operator with an Empty Spectrum

Question 29.2.1 Give an example of a densely defined closed unbounded operator A such that $\sigma(A) = \varnothing$.

29.2.2 A Densely Defined Operator A with $\sigma(A) = \mathbb{C}$

Question 29.2.2 Give an example of a densely defined operator A such that $\sigma(A) = \mathbb{C}$.

29.2.3 A Densely Defined Closed Operator A with $\sigma(A) = \mathbb{C}$

Question 29.2.3 Give an example of a densely defined and closed operator A such that $\sigma(A) = \mathbb{C}$. Deduce an example of a densely defined, unbounded, closed, nonnormal, and invertible operator.

29.2.4 Two Unbounded Strongly Commuting Self-Adjoint Operators A and B Such That $\sigma(A + B) \not\subset \sigma(A) + \sigma(B)$

Recall that if A and B are in $B(H)$ and $AB = BA$, then $\sigma(A + B) \subset \sigma(A) + \sigma(B)$ holds. More generally:

Theorem 29.2.1 ([6]) *If $B \in B(H)$ commutes with an unbounded operator A, i.e. $BA \subset AB$, then*

$$\sigma(A + B) \subset \sigma(A) + \sigma(B)$$

holds.

Remark In fact, the writers in [6] established the above result under the condition $\sigma(A) \neq \mathbb{C}$, which was imposed for other reasons. However, the inclusion $\sigma(A + B) \subset \sigma(A) + \sigma(B)$ is trivial when $\sigma(A) = \mathbb{C}$.

How about the case of two unbounded operators?

Question 29.2.4 Find two unbounded and strongly commuting self-adjoint operators A and B such that

$$\sigma(A + B) \not\subset \sigma(A) + \sigma(B).$$

29.2.5 Two Densely Closed Unbounded Operators A and B Such That $\sigma(A) + \sigma(B)$ Is Not Closed

Let H be a complex Hilbert space. Then $\sigma(A) + \sigma(B)$ is always compact when $A, B \in B(H)$.

Now, if A and B are two linear operators such that one of them is in $B(H)$ and the other is closed, then from topological considerations, we know that $\sigma(A) + \sigma(B)$, being the algebraic sum of a closed and a compact set, is closed (see e.g. [256]).

Question 29.2.5 Find two densely defined, closed, and unbounded operators A and B such that $\sigma(A) + \sigma(B)$ is not closed.

29.2.6 Two Commuting Self-Adjoint Unbounded Operators A and B Such That $\sigma(BA) \not\subset \sigma(B)\sigma(A)$

Question 29.2.6 Find two (strongly) commuting self-adjoint unbounded operators A and B such that

$$\sigma(BA) \not\subset \sigma(B)\sigma(A).$$

29.2.7 A Densely Defined Operator A with $\sigma(A^2) \neq [\sigma(A)]^2$

Question 29.2.7 Find a densely defined operator A such that

$$\sigma(A^2) \neq [\sigma(A)]^2.$$

29.2.8 A Normal Operator T with $p[\sigma(T)] \neq \sigma[p(T, T^*)]$ for Some Polynomial p

Question 29.2.8 Find an unbounded normal operator T such that

$$p[\sigma(T)] \neq \sigma[p(T, T^*)]$$

for some polynomial p of two variables.

29.2.9 Two Closed (Unbounded) Operators A and B Such That $\sigma(AB) - \{0\} \neq \sigma(BA) - \{0\}$

As we will see below, the Jacobson lemma is not satisfied in the general context of closed unbounded operators. It is, however, desirable for readers to have knowledge of the following result.

Theorem 29.2.2 ([153]) *Let A and B two closed densely defined operators such that $\sigma(AB) \neq \mathbb{C}$ and $\sigma(BA) \neq \mathbb{C}$. Then*

$$\sigma(AB) - \{0\} \neq \sigma(BA) - \{0\}.$$

Remark See [392] and [84], and the references therein for further reading.

Question 29.2.9 Find two closed operators A and B such that

$$\sigma(AB) - \{0\} \neq \sigma(BA) - \{0\}.$$

29.2.10 A (Non-closed) Densely Defined Operator A Such That $\sigma(AA^*) - \{0\} \neq \sigma(A^*A) - \{0\}$

If A is a densely defined closed operator, then both AA^* and A^*A are self-adjoint and positive, hence they both have spectra contained in $\mathbb{R}^+$. In short, $\sigma(AA^*) \neq \mathbb{C}$ and $\sigma(A^*A) \neq \mathbb{C}$. Since both A and A^* are closed, by Theorem 29.2.2 just above, we have that

$$\sigma(AA^*) - \{0\} = \sigma(A^*A) - \{0\}.$$

Is the closedness of A that important?

Question 29.2.10 Provide an unclosed densely defined operator A such that

$$\sigma(AA^*) - \{0\} \neq \sigma(A^*A) - \{0\}.$$

29.2.11   Two Self-Adjoint Operators A and B with $B \in B(H)$ Such That B Commutes with $AB - BA$ yet $\sigma(AB - BA) \neq \{0\}$

Recall the following result whose proof may be consulted on e.g. Page 5 in [295].

Proposition 29.2.3 *If $A, B \in B(H)$ and if $C = AB - BA$ is such that $AC = CA$, then $\sigma(C) = \{0\}$.*

This is a practical result whose importance is well known to operator theorists. Can we hope to have an unbounded version of it?

Question 29.2.11 Find two self-adjoint operators A and B with $B \in B(H)$ such that B commutes with $AB - BA$ yet

$$\sigma(AB - BA) \neq \{0\}.$$

29.2.12   An Unbounded, Closed, and Nilpotent Operator N with $\sigma(N) \neq \{0\}$

Recall that if $N \in B(H)$ is nilpotent, then

$$\sigma(N) = \{0\}.$$

Question 29.2.12 Show that this is not necessarily the case if N is unbounded and closed.

29.2.13 An $A \in B(H)$ and an Unbounded Closed Nilpotent Operator N with $AN = NA$ but $\sigma(A + N) \neq \sigma(A)$

It is known that if $A, N \in B(H)$ are such that $AN = NA$ and N is nilpotent, then $\sigma(A + N) = \sigma(A)$ (see e.g. Exercise 7.3.29 in [256] for an interesting proof).

Is the previous result valid in the context of one unbounded operator? Observe that there are two cases to investigate in this event:

(1) $NA \subset AN$, where $N \in B(H)$ is nilpotent.
(2) $AN \subset NA$, where $A \in B(H)$ and N is closed and nilpotent.

The first case has an affirmative answer.

Proposition 29.2.4 *Let* $N \in B(H)$ *be nilpotent, and let* A *be a densely defined closed operator such that* $NA \subset AN$. *Then*

$$\sigma(A + N) = \sigma(A).$$

Proof By Theorem 29.2.1 (just above Question 29.2.4), we know that

$$\sigma(A + N) \subset \sigma(A) + \sigma(N) = \sigma(A).$$

Conversely,

$$\sigma(A) = \sigma(A + N - N) \subset \sigma(A + N) + \sigma(-N) = \sigma(A + N)$$

if N is nilpotent and commutes with $A + N$. Therefore,

$$\sigma(A + N) = \sigma(A).$$

□

What about the case when the nilpotent operator is the unbounded one?

Question 29.2.13 Find an $A \in B(H)$ and a nilpotent closed operator N such that $AN \subset NA$ and yet

$$\sigma(A + N) \neq \sigma(A).$$

29.2.14   A Closed, Symmetric, and Positive Operator T with $\sigma(T) = \mathbb{C}$

As in the bounded case, an unbounded self-adjoint positive operator has a positive spectrum. Would closedness combined with positiveness suffice for this result to hold?

Question 29.2.14 Find a densely defined closed operator T with a domain $D(T)$ such that $\langle Tx, x \rangle \geq 0$ for all $x \in D(T)$, and yet $\sigma(T) \not\subset \mathbb{R}^+$.

29.2.15   The (Only) Four Possible Cases for the Spectrum of Closed Symmetric Operators

Theorem 29.2.5 (See e.g. Theorem 2.8, Chapter X in [66]) *Let A be a closed (densely defined) symmetric operator. Then one and only one of the following cases occurs:*

(1) $\sigma(A) = \mathbb{C}$.
(2) $\sigma(A) = \{\lambda \in \mathbb{C} : \operatorname{Im} \lambda \geq 0\}$.
(3) $\sigma(A) = \{\lambda \in \mathbb{C} : \operatorname{Im} \lambda \leq 0\}$.
(4) $\sigma(A) \subset \mathbb{R}$.

Question 29.2.15 Give examples showing that each of the four possibilities of Theorem 29.2.5 takes place.

29.2.16   Unbounded Densely Defined Closable Operators A and B Such That $A \subset B^*$, BA Is Essentially Self-Adjoint, and $\sigma(\overline{BA}) = \sigma(\overline{A}\,\overline{B}) \neq \sigma(\overline{AB})$

Question 29.2.16 (✸) Find unbounded densely defined closable operators A and B such that $A \subset B^*$, BA is essentially self-adjoint, and

$$\sigma(\overline{BA}) = \sigma(\overline{A}\,\overline{B}) \neq \sigma(\overline{AB}).$$

Answers

Answer 29.2.1 We give two examples.

First, define the closed operator A by $Af(x) = f'(x)$ on $D(A) = \{f \in L^2[0, 1] : f' \in L^2[0, 1], f(0) = 0\}$. Then $\sigma(A) = \varnothing$.

Let $f \in D(A)$. In order to find the spectrum of A, we need to solve the following differential equation:

$$\lambda f(x) - f'(x) = g(x),$$

where $g \in L^2[0, 1]$ is given, and $\lambda \in \mathbb{C}$. We find that

$$f(x) = \int_0^x e^{\lambda(x-t)} g(t) dt$$

and it is the *unique* solution that belongs to $D(A)$. This means that $\lambda I - A$ is invertible for all λ, that is,

$$\rho(A) = \mathbb{C} \text{ or } \sigma(A) = \varnothing.$$

As for the second example, define an operator $A : L^2(\mathbb{R}) \to L^2(\mathbb{R})$ by:

$$(Af)(x) = e^{-x^2} f(x - 1)$$

where $f \in L^2(\mathbb{R})$. Then it may be shown (see e.g. Exercise 10.3.23 in [256]) that $A \in B[L^2(\mathbb{R})]$, A is one-to-one with a dense range, and $\sigma(A) = \{0\}$. Let B be an (unbounded) operator with domain $D(B) = \operatorname{ran} A$, such that

$$BAf = f, \ f \in L^2(\mathbb{R}).$$

Then B is closed and $\sigma(B) = \varnothing$. We start by showing that B is closed on $D(B)$. Let (f_n) be in $D(B)$ such that $f_n \to f$ and $Bf_n \to g$. Since $f_n \in D(B) = \operatorname{ran} A$, we may write $f_n = Ag_n$ for some $g_n \in L^2(\mathbb{R})$. Hence $Ag_n \longrightarrow f$ and $BAg_n = g_n \longrightarrow g$. Thus $g \in L^2(\mathbb{R})$, and since A is continuous, we deduce that $f = Ag \in \operatorname{ran}(A) = D(B)$, i.e. $f \in D(B)$. Besides $Bf = BAg = g$ and this shows the closedness of B.

Now, we show that $\sigma(B) = \varnothing$. We need to investigate two cases:

(1) Let $\lambda = 0$. We need to show that $0 \times I - B$, i.e. B, is bijective. First, B is injective. If $Bf = 0$, where $f \in D(B) = \operatorname{ran} A$, then as $f = Ag$ for some g, we must have $BAg = 0$. But $BA = I$ and so $g = 0$ or $f = 0$. Thus, B is one-to-one.

Moreover, as $BA = I$, it is clear that B is onto. Therefore, B is bijective and so invertible by Proposition 29.1.2. Consequently, $0 \in \rho(B)$.

(2) Let $\lambda \neq 0$. We must show that $\lambda I - B$ is bijective. We start by proving injectivity. Let $f \in D(B)$. Since $\lambda \neq 0$, we may write

$$(\lambda I - B)f = 0 \Longrightarrow \lambda Af - BAf = 0$$
$$\Longrightarrow \lambda Af - f = 0$$
$$\Longrightarrow \left(A - \frac{1}{\lambda} I \right) f = 0.$$

Since $\sigma(A) = \{0\}$, $\rho(A) = \mathbb{C} \setminus \{0\}$, i.e. $A - \mu I$ is bijective for all $\mu \neq 0$. This implies that $A - \mu I$ is injective for all $\mu \neq 0$, in particular for $\mu = \frac{1}{\lambda}$ too. Hence $f = 0$, showing that $\lambda I - B$ is one-to-one.

Let us turn now to surjectivity. Let $g \in L^2(\mathbb{R})$. Since $A - \frac{1}{\lambda} I$ is surjective and $g/\lambda \in L^2(\mathbb{R})$, we have

$$\exists h \in D(A): \ \left(A - \frac{1}{\lambda} I \right) h = \frac{g}{\lambda}.$$

Now, let $f = Ah$. Then

$$(\lambda I - B)f = (\lambda I - B)Ah = \lambda Ah - BAh$$
$$= \lambda Ah - h = \lambda \left(A - \frac{1}{\lambda} I \right) h = g,$$

proving that $\lambda I - B$ is onto for all $\lambda \neq 0$. Hence $\lambda I - B$ is bijective for all $\lambda \neq 0$, that is, invertible $\lambda I - B$ for all $\lambda \neq 0$ (Proposition 29.1.2), i.e. $\mathbb{C} \setminus \{0\} \subset \rho(B)$.

Accordingly,

$$\rho(B) = \mathbb{C} \text{ or } \sigma(B) = \varnothing.$$

Answer 29.2.2 The answer is quite simple. Any non-closed operator will do! Let us prove it. If $\rho(A)$ is not empty, then it contains at least one complex λ. Hence $A - \lambda I$ is (boundedly) invertible and so $A - \lambda I$ is closed. Hence $(A - \lambda I) + \lambda I$ too is closed, that is, A is closed on $D(A)$.

Answer 29.2.3 Let A be defined by $Af(x) = -if'(x)$ on the domain $D(A) = \{f \in L^2(0, 1) : f' \in L^2(0, 1)\}$. Then A is closed. Clearly, for each $\lambda \in \mathbb{C}$, $f_\lambda(x) = e^{i\lambda x}$ is clearly in $D(A)$ and satisfies

$$Af_\lambda = \lambda f_\lambda$$

which means that $\lambda \in \sigma_p(A)$. Hence $\mathbb{C} \subset \sigma_p(A)$ and therefore

$$\sigma(A) = \sigma_p(A) = \mathbb{C}.$$

To answer the second question, consider the unbounded closed operator A as above. Since $D(A) \neq D(A^*)$, the non-normality of A follows. Since we already obtained above that $\rho(A) = \mathbb{C}$, we have that $\lambda I - A$ is invertible for any complex λ. But $\lambda I - A$ is not normal and the counterexample is therefore $\lambda I - A$ for any value of λ!

Answer 29.2.4 Let A be a self-adjoint operator with a dense domain $D(A)$. Set $B = -A$ and so B is self-adjoint, and it plainly strongly commutes with A. Since $\sigma(A), \sigma(B) \subset \mathbb{R}, \sigma(A) + \sigma(B) \subset \mathbb{R}$.

But, $A + B$ is unclosed on $D(A)$ and so $\sigma(A + B) = \mathbb{C}$. Consequently,

$$\sigma(A + B) \not\subset \sigma(A) + \sigma(B).$$

Answer 29.2.5 First, recall that any nonempty closed subset M of $\mathbb{C}$ may be regarded as the spectrum of a closed operator A on ℓ^2 (see e.g. Example 29.1.3).

Let M and N be two closed subsets of $\mathbb{C}$ such that $M + N$ is not closed (see e.g. [255]). By the above remark, we can therefore find densely defined closed operators A and B on ℓ^2 such that

$$\sigma(A) = M \text{ and } \sigma(B) = N.$$

Thus, $\sigma(A) + \sigma(B)$ is not closed.

Answer 29.2.6 Let A be a self-adjoint and boundedly invertible operator with a domain $D(A) \subset H$, then set $B = A^{-1} \in B(H)$. Clearly $BA = I_{D(A)}$ is not closed. Hence $\sigma(BA) = \mathbb{C}$. Since $\sigma(B)\sigma(A) \subset \mathbb{R}$,

$$\sigma(BA) \not\subset \sigma(B)\sigma(A).$$

Answer 29.2.7 Let H be a Hilbert space, and let $A : H \to H$ be a non-closable operator such that $A^2 = 0$ everywhere on H. Then $\sigma(A) = \mathbb{C}$ and so

$$\sigma(A^2) = \{0\} \neq \mathbb{C} = [\sigma(A)]^2.$$

Answer 29.2.8 An example already appeared in [23]. Here, we give another (perhaps simpler) example. Let $p(z, \bar{z}) = z + \bar{z}$ be a two-variable polynomial.

Let A be an unbounded self-adjoint operator with a domain $D(A)$. Set $T = iA$, where $i^2 = -1$. Then T is clearly normal. In addition, $T + T^* = 0$ on $D(A)$ and so $T + T^*$ is unclosed. This gives

$$\sigma(T + T^*) = \mathbb{C}.$$

Now, since T is skew-adjoint, we know that $\sigma(T)$ is purely imaginary and hence so is the case with $\sigma(T^*)$. Therefore, $p[\sigma(T)] \neq \mathbb{C}$, i.e.

$$p[\sigma(T)] \neq \sigma[p(T, T^*)].$$

Answer 29.2.9 Let A be an unbounded and closed operator (with a non-closed domain $D(A)$), and let $B = 0$ (everywhere defined). Then plainly

$$AB = 0 \text{ and } BA \subset 0.$$

Since BA is not closed, $\sigma(BA) = \mathbb{C}$ while clearly $\sigma(AB) = \{0\}$, that is,

$$\sigma(AB) - \{0\} \neq \sigma(BA) - \{0\}.$$

Here is another example (from [153]). Let T be a closed, unbounded, and boundedly invertible operator with domain $D(T) \subset H$. Define on $H \oplus H$

$$A = \begin{pmatrix} I & T \\ 0 & T \end{pmatrix} \text{ and } B = \begin{pmatrix} I & 0 \\ 0 & T^{-1} \end{pmatrix}$$

with $D(A) = H \oplus D(T)$ and $D(B) = H \oplus H$. Then A is closed on $D(A)$ and B is everywhere defined and bounded on $H \oplus H$. Besides,

$$AB = \begin{pmatrix} I & I \\ 0 & I \end{pmatrix} \text{ and } BA = \begin{pmatrix} I & T \\ 0 & I \end{pmatrix}.$$

Then AB is bounded and everywhere defined on $H \oplus H$. Also, BA is closed (and unbounded) on $D(BA) = H \oplus D(T)$.

Finally, it is easy to see that $\sigma(AB) = \{1\}$. Also, 1 is clearly an eigenvalue for BA and so $1 \in \sigma(BA)$. If $\lambda \neq 1$, then

$$(BA - \lambda)^{-1} = (I - \lambda)^{-1} \begin{pmatrix} I & -(I - \lambda)^{-1}T \\ 0 & I \end{pmatrix}$$

works as the unbounded inverse of $BA - \lambda$. Therefore,

$$\sigma(BA) = \mathbb{C}.$$

Answer 29.2.10 Let A be as in Answer 21.2.14, that is, A is unclosed and A^*A is self-adjoint on $H^2(\mathbb{R})$. Hence $\sigma(A^*A) \subset \mathbb{R}$.

Nevertheless, AA^* cannot be self-adjoint as this would force A to be closed (cf. [335] or [87]). In fact,

$$AA^*f(x) = -f''(x)$$

for $f \in D(AA^*) = \{f \in L^2(\mathbb{R}) : f''' \in L^2(\mathbb{R})\}$. Hence AA^* is not even closed and so $\sigma(AA^*) = \mathbb{C}$. Therefore,

$$\sigma(A^*A) - \{0\} \neq \sigma(AA^*) - \{0\},$$

as wished.

Answer 29.2.11 Let $B = 0$ everywhere on some Hilbert space H, and let A be any unbounded (self-adjoint) operator with a domain $D(A) \subsetneq H$. Then

$$BA - AB \subset 0$$

for $D(BA - AB) = D(A) \subset H$. Besides, B obviously commutes with $AB - BA$. Nonetheless,

$$\sigma(AB - BA) = \mathbb{C}$$

because $AB - BA$ is unclosed on $D(A)$.

Answer 29.2.12 Let T be any (unbounded) closed operator with a domain $D(T) \subset H$. Set $N = \begin{pmatrix} 0 & T \\ 0 & 0 \end{pmatrix}$ with $D(N) = H \oplus D(T)$.

The nilpotence of N is clear. Since N is closed, we know that $\sigma(N^2) = [\sigma(N)]^2$. Since N^2 is unclosed, $\sigma(N^2) = \mathbb{C}$. If $\sigma(N) = \{0\}$, then we would have $[\sigma(N)]^2 = \{0\}$ as well, and this is absurd. Therefore, $\sigma(N) \neq \{0\}$, as desired.

Answer 29.2.13 The simplest example perhaps is to take $A = 0$ and N as in Answer 29.2.12, and so $AN \subset NA$. By Answer 29.2.12, $\sigma(N) \neq \{0\}$. Hence

$$\sigma(A + N) = \sigma(N) \neq \{0\} = \sigma(A).$$

Another example is based on the second couple of operators A and B that appeared in Answer 29.2.9. We then found

$$BA = \begin{pmatrix} I & T \\ 0 & I \end{pmatrix}$$

(remember that T is closed and so BA too is closed) and $\sigma(BA) = \mathbb{C}$. Now, write

$$\begin{pmatrix} I & T \\ 0 & I \end{pmatrix} = \underbrace{\begin{pmatrix} I & 0 \\ 0 & I \end{pmatrix}}_{=\tilde{I}} + \underbrace{\begin{pmatrix} 0 & T \\ 0 & 0 \end{pmatrix}}_{=N}.$$

Accordingly,

$$\sigma(\tilde{I} + N) = \mathbb{C} \neq \{1\} = \sigma(\tilde{I})$$

yet $\tilde{I}$ is everywhere defined and bounded, and it commutes with N.

Answer 29.2.14 Consider again T defined by $Tf(x) = -f''(x)$ on

$$D(T) = \{f \in L^2(0, 1), f'' \in L^2(0, 1) : f(0) = f(1) = 0 = f'(0) = f'(1)\}.$$

One way of seeing that T is closed is as follows: $-d^2/dx^2$ is self-adjoint, hence closed, on both $D(SS^*)$ and $D(S^*S)$, where S is as in Answer 23.2.4. Hence T is closed on $D(T) := D(SS^*) \cap D(S^*S)$ by Example 19.1.8. However, T is not self-adjoint on $D(T)$. Now, plainly

$$\forall f \in D(T) : \langle Tf, f \rangle \geq 0.$$

Accordingly, T is closed, symmetric, and positive, and so $\sigma(T) \not\subset \mathbb{R}$ for T is not self-adjoint. In particular, $\sigma(T) \not\subset \mathbb{R}^+$.

Answer 29.2.15

(1) $\sigma(T) = \mathbb{C}$: Let $Tf(x) = -if'(x)$ be defined on

$$D(T) = \{f \in L^2(0, 1), f' \in L^2(0, 1) : f(0) = f(1) = 0\}.$$

Then T is densely defined, closed, and symmetric, and we have already found that

$$\sigma(T) = \mathbb{C}.$$

(2) $\sigma(T) = \{\lambda \in \mathbb{C} : \operatorname{Im} \lambda \leq 0\}$: (Cf. [326]) Let $Tf(x) = -if'(x)$ on the domain $D(T) = \{f \in H^1(0, \infty) : f(0) = 0\}$. Then T is densely defined, closed, and symmetric (see Example 20.1.11). To show that $\sigma(T) = \{\lambda \in \mathbb{C} : \operatorname{Im} \lambda \leq 0\}$, let $\lambda \in \mathbb{C}$ be such that $\operatorname{Im} \lambda < 0$. Set $f_\lambda(x) = e^{i\overline{\lambda}x}$ and so $f_\lambda \in D(T^*)$. Besides,

$$T^* f_\lambda = \overline{\lambda} f_\lambda,$$

i.e. $\overline{\lambda} \in \sigma_p(T^*)$. Thence, $\overline{\lambda} \in \sigma(T^*)$ or merely $\lambda \in \sigma(T)$ (see e.g. Proposition 29.1.9). Therefore

$$\{\lambda \in \mathbb{C} : \operatorname{Im} \lambda < 0\} \subset \sigma(T)$$

thereby $\{\lambda \in \mathbb{C} : \operatorname{Im} \lambda \leq 0\} \subset \sigma(T)$, and we are half way through.

For the other way inclusion, suppose that $\operatorname{Im} \lambda > 0$ and consider $h(t) = ie^{i\lambda t}$, where $t \geq 0$. Define an operator $R(\lambda, T)$ by:

$$[R(\lambda, T)g](x) = (g * h)(x) = i \int_0^x e^{i(x-t)} g(t) dt$$

where $g \in L^2(0, \infty)$. By a basic Young's inequality on convolutions (see e.g. Theorem 9.1 in [379]), we know that

$$\|R(\lambda, T)g\|_2 \leq \|g\|_2 \|h\|_1$$

as $h \in L^1(0, \infty)$ (hence $R(\lambda, T)$ is a bounded operator on $L^2(0, \infty)$). The aim is to show that $T - \lambda I$ is bijective. Set $f = R(\lambda, T)g$ and so $f \in AC[a, b]$ for all $[a, b] \subset (0, \infty)$. Clearly, $f \in L^2(0, \infty)$, $f' = i(\lambda f + g) \in L^2(0, \infty)$, and $f(0) = 0$. Therefore, $f \in D(T)$.

Now, we have

$$(T - \lambda I)f = (T - \lambda I)R(\lambda, T)g = g$$

which proves that T is onto. Also, $T - \lambda I$ is one-to-one because $\operatorname{Im} \lambda > 0$. Accordingly, $T - \lambda I$ is bijective and so it is boundedly invertible, thanks to the closedness of T.

(3) $\sigma(T) = \{\lambda \in \mathbb{C} : \operatorname{Im} \lambda \geq 0\}$: Let T be a closed and symmetric operator such that $\sigma(T) = \{\lambda \in \mathbb{C} : \operatorname{Im} \lambda \leq 0\}$ (as before). Then $-T$ remains closed and symmetric and clearly

$$\sigma(-T) = -\sigma(T) = \{\lambda \in \mathbb{C} : \operatorname{Im} \lambda \geq 0\},$$

as wished.

(4) $\sigma(T) \subset \mathbb{R}$: Just consider any unbounded self-adjoint operator T. Hence T is closed, symmetric, and $\sigma(T) \subset \mathbb{R}$.

Answer 29.2.16 ([153]) We only give the main points of the proof, leaving gaps that are to be filled in to readers.

Consider again the operators A and B introduced in Answer 19.2.13, that is,

$$A = \frac{d}{dx}, \quad D(A) = \{f \in C^1[0, 1] : f(0) = 0\}$$

and

$$B = -\frac{d}{dx}, \quad D(B) = \{f \in C^1[0, 1] : f(1) = f'(2/5) = 0\}$$

both considered in the Hilbert space $L^2(0, 1)$. Then

$$\overline{B} = -\frac{d}{dx}, \quad D(\overline{B}) = \{f \in H^1(0, 1) : f(1) = 0\}$$

and

$$\overline{A} = \frac{d}{dx}, \quad D(\overline{A}) = \{f \in H^1(0, 1) : f(0) = 0\}.$$

Clearly, $A \subset \overline{A} = B^*$. Moreover,

$$BA = -\frac{d^2}{dx^2}, \ D(BA) = \{f \in C^2[0, 1] : f'(1) = f(0) = f''(2/5) = 0\}$$

and so

$$\overline{BA} = -\frac{d^2}{dx^2}, \ D(\overline{BA}) = \{f \in H^2[0, 1] : f'(1) = f(0) = 0\}.$$

Ergo, $\overline{BA}$ is self-adjoint. Furthermore, the spectrum of $\overline{BA}$ is discrete, that is, solely constituted of eigenvalues of finite multiplicities that are isolated points of the spectrum (see e.g. Chapter XIII of [96] or [326]). To find the eigenvalues, we solve

$$-f''(x) = \lambda f(x)$$

where $\lambda > 0$ and $f \in D(T)$. We find that

$$\sigma(\overline{BA}) = \left\{ \frac{(2k-1)^2\pi^2}{4} : k \in \mathbb{N} \right\}.$$

Also, $\overline{A} \ \overline{B} = -d^2/dx^2$ is defined on

$$D(\overline{A} \ \overline{B}) = \{f \in H^2[0, 1] : f(1) = f'(0) = 0\}.$$

Finally,

$$\overline{AB} = -\frac{d^2}{dx^2}, \ D(\overline{AB}) = \{f \in H^2[0, 1] : f(1) = f'(0) = f'(2/5) = 0\}$$

and so we may find that

$$\sigma(\overline{AB}) = \left\{ \frac{25(2k-1)^2\pi^2}{4} : k \in \mathbb{N} \right\}.$$

Consequently,

$$\sigma(\overline{BA}) = \sigma(\overline{A} \ \overline{B}) \neq \sigma(\overline{AB}),$$

as required.

Remark It can be shown that the operator $\overline{A} \ \overline{B}$ is self-adjoint on $D(\overline{A} \ \overline{B}) = \{f \in H^2[0, 1] : f(1) = f'(0) = 0\}$ (for instance, $\overline{A} \ \overline{B}$ is symmetric and closed and it has a real spectrum). However, the operator $\overline{AB}$ cannot be self-adjoint. Let us posit that $\overline{AB}$ is self-adjoint. Since

$$\overline{AB} = -\frac{d^2}{dx^2}, \ D(\overline{AB}) = \{f \in H^2[0, 1] : f(1) = f'(0) = f'(2/5) = 0\},$$

it would ensue that $\overline{AB} \subset \overline{A}\ \overline{B}$, which, by the maximality of self-adjoint operators, would make us ending up having $D(\overline{AB}) = D(\overline{A}\ \overline{B})$, which is impossible.

In fact, $\overline{AB}$ is not even symmetric. Indeed, if it were, then its closedness combined with the realness of its spectrum would make it self-adjoint, and that is not the case.

Chapter 30
Matrices of Unbounded Operators

We have been hitherto using matrices of operators preponderantly as a tool to manufacture counterexamples related to other questions. Here, we confine our attention to counterexamples related to questions about basic properties of such matrices.

30.1 Questions

30.1.1 Equal Matrices and Pairwise Different Entries

Recall that if all of A, B, C, D and A', B', C', D' are in $B(H)$, then

$$T := \begin{pmatrix} A & B \\ C & D \end{pmatrix} = S := \begin{pmatrix} A' & B' \\ C' & D' \end{pmatrix}$$

yields $A = A'$, $B = B'$, $C = C'$, and $D = D'$. What about densely defined operators? The answer is negative. Indeed, an easy example is to consider an unbounded closed A with domain $D(A)$ (and hence, $D(A) \neq H$), and then let $0_{D(A)}$ be the zero operator restricted to $D(A)$. Hence,

$$\begin{pmatrix} A & 0 \\ 0_{D(A)} & 0 \end{pmatrix} = \begin{pmatrix} A & 0 \\ 0 & 0 \end{pmatrix}$$

as for all $(x, y) \in D(A) \times H$ we have

$$Ax + 0y = Ax + 0y \text{ and } 0_{D(A)}x + 0y = 0x + 0y,$$

where obviously $0_{D(A)} \neq 0$.

© The Author(s), under exclusive license to Springer Nature Switzerland AG 2022
M. H. Mortad, *Counterexamples in Operator Theory*,
https://doi.org/10.1007/978-3-030-97814-3_30

Question 30.1.1 Show that it may happen that

$$T := \begin{pmatrix} A & B \\ C & D \end{pmatrix} = S := \begin{pmatrix} A' & B' \\ C' & D' \end{pmatrix}$$

with $A \neq A'$, $B \neq B'$, $C \neq C'$, and $D \neq D'$ *simultaneously*.

30.1.2   A Bounded Matrix with All Unbounded Entries

We know that if all entries of $T := \begin{pmatrix} A & B \\ C & D \end{pmatrix}$ are bounded, then so is T. What about the converse?

Question 30.1.2 Find densely defined *unbounded* self-adjoint operators A, B, C, and D such that

$$T := \begin{pmatrix} A & B \\ C & D \end{pmatrix}$$

is bounded.

30.1.3   The Failure of the Product Formula for Some Matrices of Operators

Question 30.1.3 Find

$$A = \begin{pmatrix} A_{11} & A_{12} \\ A_{21} & A_{22} \end{pmatrix} \text{ and } B = \begin{pmatrix} B_{11} & B_{12} \\ B_{21} & B_{22} \end{pmatrix}$$

such that

$$AB \neq \begin{pmatrix} A_{11}B_{11} + A_{12}B_{21} & A_{11}B_{12} + A_{12}B_{22} \\ A_{21}B_{11} + A_{22}B_{21} & A_{21}B_{12} + A_{22}B_{22} \end{pmatrix},$$

even when B is everywhere defined and bounded, and all the products $A_{ij}B_{ij}$ and their sums are well defined.

30.1.4 A Closed Matrix Yet All Entries Are Unclosed

Question 30.1.4 Find densely defined non-closed operators A, B, C, and D such that

$$T = \begin{pmatrix} A & B \\ C & D \end{pmatrix}$$

is closed.

30.1.5 A Matrix Whose Closure Is Not Equal to the Matrix of Closures

Question 30.1.5 Find densely defined closable operators A, B, C, and D such that

$$\overline{\begin{pmatrix} A & B \\ C & D \end{pmatrix}} \neq \begin{pmatrix} \overline{A} & \overline{B} \\ \overline{C} & \overline{D} \end{pmatrix}.$$

30.1.6 A Non-closed (Essentially Self-Adjoint) Matrix Whose Entries Are All Self-Adjoint

Question 30.1.6 Find a densely defined T given by

$$T = \begin{pmatrix} A & B \\ C & D \end{pmatrix}$$

where each entry is unbounded and self-adjoint, but T is unclosed, unbounded, and essentially self-adjoint.

30.1.7   A Non-closable Matrix Yet All Entries Are Closed

Question 30.1.7 Find a densely defined T given by

$$T = \begin{pmatrix} A & B \\ C & D \end{pmatrix}$$

with all entries being closed yet T is not even closable.

30.1.8   A Densely Defined Matrix Whose Formal Adjoint Is Not Densely Defined

Question 30.1.8 Provide a densely defined $A = \begin{pmatrix} A_{11} & A_{12} \\ A_{21} & A_{22} \end{pmatrix}$ such that its formal adjoint A_*, that is,

$$A_* = \begin{pmatrix} A_{11}^* & A_{21}^* \\ A_{12}^* & A_{22}^* \end{pmatrix}$$

is not even densely defined.

30.1.9   A Closed Matrix Whose Adjoint Does Not Admit a Matrix Representation

Question 30.1.9 Find a closed and densely defined T admitting a matrix representation, but T^* does not have one.

30.1.10   A Matrix Whose Adjoint Differs from Its Formal Adjoint

Question 30.1.10 Let T be a matrix operator. Show that it may happen that $T^* \neq T_*$. Find such a T that also satisfies $\overline{T_*} = T^*$.

30.1.11   A Non-boundedly Invertible Matrix Yet All Its Entries Pairwise Commute and Its Formal Determinant Is Boundedly Invertible

Question 30.1.11 Provide a matrix $T = \begin{pmatrix} A & B \\ C & D \end{pmatrix}$ where say B is unbounded (the other operators being everywhere defined and bounded) and all of its entries pairwise commute, the formal determinant $\det T = AD - BC$ is invertible, and yet T is not boundedly invertible.

30.1.12   A Self-Adjoint Matrix Yet None of Its Entries Is Even Closed

Question 30.1.12 (⊛) Find a matrix of unbounded operators such that none of its entries is self-adjoint, yet the whole matrix is self-adjoint.

30.1.13   A Matrix of Operators A of Size n × n, Where All of Its Entries Are Unclosable and Everywhere Defined, Each A^p, $1 \leq p \leq n - 1$, Does Not Contain Any Zero Entry but $A^n = 0$

Question 30.1.13 Find a matrix of operators A of size $n \times n$, where all of its entries are everywhere defined and unclosable, each A^p, $1 \leq p \leq n - 1$, does not contain any zero entry but $A^n = 0$.

Answers

Answer 30.1.1 Define in H two unbounded closed operators A and B with domains $D(A)$ and $D(B)$, respectively (so $D(A) \neq H$ and $D(B) \neq H$). Choose $A \subset B$ with $D(A) \neq D(B)$. Now, let

$$T = \begin{pmatrix} A & 0_{D(A)} \\ 0_{D(B)} & B \end{pmatrix} \text{ and } S = \begin{pmatrix} B & 0_{D(B)} \\ 0_{D(A)} & A \end{pmatrix}.$$

Then $D(T) = D(S) = D(A) \oplus D(A)$, and for all $(x, y) \in D(A) \oplus D(A)$, we have that

$$T\begin{pmatrix} x \\ y \end{pmatrix} = \begin{pmatrix} Ax \\ By \end{pmatrix} = \begin{pmatrix} Bx \\ Ay \end{pmatrix} = S\begin{pmatrix} x \\ y \end{pmatrix},$$

meaning that $S = T$. Finally, observe that $A \neq B$ and $0_{D(A)} \neq 0_{D(B)}$, as needed.

Answer 30.1.2 An extreme but quite acceptable way is to take four unbounded and self-adjoint operators A, B, C, and D such that

$$\mathcal{D}(A) \cap \mathcal{D}(C) = \mathcal{D}(B) \cap \mathcal{D}(D) = \{0\}.$$

Then T becomes defined on $\mathcal{D}(T) = \{0\} \oplus \{0\}$, and so T is plainly bounded on $\mathcal{D}(T)$.

Answer 30.1.3 The main issue is domains. Let A be an unbounded and closed operator with domain $D(A) \neq H$. Let $I \in B(H)$ be the usual identity operator. Set

$$\mathcal{A} = \begin{pmatrix} A & 0 \\ 0 & A \end{pmatrix}, \mathcal{B} = \begin{pmatrix} I & I \\ 0 & 0 \end{pmatrix} \text{ and } S = \begin{pmatrix} A & A \\ 0 & 0 \end{pmatrix}.$$

Then

$$T := \begin{pmatrix} A & 0 \\ 0 & A \end{pmatrix} \begin{pmatrix} I & I \\ 0 & 0 \end{pmatrix} \neq \begin{pmatrix} A & A \\ 0 & 0 \end{pmatrix}$$

because

$$D(T) = \{(x, y) \in H \times H : x + y \in D(A)\} \neq D(A) \times D(A) = D(S)$$

since $(\alpha, -\alpha) \in D(T)$ but $(\alpha, -\alpha) \notin D(S)$ for some $\alpha \in H$ such that $\alpha \notin D(A)$. Consequently, $T \neq S$.

Answer 30.1.4 Let A be the Fourier transform restricted to the dense subspace $C_0^\infty(\mathbb{R}) \subset L^2(\mathbb{R})$, and let $B = I$ be the identity operator defined on ran$A$. Then, as in Answer 18.2.8

$$D(A) \cap D(B) = C_0^\infty(\mathbb{R}) \cap \text{ran}(A) = \{0\}.$$

In addition, A and B are bounded and unclosed. Therefore,

$$T := \begin{pmatrix} A & A \\ B & B \end{pmatrix}$$

is trivially closed as it is only defined at $(0_{L^2(\mathbb{R})}, 0_{L^2(\mathbb{R})})$.

Answer 30.1.5 We may just reconsider the operators A and B that appeared in Answer 30.1.4. Hence,

$$D(A) \cap D(B) = \{0\}.$$

Whence

$$T := \begin{pmatrix} A & A \\ B & B \end{pmatrix}$$

is trivially closed. At the same time, $\overline{A} = \mathcal{F}$ and $\overline{B} = I$, where $\mathcal{F}$ is the Fourier transform on $L^2(\mathbb{R})$ and I is now the identity operator on all of $L^2(\mathbb{R})$. Thus,

$$\overline{\begin{pmatrix} A & A \\ B & B \end{pmatrix}} = \begin{pmatrix} A & A \\ B & B \end{pmatrix} \neq \underbrace{\begin{pmatrix} \mathcal{F} & \mathcal{F} \\ I & I \end{pmatrix}}_{:=S} = \begin{pmatrix} \overline{A} & \overline{A} \\ \overline{B} & \overline{B} \end{pmatrix}$$

because

$$D(T) = \{(0,0)\} \neq L^2(\mathbb{R}) \oplus L^2(\mathbb{R}) = D(S).$$

Remark Another example may be found in Answer 30.1.6.

Answer 30.1.6 Let A be an unbounded self-adjoint operator with domain $D(A) \subsetneq H$; then set

$$T = \begin{pmatrix} A & A \\ A & A \end{pmatrix}$$

where $D(T) = D(A) \oplus D(A)$ (which is evidently dense in $H \oplus H$).

Let us show that T has in effect the required properties. The unboundedness of A is clear. Next, we show that such a T cannot be closed. By the density of $D(A)$, we know that if $x \in H$ (s.t. $x \notin D(A)$), then $x_n \to x$ for some $x_n \in D(A)$, thereby $(x_n, -x_n) \to (x, -x)$. The observation

$$T \begin{pmatrix} x_n \\ -x_n \end{pmatrix} = \begin{pmatrix} A & A \\ A & A \end{pmatrix} \begin{pmatrix} x_n \\ -x_n \end{pmatrix} = \begin{pmatrix} 0 \\ 0 \end{pmatrix}$$

together with $(x, -x) \notin D(T)$ amply yields the non-closedness of T. To see why T is essentially self-adjoint, we may for instance write

$$T = \underbrace{\begin{pmatrix} A & 0 \\ 0 & A \end{pmatrix}}_{L_A} + \underbrace{\begin{pmatrix} 0 & A \\ A & 0 \end{pmatrix}}_{R_A}$$

(observe in passing that both L_A and R_A are self-adjoint). Then for all $(x, y) \in D(A) \oplus D(A)$, we have

$$\|L_A(x, y)\| = \|R_A(x, y)\|$$

and so

$$\|L_A(x, y)\| \leq \|R_A(x, y)\| + b\|(x, y)\|$$

for all $b \geq 0$. By an appeal to Wüst's theorem, we obtain the essential self-adjointness of $L_A + R_A = T$, as needed.

Answer 30.1.7 Let A and B be two unbounded self-adjoint operators on a Hilbert space H such that $D(A) \cap D(B) = \{0\}$. Define

$$T = \begin{pmatrix} A & B \\ 0 & 0 \end{pmatrix}$$

on $D(A) \times D(B) \subset H \times H$. To show that T is not closable, it suffices (and it necessitates) to show that $D(T^*)$ is not dense in $H \times H$. In that regard, write:

$$T = \begin{pmatrix} A & B \\ 0 & 0 \end{pmatrix} = \underbrace{\begin{pmatrix} I & I \\ 0 & 0 \end{pmatrix}}_{R} \underbrace{\begin{pmatrix} A & 0 \\ 0 & B \end{pmatrix}}_{S},$$

which is possible for $D(T) = D(RS) = D(S) = D(A) \times D(B)$. Since $R \in B(H \times H)$, it follows that

$$T^* = S^* R^* = \begin{pmatrix} A^* & 0 \\ 0 & B^* \end{pmatrix} \begin{pmatrix} I & 0 \\ I & 0 \end{pmatrix} = \begin{pmatrix} A & 0 \\ 0 & B \end{pmatrix} \begin{pmatrix} I & 0 \\ I & 0 \end{pmatrix}$$

or merely

$$T^* = \begin{pmatrix} A & 0 \\ B & 0 \end{pmatrix}$$

for

$$D(T^*) = \{(x, y) \in H \times H : (x, x) \in D(A) \times D(B)\} = \{0\} \times H$$

because $D(A) \cap D(B) = \{0\}$. Consequently, T^* is not densely defined.

Answer 30.1.8 Let A_{11} and A_{21} be two self-adjoint operators such that $D(A_{11}) \cap D(A_{12}) = \{0\}$, and take $A_{21} = A_{22} = 0$. That is, $A = \begin{pmatrix} A_{11} & A_{12} \\ 0 & 0 \end{pmatrix}$. Then,

$$D(A_*) = \{0\} \times H,$$

i.e. $D(A_*)$ is not dense in $H \oplus H$.

Remark A more refined example may be found in [231].

Answer 30.1.9 Let A be an unbounded, densely defined, and closed operator (hence, $D(A) \neq H$). Set

$$T = \begin{pmatrix} A & 0 \\ A & 0 \end{pmatrix}$$

on $D(T) = D(A) \times H$, and so T is densely defined. To see why T is closed, take (x_n, y_n) in $D(A) \times H$ such that $(x_n, y_n) \to (x, y)$ and $T(x_n, y_n) = (Ax_n, Ax_n) \to (z, t)$. By the closedness of A, we may easily obtain that of T.

Now, write

$$T = \underbrace{\begin{pmatrix} I & 0 \\ I & 0 \end{pmatrix}}_{S} \underbrace{\begin{pmatrix} A & 0 \\ 0 & 0 \end{pmatrix}}_{\mathcal{A}}.$$

This is possible because $D(S) = H \times H$, $D(\mathcal{A}) = D(A) \times H$, and $D(T) = D(A) \times H$. Since S is everywhere defined and bounded, and $\mathcal{A}$ is densely defined, it results that

$$T^* = \begin{pmatrix} A^* & 0 \\ 0 & 0 \end{pmatrix} \begin{pmatrix} I & I \\ 0 & 0 \end{pmatrix}$$

on

$$D(T^*) = D(\mathcal{A}^* S^*) = \{(x, y) \in H \times H : x + y \in D(A^*)\}.$$

Clearly, $D(A^*) \neq H$ (why?). So, let $\alpha \in H \setminus D(A^*)$. Then

$$(\alpha, -\alpha) \in D(T^*) \text{ and } (\alpha, 0) \notin D(T^*).$$

By Proposition 18.1.3, T^* does not have a matrix representation (as $QD(T^*) \not\subset D(T^*)$).

Answer 30.1.10 Consider again the example from Answer 30.1.9, i.e.

$$T = \begin{pmatrix} A & 0 \\ A & 0 \end{pmatrix}$$

on $D(A) \times H$ where A is an unbounded, densely defined, and closed operator with domain $D(A) \neq H$. Then we already know that T is closed and that $D(T^*) = \{(x, y) \in H \times H : x + y \in D(A^*)\}$. Clearly,

$$T_* = \begin{pmatrix} A^* & A^* \\ 0 & 0 \end{pmatrix}$$

is defined on $D(T_*) = D(A^*) \times D(A^*)$. Let $\alpha \in H \setminus D(A^*)$. Then $(\alpha, -\alpha) \in D(T^*)$, but $(\alpha, -\alpha) \notin D(T_*)$. This is sufficient to declare that $T^* \neq T_*$.

Finally, we show that $\overline{T_*} = T^*$. It is plain that

$$T_* = \begin{pmatrix} A^* & A^* \\ 0 & 0 \end{pmatrix} = \begin{pmatrix} I & I \\ 0 & 0 \end{pmatrix} \begin{pmatrix} A^* & 0 \\ 0 & A^* \end{pmatrix}.$$

Hence,

$$(T_*)^* = \begin{pmatrix} A & 0 \\ 0 & A \end{pmatrix} \begin{pmatrix} I & 0 \\ I & 0 \end{pmatrix}.$$

Thus,

$$D[(T_*)^*] = \{(x, y) \in H \times H : (x, x) \in D(A) \times D(A)\} = D(A) \times H = D(T).$$

Since we already know that $T_* \subset T^*$, it follows that $T^{**} = T \subset (T_*)^*$, and so $T = (T_*)^*$. Accordingly, $\overline{T_*} = T^*$, as needed.

Answer 30.1.11 Let A be *any* unbounded operator with domain $D(A) \subset H$, and let $I \in B(H)$ be the usual identity operator. Set $T = \begin{pmatrix} I & A \\ 0 & I \end{pmatrix}$ (with $D(T) = H \oplus D(A)$). Then all entries pairwise commute. Moreover,

$$\det T = I - A0 = I$$

is obviously invertible (observe in passing that $\det T \in B(H)$ for $A0 = 0$ everywhere on H). Finally, T is invertible with the inverse given by

$$T^{-1} = \begin{pmatrix} I & -A \\ 0 & I \end{pmatrix}$$

which is clearly unbounded, that is, T is not boundedly invertible, as wished.

Answer 30.1.12 The example is inspired by one that appeared in [177]. Let A be a self-adjoint operator, and let B any *unclosed* symmetric operator that is A-bounded with a relative bound less than 1. Then both $\begin{pmatrix} A & B \\ B & A \end{pmatrix}$ and $\begin{pmatrix} B & A \\ A & B \end{pmatrix}$ are self-adjoint. Indeed, we may write

$$\begin{pmatrix} A & B \\ B & A \end{pmatrix} = \underbrace{\begin{pmatrix} A & 0 \\ 0 & A \end{pmatrix}}_{:=T_A} + \underbrace{\begin{pmatrix} 0 & B \\ B & 0 \end{pmatrix}}_{:=S_B}$$

and S_B is clearly T_A-bounded with a relative bound less than 1 (by an appeal to the Kato–Rellich theorem). Similarly, for the second matrix, we have that T_B is clearly S_A-bounded with relative bound less than 1.

Now, define T by

$$T = \begin{pmatrix} A & 0 & B & 0 \\ 0 & B & 0 & A \\ B & 0 & A & 0 \\ 0 & A & 0 & B \end{pmatrix}$$

on $D(T) = [D(A) \cap D(B)]^4 = [D(A)]^4 \subset H^4$. Then write

$$\begin{pmatrix} I & 0 & 0 & 0 \\ 0 & 0 & I & 0 \\ 0 & I & 0 & 0 \\ 0 & 0 & 0 & I \end{pmatrix} \underbrace{\begin{pmatrix} A & B & 0 & 0 \\ B & A & 0 & 0 \\ 0 & 0 & B & A \\ 0 & 0 & A & B \end{pmatrix}}_{S} \underbrace{\begin{pmatrix} I & 0 & 0 & 0 \\ 0 & 0 & I & 0 \\ 0 & I & 0 & 0 \\ 0 & 0 & 0 & I \end{pmatrix}}_{U}$$

$$= \begin{pmatrix} A & 0 & B & 0 \\ 0_{D(A)} & B_{D(A)} & 0_{D(A)} & A \\ B_{D(A)} & 0 & A & 0 \\ 0 & A & 0 & B_{D(A)} \end{pmatrix}$$

$$= T$$

as the last matrix and T share the same domain. Since S is a diagonal matrix of self-adjoint operators, it ensures that S too is self-adjoint. Since $U^* = U^{-1} = U$, we may therefore write

$$(USU)^* = (SU)^*U^* = U^*S^*U^* = USU$$

settling the question of the self-adjointness of USU, i.e. T is self-adjoint.

Now, if we regard T as an operator on $H^2 \times H^2$, then by setting $R = \begin{pmatrix} A & 0 \\ 0 & B \end{pmatrix}$ and $R' = \begin{pmatrix} B & 0 \\ 0 & A \end{pmatrix}$, we may rewrite

$$T = \begin{pmatrix} R & R' \\ R' & R \end{pmatrix}.$$

Finally, none of the entries of the self-adjoint T is closed. Indeed,

$$R^* = \begin{pmatrix} A & 0 \\ 0 & B^* \end{pmatrix} \implies \overline{R} = \begin{pmatrix} A & 0 \\ 0 & \overline{B} \end{pmatrix} \neq R$$

as $D(\overline{R}) \neq D(R)$. We reason similarly to show that R' is not closed.

Answer 30.1.13 ([263])   In an interesting preprint, I. D. Mercer [227] gave a way of constructing $n \times n$ matrices B such that none of $B, \cdots B^{n-1}$ has any zero entry yet $B^n = 0$. The general form is (though not indicated by I. D. Mercer in his preprint)

$$B = \begin{pmatrix} 2 & 2 & \cdots\cdots & 2 & 1-n \\ n+2 & 1 & \cdots\cdots & 1 & -n \\ 1 & n+2 & 1 & \cdots & 1 & \vdots \\ \vdots & & 1 & \ddots\ddots & \vdots & \vdots \\ \vdots & & \vdots & \ddots\ddots & 1 & -n \\ 1 & & 1 & \cdots & 1 & n+2 & -n \end{pmatrix}$$

By way of an example, consider the 6×6 matrix:

$$B = \begin{pmatrix} 2\,2\,2\,2\,2 & -5 \\ 8\,1\,1\,1\,1 & -6 \\ 1\,8\,1\,1\,1 & -6 \\ 1\,1\,8\,1\,1 & -6 \\ 1\,1\,1\,8\,1 & -6 \\ 1\,1\,1\,1\,8 & -6 \end{pmatrix}.$$

Then it may be checked that $B^p \neq 0_{\mathcal{M}_6}$ for all $p \leq 5$ yet $B^6 = 0_{\mathcal{M}_6}$.

The proof in the case of matrices of operators remains unchanged whether the entries are all in $B(H)$ or they are all unbounded, unclosable, and everywhere defined on H. So let T be any linear operator defined on all of H such that $T^p \neq 0$ for $1 \leq p \leq n - 1$. Then set

$$
A = \begin{pmatrix}
2T & 2T & \cdots\cdots & 2T & (1-n)T \\
(n+2)T & T & \cdots\cdots & T & -nT \\
T & (n+2)T & T \cdots & T & \vdots \\
\vdots & T & \ddots\,\ddots & \vdots & \vdots \\
\vdots & \vdots & \ddots\,\ddots & T & -nT \\
T & T & \cdots\ T\ (n+2)T & -nT
\end{pmatrix}
$$

which is defined on all $H \oplus H \oplus \cdots \oplus H$ (n copies of H). Then none of the entries of A^p with $1 \leq p \leq n - 1$ is the zero operator yet $A^n = 0$ everywhere on $H \oplus H \oplus \cdots \oplus H$.

Chapter 31
Relative Boundedness

In this chapter, we first give several recently obtained examples of relatively bounded operators (see [190] for more known and classical examples) and some corresponding counterexamples. In the second strand, we provide a few counterexamples related to some basic questions about relative boundedness.

Note that so far, we have been utilizing the latter predominantly as a tool to produce counterexamples.

31.1 Questions

31.1.1 A Function $\varphi \in L^2(\mathbb{R}^2)$ Such That $\frac{\partial^2 \varphi}{\partial x \partial y} \in L^2(\mathbb{R}^2)$ but $\varphi \notin L^\infty(\mathbb{R}^2)$

First, we give a classical example. One proof relies on elementary tools of Fourier analysis (more specifically, the Riemann–Lebesgue lemma and the Plancherel theorem), see e.g. [303], Theorem IX.28.

Theorem 31.1.1 *Let $n \leq 3$ and let f be in $D(\Delta)$, where $D(\Delta) = \{f \in L^2(\mathbb{R}^n) : \Delta f \in L^2(\mathbb{R}^n)\}$. Then f is a bounded continuous function, and for any $a > 0$, there is some b, independent of f, so that*

$$\|f\|_\infty \leq a\|\Delta f\|_2 + b\|f\|_2.$$

Corollary 31.1.2 (See e.g. Theorem X.15 in [303]) *Let M_φ be the multiplication operator by a real-valued function φ such that $\varphi \in L^2(\mathbb{R}^n) + L^\infty(\mathbb{R}^n)$ where $n \leq 3$. Then $-\Delta + M_\varphi$ is self-adjoint on $D(\Delta)$.*

© The Author(s), under exclusive license to Springer Nature Switzerland AG 2022
M. H. Mortad, *Counterexamples in Operator Theory*,
https://doi.org/10.1007/978-3-030-97814-3_31

Questions about the self-adjointness of the perturbed Laplacian have been well studied in the literature (see e.g. [303]) due to their applications in quantum mechanics. In this question and the coming two, we deal with other types of operators. The first operator treated is the wave operator.

Recall that $BMO(\mathbb{R}^2)$ designates the space of functions of bounded mean oscillation (see e.g. [98]).

Theorem 31.1.3 *([238]) Set* $\Box = \frac{\partial^2}{\partial t^2} - \frac{\partial^2}{\partial x^2}$ *(the wave operator) and let*

$$D(\Box) = \{f \in L^2(\mathbb{R}^2), \Box f \in L^2(\mathbb{R}^2)\}.$$

Then $D(\Box) \subset BMO(\mathbb{R}^2)$. *More precisely, we have*

$$\|f\|_{BMO(\mathbb{R}^2)} \leq a\|\Box f\|_2 + b\|f\|_2$$

for some positive constants a and b and all $f \in D(\Box)$.

Hence, we have the following corollary whose proof is an immediate consequence of an interpolation theorem ([98], Theorem 6.8).

Corollary 31.1.4 *Let* $2 \leq p < \infty$. *Then*

$$\|f\|_{L^p(\mathbb{R}^2)} \leq a\|\Box f\|_{L^2(\mathbb{R}^2)} + b\|f\|_{L^2(\mathbb{R}^2)}$$

where the constants a and b depend on p.

Corollary 31.1.5 *The constant "a" in the foregoing corollary may be made as small as we wish.*

To prove the previous result, we use a scaling argument. To this end, let $f_\lambda(x, y) = f(\lambda x, \lambda y)$, $\lambda > 0$. Then

$$\|\Box f_\lambda\|_2 = \lambda\|\Box f\|_2, \|f_\lambda\|_2 = \frac{1}{\lambda}\|f\|_2 \text{ and } \|f_\lambda\|_p = \frac{1}{\lambda^{\frac{2}{p}}}\|f\|_p.$$

The estimate in Corollary 31.1.4 when applied to f_λ in place of f therefore becomes

$$\|f\|_p \leq a\lambda^{\frac{2}{p}+1}\|\Box f\|_2 + b\lambda^{\frac{2}{p}-1}\|f\|_2, \lambda > 0, p \geq 2.$$

Upon taking λ small enough, the constant in front of $\|\Box f\|_2$ may be made arbitrary, as wished.

Consequently, if $\alpha > 0$, then the previous corollary allows us to say that any multiplication operator by $V \in L^{2+\alpha}(\mathbb{R}^2)$ is $\Box$-bounded. Indeed, we have by Corollary 31.1.4 that:

$$\|f\|_p \leq a\|\Box f\|_2 + b\|f\|_2$$

with $2 \leq p < \infty$. Then by (the generalized) Hölder's inequality:

$$\|Vf\|_2 \leq \|V\|_q \|f\|_p \leq a\|V\|_q \|\Box f\|_2 + b\|V\|_q \|f\|_2$$

for $\frac{1}{2} = \frac{1}{p} + \frac{1}{q}$ or merely $q = \frac{2p}{p-2}$. Since the constant ahead of $\|\Box f\|_2$ can be made small enough so that we have $a\|V\|_q < 1$, we conclude by the Kato–Rellich theorem that $\Box + V$ is self-adjoint on $D(\Box)$.

Having elaborated how to get relative boundedness from estimates as the one in Corollary 31.1.4, we content ourselves in the sequel with giving the basic estimate. The next result gives some global estimate for the wave operator in higher dimensions. For the proof that appeared in [248], I thank (once again) Professor Hart Smith for his help.

Theorem 31.1.6 *Let* $\Box = \frac{\partial^2}{\partial t^2} - \Delta_x$ *be the wave operator defined on* $L^2(\mathbb{R} \times \mathbb{R}^n)$, $n \geq 2$, *and* $p = \frac{2n+2}{n-1}$. *Then for all* $a > 0$, *there exists a* $b > 0$ *such that*

$$\|f\|_{L^p(\mathbb{R} \times \mathbb{R}^n)} \leq a\|\Box f\|_{L^2(\mathbb{R} \times \mathbb{R}^n)} + b\|f\|_{L^2(\mathbb{R} \times \mathbb{R}^n)}$$

for all $f \in L^2(\mathbb{R} \times \mathbb{R}^n)$ *such that* $\Box f \in L^2(\mathbb{R} \times \mathbb{R}^n)$.

Now, we show that the estimate in Corollary 31.1.4 fails to hold when $p = \infty$.

Question 31.1.1 (❋❋) Find $\varphi \in L^2(\mathbb{R}^2)$ such that $\frac{\partial^2 \varphi}{\partial x \partial y} \in L^2(\mathbb{R}^2)$, but φ is not essentially bounded on $\mathbb{R}^2$, i.e. $\varphi \notin L^\infty(\mathbb{R}^2)$.

31.1.2 A $u \in L^2(\mathbb{R}^{n+1})$ with $(-i\partial/\partial t - \Delta_x)u \in L^2(\mathbb{R}^{n+1})$ yet $u \notin L_t^q(L_x^r)$ for Given Values of q and r

The next classical operator on the list is the time-dependent Schrödinger operator. Recall that the time-dependent Schrödinger operator is the operator L given by

$$L = -i\frac{\partial}{\partial t} - \Delta_x,$$

where $(x, t) \in \mathbb{R}^n \times \mathbb{R} = \mathbb{R}^{n+1}$ with $n \geq 1$, which we define on

$$D(L) = \{f \in L^2(\mathbb{R}^{n+1}) : Lf \in L^2(\mathbb{R}^{n+1})\}.$$

Before stating the main estimate in this case, we define some modified mixed norms. For any integer k,

$$\|f\|_{L^q_{t,k}(L^r_x)} = \left(\int_k^{k+1} \|f(\cdot, t)\|^q_{L^r_x} dt \right)^{1/q}$$

and then

$$\|f\|_{\mathcal{L}_{p,q,r}} = \left(\sum_{k \in \mathbb{Z}} \|f\|^p_{L^q_{t,k}(L^r_x)} \right)^{1/p}.$$

Finally, let $n \geq 1$, and let q and r be the positive real numbers, possibly infinite. Define the set Ω_n as follows: for $n \neq 2$,

$$\Omega_n = \left\{ (q,r) \in \mathbb{R}^+ \times \mathbb{R}^+ : \frac{2}{q} + \frac{n}{r} \geq \frac{n}{2}, \ q \geq 2, \ r \geq 2 \right\} \tag{31.2}$$

and for $n = 2$, Ω_2 is defined in by the same expression, but omitting the point $(2, \infty)$.

The main result obtained in [239] is the following:

Theorem 31.1.7 *Let $n \geq 1$, and let $(q,r) \in \Omega_n$. Then for all $a > 0$, there exists $b > 0$ such that*

$$\|f\|_{\mathcal{L}_{2,q,r}} \leq a\|Lf\|_{L^2(\mathbb{R}^{n+1})} + b\|f\|_{L^2(\mathbb{R}^{n+1})}$$

for all $f \in D(L)$.

Using the inclusion $\mathcal{L}_{2,q,r} \subseteq L^q_t(L^r_x)$ for $q \geq 2$, we deduce:

Corollary 31.1.8 *Let $n \geq 1$, and let $(q,r) \in \Omega_n$. Then for all $a > 0$, there exists $b > 0$ such that*

$$\|u\|_{L^q_t(L^r_x)} \leq a\|Lu\|_{L^2(\mathbb{R}^{n+1})} + b\|u\|_{L^2(\mathbb{R}^{n+1})} \tag{31.3}$$

for all $u \in M^n_L$.

In particular, we get such a bound for $\|u\|_{L^q(\mathbb{R}^{n+1})}$ whenever $2 \leq q \leq (2n+4)/n$.

Question 31.1.2 (⊛⊛) Show that there *are no constants a and b* such that

$$\|u\|_{L^q_t(L^r_x)} \leq a\|Lu\|_{L^2(\mathbb{R}^{n+1})} + b\|u\|_{L^2(\mathbb{R}^{n+1})}$$

holds for all $u \in D(L)$.

31.1.3 A Function $u \in L^2(\mathbb{R}^2)$ with $(\partial/\partial t + \partial^3/\partial x^3)u \in L^2(\mathbb{R}^2)$ yet $u \notin L^p(\mathbb{R}^2)$ for Any $p > 8$

Let L be the Airy operator, that is,

$$L = \frac{\partial}{\partial t} + \frac{\partial^3}{\partial x^3}$$

defined on

$$D(L) = \{u \in L^2(\mathbb{R}^2) : Lu \in L^2(\mathbb{R}^2)\}.$$

Theorem 31.1.9 ([48]) *For all $a > 0$, there exists some $b > 0$ such that*

$$\|f\|_{L^8(\mathbb{R}^2)} \leq a\|Lf\|_{L^2(\mathbb{R}^2)} + b\|f\|_{L^2(\mathbb{R}^2)}$$

for all $f \in D(L)$.

Question 31.1.3 (⊛⊛) Let $p > 8$ (this includes the case $p = \infty$ as well). Show that there do not exist positive constants a and b such that

$$\|u\|_{L^p(\mathbb{R}^2)} \leq a\|Lu\|_{L^2(\mathbb{R}^2)} + b\|u\|_{L^2(\mathbb{R}^2)}$$

for all $u \in D(L)$.

31.1.4 A Negative V in L^2_{loc} Such That $-\Delta + V$ Is Not Essentially Self-Adjoint on C_0^∞

Based upon a certain Kato's inequality (see e.g. Theorem X.27, [303]), the following was shown in Theorem X.28 in [303]:

Theorem 31.1.10 *If V is positive and in $L^2_{loc}(\mathbb{R}^n)$, then $-\Delta + V$ is essentially self-adjoint on $C_0^\infty(\mathbb{R}^n)$.*

What about negative potentials?

Question 31.1.4 Find a negative V in $L^2_{loc}(\mathbb{R}^n)$ such that $-\Delta + V$ is not essentially self-adjoint on $C_0^\infty(\mathbb{R}^n)$.

31.1.5   A Positive V in L^2_{loc} Such That $\frac{\partial^2}{\partial t^2} - \frac{\partial^2}{\partial x^2} + V$ Is Not Essentially Self-Adjoint on C_0^∞

As mentioned just above, $-\triangle + V$ is essentially self-adjoint on $C_0^\infty(\mathbb{R}^2)$ whenever V is positive and in L^2_{loc}. It was asked in [237] whether this can be extended to the wave operator. Although the answer was somewhat expected to be negative, a full answer was not found until some anonymous referee who, even though had rejected a paper of mine, did give me the idea of how to improve a counterexample that already appeared in [237] to the case of a positive V. The example finally appeared in [248].

Question 31.1.5 (⊛) Show that there exists a positive V belonging to $L^2_{loc}(\mathbb{R}^2)$ such that $\frac{\partial^2}{\partial t^2} - \frac{\partial^2}{\partial x^2} + V$ is not essentially self-adjoint on $C_0^\infty(\mathbb{R}^2)$.

31.1.6   If B Is A-Bounded, Is B^2 A^2-Bounded? Is B^* A^*-Bounded?

Question 31.1.6 Let A be a closed operator, and let B be A-bounded. Does it follow that B^2 is A^2-bounded? Does it follow that B^* is A^*-bounded?

31.1.7   The Kato–Rellich Theorem for Three Operators?

Question 31.1.7 Find a symmetric B and two unbounded self-adjoint operators A and C (with $D(A) = D(C)$) with $D(A) \subset D(B)$ such that for some $0 < a, c < 1$ and $\alpha \geq 0$

$$\|Bx\| \leq a\|Ax\| + c\|Cx\| + \alpha\|x\|$$

for all $x \in D(A)$, yet $A + B + C$ is not self-adjoint.

31.1.8   The Kato–Rellich Theorem for Normal Operators?

Question 31.1.8 Can we expect the Kato–Rellich theorem to hold for unbounded normal operators?

Answers

Answer 31.1.1 ([238]) We are going to construct the counterexample using a linear interpolation. To prepare the ground for it, first define $(x, y) \mapsto \varphi(x, y)$ on $\mathbb{R} \times (y_n, y_{n+1})$ by

$$\varphi(x, y) = \frac{1}{y_{n+1} - y_n}[(y - y_n)f_{n+1}(x) - (y - y_{n+1})f_n(x)]$$

where $f_n(x) = \varphi(x, y_n)$, and f_n and y_n are to be carefully chosen. More precisely, we need conditions for which $\varphi \in L^2(\mathbb{R}^2)$ and $\frac{\partial^2 \varphi}{\partial x \partial y} \in L^2(\mathbb{R}^2)$ yet $\varphi \notin L^\infty(\mathbb{R}^2)$. We may write on $\mathbb{R} \times (y_1, \infty)$:

$$\|\varphi\|_2^2 = \iint_{\mathbb{R} \times (y_1, \infty)} |\varphi(x, y)|^2 dx dy = \sum_1^\infty \iint_{\mathbb{R} \times (y_n, y_{n+1})} |\varphi(x, y)|^2 dx dy$$

After some calculations, the condition making φ in L^2 is

$$\sum_1^\infty (y_{n+1} - y_n)(\|f_n\|_2^2 + \|f_{n+1}\|_2^2) < \infty.$$

Clearly,

$$\frac{\partial^2 \varphi}{\partial x \partial y} = \frac{1}{y_{n+1} - y_n}(f'_{n+1} - f'_n) \in L^2(\mathbb{R} \times (y_1, \infty))$$

whenever

$$\sum_1^\infty \frac{1}{y_{n+1} - y_n}\|\psi'_n\|_2^2 < \infty$$

where $\psi_n(x) = f_{n+1}(x) - f_n(x)$. Since $\psi_n(x) = f_{n+1}(x) - f_n(x)$, we have $f_n(x) = -\sum_n^\infty \psi_k(x)$. Finally, choosing $f_n \notin L^\infty(\mathbb{R})$ will certainly give us $\varphi \notin L^\infty(\mathbb{R}^2)$.

We are now prepared to make the appropriate choice of both f_n and y_n. Consider

$$\psi_n(x) = \begin{cases} \frac{e^n}{n}x + \frac{1}{n} & \text{if } -e^{-n} \le x \le 0, \\ -\frac{e^n}{n}x + \frac{1}{n} & \text{if } 0 \le x \le e^{-n}, \\ 0 & \text{if } |x| \ge e^{-n}. \end{cases}$$

Hence,

$$\|\psi_n'\|_2^2 \sim \frac{e^n}{n^2} \text{ and } \|\psi_n\|_2^2 \sim \frac{e^{-n}}{n^2}.$$

We also have

$$\|f_n\|_2 \le \sum_{k=n}^{\infty} \|\psi_k\|_2 = a\sum_{k=n}^{\infty} \frac{e^{-\frac{n}{2}}}{n} \simeq \int_n^{\infty} \frac{e^{-\frac{x}{2}}}{x} dx \le \frac{1}{n}\int_n^{\infty} e^{-\frac{x}{2}} dx \sim \frac{e^{-\frac{n}{2}}}{n}.$$

Choosing $y_{n+1} - y_n = e^n$, we see that the series

$$\sum_1^{\infty} \frac{1}{y_{n+1} - y_n} \|\psi_n'\|_2^2 = \sum_1^{\infty} \frac{1}{e^n} \times \frac{e^n}{n^2} = \sum_1^{\infty} \frac{1}{n^2}$$

converges and so does the series $\sum_1^{\infty}(y_{n+1} - y_n)(\|f_n\|_2^2 + \|f_{n+1}\|_2^2)$ for

$$\sum_1^{\infty}(y_{n+1} - y_n)(\|f_n\|_2^2 + \|f_{n+1}\|_2^2) \le \sum_1^{\infty} e^n \times [\frac{e^{-(n+1)}}{(n+1)^2} + \frac{e^{-n}}{n^2}] \sim \sum_1^{\infty} \frac{1}{n^2}.$$

Now φ on $\mathbb{R} \times (y_n, y_{n+1})$ is given by

$$\varphi(x, y) = e^{-n}[(y - y_n)(-\sum_{n+1}^{\infty} \psi_k(x)) - (y - y_{n+1})(-\sum_n^{\infty} \psi_k(x))].$$

Observe that this φ is only defined for $x \in \mathbb{R}$ and $y \ge y_1$. To extend it to all of $\mathbb{R}^2$, we define φ for $x \in \mathbb{R}$ and $y_1 - y_{n+1} < y < y_1 - y_n$ as follows:

$$\varphi(x, y) = \frac{1}{y_{n+1} - y_n}[(y - y_1 + y_n)f_{n+1}(x) - (y - y_1 + y_{n+1})f_n(x)].$$

The function φ is now surely in $L^2(\mathbb{R}^2)$ and such that $\frac{\partial^2 \varphi}{\partial x \partial y} \in L^2(\mathbb{R}^2)$.

Therefore, there remains to show that φ is not in $L^\infty(\mathbb{R}^2)$. Let $x > 0$ and $x \leq e^{-k}$; then $\ln x \leq -k$ or $\ln \frac{1}{x} \geq k$. So

$$f_n(x) = -x \sum_n^{\lfloor \ln \frac{1}{x} \rfloor} \frac{e^{-k}}{k} + \sum_n^{\lfloor \ln \frac{1}{x} \rfloor} \frac{1}{k} \geq (-x^2 + 1) \sum_n^{\lfloor \ln \frac{1}{x} \rfloor} \frac{1}{k}.$$

As $x \to 0$, then $\lfloor \ln \frac{1}{x} \rfloor \to \infty$; hence, $\ln \lfloor \ln \frac{1}{x} \rfloor \to \infty$ too. Thus, $f_n(x) \to \infty$, which implies that $\varphi(x, y) \to \infty$. Accordingly, $\varphi \notin L^\infty(\mathbb{R}^2)$, as coveted.

Remark There is another way of finding a counterexample, but it is not an explicit one. It reads ([237]):

First, consider $f : \mathbb{R}^2 \to \mathbb{R}$ such that

$$f(u, v) = \frac{1}{1 + |uv|}.$$

The function f is obviously positive. Besides, it does not belong to $L^2(\mathbb{R}^2)$ since

$$\iint_{\mathbb{R}^2} \frac{1}{(1 + |uv|)^2} \, du \, dv \geq \int_0^\infty \int_0^\infty \frac{1}{(1 + uv)^2} \, du \, dv = \lim_{R \to \infty} \int_0^R \int_0^R \frac{1}{(1 + uv)^2} \, du \, dv$$

But

$$\int_0^R \frac{1}{(1 + uv)^2} \, du = \frac{R}{1 + Rv}$$

and so

$$\|f\|_2^2 \geq \lim_{R \to \infty} \int_0^R \frac{R}{1 + Rv} \, dv = \infty.$$

Now by a standard result from measure theory (see e.g. Exercise 2 on Page 143 in [379]), we know that there exists a positive $\psi \in L^2(\mathbb{R}^2)$ such that $\psi f \notin L^1(\mathbb{R}^2)$.

Since $f \in L^\infty(\mathbb{R}^2)$, ψf belongs to $L^2(\mathbb{R}^2)$ and so it is legitimate to set $\varphi = \mathcal{F}^{-1}(\psi f)$ where $\mathcal{F}$ is the $L^2(\mathbb{R}^2)$-Fourier transform. By the Plancherel theorem, φ is in $L^2(\mathbb{R}^2)$. Also

$$\mathcal{F}\left(\frac{\partial^2 \varphi}{\partial x \partial y}\right) = uv \mathcal{F}\varphi = uv\psi(u, v) f(u, v).$$

Since $(u, v) \mapsto \frac{uv}{1 + |uv|} \in L^\infty(\mathbb{R}^2)$ and $\psi \in L^2(\mathbb{R}^2)$, it results that $\frac{uv}{1 + |uv|}\psi \in L^2(\mathbb{R}^2)$. Falling back once more on the Plancherel theorem yields $\frac{\partial^2 \varphi}{\partial x \partial y} \in L^2(\mathbb{R}^2)$.

Since $\mathcal{F}(\varphi) \in L^2(\mathbb{R}^2)$, since it is positive, and since $\mathcal{F}(\varphi) \notin L^1(\mathbb{R}^2)$, it finally follows, by say Proposition A.3, that $\varphi \notin L^\infty(\mathbb{R}^2)$, as wished.

Answer 31.1.2 (An Example due to Professor A. M. Davie, cf. [239]) For (q, r) to fail to be in Ω_n, one of the following 3 possibilities must occur: (i) $q < 2$ or $r < 2$; (ii) $\frac{2}{q} + \frac{n}{r} < \frac{n}{2}$; (iii) $n = 2$, $q = 2$ and $r = \infty$. We consider these cases in turn:

(i) If $q < 2$, choose a sequence $(\beta_k)_{k \in \mathbb{Z}}$ that is in l^2 but not in l^q. Let $\phi(x, t)$ be a smooth function of compact support on $\mathbb{R}^{n+1}$ that vanishes for t outside $[0, 1]$, and let $u(x, t) = \sum_{k \in \mathbb{Z}} \beta_k \phi(x, t - k)$. Then $u \in M_L^n$, but $u \notin L_t^q(L_x^r)$ for any r.

The case $r < 2$ can be dealt with similarly. We choose a sequence β_k that is in l^2 but not l^r, and a smooth ϕ that vanishes for x_1 outside $[0, 1]$, and then set $u(x, t) = \sum_{k \in \mathbb{Z}} \beta_k \phi(x - ke_1, t)$, where e_1 is the unit vector $(1, 0, \cdots, 0)$ in $\mathbb{R}^n$. Then $u \in M_L^n$, but $u \notin L_t^q(L_x^r)$ for any q.

(ii) In the case $\frac{2}{q} + \frac{n}{r} < \frac{n}{2}$ we use the scaling argument together with a cutoff to obtain a function in M_L^n.

We start with a nonzero $f \in L^2(\mathbb{R}^n)$, and let u be the solution of

$$\begin{cases} -iu_t - \triangle_x u = 0, \\ u(x, 0) = f(x) \end{cases} \tag{31.4}$$

where $f(x) = u(x, 0)$ and $g = Lu$ (an explicit example would be $f(x) = e^{-|x|^2}$ and then $u(x, t) = (1 + 4it)^{-n/2}e^{-|x|^2/(1+4it)}$). Choose a smooth function ϕ on $\mathbb{R}$ such that $\phi(0) \neq 0$ and such that ϕ and ϕ' are in L^2. Then for $\lambda > 0$, define

$$v_\lambda(x, t) = \lambda^{n/2} u(\lambda x, \lambda^2 t)\phi(t).$$

Then (using $Lu = 0$), we find $Lv_\lambda(x, t) = -i\lambda^{n/2}u(\lambda x, \lambda^2 t)\phi'(t)$. We calculate $\|v_\lambda\|_{L^2(\mathbb{R}^{n+1})} = \|f\|_{L^2(\mathbb{R}^n)}\|\phi\|_{L^2}$ and $\|Lv_\lambda\|_{L^2(\mathbb{R}^{n+1})} = \|f\|_{L^2(\mathbb{R}^n)}\|\phi'\|_{L^2}$. Also

$$\|v_\lambda\|_{L_t^q(L_x^r)} = \lambda^\beta \left\{ \int_\mathbb{R} \|u(., t)\|_{L^r(\mathbb{R}^n)}^q |\phi(\lambda^{-2}t)|^q dt \right\}^{1/q},$$

where $\beta = \frac{n}{2} - \frac{n}{r} - \frac{2}{q} > 0$. So $\lambda^{-\beta}\|v_\lambda\|_{L_t^q(L_x^r)} \to |\phi(0)|\|u\|_{L_t^q(L_x^r)}$ (note that the norm on the right may be infinite), and hence, $\|v_\lambda\|_{L_t^q(L_x^r)}$ tends to ∞ as $\lambda \to \infty$, completing the proof.

(iii) This exceptional case we treat in a similar fashion to (ii), but we have need for the result from [232] that the Strichartz inequality fails in this case. We start by fixing a smooth function ϕ on $\mathbb{R}$ such that $\phi = 1$ on $[-1, 1]$ and ϕ and ϕ' are in L^2.

Now let $M > 0$ be given, and we use [232] to find $f \in L^2(\mathbb{R}^2)$ with $\|f\|_{L^2(\mathbb{R}^2)} = 1$ such that the solution u of (31.4) satisfies $\|u\|_{L_t^2(L_x^\infty)} > M$. Then we can find $R > 0$ so that $\int_{-R}^R \|u(\cdot, t)\|_{L^\infty(\mathbb{R}^2)}^2 dt > M^2$. Let $\lambda = R^{1/2}$ and define $v(x, t) = \lambda^{n/2} u(\lambda x, \lambda^2 t) \phi(t)$. Then $\|v\|_{L^2(\mathbb{R}^3)} = \|\phi\|_{L^2}$, $\|Lv\|_{L^2(\mathbb{R}^3)} = \|\phi'\|_{L^2}$, and

$$\|v\|_{L_t^2(L_x^\infty)}^2 \geq \int_{-1}^1 \|v(\cdot, t)\|_{L^\infty(\mathbb{R}^2)}^2 dt > M^2,$$

which completes the proof, since M is arbitrary.

Answer 31.1.3 ([48]) We will show the existence of such a function u. Let $\delta > 0$. Consider

$$u_\delta(x, t) = \mathcal{F}^{-1}(g_\delta(\eta, \xi)) \text{ where } g_\delta(\eta, \xi) = \varphi(\delta\eta)V(\eta^3 + \xi)$$

and where $\mathcal{F}^{-1}$ is the inverse L^2-Fourier transform, φ is a smooth function with compact support, whereas V is a nonnegative smooth function of one variable with compact support (yet to be determined).

We want to show that $u_\delta \in D(L)$ that is equivalent (by means of the Fourier transform and the Plancherel theorem) to $\hat{u}_\delta$ and $|\xi + \eta^3|\hat{u}_\delta$ both belonging to $L^2(\mathbb{R}^2)$. This implies $(1 + |\xi + \eta^3|)\hat{u}_\delta \in L^2(\mathbb{R}^2)$.

To obtain this condition, we need an appropriate choice for support of V since $\hat{u}_\delta \in L^2(\mathbb{R}^2)$ (see below). We take the support of V to be $\{y : |y| \leq \frac{1}{2}\}$ say so that supp $\hat{u}_\delta \subset \{(\eta, \xi) \in \mathbb{R}^2 : |\xi + \eta^3| < 1\}$, whereby $(1 + |\xi + \eta^3|)\hat{u}_\delta \in L^2(\mathbb{R}^2)$. Thus we have $u_\delta \in D(L)$.

The next step is to show that u_δ does not belong to $L^p(\mathbb{R}^2)$ for any $p > 8$. We are going to show that the ratio of the L^p-norm of u_δ and the L^2-norm of u_δ goes to infinity in a suitable limit.

We first compute the L^2-norm of u_δ. We have by the Plancherel theorem

$$\|u_\delta\|_2^2 = \|\hat{u}_\delta\|_2^2 = \|g_\delta\|_2^2 = \iint_{\mathbb{R}^2} |\varphi(\delta\eta)V(\eta^3 + \xi)|^2 d\eta d\xi.$$

The change of variables $\eta = s$ and $\xi = z - s^3$ gives

$$\|u_\delta\|_2 = \delta^{-\frac{1}{2}} \|\varphi\|_{L^2(\mathbb{R})} \|V\|_{L^2(\mathbb{R})}.$$

We also have

$$u_\delta(x, t) = \mathcal{F}^{-1}(g_\delta(\eta, \xi)) = \iint_{\mathbb{R}^2} \varphi(\delta\eta)V(\eta^3 + \xi)e^{i\eta x + it\xi} d\eta d\xi.$$

By the same change of variables as above, one gets

$$u_\delta(x, t) = \int_{\mathbb{R}} \int_{-\frac{1}{2}}^{\frac{1}{2}} \varphi(\delta s) V(z) e^{isx - is^3 t + izt} dz ds.$$

Setting $\delta s = r$ gives us

$$u_\delta(x, t) = \delta^{-1} \check{V}(t) \int_{\mathbb{R}} \varphi(r) e^{ir\frac{x}{\delta} - ir^3 \frac{t}{\delta^3}} dr$$

which is equal to

$$u_\delta(x, t) = \delta^{-1} \check{V}(t) H\left(\frac{x}{\delta}, \frac{t}{\delta^3}\right),$$

where H is some function of two variables. Therefore,

$$\|u_\delta\|_p^p = \delta^{4-p} \iint_{\mathbb{R}^2} |H(\mu, \tau) \check{V}(\delta^3 \tau)|^p d\mu d\tau.$$

We need to investigate how the integral on the right hand side of the last equation behaves as $\delta \to 0$. Since $\check{V}$ is continuous,

$$\lim_{\delta \to 0} \check{V}(\delta^3 \tau) = \check{V}(0) = \|V\|_{L^1(\mathbb{R})}$$

as V is nonnegative. So

$$\lim_{\delta \to 0} |H(\mu, \tau) \check{V}(\delta^3 \tau)|^p = |H(\mu, \tau)|^p \|V\|_{L^1(\mathbb{R})}^p.$$

Fatou's lemma then yields

$$\liminf_{\delta \to 0} \iint_{\mathbb{R}^2} |H(\mu, \tau) \check{V}(\delta^3 \tau)|^p d\mu d\tau \geq \|H\|_{L^p(\mathbb{R}^2)}^p \|V\|_{L^1(\mathbb{R})}^p.$$

In the end, since

$$\frac{\|u_\delta\|_p}{\|u_\delta\|_2} = \delta^{\frac{4}{p} - \frac{1}{2}} \frac{\sqrt[p]{\iint_{\mathbb{R}^2} |H(\mu, \tau) \check{V}(\delta^3 \tau)|^p d\mu d\tau}}{\|\varphi\|_{L^2(\mathbb{R})} \|V\|_{L^2(\mathbb{R})}},$$

using the argument above and sending $\delta \to 0$ yield

$$\frac{\|u_\delta\|_p}{\|u_\delta\|_2} \longrightarrow +\infty \text{ for } p > 8,$$

as needed.

Answer 31.1.4 ([303]) Let $V(x) = -2/|x|^2$ on $L^2(\mathbb{R}^5)$. Then $-\Delta + V$ is not essentially self-adjoint on $C_0^\infty(\mathbb{R}^5)$ by Theorem X.11 in [303] where more details may be found.

Answer 31.1.5 The proof exploits the non-essential self-adjointness of the one-dimensional Laplacian perturbed by some *negative* potential.

The operator $-\frac{d^2}{dt^2} - t^4$ is not essentially self-adjoint on $C_0^\infty(\mathbb{R})$ (details may be found in [303]) meaning that

$$\left(-\frac{d^2}{dt^2} - t^4\right) f(t) = -if(t) \text{ or } \left(\frac{d^2}{dt^2} + t^4\right) f(t) = if(t)$$

has a nonzero solution in $L^2(\mathbb{R})$. Now the perturbed one-dimensional Laplacian by x^2 has as eigenvalue $\frac{1}{2}$, i.e. there is a nonzero $g \in L^2(\mathbb{R})$ such that

$$\left(-\frac{d^2}{dx^2} + x^2\right) g(x) = \frac{1}{2} g(x)$$

By adding up the last two displayed equations, we get

$$\left(\frac{\partial^2}{\partial t^2} - \frac{\partial^2}{\partial x^2}\right) f(x)g(t) + (t^4 + x^2)f(x)g(t) = \left(i + \frac{1}{2}\right) f(x)g(t). \qquad (31.5)$$

Set $\varphi(x, t) = f(x)g(t)$. Since f, g are both in $L^2(\mathbb{R})$, φ will be in $L^2(\mathbb{R}^2)$, and Eq. (31.5) will have a nonzero solution in $L^2(\mathbb{R}^2)$, and yet $V(x, t) = t^4 + x^2$ is *nonnegative* and it belongs to $L^2_{loc}(\mathbb{R}^2)$.

Answer 31.1.6 The answer is negative for both questions. For the first question, let T be a closed densely defined operator with a domain $D(T)$. Then clearly T is $|T|$-bounded. Choose T as in Question 21.2.46 (we may even sophisticate the example by assuming that T is symmetric with the aid of Question 20.2.23), and so $D(T^2) = \{0\}$. Hence, T^2 cannot be $|T|^2$-bounded for $|T|^2$ is self-adjoint and so

$$D(|T|^2) \not\subset D(T^2) = \{0\}.$$

As for the second question, consider any closed densely defined operator T that is not symmetric. Then T is $|T|$-bounded. However,

$$D(|T^*|) = D(|T|) = D(T) \not\subset D(T^*),$$

i.e. T^* cannot be $|T|^*$-bounded.

Answer 31.1.7 Let A be any unbounded self-adjoint operator, and let B be a symmetric operator that is A-bounded with relative bound $a < 1$. Then $A + B$ is self-adjoint on $D(A)$. Now, for any self-adjoint C with $D(C) = D(A)$, we have

$$\|Bx\| \leq a\|Ax\| + c\|Cx\| + \alpha\|x\|$$

for any $c \geq 0$ and all $x \in D(A)$. In particular, the previous inequality holds for $C = -(A + B)$, and we clearly see that $A + B + C$, which is the restriction of 0 to $D(A)$, is not even closed and so it cannot be self-adjoint.

Answer 31.1.8 ([251]) Consider the operators

$$A = -\frac{d^2}{dx^2} \text{ and } B = iV,$$

where V is a real-valued function in $L^2(\mathbb{R}) + L^\infty(\mathbb{R})$, defined on $D(A) = H^2(\mathbb{R})$ and $D(B) = \{f \in L^2(\mathbb{R}) : Vf \in L^2(\mathbb{R})\}$, respectively. Obviously, A is normal (it is self-adjoint) and so is B. It is also well known (see e.g. Theorem X.15 in [303]) that B is A-bounded with an arbitrary relative bound.

Set $N = A + B$. We have for any $f \in D(N) = D(A)$

$$Nf(x) = -f''(x) + iV(x)f(x)$$

and applying the Hess–Kato theorem, say, yields

$$N^*f(x) = -f''(x) - iV(x)f(x).$$

Doing some easy calculations gives us

$$NN^*f(x) = f^{(4)}(x) + iV''(x)f(x) + 2iV'(x)f'(x) + V^2(x)f(x)$$

and

$$N^*Nf(x) = f^{(4)}(x) - iV''(x)f(x) - 2iV'(x)f'(x) + V^2(x)f(x)$$

for all $f \in D(A^2)$. Therefore, N is not normal on $D(A)$.

Chapter 32
More Questions and Some Open Problems II

32.1 More Questions

Question 32.1.1 (Can you) Find a densely defined right invertible operator T and a densely defined operator S such that ST is densely defined and

$$(ST)^* \neq T^* S^*.$$

Question 32.1.2 (See e.g. [314]) Give an example of a densely defined, maximally symmetric, closed operator that is not self-adjoint.

Question 32.1.3 ([375]) Let A be a closed, symmetric, *and non-self-adjoint* operator with domain $D(A) \subset H$. Is $|A| + A$ always unclosed? What about $|A| - A$?

Question 32.1.4 (See [137]) On a Hilbert space H, find two skew-adjoint operators A and B with $B \in B(H)$ such that the restriction of $AB - BA$ to $D(AB) \cap D(BA) \cap D(B^2) \cap D(A^2)$ has no skew-adjoint extension.

© The Author(s), under exclusive license to Springer Nature Switzerland AG 2022
M. H. Mortad, *Counterexamples in Operator Theory*,
https://doi.org/10.1007/978-3-030-97814-3_32

Question 32.1.5 ([172]) Construct an unbounded quasinormal operator T such that

$$(T^n)^* \neq T^{*n}$$

for every $n \geq 2$.

Question 32.1.6 (Cf. [48, 239] and [248]) Let $n \geq 2$, and let $p > \frac{2n+2}{n-1}$. Show that there are no constants a and b such that the estimate in Theorem 31.1.6 (just above Question 31.1.1) holds for all $u \in L^2(\mathbb{R} \times \mathbb{R}^n)$ such that $\Box u \in L^2(\mathbb{R} \times \mathbb{R}^n)$.

32.2   Some Open Problems

Question 32.2.1 (Cf. Question 21.2.42) Can you find unbounded self-adjoint operators S and T such that

$$D(ST) = D(TS) = D(S+T) = \{0\}?$$

Question 32.2.2 (Cf. [18]) Let $B \in B(H)$ be normal, and let A be an unbounded self-adjoint operator such that BA is self-adjoint. Does it follow that B^*A is self-adjoint? In other language, do we have

$$BA = AB^* \implies B^*A = AB?$$

Remark The following observation might be useful: BA is closed if and only if B^*A is closed.

Question 32.2.3 Let $B \in B(H)$ be normal, and let A be an unbounded closed operator. Do we have

$$BA = AB \implies B^*A = AB^*?$$

Question 32.2.4 (Cf. [258]) Let A be an unbounded self-adjoint operator A. Can we always find a self-adjoint operator $B \in B(H)$ such that $AB - BA$ is unbounded?

Question 32.2.5 (Cf. Question 28.2.11) Can you find two self-adjoint positive operators A and B such that $AB - BA$ is bounded, whereas $A^\alpha B^\alpha - B^\alpha A^\alpha$ is unbounded for all $\alpha \in (0, 1)$? What about the same question when $\alpha > 1$?

Question 32.2.6 The following result was shown in [31]: *Let A and B be two strongly commuting unbounded normal operators. When $D(\overline{A + B}) \subset D(\overline{|A| - |B|})$, then*

$$\left\| \overline{|A| - |B|} \right\| \leq \overline{|A + B|}.$$

Can we prove

$$\left\| \overline{|A| - |B|} \right\| \leq \overline{|A + B|}$$

by solely assuming that A and B are strongly commuting normal operators?

Question 32.2.7 Is it possible to find two normal operators A and B, with $B \in B(H)$ such that $BA \subset \lambda AB \neq 0$ with $|\lambda| \neq 1$?

Question 32.2.8 Since a densely defined paranormal operator is not necessarily closable as we already observed in Question 25.1.3, it would be interesting to find an unbounded paranormal operator that is *everywhere defined*. Can we find such an example?

Question 32.2.9 Is a densely defined closed paranormal operator with a real spectrum necessarily self-adjoint?

Question 32.2.10 *It was shown in [85] that if A is a closed densely defined hyponormal operator, then*

$$AA^* \leq A^*A.$$

Does a densely defined closed hyponormal operator A have to satisfy $D(A^*A) \subset D(AA^*)$?

Question 32.2.11 (Cf. [292]) Let A be a closed densely defined subnormal operator (see e.g. [353] for their definition) such that A^n is quasinormal for some integer $n \geq 2$. Then A is quasinormal, and this was obtained in [293]. What about the same question for closed hyponormal or closed paranormal operators?

Question 32.2.12 (Cf. [350], and [233]) Let A and B be two closed hyponormal operators such that $TA \subset B^*T$, where $T \in B(H)$. Is it true that $TA^* \subset BT$?

Appendix A: A Quick Review of the Fourier Transform

Definition A.1 Let f be in $L^1(\mathbb{R})$. The Fourier transform of f is defined by

$$\mathcal{F}f(t) = \hat{f}(t) = \frac{1}{\sqrt{2\pi}} \int_{-\infty}^{\infty} f(x)e^{-itx}dx = \int_{-\infty}^{\infty} f(x)e^{-itx}d\mu(x)$$

where $i = \sqrt{-1}$ and where $d\mu(x) = \frac{1}{\sqrt{2\pi}}dx$.

Remark We continue to write $L^1(\mathbb{R})$ even with respect to $d\mu(x)$.

Remark In case of $f \in L^1(\mathbb{R}^n)$, we define $\hat{f}$ as

$$\hat{f}(t) = \frac{1}{(2\pi)^{\frac{n}{2}}} \int_{\mathbb{R}^n} f(x)e^{-it\cdot x}dx$$

where $t \cdot x = t_1x_1 + \cdots + t_nx_n$.

The next theorem gathers a few basic properties of the Fourier transform:

Theorem A.1 *(see [313])*

(1) $\hat{f}$ *is linear.*
(2) $\hat{f}$ *is continuous from* $L^1(\mathbb{R})$ *into* $L^\infty(\mathbb{R})$, *i.e.*

$$\exists M > 0 : \|\hat{f}\|_{L^\infty(\mathbb{R})} \leq M\|f\|_{L^1(\mathbb{R})}$$

for all $f \in L^1(\mathbb{R})$.
(3) *If* $f \geq 0$, *then*

$$\|\hat{f}\|_{L^\infty(\mathbb{R})} = \|f\|_{L^1(\mathbb{R})}.$$

(4) If $g(x) = -ixf(x)$ and $g \in L^1(\mathbb{R})$, then $\hat{f}$ is differentiable and $\hat{f}'(t) = \hat{g}(t)$. Also, if $f, f' \in L^1(\mathbb{R})$, then $\widehat{f'} = it\hat{f}$.

We may also extend the definition of the Fourier transform to all of $L^2(\mathbb{R})$ (see e.g. [215] or [313]).

Theorem A.2 (Plancherel Theorem) *Let f be a function in $L^1(\mathbb{R}) \cap L^2(\mathbb{R})$. Then $\hat{f} \in L^2(\mathbb{R})$, and we have $\|\hat{f}\|_{L^2(\mathbb{R})} = \|f\|_{L^2(\mathbb{R})}$. Then $f \mapsto \hat{f}$ has a unique extension to $L^2(\mathbb{R}) \to L^2(\mathbb{R})$ such that for all $f \in L^2(\mathbb{R})$*

$$\|\hat{f}\|_{L^2(\mathbb{R})} = \|f\|_{L^2(\mathbb{R})}.$$

Furthermore, $f \mapsto \hat{f}$ is a surjective isometry from $L^2(\mathbb{R})$ onto $L^2(\mathbb{R})$.

Remarks

(1) We may continue to denote the Fourier transform by $\hat{f}$ when $f \in L^1(\mathbb{R}) \cup L^2(\mathbb{R})$.
(2) If $f \in L^2(\mathbb{R})$, and $f \notin L^1(\mathbb{R})$, then $\hat{f}$ is to be understood as an $L^2(\mathbb{R})$-limit as follows [313]: If

$$g_k(t) = \int_{-k}^{k} f(x)e^{-ixt}d\mu(x) \text{ and } h_k(t) = \int_{-k}^{k} \hat{f}(x)e^{ixt}d\mu(x),$$

then as k tends to ∞, we have

$$\|g_k - \hat{f}\|_2 \to 0 \text{ and } \|h_k - f\|_2 \to 0.$$

(3) All the previous results remain true in $\mathbb{R}^n$.
(4) We still have on $L^2(\mathbb{R})$ the important property that the Fourier transform of f' becomes a product by $t\hat{f}$ up to a factor. This also applies to partial derivatives.

Proposition A.3 *Let $n \geq 1$. If $f \in L^2(\mathbb{R}^n)$ and $\hat{f} \notin L^1(\mathbb{R}^n)$ with $\hat{f} \geq 0$, then $f \notin L^\infty(\mathbb{R}^n)$.*

Theorem A.4 *If $f \in C_0^\infty(\mathbb{R})$ such that $\hat{f} \in C_0^\infty(\mathbb{R})$, then $f = 0$.*

The preceding theorem may also be viewed as a special case of the following L^2-version of a famous Hardy theorem (see [154]) that was obtained in [75]. See also [202].

Theorem A.5 *Let $f \in L^2(\mathbb{R})$. Assume also that*

$$\int_{\mathbb{R}} |f(x)|^2 e^{\alpha x^2} dx < \infty \text{ and } \int_{\mathbb{R}} |\hat{f}(y)|^2 e^{\beta y^2} dy < \infty$$

for some $\alpha, \beta > 0$. If $\alpha\beta \geq 1$, then $f = 0$.

One may wonder whether Theorem A.4 remains valid for the so-called Fourier cosine transform. The answer is obviously no as any nonzero odd function provides a counterexample. However, the same idea of the proof of Theorem A.4 works to establish the following:

Theorem A.6 *If $f \in C_0^\infty(\mathbb{R})$ is even and such that its Fourier cosine transform too is in $C_0^\infty(\mathbb{R})$, then $f = 0$.*

Appendix B: A Word on Distributions and Sobolev Spaces

Distribution theory is such a vast domain in mathematical analysis. It is treated in many textbooks. We could cite [116, 215, 302]. In this subsection, we recall briefly a few facts needed from this theory. A good portion of the material here is borrowed from [326].

As is customary in distribution theory, we shall denote $C_0^\infty(\mathbb{R}^n)$ by $\mathcal{D}(\mathbb{R}^n)$. We also introduce the following notations: $\partial_k = \frac{\partial}{\partial x_k}$,

$$\partial^\alpha = \left(\frac{\partial}{\partial x_1}\right)^{\alpha_1} \cdots \left(\frac{\partial}{\partial x_n}\right)^{\alpha_n}$$

and $|\alpha| = \alpha_1 + \cdots + \alpha_n$ for $\alpha = (\alpha_1, \cdots, \alpha_n)$, and each α_k is in $\mathbb{Z}^+$.

Definition B.2 A linear functional T on $\mathcal{D}(\mathbb{R}^n)$ is called a distribution on $\mathbb{R}^n$ if for each compact set K of $\mathbb{R}^n$, there are $n_K \in \mathbb{Z}^+$ and $C_K > 0$ such that

$$|T(\varphi)| \leq C_K \sup\{|\partial^\alpha \varphi(x)| : x \in K, \ |\alpha| \leq n_K\}$$

for all $\varphi \in \mathcal{D}(\mathbb{R}^n)$ with supp $\varphi \subset K$.

The vector space of distributions on $\mathbb{R}^n$ is designated by $\mathcal{D}'(\mathbb{R}^n)$.

Remark Obviously, the continuity of a distribution T may be interpreted as: Whenever $\varphi_m \in \mathcal{D}(\mathbb{R}^n)$ with $\varphi_m \to \varphi$ in $\mathcal{D}(\mathbb{R}^n)$, then (in $\mathbb{C}$)

$$T(\varphi_m) \to T(\varphi),$$

where $\varphi_m \to \varphi$ in $\mathcal{D}(\mathbb{R}^n)$ means that there is some fixed compact set $K \subset \mathbb{R}^n$ such that $\text{supp}(\varphi_m - \varphi) \subset K$ and for each α (as defined above)

$$\partial^\alpha \varphi_m \longrightarrow \partial^\alpha \varphi$$

as $m \to \infty$ uniformly on K.

Therefore, a distribution is a continuous linear functional on $\mathcal{D}(\mathbb{R}^n)$.

All $L^p(\mathbb{R}^n)$-functions may be regarded as distributions. To see this, we introduce the following space:

$$L^p_{\text{loc}}(\mathbb{R}^n) = \{ f : \mathbb{R}^n \to \mathbb{C} \text{ measurable,} \int_K |f|^p < \infty \text{ for each compact } K \subset \mathbb{R}^n \}$$

where $p \geq 1$.

Now if $f \in L^1_{\text{loc}}(\mathbb{R}^n)$, then for any $\varphi \in \mathcal{D}(\mathbb{R}^n)$, it can be shown that

$$T_f(\varphi) = \int_{\mathbb{R}^n} f \varphi \, dx$$

defines a continuous linear functional on $\mathcal{D}(\mathbb{R}^n)$, i.e. $T_f \in \mathcal{D}'(\mathbb{R}^n)$.

It is well known that $L^p(\mathbb{R}^n)$ and $L^q(\mathbb{R}^n)$ are not comparable. Nonetheless, $L^p_{\text{loc}}(\mathbb{R}^n)$ and $L^q_{\text{loc}}(\mathbb{R}^n)$ are always comparable. More precisely,

$$q \geq p \implies L^q_{\text{loc}}(\mathbb{R}^n) \subset L^p_{\text{loc}}(\mathbb{R}^n)$$

by a simple application of Hölder's inequality. Hence, $L^p_{\text{loc}}(\mathbb{R}^n) \subset L^1_{\text{loc}}(\mathbb{R}^n)$. Finally, since clearly $L^p(\mathbb{R}^n) \subset L^p_{\text{loc}}(\mathbb{R}^n)$, we have that any $L^p(\mathbb{R}^n)$-function yields a distribution.

One of the main virtues of distributions is that we can always differentiate them in the following sense:

Definition B.3 Let $T \in \mathcal{D}'(\mathbb{R}^n)$. We define $\partial^\alpha T$ by its action on every $\varphi \in \mathcal{D}(\mathbb{R}^n)$ as

$$(\partial^\alpha T)(\varphi) = (-1)^{|\alpha|} T(\partial^\alpha \varphi).$$

We call $\partial^\alpha T$ the distributional derivative of T.

Remark In terms of $L^1_{\text{loc}}(\mathbb{R}^n)$-functions, we have:

We say that a function $g \in L^1_{\text{loc}}(\mathbb{R}^n)$ is the weak derivative of $f \in L^1_{\text{loc}}(\mathbb{R}^n)$ if the distributional derivative $\partial^\alpha f$ is given by the function g, i.e.

$$(-1)^{|\alpha|} \int_{\mathbb{R}^n} f \partial^\alpha(\varphi) dx = \int_{\mathbb{R}^n} g \varphi \, dx$$

for all $\varphi \in \mathcal{D}(\mathbb{R}^n)$.

Remark All definitions and results still make sense if $\mathbb{R}^n$ is replaced by an open set $\Omega \subset \mathbb{R}^n$.

All that to define Sobolev spaces (readers may wish to consult [157, 215], or [303] for further reading).

Definition B.4 Let $m \in \mathbb{Z}_+$. Set

$$H^m(\mathbb{R}^n) = \{f \in L^2(\mathbb{R}^n) : \partial^\alpha f \in L^2(\mathbb{R}^n) \text{ for all } \alpha \in \mathbb{Z}_+^n, |\alpha| \le m\},$$

where derivatives are taken in the distributional sense. Then $H^m(\mathbb{R})$ is called a Sobolev space.

Remark The vector space $H^m(\mathbb{R}^n)$ is a Hilbert space with respect to the inner product:

$$\langle f, g \rangle_{H^m(\mathbb{R}^n)} = \sum_{|\alpha| \le m} \langle \partial^\alpha f, \partial^\alpha g \rangle_{L^2(\mathbb{R}^n)}.$$

Using the Fourier transform, we may describe Sobolev spaces as follows:

Theorem B.7

$$H^m(\mathbb{R}^n) = \{f \in L^2(\mathbb{R}^n) : |x|^{\frac{m}{2}} \hat{f}(t) \in L^2(\mathbb{R}^n)\}$$

(where $|x|^2 = x_1^2 + \cdots + x_n^2$). Besides, if $p \in \mathbb{Z}_+$ and $m > p + \frac{n}{2}$, then $H^m(\mathbb{R}^n) \subset C^p(\mathbb{R}^n)$.

The last stop here is at absolutely continuous functions (more on this topic may be found in [379]).

Definition B.5 Let $a, b \in \mathbb{R}$ with $a < b$. A function f on $[a, b]$ is said to be absolutely continuous if for each $\varepsilon > 0$, there exists $\delta > 0$ such that for any collection of pairwise disjoint subintervals (a_k, b_k)

$$\sum_{k=1}^n |f(b_k) - f(a_k)| < \varepsilon \text{ if } \sum_{k=1}^n (b_k - a_k) < \delta.$$

We denote the set of all absolutely continuous functions on $[a, b]$ by $\mathrm{AC}[a, b]$.

Remark Absolute continuous functions are continuous. Are they differentiable? The answer is "almost yes":

Theorem B.8 *A function f is absolutely continuous on $[a, b]$ if and only if f' exists almost everywhere on $[a, b]$, $f' \in L^1[a, b]$, and*

$$f(x) - f(a) = \int_a^x f'(t)dt$$

for all $x \in [a, b]$.

Finally, we give some practical results about Sobolev spaces (mainly from [326]):

Proposition B.9

(1) If $f \in H^1(0, \infty)$, then $\lim_{y \to \infty} f(y) = 0$.
(2) If $f \in H^1(\mathbb{R})$, then

$$\lim_{x \to -\infty} f(x) = \lim_{y \to \infty} f(y) = 0.$$

Proposition B.10 *Let I be an open interval, and let c be in the closure of I. For any $\varepsilon > 0$, there is a constant $b_\varepsilon > 0$ such that:*

$$|f(c)| \leq \varepsilon \|f'\|_2 + b_\varepsilon \|f\|_2, \quad \forall f \in H^1(I)$$

where $H^1(I) = \{f \in AC(I) : f' \in L^2(I)\}$.

Remark We use $H^1(a, b)$ for $\{f \in AC[a, b] : f' \in L^2(a, b)\}$, and sometimes as $\{f \in L^2(a, b) : f' \in L^2(a, b)\}$, where in the latter the derivative is a distributional one. Similarly, $H^2(a, b)$ may designate $\{f \in L^2(a, b) : f'' \in L^2(a, b)\}$, and also $\{f \in C^1[a, b] : f' \in H^1(a, b)\}$.

Bibliography

1. Abrahamse, M.B.: Commuting subnormal operators. Illinois J. Math. **22**(1), 171–176 (1978)
2. Albrecht, E., Spain, P.G.: When products of self-adjoints are normal. Proc. Am. Math. Soc. **128**(8), 2509–2511 (2000)
3. Alpay, D.: An advanced complex analysis problem book.In: Topological vector spaces, functional analysis, and Hilbert spaces of analytic functions. Birkhäuser/Springer, Cham (2015)
4. Alpay, D.: A Complex Analysis Problem Book, 2nd edn. Birkhäuser/Springer, Cham (2016)
5. Ando, T.: Operators with a norm condition. Acta Sci. Math. (Szeged) **33**, 169–178 (1972)
6. Arendt, W., Räbiger, F., Sourour, A.: Spectral properties of the operator equation $AX + XB = Y$. Quart. J. Math. Oxford Ser.(2) **45**(178), 133–149 (1994)
7. Arlinskiĭ, Y., Tretter, Ch.: Everything is possible for the domain intersection $\mathrm{dom}T \cap \mathrm{dom}T^*$. Adv. Math. **374**, 107383 (2020)
8. Arlinskiĭ, Y., Zagrebnov, V.A.: Around the van Daele-Schmüdgen theorem. Integr. Equ. Oper. Theory **81**(1), 53–95 (2015)
9. Axler, S.: Linear Algebra Done Right. 3rd edn. Undergraduate Texts in Mathematics. Springer, Cham (2015)
10. Axler, S.: Measure, integration and real analysis. In: Graduate Texts in Mathematics, vol. 282. Springer, Cham (2020)
11. Bagby, R.J.: An elementary proof of the spectral theorem for unbounded operators. Thesis (Master of arts)-Rice University (USA) (1965)
12. Bakić, D., Guljaš, B.: A note on compact operators and operator matrices. Math. Commun. **4**(2), 159–165 (1999)
13. Bala, A.: Binormal operators. Indian J. Pure Appl. Math. **8**(1), 68–71 (1977)
14. Barraa, M., Boumazghour, M.: Numerical range submultiplicity. Linear Multilinear Algebra **63**(11), 2311–2317 (2015)
15. Beauzamy, B.: Un opérateur sans sous-espace invariant: simplification de l'exemple de P. Enflo. (French) [An operator with no invariant subspace: simplification of the example of P. Enflo]. Integr. Equ. Oper. Theory **8**(3), 314–384 (1985)
16. Beck, W.A., Putnam, C.R.: A note on normal operators and their adjoints. J. London Math. Soc. **31**, 213–216 (1956)
17. Benali, A., Mortad, M.H.: Generalizations of Kaplansky's theorem involving unbounded linear operators. Bull. Pol. Acad. Sci. Math. **62**(2), 181–186 (2014)
18. Bensaid, I.F.Z., Dehimi, S., Fuglede, B., Mortad, M.H.: The Fuglede theorem and some intertwining relations. Adv. Oper. Theory **6**(1), Paper No. 9, 8 pp. (2021)

579

19. Berberian, S.K.: Note on a theorem of Fuglede and Putnam. Proc. Am. Math. Soc. **10**, 175–182 (1959)
20. Berberian, S.K.: A note on operators unitarily equivalent to their adjoints. J. London Math. Soc. **37**, 403–404 (1962)
21. Berberian, S.K.: Introduction to Hilbert space. Reprinting of the 1961 original. With an addendum to the original. Chelsea Publishing Co., New York (1976)
22. Bernau, S.J.: The square root of a positive self-adjoint operator. J. Austral. Math. Soc. **8**, 17–36 (1968)
23. Bernau, S.J.: An unbounded spectral mapping theorem. J. Austral. Math. Soc. **8**, 119–127 (1968)
24. Bernau, S.J., Smithies, F.: A note on normal operators. Proc. Cambridge Philos. Soc. **59**, 727–729 (1963)
25. Bhatia, R.: Modulus of continuity of the matrix absolute value. Indian J. Pure Appl. Math. **41**(1), 99–111 (2010)
26. Bhatia, R., Kittaneh, F.: The matrix arithmetic-geometric mean inequality revisited. Linear Algebra Appl. **428**(8–9), 2177–2191 (2008)
27. Bhatia, R., Rosenthal, P.: How and why to solve the operator equation $AX - XB = Y$. Bull. London Math. Soc. **29**(1), 1–21 (1997)
28. Birman, M. Sh., Solomjak, M.Z.: Spectral theory of selfadjoint operators in Hilbert space. Translated from the 1980 Russian original by Khrushchëv, S., Peller, V. Mathematics and its Applications (Soviet Series). D. Reidel Publishing Co., Dordrecht (1987)
29. Bishop, E.: Spectral theory for operators on a Banach space. Trans. Am. Math. Soc. **86**, 414–445 (1957)
30. Borwein, J.M., Richmond, B.: How many matrices have roots?. Canad. J. Math. 36(2), 286–299 (1984)
31. Boucif, I., Dehimi, S., Mortad, M.H.: On the absolute value of unbounded operators. J. Operator Theory **82**(2), 285–306 (2019)
32. Bourgeois, G.: On commuting exponentials in low dimensions. Linear Algebra Appl. **423**(2–3), 277–286 (2007)
33. Bram, J.: Subnormal operators. Duke Math. J. **22**, 75–94 (1955)
34. Brasche, J.F., Neidhardt, H.: Has every symmetric operator a closed symmetric restriction whose square has a trivial domain? Acta Sci. Math. (Szeged) **58**(1–4), 425–430 (1993)
35. Brooke, J.A., Busch, P., Pearson, D.B.: Commutativity up to a factor of bounded operators in complex Hilbert space. R. Soc. Lond. Proc. Ser. A Math. Phys. Eng. Sci. **458**(2017), 109–118 (2002)
36. Browder, F.E.: Commutators with powers of an unbounded operator in Hilbert space. Proc. Am. Math. Soc. **16**, 1211–1213 (1965)
37. Brown, A.: On a class of operators. Proc. Am. Math. Soc. **4**, 723–728 (1953)
38. Brown, A., Halmos, P.R., Shields, A.L.: Cesàro operators. Acta Sci. Math. (Szeged) **26**, 125–137 (1965)
39. Brown, A., Pearcy, C.: Spectra of tensor products of operators. Proc. Am. Math. Soc. **17**, 162–166 (1966)
40. Calderón, A.-P.: Commutators of singular integral operators. Proc. Nat. Acad. Sci. U.S.A. **53**, 1092–1099 (1965)
41. Campbell, S.L.: Linear operators for which T^*T and TT^*. Proc. Am. Math. Soc. **34**, 177–180 (1972)
42. Campbell, S.L.: Linear operators for which T^*T and TT^* commute II. Pacific J. Math. **53**, 355–361 (1974)
43. Campbell, S.L.: Operator-valued inner functions analytic on the closed disc. II. Pacific J. Math. **60**(2), 37–49 (1975)
44. Campbell, S.L.: Linear operators for which T^*T and $T + T^*$ commute. Pacific J. Math. **61**(1), 53–57 (1975)
45. Campbell, S.L.: Linear operators for which T^*T and $T + T^*$ commute. III. Pacific J. Math. **76**(1), 17–19 (1978)

46. Campbell, S.L., Gellar, R.: Linear operators for which T^*T and $T + T^*$ commute. II. Trans. Am. Math. Soc. **226**, 305–319 (1977)
47. Cavaretta, A.S., Smithies, L.: Lipschitz-type bounds for the map $A \to |A|$ on $\mathcal{L}(\mathcal{H})$. Linear Algebra Appl. **360**, 231–235 (2003)
48. Chaban, A., Mortad, M.H.: Global space-time L^p-estimates for the Airy operator on $L^2(\mathbb{R}^2)$ and some applications. Glas. Mat. Ser. III **47**(67), 373–379 (2012)
49. Chaban, A., Mortad, M.H.: Exponentials of bounded normal operators. Colloq. Math. **133**(2), 237–244 (2013)
50. Chaichian, M., Demichev, A.: Introduction to Quantum Groups. World Scientific Publishing Co., Inc., River Edge, NJ (1996)
51. Chan, N.N., Kwong, M.K.: Hermitian matrix inequalities and a conjecture. Am. Math. Monthly **92**(8), 533–541 (1985)
52. Charles, J., Mbekhta, M., Queffélec, H.: Analyse fonctionnelle et théorie des opérateurs (French). Dunod, Paris (2010)
53. Chavan, S., Jabłoński, Z.J., Jung, I.B., Stochel, J.: Taylor spectrum approach to Brownian-type operators with quasinormal entry. Ann. Mat. Pura Appl. (4) **200**(3), 881–922 (2021)
54. Chellali, Ch., Mortad, M.H.: Commutativity up to a factor for bounded and unbounded operators. J. Math. Anal. Appl. **419**(1), 114–122 (2014)
55. Chen, K.Y., Herrero, D.A., Wu, P.Y.: Similarity and quasisimilarity of quasinormal operators. J. Operator Theory **27**(2), 385–412 (1992)
56. Chernoff, P.R.: Semigroup product formulas and addition of unbounded operators, Thesis (Ph.D.)-Harvard University. ProQuest LLC, Ann Arbor, MI (1968)
57. Chernoff, P.R.: Product formulas, nonlinear semigroups, and addition of unbounded operators. In: Memoirs of the American Mathematical Society, No. 140. American Mathematical Society, Providence, R. I. (1974)
58. Chernoff, P.R.: A semibounded closed symmetric operator whose square has trivial domain. Proc. Am. Math. Soc. **89**(2), 289–290 (1983)
59. Chō, M., Lee, J.I., Yamazaki, T.: On the operator equation $AB = zBA$. Sci. Math. Jpn. **69**(2), 257–263 (2009)
60. Choi, M.D.: Adjunction and inversion of invertible Hilbert-space operators. Indiana Univ. Math. J. **23**, 413–419 (1973/74)
61. Cimprič, J., Savchuk, Y., Schmüdgen, K.: On q-normal operators and the quantum complex plane. Trans. Am. Math. Soc. **366**(1), 135–158 (2014)
62. Clancey, K.F.: Examples of nonnormal seminormal operators whose spectra are not spectral sets. Proc. Am. Math. Soc. **24**, 797–800 (1970)
63. Clary, S.: Equality of spectra of quasi-similar hyponormal operators. Proc. Am. Math. Soc. **53**(1), 88–90 (1975)
64. Coddington, E.A.: Formally normal operators having no normal extensions. Canad. J. Math. **17**, 1030–1040 (1965)
65. Cohen, P.J.: A counterexample to the closed graph theorem for bilinear maps. J. Functional Analysis **16**, 235–240 (1974)
66. Conway, J.B.: A Course in Functional Analysis, 2nd edn. Springer, New York (1990)
67. Conway, J.B.: The theory of subnormal operators. In: Mathematical Surveys and Monographs, vol. 36. American Mathematical Society, Providence, RI (1991)
68. Conway, J.B., Feldman, N.S.: The state of subnormal operators: A glimpse at Hilbert space operators. Oper. Theory Adv. Appl. **207**, 177–194 (2010). Birkhäuser Verlag, Basel
69. Conway, J.B., Hadwin, D.W.: Strong limits of normal operators. Glasgow Math. J. **24**(1), 93–96 (1983)
70. Conway, J.B., Morrel, B.B.: Roots and logarithms of bounded operators on Hilbert space. J. Funct. Anal. **70**(1), 171–193 (1987)
71. Conway, J.B., Prăjitură, G.: On λ-commuting operators. Studia Math. **166**(1), 1–9 (2005)
72. Conway, J.B., Prăjitură, G., Rodríguez-Martínez, A.: Powers and direct sums. J. Math. Anal. Appl. **413**(2), 880–889 (2014)

73. Conway, J.B., Szymański, W.: Linear combinations of hyponormal operators. Rocky Mountain J. Math. **18**(3), 695–705 (1988)

74. Cowen, C.C.: Subnormality of the Cesàro operator and a semigroup of composition operators. Indiana Univ. Math. J. **33**(2), 305–318 (1984)

75. Cowling, M., Price, J.F.: Generalisations of Heisenberg's inequality. In: Harmonic analysis (Cortona, 1982). Lecture Notes in Mathematics, vol. 992, pp. 443–449. Springer, Berlin (1983)

76. Crabb, M.J., Spain, P.G.: Commutators and normal operators. Glasgow Math. J. **18**(2), 197–198 (1977)

77. Curto, R.E., Muhly, P.S., Xia, J.: Hyponormal pairs of commuting operators. In: Contributions to Operator Theory and Its Applications (Mesa, AZ, 1987). Operator Theory: Advances and Applications, vol. 35, pp. 1–22. Birkhäuser, Basel (1988)

78. Curto, R.E., Putinar, M.: Existence of nonsubnormal polynomially hyponormal operators. Bull. Am. Math. Soc.(N.S.) **25**(2), 373–378 (1991)

79. Curto, R.E., Lee, S.H., Yoon, J.: Quasinormality of powers of commuting pairs of bounded operators. J. Funct. Anal. **278** (3), 108342, 23 pp. (2020)

80. Cycon, H.L., Froese, R.G., Kirsch, W., Simon, B.: Schrödinger operators with application to quantum mechanics and global geometry. In: Texts and Monographs in Physics. Springer Study Edition. Springer, Berlin (1987)

81. Daniluk, A.: On the closability of paranormal operators. J. Math. Anal. Appl. **376**(1), 342–348 (2011)

82. Deckard, D., Pearcy, C.: Another class of invertible operators without square roots. Proc. Am. Math. Soc. **14**, 445–449 (1963)

83. Dehimi, S., Mortad, M.H.: Bounded and unbounded operators similar to their adjoints. Bull. Korean Math. Soc. **54**(1), 215–223 (2017)

84. Dehimi, S., Mortad, M.H.: Right (or left) invertibility of bounded and unbounded operators and applications to the spectrum of products. Complex Anal. Oper. Theory **12**(3), 589–597 (2018)

85. Dehimi, S., Mortad, M.H.: Generalizations of Reid inequality. Mathematica Slovaca **68**(6), 1439–1446 (2018)

86. Dehimi, S., Mortad, M.H.: Chernoff-like counterexamples related to unbounded operators. Kyushu J. Math. **74**(1), 105–108 (2020)

87. Dehimi, S., Mortad, M.H.: Unbounded operators having self-adjoint or normal powers and some related results (2020). arXiv:2007.14349

88. Dehimi, S., Mortad, M.H., Tarcsay, Z.: On the operator equations $A^n = A^*A$. Linear Multilinear Algebra **69**(9), 1771–1778 (2021). https://doi.org/10.1080/03081087.2019.1641463

89. Devinatz, A., Nussbaum, A.E.: On the permutability of normal operators. Ann. of Math. (2) **65**, 144–152 (1957)

90. Devinatz, A., Nussbaum, A.E., von Neumann, J.: On the permutability of self-adjoint operators. Ann. of Math. (2) **62**, 199–203 (1955)

91. Dixmier, J.: L'adjoint du produit de deux opérateurs fermés (French). Annales de la Faculté des Sciences de Toulouse 4è Série **11**, 101–106 (1947)

92. Douglas, R.G.: On majorization, factorization, and range inclusion of operators on Hilbert space. Proc. Am. Math. Soc. **17**, 413–415 (1966)

93. Douglas, R.G.: On the operator equation $S^*XT = X$ and related topics. Acta Sci. Math. (Szeged) **30**, 19–32 (1969)

94. Dragomir, S.S., Moslehian, M.S.: Some inequalities for (α, β)-normal operators in Hilbert spaces. Facta Univ. Ser. Math. Inform. **23**, 39–47 (2008)

95. Dunford, N., Schwartz, J.T.: Linear operators. Part I. General theory. In: With the assistance of William G. Bade and Robert Bartle, G.: Reprint of the 1958 original. Wiley Classics Library. A Wiley-Interscience Publication. Wiley, New York (1988)

96. Dunford, N., Schwartz, J.T.: Linear operators. Part II. Spectral theory. Selfadjoint operators in Hilbert space. In: With the assistance of William G. Bade and Robert Bartle, G.: Reprint of the 1963 original. Wiley Classics Library. A Wiley-Interscience Publication. Wiley, New York (1988)

97. Dunford, N., Schwartz, J.T.: Linear operators. Part III. Spectral operators. In: With the assistance of William G. Bade and Robert G. Bartle. Reprint of the 1971 original. Wiley Classics Library. A Wiley-Interscience Publication. Wiley, New York (1988)

98. Duoandikoetxea, J.: Fourier analysis, Translated and revised from the 1995 Spanish original by David Cruz-Uribe. G.S.M., vol. 29. American Mathematical Society, Providence, RI (2001)

99. Elsner, L., Ikramov, Kh.D.: Normal matrices: an update. Linear Algebra Appl. **285**(1–3), 291–303 (1998)

100. Embry, M.R.: Conditions implying normality in Hilbert space. Pacific J. Math. **18**, 457–460 (1966)

101. Embry, M.R.: Similarities involving normal operators on Hilbert space. Pacific J. Math. **35**, 331–336 (1970)

102. Embry, M.R.: A connection between commutativity and separation of spectra of operators. Acta Sci. Math. (Szeged) **32**, 235–237 (1971)

103. Embry, M.R.: A generalization of the Halmos-Bram criterion for subnormality. Acta Sci. Math. (Szeged) **35**, 61–64 (1973)

104. Enflo, P.: On the invariant subspace problem for Banach spaces. Acta Math. **158**(3–4), 213–313 (1987)

105. Engel, K.-J.: Operator matrices and systems of evolution equations, book manuscript (unpublished)

106. Ezzahraoui, H., Mbekhta, M., Salhi, A., Zerouali, E.H.: A note on roots and powers of partial isometries. Arch. Math. (Basel) **110**(3), 251–259 (2018)

107. Fillmore, P.A., Williams, J.P.: On operator ranges. Adv. Math. **7**, 254–281 (1971)

108. Flanders, H., Wimmer, H.K.: On the matrix equations $AX - XB = C$ and $AX - YB = C$. SIAM J. Appl. Math. **32**(4), 707–710 (1977)

109. Foiaş, C., Williams, J.P.: Some remarks on the Volterra operator. Proc. Am. Math. Soc. **31**, 177–184 (1972)

110. Folland, G.F., Sitaram, A.: The uncertainty principle: a mathematical survey. J. Fourier Anal. Appl. **3**(3), 207–238 (1997)

111. Fong, C.K., Istrăţescu, V.I.: Some characterizations of Hermitian operators and related classes of operators. Proc. Am. Math. Soc. **76**, 107–112 (1979)

112. Fong, C.K., Tsui, S.K.: A note on positive operators. J. Operator Theory **5**(1), 73–76 (1981)

113. Frid, N., Mortad, M.H.: When nilpotence implies normality of bounded linear operators. arXiv:1901.09435

114. Frid, N., Mortad, M.H., Dehimi, S.: On nilpotence of bounded and unbounded linear operators. arXiv preprint arXiv:2008.09509 (2020)

115. Friedland, S.: A characterization of normal operators. Israel J. Math. **42**(3), 235–240 (1982)

116. Friedlander, F.G.: Introduction to the Theory of Distributions, 2nd edn. With additional material by Joshi, M. Cambridge University, Cambridge (1998)

117. Fuglede, B.: A commutativity theorem for normal operators. Proc. Nati. Acad. Sci. **36**, 35–40 (1950)

118. Fuglede, B.: Solution to Problem 3. Math. Scand. **2**, 346–347 (1954)

119. Fuglede, B.: Conditions for two self-adjoint operators to commute or to satisfy the Weyl relation. Math. Scand. **51**(1), 163–178 (1982)

120. Furuta, T.: An extension of the Fuglede-Putnam theorem to subnormal operators using a Hilbert-Schmidt norm inequality. Proc. Am. Math. Soc. **81**(2), 240–242 (1981)

121. Furuta, T.: Applications of the polar decomposition of an operator. Yokohama Math. J. **32**(1–2), 245–253 (1984)

122. Furuta, T.: A counterexample to a conjectured Hermitian matrix inequality. Am. Math. Monthly **94**(3), 271–272 (1987)

123. Furuta, T.: Invitation to Linear Operators: From Matrices to Bounded Linear Operators on a Hilbert Space. Taylor and Francis Ltd., London (2001)

124. Furuta, T.: Applications of polar decompositions of idempotent and 2-nilpotent operators. Linear Multilinear Algebra **56**(1–2), 69–79 (2008)

125. Garcia, S.R., Horn, R.A.: A second course in linear algebra. In: Cambridge Mathematical Textbooks. Cambridge University, Cambridge (2017)

126. Garcia, S.R., Tener, J.E.: Unitary equivalence of a matrix to its transpose. J. Operator Theory **68**(1), 179–203 (2012)

127. Garcia, S.R., Sherman, D., Weiss, G.: On the similarity of AB and BA for normal and other matrices. Linear Algebra Appl. **508**, 14–21 (2016)

128. Gelbaum, B.R., Olmsted, J.M.H.: Counterexamples in Analysis. Corrected reprint of the second (1965) edition. Dover Publications Inc., Mineola (2003)

129. George, A., Ikramov, Kh.D.: Unitary similarity of matrices with quadratic minimal polynomials. Linear Algebra Appl. **349**, 11–16 (2002)

130. Gerasimova, T.G.: Unitary similarity to a normal matrix. Linear Algebra Appl. **436**(9), 3777–3783 (2012)

131. Gesztesy, F., Schmüdgen, K.: On a theorem of Z. Sebestyén and Zs. Tarcsay. Acta Sci. Math. (Szeged) **85**(1–2), 291–293 (2019)

132. Gheondea, A.: When are the products of normal operators normal? Bull. Math. Soc. Sci. Math. Roumanie (N.S.) **52**(100)/2, 129–150 (2009)

133. Goldberg, S.: Unbounded Linear Operators: Theory and Applications. Dover Publications Inc., Mineola (2006)

134. Gohberg, I., Krupnik, N.: One-Dimensional Linear Singular Integral Equations. I. Introduction. Translated from the 1979 German translation by Bernd Luderer and Steffen Roch and revised by the authors. In: Operator Theory: Advances and Applications, vol. 53. Birkhäuser, Basel (1992)

135. Goldberg, M.: Zwas, G.: On matrices having equal spectral radius and spectral norm. Linear Algebra Appl. **8**, 427–434 (1974)

136. Gohberg, I., Goldberg, S., Kaashoek, M.A.: Basic Classes of Linear Operators. Birkhäuser, Basel (2003)

137. Goldstein, J.A.: Some counterexamples involving selfadjoint operators. Rocky Mountain J. Math. **2**(1), 143–149 (1972)

138. Gong, W.: A simple proof of an extension of the Fuglede-Putnam theorem. Proc. Am. Math. Soc. **100**(3), 599–600 (1987)

139. Gong, W.B., Han, D.G.: Spectrum of the products of operators and compact perturbations. Proc. Am. Math. Soc. **120**(3), 755–760 (1994)

140. Gray, L.J.: Products of Hermitian operators. Proc. Am. Math. Soc. **59**(1), 123–126 (1976)

141. Grivaux, S., Roginskaya, M.: Multiplicity of direct sums of operators on Banach spaces. Bull. Lond. Math. Soc. **41**(6), 1041–1052 (2009)

142. Gustafson, K.: On projections of self-adjoint operators and operator product adjoints. Bull. Am. Math. Soc. **75**, 739–741 (1969)

143. Gustafson, K.: On operator sum and product adjoints and closures. Canad. Math. Bull. **54**(3), 456–463 (2011)

144. Gustafson, K., Mortad, M.H.: Unbounded products of operators and connections to Dirac-type operators. Bull. Sci. Math. **138**(5), 626–642 (2014)

145. Gustafson, K., Mortad, M.H.: Conditions implying commutativity of unbounded self-adjoint operators and related topics. J. Operator Theory **76**(1), 159–169 (2016)

146. Halmos, P.R.: Normal dilations and extensions of operators. Summa Brasil. Math. **2**, 125–134 (1950)

147. Halmos, P.R.: Commutativity and spectral properties of normal operators. Acta Sci. Math. Szeged **12**, 153–156 (1950). Leopoldo Fejér Frederico Riesz LXX annos natis dedicatus, Pars B

148. Halmos, P.R.: A Hilbert Space Problem Book, 2nd edn. Springer, Berlin (1982)

149. Halmos, P.R.: Linear algebra problem book. In: The Dolciani Mathematical Expositions, vol. 16. Mathematical Association of America, Washington (1995)
150. Halmos, P.R., Lumer, G.: Square roots of operators II. Proc. Am. Math. Soc. **5**, 589–595 (1954)
151. Halmos, P.R., Lumer, G., Schäffer, J.J.: Square roots of operators. Proc. Am. Math. Soc. **4**, 142–149 (1953)
152. Han, J.K., Lee, H.Y., Lee, W.Y.: Invertible completions of 2×2 upper triangular operator matrices. Proc. Am. Math. Soc. **128**(1), 119–123 (2000)
153. Hardt, V., Konstantinov, A., Mennicken, R.: On the spectrum of the product of closed operators. Math. Nachr. **215**, 91–102 (2000)
154. Hardy, G.H.: A theorem concerning Fourier transforms. J. London Math. Soc. **8**(3), 227–231 (1933)
155. Hassi, S., Sebestyén, Z., de Snoo, H.S.V.: On the nonnegativity of operator products. Acta Math. Hungar. **109**(1–2), 1–14 (2005)
156. Hess, P., Kato, T.: Perturbation of closed operators and their adjoints. Comment. Math. Helv. **45**, 524–529 (1970)
157. Hirsch, F., Lacombe, G.: Elements of functional analysis. Translated from the 1997 French original by Silvio Levy. In: Graduate Texts in Mathematics, vol. 192. Springer, New York (1999)
158. Hladnik, M., Omladič, M.: Spectrum of the product of operators. Proc. Am. Math. Soc. **102**(2), 300–302 (1988)
159. Hoffman, K.: Banach Spaces of Analytic Functions. Prentice-Hall Series in Modern Analysis Prentice-Hall, Inc., Englewood Cliffs, N. J. (1962)
160. Holland Jr., S.S.: On the adjoint of the product of operators. J. Functional Analysis **3**, 337–344 (1969)
161. Hoover, T.B.: Quasi-similarity of operators. Illinois J. Math. **16**, 678–686 (1972)
162. Horn, R.A., Johnson, C.R.: Matrix Analysis, 2nd edn. Cambridge University, Cambridge (2013)
163. Horowitz, Ch.: An elementary counterexample to the open mapping principle for bilinear maps. Proc. Am. Math. Soc. **53**(2), 293–294 (1975)
164. Hou, J.: Solution of operator equations and tensor products. J. Math. Res. "Sr Exposition" **12**(4), 479–486 (1992)
165. Hou, J.: On the tensor products of operators. Acta Math. Sinica **37**(3), 432 (1994). Acta Math. Sinica (N.S.) **9**(2), 195–202 (1993)
166. Huff, Ch.W.: On pairs of matrices (of order two) A, B satisfying the condition $e^A e^B = e^{A+B} \neq e^B e^A$. Rend. Circ. Mat. Palermo (2), **2**(1953), 326–330 (1954)
167. Ichinose, W., Iwashita, K.: On the uniqueness of the polar decomposition of bounded operators in Hilbert spaces. J. Operator Theory **70**(1), 175–180 (2013)
168. Ikramov, Kh.D.: Binormal Matrices. (Russian); Translated from Zap. Nauchn. Sem. S.-Peterburg. Otdel. Mat. Inst. Steklov. (POMI) **463** (2017), Chislennye Metody i Voprosy Organizatsii Vychisleniĭ. XXX, 132–141. J. Math. Sci. (N.Y.) **232**(6), 830–836 (2018)
169. Istrăţescu, V., Saitô, T., Yoshino, T.: On a class of operators. Tôhoku Math. J. (2), **18**, 410–413 (1966)
170. Ito, T., Wong, T.K.: Subnormality and quasinormality of Toeplitz operators. Proc. Am. Math. Soc. **34**, 157–164 (1972)
171. Ito, M., Yamazaki, T., Yanagida, M.: On the polar decomposition of the product of two operators and its applications. Integr. Equ. Oper. Theory **49**(4), 461–472 (2004)
172. Jabłoński, Z.J., Jung, Il.B., Stochel, J.: Unbounded quasinormal operators revisited. Integr. Equ. Oper. Theory **79**(1), 135–149 (2014)
173. Jeon, I.H., Kim, S.H., Ko, E., Park, J.E.: On positive-normal operators. Bull. Korean Math. Soc. **39**(1), 33–41 (2002)
174. Jeon, I.H., Kim, I.H., Tanahashi, K., Uchiyama, A.: Conditions implying self-adjointness of operators. Integr. Equ. Oper. Theory **61**(4), 549–557 (2008)

175. Jiang, Ch.-C.: On products of two Hermitian operators. Linear Algebra Appl. **430**(1), 553–557 (2009)
176. Jibril, A.A.S.: On operators for which $T^{*^2}T^2 = (T^*T)^2$. Int. Math. Forum **5**(45–48), 2255–2262 (2010)
177. Jin, G., Chen, A.: Some basic properties of block operator matrices. arXiv:1403.7732
178. Johnson, C.R.: Normality and the numerical range. Linear Algebra Appl. **15**(1), 89–94 (1976)
179. Johnson, C.R., Zhang, F.: An operator inequality and matrix normality. Linear Algebra Appl. **240**, 105–110 (1996)
180. Jørgensen, P.E.T.: Unbounded operators: perturbations and commutativity problems. J. Funct. Anal. **39**(3), 281–307 (1980)
181. Jorgensen, P., Tian, F.: Non-commutative analysis. With a foreword by Wayne Polyzou. World Scientific Publishing Co. Pte. Ltd., Hackensack, NJ (2017)
182. Jung, Il.B., Stochel, J.: Subnormal operators whose adjoints have rich point spectrum. J. Funct. Anal. **255**(7), 1797–1816 (2008)
183. Jung, Il.B., Mortad, M.H., Stochel, J.: On normal products of selfadjoint operators. Kyungpook Math. J. **57**, 457–471 (2017)
184. Kadison, R.V., Ringrose, J.R.: Fundamentals of the Theory of Operator Algebras, vol. I. Elementary theory. Reprint of the 1983 original, G.S.M., vol. 15. American Mathematical Society, Providence, RI (1997)
185. Kadison, R.V., Singer, I.M.: Three test problems in operator theory. Pacific J. Math. **7**, 1101–1106 (1957)
186. Kalisch, G.K.: On similarity, reducing manifolds, and unitary equivalence of certain Volterra operators. Ann. of Math. (2) **66**, 481–494 (1957)
187. Kamowitz, H.: On operators whose spectrum lies on a circle or a line. Pacific J. Math. **20**, 65–68 (1967)
188. Kaplansky, I.: Products of normal operators. Duke Math. J. **20**(2), 257–260 (1953)
189. Kato, T.: Continuity of the map $S \mapsto |S|$ for linear operators. Proc. Japan Acad. **49**, 157–160 (1973)
190. Kato, T.: Perturbation Theory for Linear Operators, 2nd edn. Springer, New York (1980)
191. Kato, Y.: Some examples of θ-operators. Kyushu J. Math. **48**(1), 101–109 (1994)
192. Khasbardar, S., Thakare, N.: Some counterexamples for quasinormal operators and related results. Indian J. Pure Appl. Math. **9**(12), 1263–1270 (1978)
193. Kittaneh, F.: On generalized Fuglede-Putnam theorems of Hilbert-Schmidt type. Proc. Am. Math. Soc. **88**(2), 293–298 (1983)
194. Kittaneh, F.: On normality of operators. Rev. Roumaine Math. Pures Appl. **29**(8), 703–705 (1984)
195. Kittaneh, F.: Notes on some inequalities for Hilbert space operators. Publ. Res. Inst. Math. Sci. **24**(2), 283–293 (1988)
196. Kittaneh, F.: On some equivalent metrics for bounded operators on Hilbert space. Proc. Am. Math. Soc. **110**(3), 789–798 (1990)
197. Kittaneh, F.: On the normality of operator products. Linear and Multilinear Algebra **30**(1–2), 1–4 (1991)
198. Kittaneh, F.: Inequalities for commutators of positive operators. J. Funct. Anal. **250**(1), 132–143 (2007)
199. Kittaneh, F.: Norm inequalities for commutators of positive operators and applications. Math. Z. **258**(4), 845–849 (2008)
200. Klimyk, A., Schmüdgen, K.: Quantum groups and their representations. In: Texts and Monographs in Physics. Springer, Berlin (1997)
201. Kosaki, H.: Unitarily invariant norms under which the map $A \to |A|$ is Lipschitz continuous. Publ. Res. Inst. Math. Sci. **28**(2), 299–313 (1992)
202. Kosaki, H.: On intersections of domains of unbounded positive operators. Kyushu J. Math. **60**(1), 3–25 (2006)

203. Koshmanenko, V.D., Ôta, S.: On the characteristic properties of singular operators. (Ukrainian); translated from Ukraïn. Mat. Zh. **48**(11), 1484–1493 (1996). Ukrainian Math. J. **48**(11), 1677–1687 (1997)
204. Kriete III, T.L., Trutt, D.: The Cesàro operator in ℓ^2 is subnormal. Am. J. Math. **93**, 215–225 (1971)
205. Kubrusly, C.S.: An Introduction to Models and Decompositions in Operator Theory. Birkhäuser Boston, Inc., Boston, MA (1997)
206. Kubrusly, C.S.: Hilbert Space Operators, A Problem Solving Approach. Birkhäuser Boston, Inc., Boston, MA (2003)
207. Kubrusly, C.S.: The Elements of Operator Theory. 2nd edn. Birkhäuser/Springer, New York (2011)
208. Kubrusly, C.S., Vieira, P.C.M., Zanni, J.: Powers of posinormal operators. Oper. Matrices **10**(1), 15–27 (2016)
209. Kulkarni, S.H., Nair, M.T., Ramesh, G.: Some properties of unbounded operators with closed range. Proc. Indian Acad. Sci. Math. Sci. **118**(4), 613–625 (2008)
210. Laberteux, K.R.: Problem 10377. Am. Math. Monthly **101**, 362 (1994)
211. Lee, M.Y., Lee, S.H.: On a class of operators related to paranormal operators. J. Korean Math. Soc. **44**(1), 25–34 (2007)
212. Lee, S.H., Lee, W.Y., Yoon, J.: An answer to a question of A. Lubin: the lifting problem for commuting subnormals. Israel J. Math. **222**(1), 201–222 (2017)
213. Li, Ch.-K., Poon, Y.-T.: Spectral inequalities and equalities involving products of matrices. Linear Algebra Appl. **323**(1–3), 131–143 (2001)
214. Li, C.K., Poon, Y.T.: Spectrum, numerical range and Davis-Wielandt shell of a normal operator. Glasg. Math. J. **51**(1), 91–100 (2009)
215. Lieb, E.H., Loss, M.: Analysis, 2nd edn. In: Graduate Studies in Mathematics, vol. 14. American Mathematical Society, Providence, RI (2001)
216. Lubin, A.: Weighted shifts and products of subnormal operators. Indiana Univ. Math. J. **26**(5), 839–845 (1977)
217. Lubin, A.: A subnormal semigroup without normal extension. Proc. Am. Math. Soc. **68**(2), 176–178 (1978)
218. Lubin, A.: Extensions of commuting subnormal operators. In: Hilbert space operators (Proceedings of the Conference California State University, Long Beach, California, 1977), pp. 115–120. Lecture Notes in Mathematical, vol. 693. Springer, Berlin (1978). Proceedings of the Conference of Army Physicians, Central Mediterranean Forces
219. Lumer, G., Rosenblum, M.: Linear operator equations. Proc. Am. Math. Soc. **10**, 32–41 (1959)
220. Markin, M.V.: On a characterization of finite-dimensional vector spaces. arXiv:2008.09270
221. Martínez Carracedo, C., Sanz Alix, M.: The theory of fractional powers of operators. In: North-Holland Mathematics Studies, vol. 187. North-Holland Publishing Co., Amsterdam (2001)
222. Mbekhta, M., Suciu, L.: Partial isometries and the conjecture of CK Fong and SK Tsui. J. Math. Anal. Appl. **437**(1), 431–444 (2016)
223. McCarthy, C.A.: On a theorem of Beck and Putnam. J. London Math. Soc. **39**, 288–290 (1964)
224. McCullough, S.A., Rodman, L.: Hereditary classes of operators and matrices. Am. Math. Monthly **104**(5), 415–430 (1997)
225. McIntosh, A.: Counterexample to a question on commutators. Proc. Am. Math. Soc. **29**, 337–340 (1971)
226. Mecheri, S., Bachir, A.: Positive answer to the conjecture by Fong and Istratescu. Bull. Korean Math. Soc. **42**(4), 869–873 (2005)
227. Mercer, I.D.: Finding "nonobvious" nilpotent matrices (2005). http://people.math.sfu.ca/~idmercer/nilpotent.pdf
228. Messirdi, B., Mortad, M.H.: On different products of closed operators. Banach J. Math. Anal. **2**(1), 40–47 (2008)

229. Meziane, M., Mortad, M.H.: Maximality of linear operators. Rend. Circ. Mat. Palermo, Ser II. **68**(3), 441–451 (2019)
230. Möller, M.: On the essential spectrum of a class of operators in Hilbert space. Math. Nachr. **194**, 185–196 (1998)
231. Möller, M., Szafraniec, F.H.: Adjoints and formal adjoints of matrices of unbounded operators. Proc. Am. Math. Soc. **136**(6), 2165–2176 (2008)
232. Montgomery-Smith, S.J.: Time decay for the bounded mean oscillation of solutions of the Schrödinger and wave equations. Duke Math. J. **91**, 393–408 (1998)
233. Moore, R.L., Rogers, D.D., Trent, T.T.: A note on intertwining M-hyponormal operators. Proc. Am. Math. Soc. **83**(3), 514–516 (1981)
234. Morinaga, K., Nôno, T.: On the non-commutative solutions of the exponential equation $e^x e^y = e^{x+y}$. II. J. Sci. Hiroshima Univ. Ser. A. **17**, 345–358 (1954)
235. Morinaga, K., Nôno, T.: On the non-commutative solutions of the exponential equation $e^x e^y = e^{x+y}$. II. J. Sci. Hiroshima Univ. Ser. A. **18**, 137–178 (1954)
236. Mortad, M.H.: An application of the Putnam-Fuglede theorem to normal products of self-adjoint operators. Proc. Am. Math. Soc. **131**(10), 3135–3141 (2003)
237. Mortad, M.H.: Normal products of self-adjoint operators and self-adjointness of the perturbed wave operator on $L^2(\mathbb{R}^n)$. Thesis (Ph.D.)-The University of Edinburgh (United Kingdom). ProQuest LLC, Ann Arbor, MI (2003)
238. Mortad, M.H.: Self-adjointness of the perturbed wave operator on $L^2(\mathbb{R}^n)$, $n \geq 2$. Proc. Am. Math. Soc. **133**(2), 455–464 (2005)
239. Mortad, M.H.: On L^p-estimates for the time-dependent Schrödinger operator on L^2. J. Ineq. Pure Appl. Math. **8**(3), Art. 80, 8pp. (2007)
240. Mortad, M.H.: On some product of two unbounded self-adjoint operators. Integr. Equ. Oper. Theory **64**(3), 399–408 (2009)
241. Mortad, M.H.: Yet more versions of the Fuglede-Putnam theorem. Glasg. Math. J. **51**(3), 473–480 (2009)
242. Mortad, M.H.: Commutativity up to a factor: more results and the unbounded case. Z. Anal. Anwendungen: Journal for Analysis and its Applications **29**(3), 303–307 (2010)
243. Mortad, M.H.: On a Beck-Putnam-Rehder theorem. Bull. Belg. Math. Soc. Simon Stevin **17**(4), 737–740 (2010)
244. Mortad, M.H.: Similarities involving unbounded normal operators. Tsukuba J. Math. **34**(1), 129–136 (2010)
245. Mortad, M.H.: Products and sums of bounded and unbounded normal operators: Fuglede-Putnam versus Embry. Rev. Roumaine Math. Pures Appl. **56**(3), 195–205 (2011)
246. Mortad, M.H.: On the adjoint and the closure of the sum of two unbounded operators. Canad. Math. Bull. **54**(3), 498–505 (2011)
247. Mortad, M.H.: An implicit division of bounded and unbounded linear operators which preserves their properties. Glas. Mat. Ser. III **46**(66-2), 433–438 (2011)
248. Mortad, M.H.: Global space-time L^p estimates for the wave operator on L^2. Rend. Semin. Mat. Univ. Politec. Torino **69**(1), 91–96 (2011)
249. Mortad, M.H.: Exponentials of normal operators and commutativity of operators: A new approach. Colloq. Math. **125**(1), 1–6 (2011)
250. Mortad, M.H.: An all-unbounded-operator version of the Fuglede-Putnam theorem. Complex Anal. Oper. Theory **6**(6), 1269–1273 (2012)
251. Mortad, M.H.: On the normality of the sum of two normal operators. Complex Anal. Oper. Theory **6**(1), 105–112 (2012)
252. Mortad, M.H.: A contribution to the Fong-Tsui conjecture related to self-adjoint operators (2012). arXiv:1208.4346
253. Mortad, M.H.: Commutativity of unbounded normal and self-adjoint operators and applications. Oper. Matrices **8**(2), 563–571 (2014)
254. Mortad, M.H.: A criterion for the normality of unbounded operators and applications to self-adjointness. Rend. Circ. Mat. Palermo (2) **64**(1), 149–156 (2015)

255. Mortad, M.H.: Introductory Topology: Exercises and Solutions, vol. xvii, 2nd edn. (English). World Scientific, Hackensack, NJ (ISBN 978-981-3146-93-8/hbk; 978-981-3148-02-4/pbk), 356 p. (2017)

256. Mortad, M.H.: An Operator Theory Problem Book. World Scientific Publishing Co., New York (2018). https://doi.org/10.1142/10884. ISBN: 978-981-3236-25-7 (hardcover)

257. Mortad, M.H.: On the absolute value of the product and the sum of linear operators. Rend. Circ. Mat. Palermo, II. Ser **68**(2), 247–257 (2019)

258. Mortad, M.H.: Counterexamples related to commutators of unbounded operators. Results Math. **74**(4), Paper No. 174 (2019)

259. Mortad, M.H.: On the triviality of domains of powers and adjoints of closed operators. Acta Sci. Math. (Szeged) **85**, 651–658 (2019)

260. Mortad, M.H.: On the invertibility of the sum of operators. Anal. Math. **46**(1), 133–145 (2020)

261. Mortad, M.H.: On the existence of normal square and nth roots of operators. J. Anal. **28**(3), 695–703 (2020)

262. Mortad, M.H.: Yet another generalization of the Fuglede-Putnam theorem to unbounded operators (2020). arXiv:2003.00339

263. Mortad, M.H.: Unbounded operators: (square) roots, nilpotence, closability and some related invertibility results (2020). arXiv:2007.12027

264. Mortad, M.H.: Counterexamples related to unbounded paranormal operators. Examples and Counterexamples **1**, 100017 (2021)

265. Mortad, M.H.: The Fuglede-Putnam Theory (submitted monograph)

266. Moslehian, M.S., Nabavi Sales, S.M.S.: Fuglede-Putnam type theorems via the Aluthge transform. Positivity **17**(1), 151–162 (2013)

267. Moyls, B.N., Marcus, M.D.: Field convexity of a square matrix. Proc. Am. Math. Soc., **6**, 981–983 (1955)

268. Murnaghan, F.D.: On the unitary invariants of a square matrix. Proc. Nati. Acad. Sci. **18**(2), 185–189 (1932)

269. Nagel, R.: Towards a "matrix theory" for unbounded operator matrices. Math. Z. **201**(1), 57–68 (1989)

270. Nagel, R.: The spectrum of unbounded operator matrices with nondiagonal domain. J. Funct. Anal. **89**(2), 291–302 (1990)

271. Naimark, M.: On the square of a closed symmetric operator. Dokl. Akad. Nauk SSSR **26**, 866–870 (1940); ibid. **28**, 207–208 (1940)

272. Nelson, E.: Analytic vectors. Ann. of Math. (2), **70**, 572–615 (1959)

273. Nussbaum, A.E.: A commutativity theorem for semibounded operators in Hilbert space. Proc. Am. Math. Soc. **125**(12), 3541–3545 (1997)

274. Okazaki, Y.: Boundedness of closed linear operator T satisfying $R(T) \subset D(T)$. Proc. Japan Acad. Ser. A Math. Sci. **62**(8), 294–296 (1986)

275. Olagunju, A.: A note on closed operators. Proc. Cambridge Philos. Soc. **57**, 426 (1961)

276. Olson, B., Shaw, S., Shi, Ch., Pierre, Ch., Parker, R.: Circulant matrices and their application to vibration analysis. Appl. Mech. Rev. **66**(4) (2014). American Society of Mechanical Engineers

277. Ó Searcóid, M.: A note on a comment of S. Axler and J. H. Shapiro: "Putnam's theorem, Alexander's spectral area estimate and VMO". Irish Math. Soc. Newslett. No. **15**, 52–56 (1985)

278. Ôta, S.: Closed linear operators with domain containing their range. Proc. Edinburgh Math. Soc. (2) **27**(2), 229–233 (1984)

279. Ôta, S.: Unbounded nilpotents and idempotents. J. Math. Anal. Appl. **132**(1), 300–308 (1988)

280. Ôta, S.: On a singular part of an unbounded operator. Z. Anal. Anwendungen **7**(1), 15–18 (1988)

281. Ôta, S.: On normal extensions of unbounded operators. Bull. Polish Acad. Sci. Math. **46**(3), 291–301 (1998)

282. Ôta, S.: Some classes of q-deformed operators. J. Operator Theory **48**(1), 151–186 (2002)

283. Ôta, S.: On q-deformed hyponormal operators. Math. Nachr. **248**(249), 144–150 (2003)

284. Ôta, S., Schmüdgen, K.: On some classes of unbounded operators. Integr. Equ. Oper. Theory **12**(2), 211–226 (1989)

285. Ôta, S., Schmüdgen, K.: Some selfadjoint 2×2 operator matrices associated with closed operators. Integr. Equ. Oper. Theory **45**(4), 475–484 (2003)

286. Ôta, S., Szafraniec, F.H.: Notes on q-deformed operators. Studia Math., **165**(3), 295–301 (2004)

287. Ozawa, T.: Nonexistence of positive commutators. Hiroshima Math. J. **20**(1), 209–211 (1990)

288. Patel, A.B., Ramanujan, P.B.: On sum and product of normal operators. Indian J. Pure Appl. Math. **12**(10), 1213–1218 (1981)

289. Pedersen, S.: Anticommuting self-adjoint operators. J. Funct. Anal. **89**(2), 428–443 (1990)

290. Pietrzycki, P.: The single equality $A^{*n} A^n = (A^*A)^n$ does not imply the quasinormality of weighted shifts on rootless directed trees. J. Math. Anal. Appl. **435**(1), 338–348 (2016)

291. Pietrzycki, P.: Reduced commutativity of moduli of operators. Linear Algebra Appl. **557**, 375–402 (2018)

292. Pietrzycki, P., Stochel, J.: Subnormal nth roots of quasinormal operators are quasinormal. J. Funct. Anal. **280**(12), Article 109001 (2021). https://doi.org/10.1016/j.jfa.2021.109001

293. Pietrzycki, P., Stochel, J.: On nth roots of bounded and unbounded quasinormal operators. arxiv: 2103.09961v2

294. Putnam, C.R.: On normal operators in Hilbert space. Am. J. Math. **73**, 357–362 (1951)

295. Putnam, C.R.: Commutation Properties of Hilbert Space Operators and Related Topics. Springer, New York (1967)

296. Putnam, C.R.: Almost normal operators, their spectra and invariant subspaces. Bull. Am. Math. Soc. **79**, 615–624 (1973)

297. Putnam, C.R.: Spectra of polar factors of hyponormal operators. Trans. Am. Math. Soc. **188**, 419–428 (1974)

298. Putnam, C.R.: Normal operators and strong limit approximations. Indiana Univ. Math. J. **32**(3), 377–379 (1983)

299. Radjavi, H., Rosenthal, P.: Hyperinvariant subspaces for spectral and n-normal operators. Acta Sci. Math. (Szeged) **32**, 121–126 (1971)

300. Radjavi, H., Williams, J.P.: Products of self-adjoint operators. Michigan Math. J. **16**, 177–185 (1969)

301. Read, C.J.: A solution to the invariant subspace problem. Bull. London Math. Soc. **16**(4), 337–401 (1984)

302. Reed, M., Simon, B.: Methods of modern mathematical physics. In: Functional Analysis, vol. 1. Academic Press, New York (1972)

303. Reed, M., Simon, B.: Methods of modern mathematical physics. Fourier Analysis, Self-Adjointness, vol. 2. Academic Press, New York (1975)

304. Rehder, W.: On the adjoints of normal operators. Arch. Math. (Basel) **37**(2), 169–172 (1981)

305. Rehder, W.: On the product of self-adjoint operators. Internat. J. Math. and Math. Sci. **5**(4), 813–816 (1982)

306. Reid, W.T.: Symmetrizable completely continuous linear transformations in Hilbert space. Duke Math. J. **18**, 41–56 (1951)

307. Rhaly Jr., H.C.: Posinormal operators. J. Math. Soc. Japan **46**(4), 587–605 (1994)

308. Rosenblum, M.: On the operator equation $BX - XA = Q$. Duke Math. J. **23**, 263–269 (1956)

309. Rosenblum, M.: On a theorem of Fuglede and Putnam. J. London Math. Soc. **33**, 376–377 (1958)

310. Rosenblum, M.: The operator equation $BX - XA = Q$ with self-adjoint A and B. Proc. Am. Math. Soc. **20**, 115–120 (1969)

311. Roth, W.E.: The equations $AX - YB = C$ and $AX - XB = C$ in matrices. Proc. Am. Math. Soc. **3**, 392–396 (1969)

312. Rudin, W.: Function theory in polydiscs. W. A. Benjamin, Inc., Amsterdam (1969)

313. Rudin, W.: Real and Complex Analysis, 3rd edn. McGraw-Hill Book Co., New York (1987)

314. Rudin, W.: Functional Analysis. In: McGraw-Hill Book Co., International Series in Pure and Applied Mathematics, 2nd edn. McGraw-Hill, Inc., New York (1991)

315. Rynne, B.P., Youngson, M.A.: Linear functional analysis. In: Springer Undergraduate Mathematics Series. Springer-Verlag London, Ltd., London (2000)
316. Schäffer, J.J.: More about invertible operators without roots. Proc. Am. Math. Soc. **16**, 213–219 (1965)
317. Schechter, M.: The conjugate of a product of operators. J. Functional Analysis **6**, 26–28 (1970)
318. Schmincke, U.-W.: Distinguished selfadjoint extensions of Dirac operators. Math. Z. **129**, 335–349 (1972)
319. Schmoeger, C.: On normal operator exponentials. Proc. Am. Math. Soc. **130**(3), 697–702 (2001)
320. Schmoeger, C.: Operator Exponentials on Hilbert Spaces. Mathematisches Institut I, Universität Karlsruhe, Karlsruhe (2001)
321. Schmüdgen, K.: On domains of powers of closed symmetric operators. J. Operator Theory **9**(1), 53–75 (1983)
322. Schmüdgen, K.: On restrictions of unbounded symmetric operators. J. Operator Theory **11**(2), 379–393 (1984)
323. Schmüdgen, K.: On commuting unbounded selfadjoint operators. I. Acta Sci. Math. (Szeged) **47**(1–2), 131–146 (1984)
324. Schmüdgen, K.: A formally normal operator having no normal extension. Proc. Am. Math. Soc. **95**(3), 503–504 (1985)
325. Schmüdgen, K.: Unbounded operator algebras and representation theory. In: Operator Theory: Advances and Applications, vol. 37. Birkhäuser, Basel (1990)
326. Schmüdgen, K.: Unbounded Self-Adjoint Operators on Hilbert Space, vol. 265. Springer, Berlin (2012). GTM
327. Schwaiger, J.: More on rootless matrices. Anz. Österreich. Akad. Wiss. Math.-Natur. Kl. **141**(3–8) (2005/2006)
328. Schweinsberg, A.: The operator equation $AX - XB = C$ with normal A and B. Pacific J. Math. **102**(2), 447–453 (1982)
329. Schwenk, J., Wess, J.A.: q-deformed quantum mechanical toy model. Phys. Lett. B **291**(3), 273–277 (1992)
330. Sebestyén, Z.: Positivity of operator products. Acta Sci. Math. (Szeged) **66**(1–2), 287–294 (2000)
331. Sebestyén, Z., Stochel, J.: Restrictions of positive selfadjoint operators. Acta Sci. Math. (Szeged) **55**(1–2), 149–154 (1991)
332. Sebestyén, Z., Stochel, J.: On products of unbounded operators. Acta Math. Hungar. **100**(1–2), 105–129 (2003)
333. Sebestyén, Z., Stochel, J.: On suboperators with codimension one domains. J. Math. Anal. Appl. **360**(2), 391–397 (2009)
334. Sebestyén, Z., Tarcsay, Zs.: T^*T always has a positive selfadjoint extension. Acta Math. Hungar. **135**(1–2), 116–129 (2012)
335. Sebestyén, Z., Tarcsay, Zs.: A reversed von Neumann theorem. Acta Sci. Math. (Szeged) **80**(3–4), 659–664 (2014)
336. Sebestyén, Z., Tarcsay, Zs.: Adjoint of sums and products of operators in Hilbert spaces. Acta Sci. Math. (Szeged) **82**(1–2), 175–191 (2016)
337. Sebestyén, Z., Tarcsay, Zs.: On the square root of a positive selfadjoint operator. Period. Math. Hungar. **75**(2), 268–272 (2017)
338. Sebestyén, Z., Tarcsay, Zs.: Range-kernel characterizations of operators which are adjoint of each other (2020). arXiv:2002.01213
339. Sheth, I.H.: On hyponormal operators. Proc. Am. Math. Soc. **17**, 998–1000 (1966)
340. Sibirskiĭ, K.S.: A minimal polynomial basis of unitary invariants of a square matrix of order three. (Russian). Mat. Zametki **3**, 291–295 (1968)
341. Simon, B.: Operator Theory: A Comprehensive Course in Analysis, Part 4. American Mathematical Society, Providence, RI (2015)

342. Sirotkin, G.: Infinite matrices with "few" non-zero entries and without non-trivial invariant subspaces. J. Funct. Anal. **256**(6), 1865–1874 (2009)
343. So, W.: Equality cases in matrix exponential inequalities. SIAM J. Matrix Anal. Appl. **13**(4), 1154–1158 (1992)
344. Stampfli, J.G.: Hyponormal operators. Pacific J. Math. **12**, 1453–1458 (1962)
345. Stampfli, J.G.: Hyponormal operators and spectral density. Trans. Am. Math. Soc. **117**, 469–476 (1965)
346. Stampfli, J.G.: Which weighted shifts are subnormal? Pacific J. Math. **17**, 367–379 (1966)
347. Steen, L.A., Seebach Jr, J.A.: Counterexamples in topology. Reprint of the second (1978) edition. Dover Publications, Inc., Mineola, NY (1995)
348. Stenger, W.: On the projection of a selfadjoint operator. Bull. Am. Math. Soc. **74**, 369–372 (1968)
349. Stochel, J.: Seminormality of operators from their tensor product. Proc. Am. Math. Soc., **124**(1), 135–140 (1996)
350. Stochel, J.: An asymmetric Putnam-Fuglede theorem for unbounded operators. Proc. Am. Math. Soc. **129**(8), 2261–2271 (2001)
351. Stochel, J.: Lifting strong commutants of unbounded subnormal operators. Integr. Equ. Oper. Theory **43**(2), 189–214 (2002)
352. Stochel, J., Stochel, J.B.: Composition operators on Hilbert spaces of entire functions with analytic symbols. J. Math. Anal. Appl. **454**(2), 1019–1066 (2017)
353. Stochel, J., Szafraniec, F.H.: On normal extensions of unbounded operators I. J. Operator Theory **14**(1), 31–55 (1985)
354. Stochel, J., Szafraniec, F.H.: On normal extensions of unbounded operators. II. Acta Sci. Math. (Szeged) **53**(1–2), 153–177 (1989)
355. Stochel, J., Szafraniec, F.H.: On normal extensions of unbounded operators. III. Spectral properties. Publ. Res. Inst. Math. Sci. **25**(1), 105–139 (1989)
356. Stochel, J., Szafraniec, F.H.: Domination of unbounded operators and commutativity. J. Math. Soc. Japan **55**(2), 405–437 (2003)
357. Sylvester, J.J.: Sur l'équation en matrices $px = xq$ (French). C. R. Acad. Sci. Paris **99**, 67–71, 115–116 (1884)
358. Szafraniec, F.H.: On normal extensions of unbounded operators. IV. A matrix construction. In: Operator Theory and Indefinite Inner Product Spaces, pp. 337–350, Operator Theory: Advances and Applications, vol. 163. Birkhäuser, Basel (2006)
359. Sz.-Nagy, B.: Perturbations des transformations linéaires fermées (French). Acta Sci. Math. Szeged **14**, 125–137 (1951)
360. Tarcsay, Zs.: Operator extensions with closed range. Acta Math. Hungar. **135**, 325–341 (2012)
361. Taylor, A.E., Lay, D.C.: Introduction to functional analysis. Reprint of the 2nd edn. Robert E. Krieger Publishing Co., Inc., Melbourne, FL (1986)
362. ter Elst, *A.F.M., Sauter, M.: Nonseparability and von Neumann's theorem for domains of unbounded operators. J. Operator Theory **75**(2), 367–386 (2016)
363. Tian, F.: On commutativity of unbounded operators in Hilbert space. Thesis (Ph.D.) The University of Iowa, Iowa (2011). http://ir.uiowa.edu/etd/1095/
364. Tretter, Ch.: Spectral theory of block operator matrices and applications. Imperial College Press, London (2008)
365. Uchiyama, M.: Operators which have commutative polar decompositions. In: Contributions to Operator Theory and Its Applications, pp. 197–208. Operator Theory: Advances and Applications, vol. 62. Birkhäuser, Basel (1993)
366. Uchiyama, M.: Powers and commutativity of selfadjoint operators. Math. Ann. **300**(4), 643–647 (1994)
367. van Casteren, J.A.W.: Adjoints of products of operators in Banach space. Arch. Math. (Basel) **23**, 73–76 (1972)
368. van Casteren, J.A.W., Goldberg, S.: The Conjugate of the product of operators. Studia Math. **38**, 125–130 (1970)
369. van Daele, A.: On pairs of closed operators. Bull. Soc. Math. Belg. Sér. B **34**(1), 25–40 (1982)

370. Van Daele, A., Kasparek, A.: On the strong unbounded commutant of an $\mathcal{O}^*$-algebra. Proc. Am. Math. Soc. **105**(1), 111–116 (1989)

371. Vasilescu, F.-H.: Anticommuting self-adjoint operators. Rev. Roumaine Math. Pures Appl. **28**(1), 76–91 (1983)

372. von Neumann, J.: Zur Theorie der unbeschränkten Matrizen (German). J. Reine Angew. Math. **161**, 208–236 (1929)

373. von Neumann, J.: Approximative properties of matrices of high finite order. Portugaliae Math. **3**, 1–62 (1942)

374. Wang, B.-Y., Zhang, F.: Words and normality of matrices. Linear and Multilinear Algebra **40**(2), 111–118 (1995)

375. Weidmann, J.: Linear Operators in Hilbert Spaces. Springer, Berlin (1980)

376. Weiss, G.: The Fuglede commutativity theorem modulo the Hilbert-Schmidt class and generating functions for matrix operators. II. J. Operator Theory **5**(1), 3–16 (1981)

377. Wermuth, E.M.E.: Two remarks on matrix exponentials. Linear Algebra Appl. **117**, 127–132 (1989)

378. Wermuth, E.M.E.: A remark on commuting operator exponentials. Proc. Am. Math. Soc. **125**(6), 1685–1688 (1997)

379. Wheeden, R.L., Zygmund, A.: Measure and integral, An introduction to real analysis. In: Pure and Applied Mathematics, vol. **43**. Marcel Dekker, Inc., Basel (1977)

380. Wiegmann, N.A.: Normal products of matrices. Duke Math. J. **15**, 633–638 (1948)

381. Wiegmann, N.A.: A note on infinite normal matrices. Duke Math. J. **16**, 535–538 (1949)

382. Williams, D.P.: Lecture notes on the spectral theorem. https://www.math.dartmouth.edu/~dana/bookspapers/ln-spec-thm.pdf

383. Williams, J.P.: Spectra of products and numerical ranges. J. Math. Anal. Appl. **17**, 214–220 (1967)

384. Williams, J.P.: Operators Similar to Their Adjoints. Proc. Am. Math. Soc. **20**, 121–123 (1969)

385. Wogen, W.R.: Subnormal roots of subnormal operators. Integr. Equ. Oper. Theory **8**(3), 432–436 (1985)

386. Wright, J.D.M.: All operators on a Hilbert space are bounded. Bull. Am. Math. Soc. **79**, 1247–1250 (1973)

387. Wu, P.Y.: All (?) about quasinormal operators. In: Operator Theory and Complex Analysis (Sapporo, 1991), pp. 372–398. Operator Theory: Advances and Applications, vol. 59. Birkhäuser, Basel (1992)

388. Wu, D.Y., Chen, A.: On the adjoint of operator matrices with unbounded entries II. Acta Math. Sin. (Engl. Ser.) **31**(6), 995–1002 (2015)

389. Xia, D.X.: On the nonnormal operators-semihyponormal operators. Sci. Sinica **23**(6), 700–713 (1980)

390. Xu, H.: Two results about the matrix exponential. Linear Algebra Appl. **262**, 99–109 (1997)

391. Ya Azizov, T., Dijksma, A.: Closedness and adjoints of products of operators, and compressions. Integr. Equ. Oper. Theory **74**(2), 259–269 (2012)

392. Ya Azizov, T., Denisov, M., Philipp, F.: Spectral functions of products of selfadjoint operators. Math. Nachr. **285**(14–15), 1711–1728 (2012)

393. Yamazaki, T., Yanagida, M.: Relations between two operator inequalities and their applications to paranormal operators. Acta Sci. Math. (Szeged) **69**(1–2), 377–389 (2003)

394. Yang, J., Du, H.-K.: A note on commutativity up to a factor of bounded operators. Proc. Am. Math. Soc. **132**(6), 1713–1720 (2004)

395. Yood, B.: Rootless Matrices. Math. Mag. **75**(3), 219–223 (2002)

396. Yoshino, T.: A counterexample to a conjecture of Friedland. Israel J. Math. **75**(2–3), 273–275 (1991)

397. Youm, D.: q-deformed conformal quantum mechanics. Phys. Rev. D **62**, 095009 (2000). https://doi.org/10.1103/PhysRevD.62.095009

Index

Symbols
λ-commutativity, 163
(BN)-class, 226
(WN)-class, 97
L_{loc}^{p}-functions, 576

A
Absolutely continuous (function), 577
Anti-commuting (bounded) operators, 25
Anti-commuting (unbounded) operators, 473
Approximate eigenvalue, 122

B
Bilateral backward shift, 6
Bilateral (forward) shift, 6
Bilateral left shift, 6
Bilateral (right) shift, 6
Block circulant matrix, 67

C
Cartesian decomposition of a bounded
    operator, 22
Cartesian decomposition of unbounded
    operators, 474
Cesàro matrix, 260
Circulant matrix, 67
Commutativity (bounded case), 5
Commutativity (unbounded case), 297
Commutator, 9
Commutator (unbounded), 503
Compression spectrum, 253
Convergence

norm, 39
strong, 39
uniform, 39
weak, 39
Core, 347

D
Deficiency indices, 409
Densely defined (operator), 296
Devinatz-Nussbaum-von-Neumann theorem,
    471
Diagonal matrix of operators, 298
Direct sum of Hilbert spaces, 59
Direct sum (infinite) of the operators, 61
Direct sum of operators, 61
Distribution, 575
Distributional derivative, 576
Double commuting operators, 180

E
Eigenspace, 122
Eigenvalue, 122
Eigenvector, 122
Embry Theorem, 185
Exponential of an operator, 191
Extension (of unbounded operators), 295

F
Fong-Tsui conjecture, 289
Formal adjoint (of a matrix of unbounded
    operators), 348
Formal determinant, 63

© The Author(s), under exclusive license to Springer Nature Switzerland AG 2022
M. H. Mortad, *Counterexamples in Operator Theory*,
https://doi.org/10.1007/978-3-030-97814-3

Fourier transform, 571
Fuglede-Putnam Theorem, 179, 489
Fuglede Theorem, 179
Fundamental symmetry, 21

G
Gelfand-Beurling Formula, 145
Generalized Cauchy-Schwarz inequality, 103
Graph inner product, 308
Graph norm, 308
Graph (of an operator), 3, 295

H
Heisenberg canonical commutation relation, 503
Hess-Kato Theorem, 379

I
Identity (matrix of operators), 60
Image, 3, 295
Imaginary part of a bounded operator, 22
Imaginary part of an unbounded operator, 474
Index of nilpotence, 10
Intertwine, 180, 490
Invertible, 10
Invertible (matrix of operators), 62
Iterated operator, 7

J
Jacobson Lemma, 127, 525

K
Kato-Rellich Theorem, 379
Kernel, 3, 295

L
Linear functional, 4

M
Matrix of unbounded operators, 297
Metrically equivalent operators, 264
Monotone shift, 220

N
Nelson's example, 469
Neumann series, 11

Norm
  of an operator, 4
Nullity, 3
Null-space, 3, 295
Numerical range, 145
Numerical range (of an unbounded operator), 493

O
Off-diagonal matrix of operators, 298
Operator
  absolute value, 95
  absolute value (of a closed operator), 379
  adjoint, 5
  adjoint (of densely defined operators), 345
  (α, β)-normal, 230
  Airy, 557
  binormal, 226
  bounded, 3
  bounded below, 21
  bounded (not everywhere defined), 293
  boundedly invertible, 308
  Cesàro, 260
  closable, 307
  closed, 307
  closure, 307
  co-hyponormal, 208
  co-isometry, 21
  co-posinormal, 230
  compact, 159
  congruent, 267
  cube root, 75
  diagonal, 146
  essentially self-adjoint, 375
  finite rank, 159
  formally normal, 477
  hermitian, 21
  Hilbert-Schmidt, 161
  hyponormal, 207
  hyponormal (unbounded), 457
  idempotent, 21
  idempotent (unbounded), 321
  identity, 4
  interrupter, 230
  inverse, 10
  invertible, 10
  invertible (unbounded), 308
  isometry, 21
  left invertible, 12
  left invertible (unbounded), 317
  lower semibounded, 346
  modulus, 95
  multiplication, 4, 294

negative, 47
nilpotent, 10
nilpotent (unbounded), 320
nonnegative, 47
normal, 21
normal (unbounded), 441
normaloid, 147
nth root, 75
orthogonal projection, 21
paranormal, 209
paranormal (unbounded), 458
partial isometry, 21
p-hyponormal, 213
posinormal, 230
positive, 47
positive (unbounded), 347
projection, 21
projection (unbounded), 321
q-normal, 53, 460
quasinilpotent, 146
quasinormal, 208
quasinormal (unbounded), 458
quasi-invertible, 267
quasi-similar, 268
resolvent, 519
right invertible, 12
right invertible (unbounded), 317
self-adjoint, 21
self-adjoint (unbounded), 375
semibounded, 346
semi-hyponormal, 213
semi-normal, 208
shift, 5
skew-adjoint, 21
skew-adjoint (unbounded), 375
skew-Hermitian, 21
skew-symmetric, 21
square root, 75
square root (positive and unbounded), 378
square root (unbounded), 435
strictly positive, 47
subnormal, 208
symmetric, 21
symmetric (unbounded), 346
θ-class, 228
time-dependent Schrödinger, 555
unbounded, 3, 293
unitary, 21
upper semibounded, 346
Volterra, 4
wave, 554, 555
weakly normal, 97
weighted shift, 6
Operator matrix, 59

P
Plancherel theorem, 572
Polar decomposition, 96
Polar decomposition (of a closed operator),
    451
Polar decomposition (of unbounded self-
    adjoint or normal operators,
    451
Product of operators, 5
Product (unbounded operators), 295
Pure contraction, 11

R
Range, 3, 295
Rank, 3
Real part of a bounded operator, 22
Real part of an unbounded operator, 474
Reid's inequality, 222
Relative bound, 309
Resolvent set, 121, 519
Restriction of a linear operator, 4
Rootless matrix, 75

S
Similar operators, 21
Sobolev space, 577
Spectral mapping theorem, 124, 169
Spectral measure, 169
Spectral radius, 145
Spectral radius theorem, 145
Spectral Theorem for Self-adjoint Operators,
    169, 170
Spectral Theorem for Unbounded Self-adjoint
    Operators, 378
Spectrum, 121, 519
    continuous, 122
    residual, 122
    approximate point, 122
    point, 122
Strictly increasing norm, 50
Strong commutativity, 465
Strongly anticommuting (unbounded)
    operators, 473
Strong operator topology, 39
Sum (unbounded operators), 296
Sylvester equation, 281

T
Tensor product (of operators), 217
Toeplitz decomposition, 22
Toeplitz-Hausdorff theorem, 145

Topology of the operator norm, 39
Trace of a matrix, 49

U
Uniform topology, 39
Unitarily equivalent operators, 22
Universally commutable operators in the
    classical sense, 509

V
van Daele-Schmüdgen theorem, 414

W
Wüst theorem, 379
Weak derivative, 576
Weak operator topology, 39
Weight, 7

Printed in the United States
by Baker & Taylor Publisher Services